BIOMEDICAL
ELECTRON
MICROSCOPY

BIOMEDICAL ELECTRON MICROSCOPY

ILLUSTRATED METHODS AND INTERPRETATIONS

Arvid B. Maunsbach

Department of Cell Biology
Institute of Anatomy
University of Aarhus
Aarhus, Denmark

Björn A. Afzelius

Department of Ultrastructure Research
The Arrhenius Laboratories
Stockholm University
Stockholm, Sweden

ACADEMIC PRESS

San Diego   London   Boston   New York   Sydney   Tokyo   Toronto

Copyright © 1999 by ACADEMIC PRESS

Academic Press
a division of Harcourt Brace & Company
525 B Street, Suite 1900, San Diego, California 92191-4495, USA
http://www.apnet.com

Academic Press
24-28 Oval Road, London NW1 7DX, UK
http://www.hbuk.co.uk/ap/

Library of Congress Catalog Card Number: 98-85235

International Standard Book Number: 0-12-480610-4

Printed and bound by CPI Group (UK) Ltd, Croydon, CR0 4YY

Transferred to Digital Print 2011

CONTENTS

APPENDIX

PRACTICAL METHODS

FOREWORD

Professors Arvid Maunsbach and Björn Afzelius are uniquely qualified to author a book on methods of specimen preparation for biomedical electron microscopy. They are not newcomers to this field. They were among the pioneers in the application of the electron microscope to the study of cells and tissues. In distinguished research careers spanning some 40 years, they have witnessed the development of a broad range of methods of specimen preparation for transmission electron microscopy, and have made important contributions to the improvement of some of those methods. The electron micrographs in their classic papers on the kidney, liver, and spermatozoa have set a high standard of quality that others have sought to attain.

There is a real need for this book. Regrettably, the current preoccupation of biomedical scientists with the newer field of molecular biology, and the pressures to publish early, and often, to compete for dwindling grant funds have led to a decline in the quality of published electron micrographs. Meticulous preparation of tissues is essential to avoid artifacts that may lead to interpretations of questionable validity. This well-written, abundantly illustrated, book describes in detail the procedures required to obtain images of cells and cell organelles free of artifact, and provides guidance for image processing and interpretation. It will be a rich source of information, and inspiration, for younger investigators struggling to attain optimal results with traditional methods and for older investigators desirous of applying some of the newer techniques to their own research.

Don W. Fawcett, M.D.
Hersey Professor of Anatomy and Cell Biology, Emeritus
Harvard Medical School

PREFACE

Electron microscopy has fundamentally changed the knowledge about cell structure and function. It is now an indispensable tool in many fields of cell biology and medicine and includes a large variety of preparatory methods. Each of these has its own sets of information possibilities along with interpretative difficulties and pitfalls. A profound understanding of the methods is therefore required for an in-depth analysis and evaluation of structure–function relationships. In this volume we illustrate the basic preparatory methods for transmission electron microscopy in biology and medicine, and we hope that this monograph will be a guide through the broad spectrum of methods.

From discussions with our research students we have realized that the interpretation or "reading" of electron micrographs often presents more of a problem than the actual preparation and microscopy of the biological specimens. We have found that the best way to help students in critically evaluating their micrographs is to show examples of their—and of our—good and failed preparations. These comparisons have then formed the basis of discussions of the biological significance of various structures in electron micrographs.

This monograph should be useful for investigators working with transmission electron microscopy in biology and medicine—beginners as well as more experienced researchers. Investigators in other fields of biology may use it as a guide in evaluating and interpreting electron micrographs. Research technicians may find it useful for choosing optimal preparation methods and identifying artifacts.

The first part of this monograph presents the basic procedures in biological electron microscopy, ranging from fixation through microtomy and microscopy to photographic procedures. The second part exemplifies more special preparation techniques, including autoradiography, cytochemistry, immunoelectron microscopy, and computer-assisted image analysis of electron micrographs. Each chapter begins with an introduction, which we have kept quite short, but which includes a note on the early development in the field. The presentations of the micrographs start with a short section on the preparation method. Next to this is a description of what can be observed and finally we have some comments on the results. Details of the preparatory procedures used for most of the illustrations are included in the Appendix (Practical Methods) or can be found in the literature listed at the end of each chapter. These reference lists include both classical studies and more recent key publications.

Most illustrations come from preparations that we have performed specifically for this monograph. Some originate from our own published papers (see Acknowledgments for Reproduction of Figures). As experimental material we have mainly used tissues and cell types that have been extensively studied in our laboratories, particularly kidney tubule cells, hepatocytes, and spermatozoa. To a large extent, the content of this volume therefore reflects our own hands-on experiences. This focus on a few types of cells and tissues may appear to be a limitation, but will facilitate comparisons of different procedures on the same tissue. Moreover, many methods apply almost equally well to different cell types.

ACKNOWLEDGMENTS

Much of the inspiration for the present work can be traced back to our early years of training in biological electron microscopy under the critical scientific guidance of Professor Fritiof S. Sjöstrand.

Plans for this monograph were already laid in the 1970s in connection with a series of postgraduate courses in biological electron microscopy organized by José David-Ferreira, then director of the Laboratorio Biologia Celular at the Gulbenkian Institute, Oeiras, Portugal. Over the following years, we continued to assemble material during our postgraduate courses in Denmark and Sweden until we decided to finalize the work.

We are indebted to the late Sven-Olof Bohman, to Hans Hebert and Bjørn Johansen for very useful advice on various chapters of this book; to José and Karin David-Ferreira for allowing us to include some micrographs; and to our co-authors over the years for inspiring discussions. We also thank several colleagues and friends who generously shared their specimens or antibodies with us, helped to prepare specimens or micrographs, and/or gave us their constructive comments: Ulla Afzelius, Peter Agre, Pier Luigi Bellon, Emile L. Boulpaep, Erik Ilsø Christensen, Gunna Christiansen, Romano Dallai, Carl Christian Danielsen, Jens Dørup, Hans Jørgen Gundersen, Anders Höög, Kaj Josephsen, Peter Leth Jørgensen, Lars Kihlborg, Salvatore Lanzavecchia, Else-Merete Løcke, Søren Mogensen, Jesper Vuust Møller, Søren Nielsen, T. Steen Olsen, Peter Ottosen, Kaarina Pihakaski-Maunsbach, Finn Reinholt, Elisabeth Skriver, Margaret Söderholm, Karen Thomsen, Pierre Verroust, Hans Ørskov, and several others, who may not always have been aware of how valuable their comments were to us.

This monograph would not have been completed without the technical help of several persons, particularly Karen Thomsen and Else-Merete Løcke in Aarhus and Ulla Afzelius in Stockholm, who participated in this venture with great skill and enthusiasm. We also acknowledge the outstanding photographic work of Albert Meier and thank him for many valuable suggestions. At previous stages of this project we also received important help from Margrethe Aarup, Gunilla Almesjö, Christina Bohman, Poul Boldsen, Marianne Ellegaard, Maj Hasselgren, and Inger Kristoffersen. Stig Sundelin in Stockholm and Arne Christensen, Ole Moeskjær, and the late Ejnar Hansen in Aarhus carefully maintained microscopes and computer equipment.

The task of bringing this volume into readable shape was performed by Dorrit Ipsen, Jytte Kragelund, and Bente Kragh. We thank them for very competent secretarial help and editing and for being enormously patient with our constant rewriting of the text.

Last but not least we thank the staff at Academic Press in San Diego for very constructive and stimulating cooperation: Craig Panner for advice and planning, Lori Asbury for production, Michael Remener for design, and Debby Bicher for artwork.

This project was financially supported in part by the Danish Medical Research Council, the Swedish Natural Science Research Council, the Danish Biomembrane Research Center, the Karen Elise Jensen Foundation, the Novo Nordisk Foundation, and the Aarhus University Research Foundation.

A.B.M. and B.A.A.
Aarhus and Stockholm
February 1998

MICROGRAPH INTERPRETATION

Present-day ultrastructure research aims at the understanding of biological structure–function relationships at cellular and molecular levels using a wide variety of preparatory procedures, such as cytochemistry, immunocytochemistry, negative staining, and image analysis. These methods all depend on a series of basic procedures that are crucial for the outcome of the analyses, e.g., fixation procedures, ultramicrotomy, and, not the least, microscopy.

For optimal results a number of methodological aspects have to be considered: What preparation method should be used? Should different methods be applied in parallel? In what ways will the preparation procedure influence the object? Is the object frequent or rare? Which magnification should be used for optimal results? What quantitative aspects should be considered? What is the appearance of a "golden standard" preparation?

The first part of this chapter serves to illustrate briefly some of these questions. It starts with illustrations of kidney and liver cells, which we regard to be almost optimally prepared by present-day standards. Comparisons are then made between entirely different methods. The necessity to use the right magnifications for the object under study is emphasized and we raise the important question of whether an abnormal pattern is due to a preparatory artifact or represents a pathological change of the biological object.

The last part of this chapter deals with the fact that electron micrographs represent scientific raw data, the significance of which may not be readily apparent. Interpretation requires an understanding of the principles of the techniques as well as experience in detecting methodological errors. The electron microscope image is characterized by its wealth of information. However, only part of the information is directly related to the biological object itself; other parts depend on various preparatory steps and to instrument characteristics. As a consequence, electron micrographs are far less accessible to interpretation than they appear to be.

The process of electron micrograph interpretation can intentionally or unintentionally be divided into a series of consecutive steps or levels of analysis, for example:

- detection of objects
- identification of artifacts
- analysis of geometry
- biological identification
- analysis of dynamics

Examples of these five steps of analysis are given in the following. The late steps of interpretation usually depend on the early ones. An ability to master the initial sequence of interpretation is required for the analysis of the late sequence to be meaningful.

The validity of the final biological conclusions is invariably correlated to the quality of the information obtained. Thus, emphasis on high technical standards and methodological insight is not *l'art pour l'art* in electron microscopy but a requirement for the analysis.

1. Classical Preparation Method

FIGURE 1.1 A transmission electron micrograph showing part of the cytoplasm in a rat liver cell. The anesthetized animal was perfusion fixed through the abdominal aorta with 1% glutaraldehyde in 0.1 M cacodylate buffer. Excised blocks of liver tissue were postfixed, first in the same glutaraldehyde fixative and then in 1% osmium tetroxide, dehydrated in ethanol, and embedded in Epon 812. Ultrathin sections were cut on a diamond knife and stained with uranyl acetate and lead citrate. × 36,000.

This liver cell shows:

- a cytoplasmic ground substance that is uniformly stained,
- a nucleus with euchromatin, heterochromatin, and a surrounding nuclear envelope,
- mitochondria with a uniformly dense matrix, clearly visible mitochondrial cristae, and occasional matrix granules,
- peroxisomes,
- ribosomes that are seen as free particles in the cytoplasm,
- rough-surfaced endoplasmic reticulum consisting of elongated cisternae with associated ribosomes,
- smooth-surfaced endoplasmic reticulum surrounded by glycogen particles, and
- cisternae and vacuoles of the Golgi apparatus with stainable contents.

COMMENTS

This micrograph is representative of a hepatocyte prepared by a conventional chemical fixation and embedding procedure. This micrograph is considered as approaching the "state-of-the-art" preparation. It is free of obvious artifacts, such as disruption of membranes, extracted regions of ground substance, or organelles, and shows no instrumental errors, such as astigmatism or specimen drift, yet there is no way to know the "true" structural appearance of a liver cell or any other cell in a transmission electron micrograph. For example, one may question whether the membranes of the RER in reality have a slightly wavy conformation, as seen here, or are more planar. Have the mitochondria retained the same diameters and positions as they had in the living cell? Are the separate mitochondrial profiles seen in this figure in reality in continuity outside the plane of the section?

The electron micrograph is to be regarded as representing a snapshot of a moment in the life of the cell. When interpreting the micrograph the investigator must also take the dynamics and the movements of the organelles in consideration. Mitochondria may divide or fuse with other mitochondria, various organelles may be transported along the microtubules, which themselves may grow or shorten, and secretory vesicles open at the cell surface.

The ultrastructure of cells and tissues as they appear in conventionally fixed and resin-embedded specimens has not changed appreciably since the early 1960s, despite the dramatic improvements of the transmission electron microscope itself, which now features increased flexibility in handling and extended automation due to computerization. Indeed, the major improvement in transmission electron microscopy of noncryo-preparations has been the result of an expanded repertoire of preparatory methods such as cytochemistry, autoradiography, and, not the least, immunoelectron microscopy.

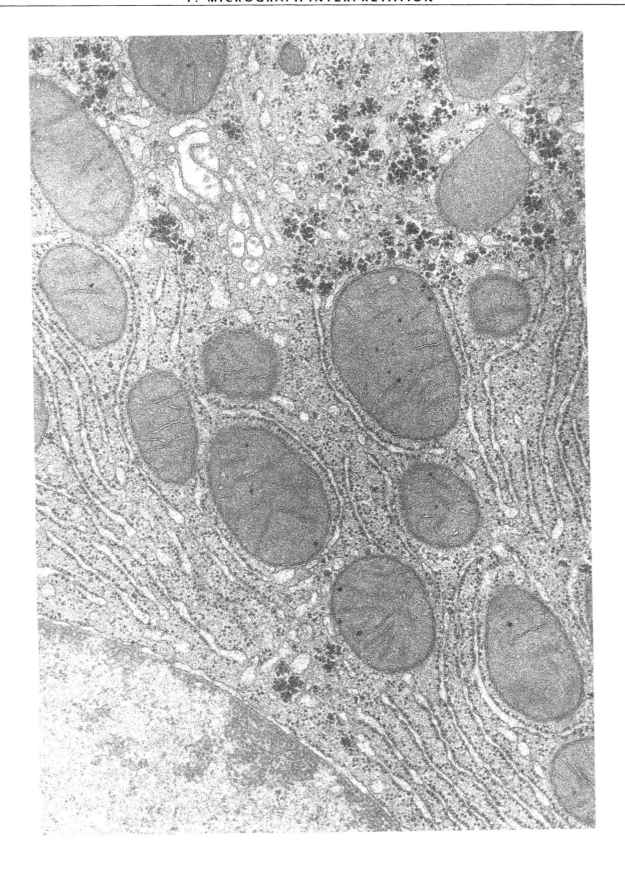

2. Low Temperature Approach

FIGURE 1.2A A transmission electron micrograph showing part of the cytoplasm in a rat liver cell. The anesthetized animal was perfusion fixed with 1% glutaraldehyde in 0.1 M cacodylate buffer. The excised tissue blocks were rinsed in buffer and infiltrated for 2 hr with 2.3 M sucrose, frozen by rapid immersion in Freon cooled with liquid nitrogen, and cryosectioned at $-115°C$ on a diamond knife. The cryosections were picked up with a droplet consisting of a 1:1 mixture of 2.3 M sucrose and 2% methylcellulose. The sections were rinsed on phosphate-buffered saline (PBS), transferred to Formvar-coated grids, and dried in the presence of 1.8% methylcellulose containing 0.3% uranyl acetate. $\times$ 30,000.

This liver cell contains a densely stained cytoplasmic ground substance of uniform appearance, well-preserved mitochondria, and other organelles. Membranes of all organelles appear as light lines contrasting the densely stained cytoplasmic matrix. Glycogen has been extracted and its sites appear as electron-lucid areas.

FIGURE 1.2B A freeze-fracture replica of a glutaraldehyde-fixed liver cell. The liver was perfusion fixed in 2% glutaraldehyde in 0.1 M cacodylate buffer, rinsed in buffer, and transferred to 30% glycerol for 25 hr. The tissue was then freeze-fractured in a Balzers freeze-fracture unit BA 300. $\times$ 30,000.

Part of the cell nucleus is seen at the upper left. As the plane of fracture has passed along the nuclear envelope, several nuclear pore complexes are observed. There are many cisternae of the rough-surfaced endoplasmic reticulum in the cytoplasm, although their ribosomes are not visible. Most of the rounded organelles in the cytoplasm are mitochondria, although some might be lysosomes or peroxisomes.

COMMENTS

In Fig. 1.2A the cytoplasmic ground substance and mitochondrial matrix appear intensely stained, whereas membranes, due to the absence of osmium tetroxide fixation, stand out against the surrounding cytoplasmic or mitochondrial matrix by being electron translucent.

In Fig. 1.2B the general architecture of a liver cell can be recognized, although there are several differences, such as the more three-dimensional appearance of the freeze-fractured specimen. The freeze-fracture replica also provides information of membrane surfaces that cannot be retrieved by other methods. The appearance of these liver cells is considerably different from that in Fig. 1.1, yet it must be considered as equally representative of the cytoplasmic architecture of a liver cell. Many of the cytoplasmic components seen in Fig. 1.1 can be recognized, although they differ in appearance because of the different techniques.

The fact that the general architecture of the cells is similar in all three preparations (Figs. 1.1 and 1.2) provides evidence that they reflect life-like appearances of a liver cell. The implicit conclusion is that the application of different procedures in parallel is essential for the analysis of cellular fine structure. In the following chapters, liver cells will be shown after many different types of preparatory procedures and may be compared with Figs. 1.1 and 1.2.

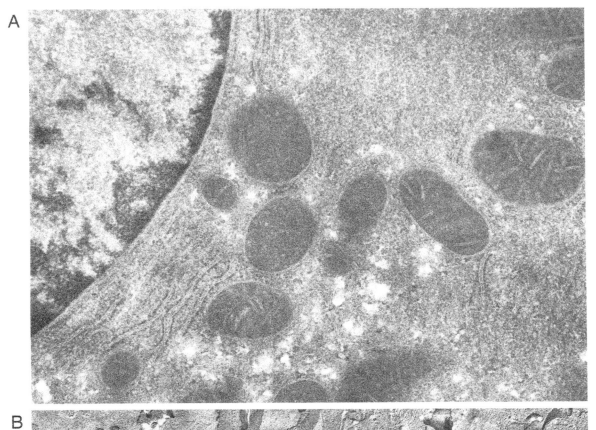

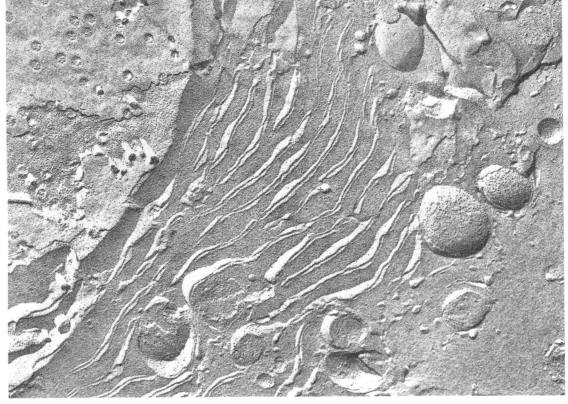

3. A Common Test Specimen

FIGURE 1.3 A survey electron micrograph of proximal tubule epithelium from a rat kidney. Following perfusion fixation with 1% glutaraldehyde in 0.1 M cacodylate buffer, the tissue was treated with potassium ferrocyanide for contrast enhancement and embedded in Epon. Ultrathin sections were stained with lead citrate. $\times$ 17,500.

This micrograph shows proximal tubule cells extending from the tubule lumen (LU) to the basement membrane (BM). The cells interdigitate extensively, and the cell membranes (arrowheads) that border the lateral intracellular spaces have a winding course. Tight junctions (arrows) close the lateral intracellular spaces toward the lumen. The cytoplasm is dominated by numerous mitochondria (M) as well as by lysosomes (L) and peroxisomes (P). The cells have a prominent Golgi region (G) often located equatorially in relation to the cell nucleus (N). In the apical cytoplasm there are numerous small and large endocytic vacuoles (E) as well as dense apical tubules (DAT), which participate in the membrane recycling process in the apical cytoplasm. The brush border (BB) is composed of numerous microvilli, which in this micrograph are sectioned somewhat obliquely.

COMMENTS

This micrograph of extensively interdigitating proximal tubule cells is to be compared with corresponding cells following cryosectioning (Fig. 8.21A) and freeze-fracture (Fig. 17.8C). Although liver cells may be the most commonly used test object when exploring new preparation methods in biological electron microscopy, the proximal tubule cells are also suitable test objects because they are very sensitive to variations in preparation procedures, including type of fixative, osmolality of the fixative vehicle, mode of application of the fixative to the tissue, hydrostatic pressure during perfusion, handling during immersion fixation, and subsequent steps during dehydration. Much is known about their functional characteristics and they are therefore suitable for correlations between structure and function. Additionally, as with many other cell types, rat proximal tubule cells show basically the same cellular architecture as human proximal tubule cells, although human biopsy material usually exhibits more preparatory artifacts than samples from experimental animals.

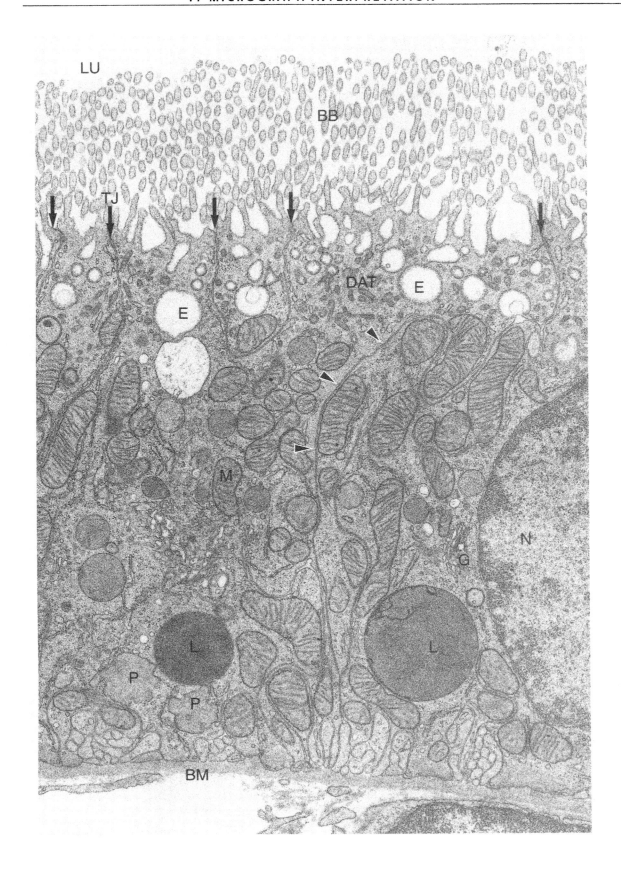

4. Detection of Objects

FIGURE 1.4A A lysosome in a cell in the renal proximal tubule. After fixation with glutaraldehyde, cytochemical incubation for acid phosphatase, postfixation in osmium tetroxide, and embedding in Epon, the section was stained with uranyl acetate and lead citrate. × 39,000.

The lysosome is identified through its dense reaction product for acid phosphatase. The center of the lysosome is, however, devoid of reaction product but seems to contain a homogeneous material.

FIGURE 1.4B A higher magnification of the same negative showing the central part of the lysosome in Fig. 14A. × 270,000.

At this magnification the material inside the lysosome shows a periodic structure.

FIGURE 1.4C Part of the apical region of a cell in the renal proximal tubule. The tissue was fixed in glutaraldehyde and tannic acid, postfixed in uranyl acetate, and embedded in melamine. The section was stained with lead citrate. × 70,000.

The cytoplasm contains many small membrane-limited vacuoles, and the apical plasma membrane shows invaginations into the cytoplasm. None of these show an extracellular or cytoplasmic coating.

FIGURE 1.4D A similar cell region as in Fig. 1.4C, except that the tissue was fixed with glutaraldehyde and embedded in Epon and the section was stained with uranyl acetate and lead citrate. × 60,000.

A plasma membrane invagination exhibits a distinct cytoplasmic (arrowheads) as well as extracellular coating.

FIGURE 1.4E Mitochondria in the cytoplasm of a rat bone marrow cell fixed in 2% osmium tetroxide, embedded in Epon, and section stained with uranyl acetate. × 80,000.

Light areas exist within the mitochondrial matrix; some fibers are contained in these otherwise "empty" regions (arrows).

COMMENTS

The possibility to detect an object is dependent on its size, frequency, distribution, and electron density within the electron micrograph. Very small objects, in the nanometer range, may not be detected against the "background noise" in the image, despite having high electron density. At low magnification, fine details go undetected.

Cellular components may be difficult, if not impossible, to detect when contrast is low. The distinct coating of the cell membrane seen in Fig. 1.4D represents clathrin, which is a protein characteristic of endocytic invaginations and vacuoles. Clathrin is visible in stained sections (Fig. 1.4D), but is often not discerned in cells with densely contrasted ground cytoplasm, such as Fig. 1.4C, or in unstained sections.

Even if an object is large enough to be seen and has a contrast that differs from its background, it often goes undetected simply because the investigator searches for other features; when inspecting Fig. 1.4E, attention may be on the cristae of the mitochondria rather than on its matrix with fibers now known to be DNA (see Nass and Nass, 1963; Nass *et al.*, 1965).

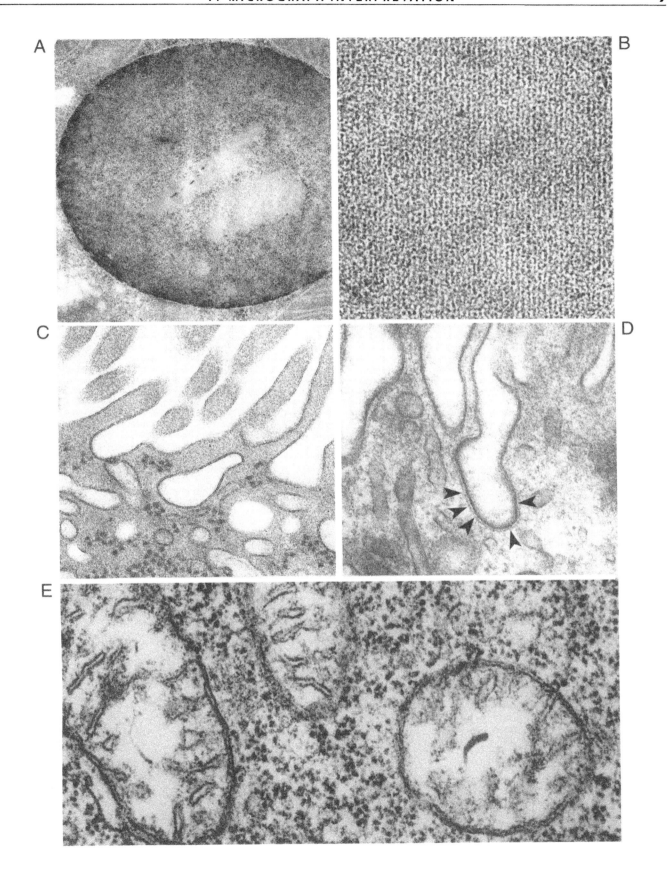

5. Identification of Artifacts

FIGURE 1.5A Cross-sectioned human cilia in biopsy from the respiratory tract. The biopsy was fixed in 1% glutaraldehyde, postfixed in osmium tetroxide, and embedded in Epon. The section was stained with uranyl acetate and lead citrate. × 110,000.

Cilia show a normal fine structure with a well-defined cell membrane and an axoneme with nine double microtubules surrounding two central microtubules.

FIGURE 1.5B Cross-sectioned human cilia from the respiratory tract in biopsy taken about 9 hr post mortem. × 110,000.

The two central microtubules are absent and the cell membrane is broken.

FIGURE 1.5C Apatite crystals in globules of the dental anlage in rat. The tissue was fixed in glutaraldehyde and osmium tetroxide and was observed unstained. × 75,000.

The crystallites appear as double lines.

FIGURE 1.5D Same object as in Fig. 1.5C, but the electron micrograph was taken at another focus setting. × 75,000.

The same crystallites can be observed as in Fig. 1.5C, but each of them appears as a single sharp needle.

FIGURE 1.5E Rat renal cortex fixed by dripping osmium tetroxide solution on the surface of the still functioning kidney. The tissue was embedded in Epon, and section stained. × 15,000.

All components of the tissue appear well preserved. However, there is an unusually large number of thrombocytes in the capillary.

COMMENTS

Following the detection of a pattern, the investigator has to judge the extent to which it may be influenced by preparation artifacts. This judgement is the second level of analysis and requires a knowledge of the particular preparation procedure, as well as of the influence of the individual steps in the procedure. Patterns, which are referred to as artifacts, usually have three characteristics: (1) They alter the structures or distort the electron optical images, (2) They can be traced to a specific step in the preparation procedure, and (3) they can be avoided or minimized by altering the technique.

In Fig. 1.5B the artifactual absence of central microtubules can be traced to the pronounced autolytic changes post mortem. Such changes are absent in fresh biopsies, as in Fig. 1.5A.

Figure 1.5D shows a focused image of the apatite crytals in the dental anlage, whereas Fig. 1.5C is overfocused, which leads to an apparent doubling of the crystallites. Awareness of focusing effects on image formation is therefore essential in analyses of small biological objects (see Chapter 9). In Fig. 1.5E the cells appear well preserved, but the slow penetration of the fixative into the functioning kidney presumably led to the formation of small blood clots.

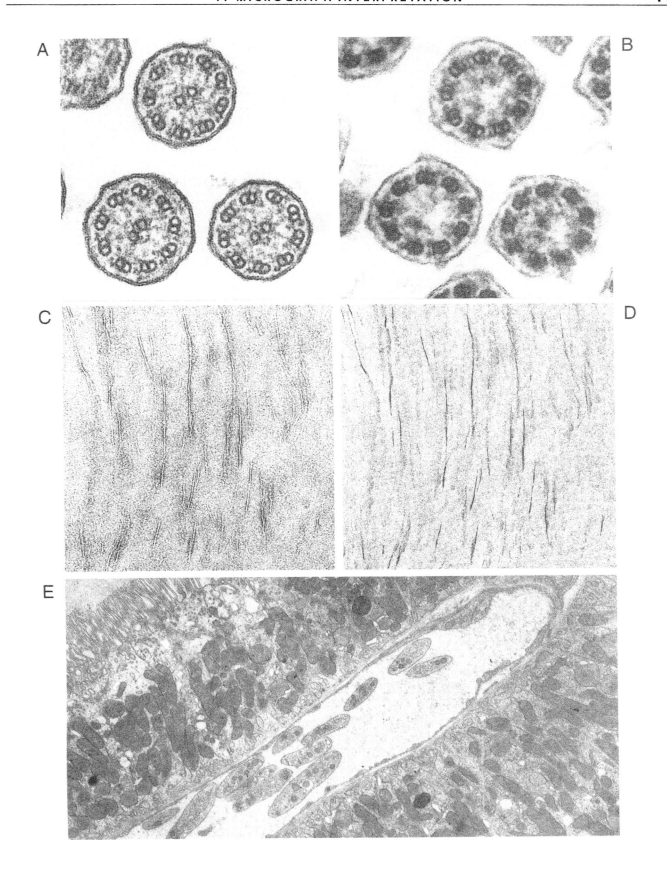

6. Analysis of Geometry

FIGURE 1.6A Red cells in a preparation of fixed and resin embedded human blood. × 6000.

The profiles of these three sectioned erythrocytes differ greatly.

FIGURE 1.6B Part of a cell in the renal proximal tubule. The tissue was fixed with glutaraldehyde, and small pieces of tissue were subsequently immersed in a solution of ruthenium red before resin embedding, microtomy, and microscopy. × 24,000.

In the apical part of the cell some vacuoles appear to be empty whereas others are filled with a strongly electron dense material, similar to that observed outside the cell.

FIGURES 1.6C AND 1.6D Mitochondria in a liver cell before (Fig. 1.6C) and after (Fig. 1.6D) tilting the section 33° with the aid of a goniometer in the microscope. The tilting axis is oriented vertically. × 45,000.

Membranes (horizontal arrows), which stand out distinctly in Fig. 1.6C, appear unsharp in Fig. 1.6D, Membranes, which are oriented perpendicular to the tilt axis, show no change (vertical arrows).

FIGURES 1.6E AND 1.6F Section of muscle fibrils from arrow-worm tilted 35° in one direction (Fig. 1.6E) and 25° in the opposite direction (Fig. 1.6F) around an axis oriented at right angle to the elongated profiles in Fig. 1.6E. × 63,000.

Figure 1.6E shows mainly elongated linear profiles in the muscle fibrils, whereas in Fig. 1.6F the muscle filaments appear as small circles and dots.

COMMENTS

Electron micrographs are two-dimensional records of three-dimensional objects. The appearance of various details within the specimen depends on their orientation relative to the electron beam. The three-dimensional shape is usually deduced from an inspection of many similar objects viewed from different angles. Tilting the object in a goniometer stage may also be very useful. Using this means, a cross-sectioned filament can be distinguished from a small granule (Figs. 1.6E and 1.6F) and a cross-sectioned membrane from a filament (Figs. 1.6C and 1.6D).

The well-known shape of a red cell would indeed be difficult to deduce immediately from a random section such as that in Fig. 1.6A. The three profiles are, however, fully compatible with that of a biconcave disc, a structure well known from light microscopy. It is also impossible to ascertain whether a vacuole-like structure within the cell is continuous with the cell membrane at a level above or below the section. This problem may be successfully attacked by studying serial sections or by adding tracers. For example, in Fig. 1.6B the ruthenium red tracer has gained access to some "vacuolar profiles," which actually represent the invaginations of the cell membrane.

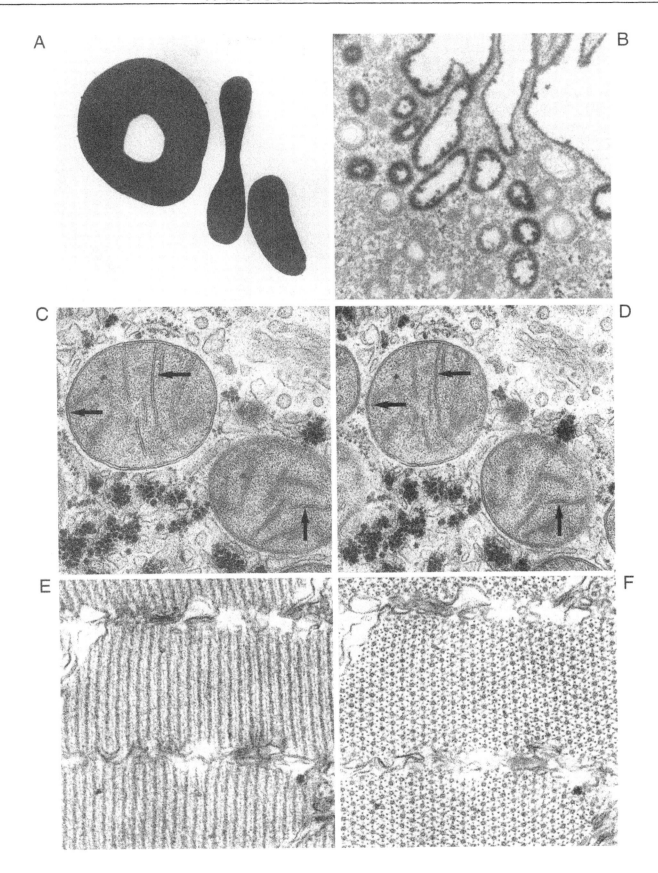

7. BIOLOGICAL IDENTIFICATION

FIGURE 1.7A Part of human kidney cells fixed in glutaraldehyde, freeze-substituted in Lowicryl HM20, and immunolabeled for aquaporin-1 using polyclonal anti-aquaporin-1 antibodies. The primary antibodies were detected with protein A coupled to 10-nm colloidal gold particles. × 60,000.

The cell membrane, in contrast to the intracellular structures, is intensely labeled with colloidal gold particles.

FIGURE 1.7B Cell in the renal proximal tubule following glutaraldehyde fixation, cytochemical incubation for acid phosphatase, osmium tetroxide fixation, Epon embedding, and section staining with uranyl acetate. × 35,000.

This part of the cytoplasm contains three cytoplasmic bodies that have reacted to acid phosphatase.

FIGURE 1.7C Respiratory epithelium from rabbit following glutaraldehyde and osmium tetroxide fixations, resin embedding, and section staining with uranyl acetate and lead citrate. × 80,000.

The image shows round structures, some with the structural characteristics of cilia and others with more or less opaque internal contents.

FIGURE 1.7D Cytoplasm of a lymphocyte from a HIV-positive patient. The cell was fixed in glutaraldehyde and osmium tetroxide, resin embedded, and section stained with uranyl acetate and lead citrate. × 40,000.

The cytoplasm contains a number of vesicular components, some with and some without stainable contents. Two such structures have particularly prominent electron-dense cores.

COMMENTS

The identification of a biological structure requires a correlation to biochemical, physiological, or other data. Nowadays it is simple enough to identify a mitochondrion in an electron micrograph. However, initially the identification was far from trivial, it had to be based on a series of studies that eventually resulted in the association of a certain characteristic structural pattern with organelles containing the respiratory chain. Indeed, the biological identification of structural components is almost invariably based on electron microscopy in combination with vital microscopy, cytochemistry, biochemistry, or physiology.

Certain structures cannot be identified unequivocally in the electron micrograph because they lack specific ultrastructural characteristics. For instance, decisive identification criteria may be enzymatic rather than structural, as is the case with lysosomes (Fig. 1.7B). The identification and localization of proteins can be performed with immunoelectron microscopy (Fig. 1.7A). In other instances, biological structures can be identified after consideration of their purely structural appearances, such as in Fig. 1.7C, where the objects among the cilia can be recognized as cross-sectioned bacteria.

In some cases the structure of a suspected biological identity is unknown. Such was the situation in 1983 when the micrograph in Fig. 1.7D was taken. Only later was the fine structure of the HIV virus defined, and it was realized that the two vesicles with electron-dense contents, although probably viruses, may not have been HIV viruses.

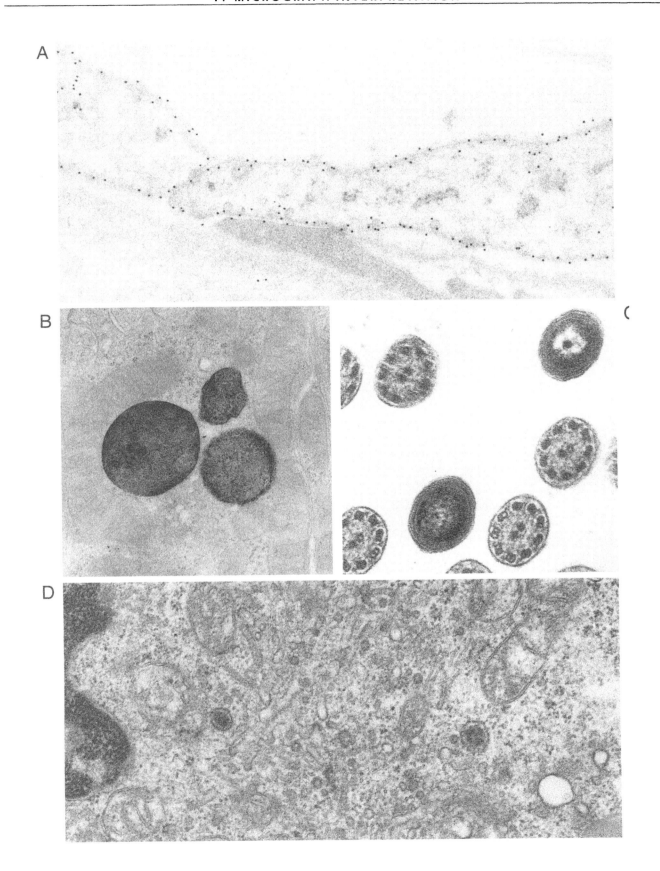

8. Biological Diversity

FIGURES 1.8A–1.8F Human sperm flagella from six different men. Spermatozoa were fixed in glutaraldehyde, followed by osmium tetroxide, embedded in Epon, section stained, and recorded in the electron microscope at primary magnifications around 30,000. The sperm tail in Fig. 1.8A, but not those in Figs. 1.8B–1.8F, originates from a man with normally motile spermatozoa. × 150,000.

A close inspection of these sperm tails reveals that they are all different. As compared to Fig. 1.8A, the sperm tail in Fig. 1.8B lacks outer dynein arms, whereas the one in Fig. 1.8C lacks both outer and inner dynein arms. The sperm cell in Figure 1.8D has only one central microtubule and no so-called central sheath surrounding it, whereas the sperm tail in Fig. 1.8E lacks the central sheath but is otherwise normal. The sperm tails in Fig. 1.8F lack three microtubular doublets.

FIGURE 1.8G Cytoplasm of a muscle cell from a patient with a genetic hypermetabolic condition called Luft's syndrome. The biopsy was taken from a leg muscle, fixed in glutaraldehyde and then osmium tetroxide, and embedded in Epon. × 20,000.

Mitochondria show abnormal inclusions in the form of crystalline arrays.

FIGURE 1.8H AND 1.8I Part of glomerulus in a kidney biopsy from a patient with Fabry's disease. The tissue was fixed in glutaraldehyde and osmium tetroxide and embedded in Epon. × 12,000 (Fig. 1.8H); × 350,000 (Fig. 1.8I).

The glomerular epithelial cells contain densely stained, myelin-like bodies that are also present in the urinary space. A small region of cell cytoplasm is further enlarged in Fig. 1.8I, where the material has a distinctly layered appearance.

COMMENTS

Spermatozoa from the six men had different degrees of motility, ranging from normal (Fig. 1.8A) to completely immotile (Figs. 1.8C–1.8E), whereas the spermatozoa in Fig. 1.8B showed reduced motility. These observations illustrate that changes in the architecture of the microtubule apparatus of sperm tails may be related directly to the motility of the spermatozoa and hence to the fertility of the man.

The human cells illustrated here (six different sperm tails, a muscle biopsy in Luft's syndrome, and a kidney biopsy in Fabry's disease) all illustrate structural changes that can be linked directly to specific diseases or genetic syndromes. These structural changes may be so specific that they are diagnostic. In fact, the diagnosis of Fabry's disease was initially suspected on the basis of this particular biopsy (Malmqvist *et al.*, 1971). The characteristic abnormalities of the sperm tail axonemes in Figs. 1.8B, 1.8C, 1.8E, and 1.8F also explain the infertility of the patients. Finally, it should be emphasized that the structural modifications of the cells illustrated here are highly reproducible and unrelated to any form of preparatory shortcomings or artifacts.

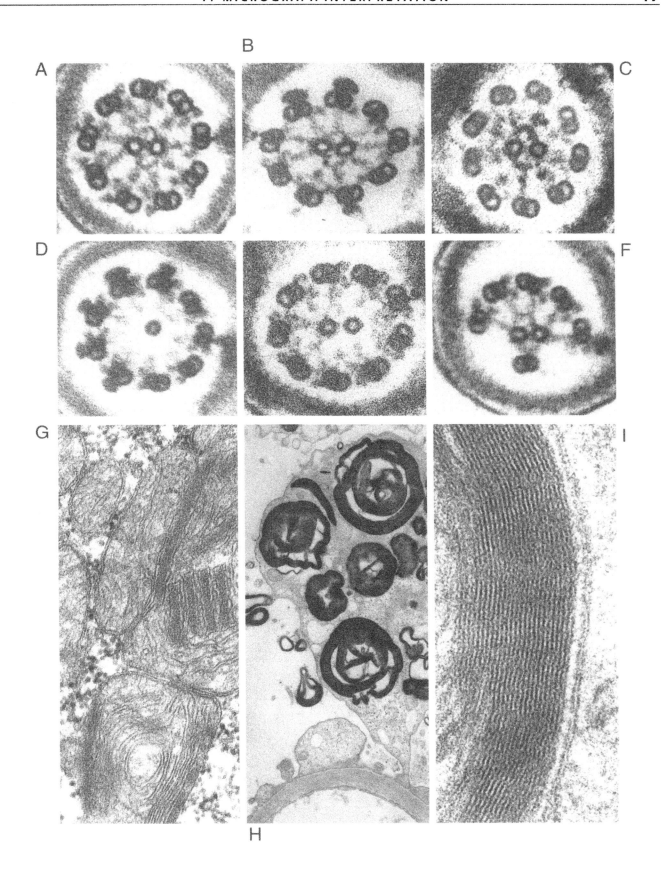

9. Analysis of Dynamics: Endocytosis

FIGURE 1.9A An electron microscope autoradiograph of cells that have endocytosed ^{125}I-labeled homologous albumin. The radioactive albumin was microperfused into the lumen of a proximal tubule of rat kidney. After 10 min the tubule was microperfusion fixed with 3% glutaraldehyde and postfixed in the same glutaraldehyde solution and then in osmium tetroxide. Following Epon embedding, thin sections were prepared for electron microscope autoradiography using Ilford L4 Nuclear Research Emulsion. × 12,000.

The autoradiographic grains are confined to the apical part of the cells and are predominantly associated with the small and large endocytic vacuoles, whereas lysosomes and other cell components deeper in the cytoplasm are unlabeled.

FIGURE 1.9B An electron micrograph of the apical part of the cell in the renal proximal tubule of *Necturus maculosus* following microperfusion with a solution of ferritin. The tubule was fixed with glutaraldehyde after 45 min and processed for electron microscopy. × 25,000.

Ferritin molecules are present in the tubule lumen where they are mainly associated with the outside of the cell membrane and endocytic invaginations (short arrows). Ferritin is also present in endocytic vacuoles, but is almost completely absent in the dense apical tubules, which are typical structures in this part of the cells (long arrows).

COMMENTS

These figures illustrate the dynamics of endocytosis, including time sequence, direction of transport, and identification of involved cell components. The absorption of homologous albumin, which has been lightly iodinated without denaturing the protein, demonstrates the physiological endocytic pathway in these cells (Maunsbach, 1966). Ferritin provides additional information with respect to the identification of the cell organelles that participate in endocytosis. In particular the ferritin experiments show that the dense apical tubules do not participate in the transport of ferritin from endocytic invaginations to endocytic vacuoles. Instead the analysis of ferritin absorption led to the conclusion that the dense apical tubules return excess membrane material from the endocytic vacuoles to the plasma membrane and thus participate in the recycling of membrane from the endocytic vacuoles to the plasma membrane in the proximal tubule (Maunsbach, 1976).

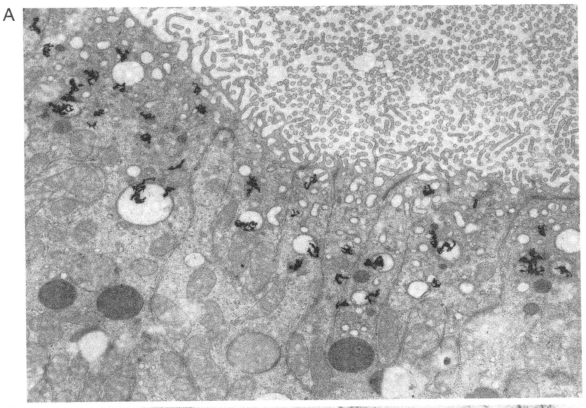

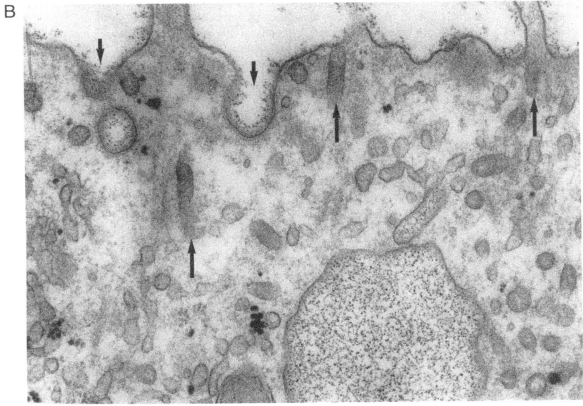

10. Analysis of Dynamics: Synthesis

FIGURE 1.10A Electron microscope autoradiographs of duck reticulocytes incubated for 45 min *in vitro* with ^{3}H-leucine (Miller and Maunsbach, 1966). After incubation with cold leucine the cells were fixed and further processed for electron microscope autoradiography. × 8000.

Erythrocytes and reticulocytes are shown. Some cells are labeled extensively whereas others are unlabeled or show only a few autoradiographic grains.

FIGURE 1.10B A higher magnification of parts of two cells in a similar preparation. × 26,000.

The heavily labeled reticulocyte is characterized by abundant polyribosomes, in contrast to the unlabeled reticulocyte (lower right), which contains only single ribosomes, and the erythrocytes (upper left) that are not active in protein synthesis.

COMMENTS

Electron microscope autoradiography can be used in many biological systems to follow the dynamics of protein synthesis with respect to the site of synthesis and the further processing of the proteins. Figure 1.10A illustrates that in a mixture of reticulocytes and erythrocytes, only reticulocytes are capable of synthesizing hemoglobin on *in vitro* incubation. Actively synthesizing cells are characterized by an abundance of polyribosomes (Fig. 1.10B), whereas erythrocytes or reticulocytes with only single ribosomes are inactive. The process of protein synthesis can also be analyzed within cells. Thus in the exocrine pancreas cell, autoradiography not only helps reveal the site of synthesis, but also the main secretory pathway and the time sequence of the intracellular processing (Palade, 1975).

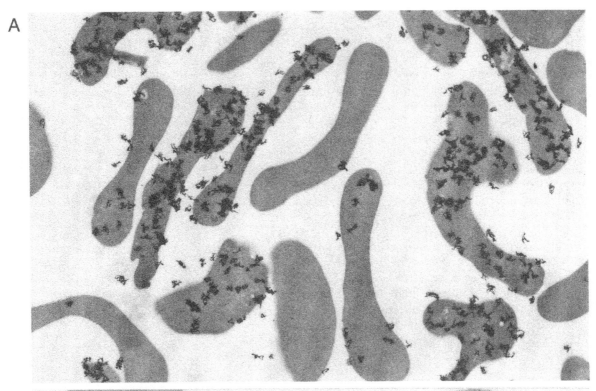

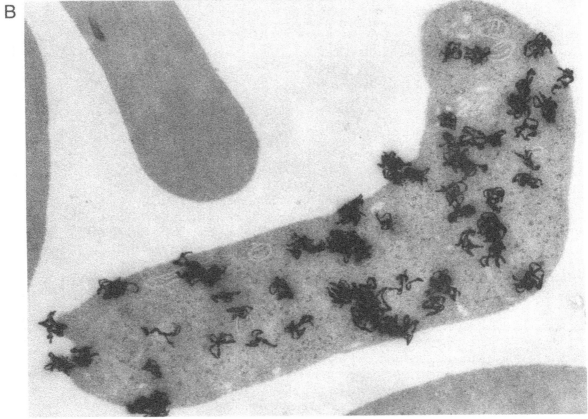

11. Comparison of Methods

FIGURES 1.11A–1.11E Membrane fragments containing Na,K-ATPase, the sodium–potassium ion pump. The membranes were purified from the outer renal medulla of the pig kidney and prepared for transmission electron microscopy in different ways. Figure 1.11A (× 105,000) shows a sectioned pellet of the isolated membranes after conventional glutaraldehyde/osmium tetroxide fixation and epoxy embedding, Fig. 1.11B (× 120,000) shows a membrane fragment after freeze-fracture, Fig. 1.11C (× 85,000) shows a membrane fragment after negative staining with uranyl acetate, Fig. 1.11D (× 85,000) shows a membrane fragment following incubation with anti-Na,K-ATPase antibodies that were detected with protein A conjugated to 5 nm colloidal gold, and Fig. 1.11E (× 175,000) shows double immunolabeling of a Na,K-ATPase membrane with two different sequence-specific antibodies detected with protein A conjugated to 10 and 5 nm colloidal gold, respectively.

These five electron micrographs illustrate the characteristics of different preparation methods as applied to the same biological object. Each method reveals specific biological information: the general structure of the membranes (Fig. 1.11A), small surface objects are observable with negative staining (Fig. 1.11C), intramembrane components can be detected in freeze-fracture replicas (Fig. 1.11B), and the distribution of proteins in the membranes can be monitored with immunoelectron microscopy (Figs. 1.11D and 1.11E).

FIGURE 1.11F Atomic force microscope image of the lipid region of an isolated Na,K-ATPase membrane, corresponding to the central smooth area of Fig. 1.11C. The image was recorded with a silicone nitride cantilever in aqueous solution according to Maunsbach and Thomsen (1994). Magnification about 4,000,000. [*See also color insert.*]

This image shows a flattened surface with small elevations arranged in a nearly crystalline array. The resolution in this image is estimated to be better than 0.5 nm (5 Å). The average distance between the centers of the elevations is 6–7 Å.

COMMENTS

In biological electron microscopy it is often essential to analyze an object with two or more techniques in parallel. This provides complementary information regarding the structure and may provide enzymatic or immunological information as well.

The Na,K-ATPase membranes shown here represent a simple biological object, as it contains lipids and only one protein, the Na,K-ATPase. Based on biochemical and chemical information it can be deduced that the surface particles observed by negative staining in Fig. 1.11C represent the protomer of the enzyme and that the intramembrane particles observed by freeze-fracture (Fig. 1.11B) represent dimeric enzyme units (Deguchi *et al.*, 1977).

The smooth membrane regions observed by negative staining as well as by freeze-fracturing reveal different aspects of the lipid bilayer. The thickness of the lipid region of the membrane, as determined in the atomic force microscope, is close to 5 nm corresponding to a lipid bilayer. The surface of the lipid bilayer, as imaged by atomic force microscopy, exhibits regular arrays of elevations likely to represent the hydrophilic ends of individual phospholipid molecules seen head on (Fig. 1.11F). It is evident that information about an object can be gathered with the aid of different electron microscopy methods and that other types of microscopy may provide supplementary information.

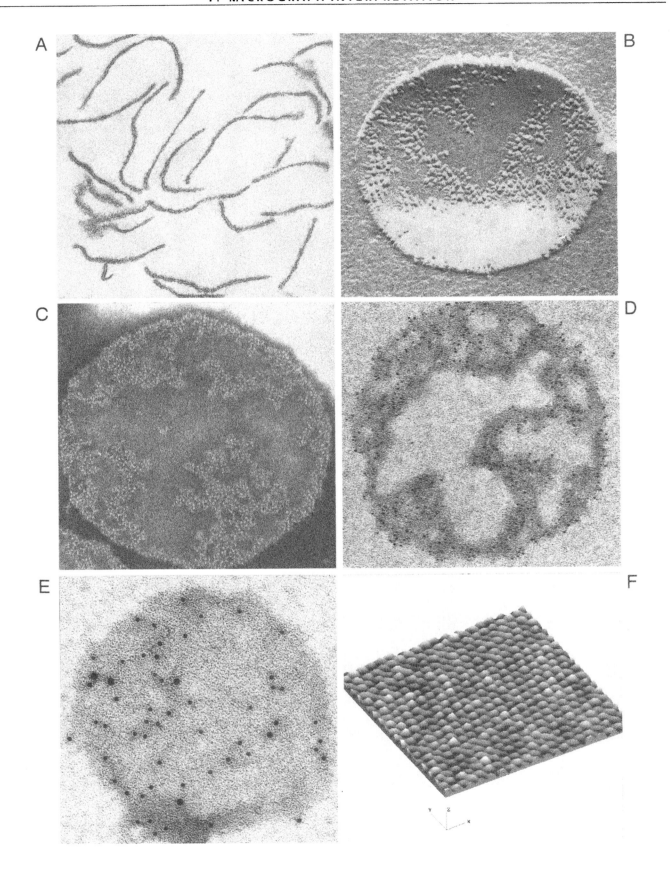

12. Variations in Magnifications

FIGURES 1.12A–1.12F Electron micrographs of one and the same section of human spermatozoa. The semen was fixed in a mixture of 2% glutaraldehyde, 1% tannic acid, and 1.8% sucrose in phosphate buffer, postfixed with 1% uranyl acetate, and embedded in Epon. The section was stained with uranyl acetate and lead citrate, and electron micrographs were recorded at original magnifications ranging from 500 to 50,000 times without changing apertures in the microscope. Final magnifications × 1000, × 3500, × 10,000, × 35,000, × 100,000, and × 350,000 for Figs. 1.12A–1.12F, respectively.

At low magnification the variety of spermatozoa—mature and immature—can be appreciated. The same sperm head is seen approximately in the center of Figs. 1.12A–1.12C, and an adjacent cross-sectioned sperm tail is shown in the center of Figs. 1.12D–1.12F.

COMMENTS

This series of micrographs shows the great range of magnification that can be obtained with a transmission electron microscope. Thus, at the low magnification in Fig. 1.12A, a survey of the specimen with its many cell types can be appreciated. At gradually increasing magnifications, more structural details become observable, and at the highest magnifications the protofilament organization of the microtubules becomes evident. The conclusion is that the magnification must be chosen according to the aim of the study. It is usually advisable to start visual inspection at a low mangification and gradually shift to high magnifications. When recordings at different magnifications are to be made, it is, however, better to start at the high magnification because some deterioration in section quality may occur in the electron beam.

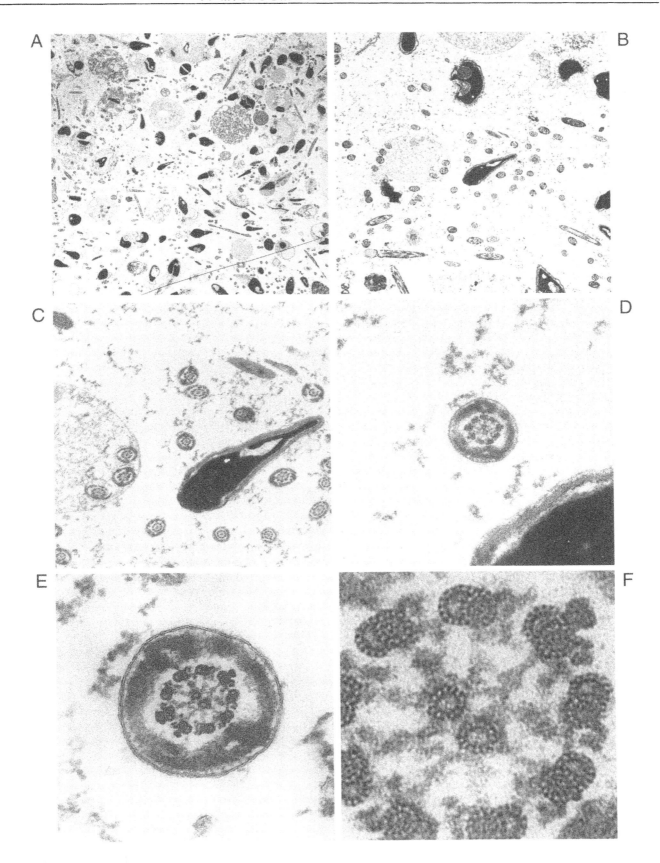

13. Interpretation Difficulties

FIGURE 1.13A Liver tissue fixed in glutaraldehyde and osmium tetroxide and embedded in Epon. × 11,000.

A lymphocyte located in an opening between the endothelial cells separating the space of Disse (left) and the hepatic sinusoid (right) in rat liver.

FIGURE 1.13B A subepithelial cell, possibly an endothelial cell, in the trachea from a rat. The tissue was fixed in 2% glutaraldehyde, 1% tannic acid, and 1.8% sucrose in 0.1 mM phosphate buffer. The tissue was postfixed in 1% uranyl acetate and embedded in melamine. Ultrathin sections were stained with uranyl acetate and lead citrate. × 155,000.

This electron micrograph shows a flattened cell in the middle where the plasma membrane shows invaginations (asterisks) as well as internal vesicles. One of them shows an external electron-dense layer. Both cells also show an endoplasmic reticulum with densely stained ribosomes.

COMMENTS

These two figures illustrate interpretation difficulties that may appear when trying to determine the movement of a cell through an endothelial barrier (Fig. 1.13A) or the possible movement of vesicles through an endothelium-like cell (Fig. 1.13B). In neither case is it possible to arrive at firm conclusions solely on the basis of the observed structural pattern. The cell within an opening of the endothelium, possibly a lymphocyte, may be in transit between the space of Disse (left) to the sinusoid (right), as the cell appears to have pseudopodia on its right part. However, this is only a guess, and further experimental studies using radioactively labeled cell or vital microscopy may answer this question.

Vesicles in contact with the cell membrane may be interpreted as showing exocytosis or endocytosis, or they may be stationary caveolae. However, only by properly designed experiments is it possible to arrive at valid conclusions. It is also not known if the system of membrane invaginations and vesicles in fact forms a continuous channel through the cell and whether the largest vesicle with an external stained coat is indeed a vesicle rimmed by clathrin. Tracer experiments may be the only possible means of determining the role of these cell components.

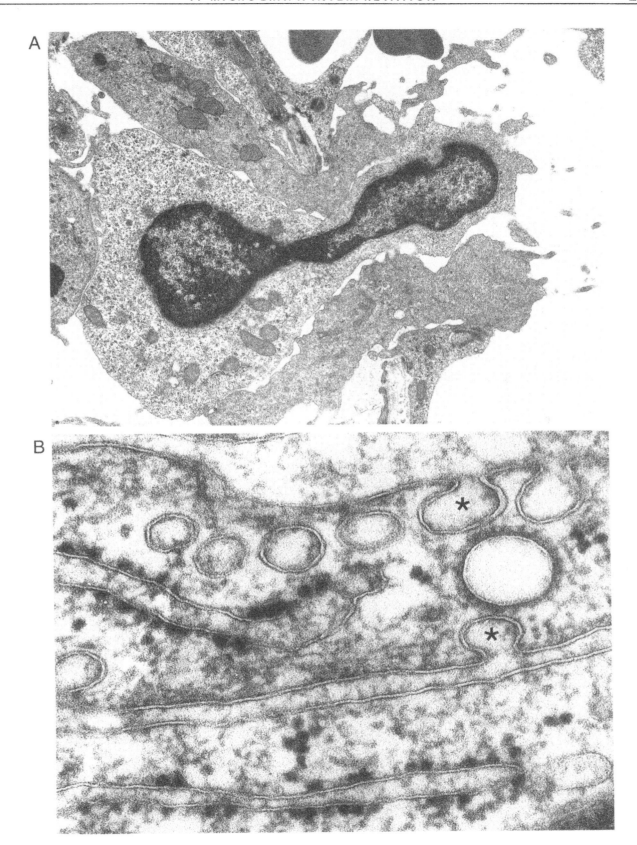

14. Diagnostic Pathology

FIGURE 1.14A A biopsy of human nasal mucosa fixed in a "routine" 10% formalin solution. After 1 week in the fixative the specimen was rinsed, postfixed in osmium tetroxide, dehydrated in ethanol, embedded in Epon, sectioned, and section stained. × 20,000.

Cytoplasm of epithelial cell with cilia showing swollen endoplasmic reticulum and mitochondria that often appear "exploded." Cilia and microvilli are essentially normal.

FIGURE 1.14B A cell from a human adrenal tumor. The specimen was obtained during surgery and fixed in "routine" 10% formalin solution. It was then dehydrated in acetone and embedded in paraffin. Some months later the specimen was deparaffinized in toluene, hydrated via acetone, fixed with glutaraldehyde and osmium tetroxide, and embedded in Epon. Sections were double stained. × 8000.

This cell shows an extensively indented nucleus and cytoplasm with many mitochondria. The general cell architecture is well preserved.

FIGURE 1.14C A higher magnification of cytoplasm from an adjacent cell to that in Fig. 1.14B. × 30,000.

Mitochondrial cristae have a round cross section, so-called tubular cristae (arrowheads). There are no small stainable granules in the cytoplasm.

COMMENTS

For diagnostic pathology at the light microscopic level, most specimens are fixed in formalin and embedded in paraffin. This procedure, in combination with staining methods, including immunohistochemical procedures, provides a firm basis for much diagnostic work. However, in some cases, diagnostic characteristics are associated with subcellular components below the resolution level of the light microscope.

Figure 1.14A shows that cell fine structure can be reasonably well preserved even after fixation in a "routine" formalin solution. Cilia are easily identified and appear normal, which in this case was important for the diagnosis. The cytoplasm, however, shows changes, mainly involving the deformation of mitochondria and endoplasmic reticulum. One reason why a 10% formalin solution (which contains 4% formaldehyde) is a less than optimal fixative for electron microscopy is that it contains formic acid and other impurities and that it often does not contain solutes that protect cells from osmotic distortions. A better result can be expected from a buffered and osmotically balanced formaldehyde solution.

In Figs. 1.14B and 1.14C the cell originates from an adrenal tumor, and the question was whether the tumor is a pheochromocytoma from the adrenal medulla or a cancer of the adrenal cortex. Because the former contains small, dense secretion granules, which are usually observable by electron microscopy, also in originally paraffin-embedded material, the cells in Figs. 1.14B and 1.14C, which are devoid of granules, more likely originate from a cancer of the adrenal cortex. This conclusion is further supported by the observation (Fig. 1.14C) that the mitochondria have tubular cristae—a feature characteristic of cells from normal adrenal cortex and other steroid-synthesizing cells. Although as a general rule only optimally prepared tissues should be analyzed by electron microscopy, this rule can be violated occasionally and useful information obtained.

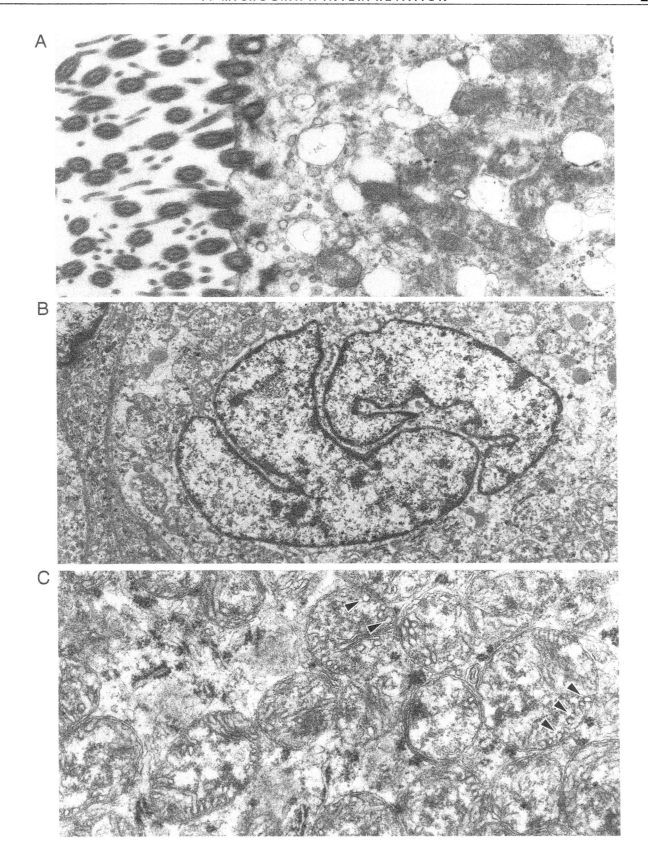

References

Afzelius, B. A. (1976). A human syndrome caused by immotile cilia. *Science* **193**, 317–319.

Afzelius, B. A. (1987). Interpretation of electron micrographs. *Scan. Microsc.* **1**, 1157–1165.

Birn, H., Christensen, E. I., and Nielsen, S. (1993). Kinetics of endocytosis in renal proximal tubule studied with ruthenium red as membrane marker. *Am. J. Physiol.* **264** (*Renal Fluid Electrolyte Physiol.* **33**, F239–F250.

Blowers, B., and Maser, M. (1988). Artifacts in fixation for transmission electron microscopy. *In* "Artifacts in Biological Electron Microscopy" (R. F. E. Crang and K. L. Klomparens, eds.), pp. 13–42. Plenum Press, New York London.

Cheville, N. F. (1994). "Ultrastructural Pathology. An Introduction to Interpretation." Iowa State Univ. Press, Ames, IA.

Crang, R. F. E., and Klomparens, K. L. (eds.) (1988). "Artifacts in Biological Electron Microscopy." Plenum Press, New York/London.

Deguchi, N., Jørgensen, P. L., and Maunsbach, A. B. (1977). Ultrastructure of the sodium pump: Comparison of thin sectioning, negative staining, and freeze-fracture of purified, membrane-bound (Na$^+$, K$^+$)-ATPase. *J. Cell Biol.* **75**, 619–634.

Dallai, R., and Afzelius, B. (1994). Three-dimensional reconstructions of accessory tubules observed in the sperm axonemes of two insect species. *J. Struct. Biol.* **113**, 225–237.

Dvorak, A. M. (1987). Procedural guide to specimen handling for the ultrastructural pathology service laboratory. *J. Electron Microsc. Tech.* **6**, 255–301.

Fawcett, D. W. (1981) "The Cell." Saunders, Philadelphia.

Harb, J. M. (1993). Interpretation of TEM micrographs for human diagnosis. *Microsc. Soc. Am. Bull.* **23**, 206–218.

Hockley, D. J., Wood, R. D., Jacobs, J. P., and Garrett, A. J. (1988). Electron microscopy of human immunodeficiency virus. *J. Gen. Virol.* **69**, 2455–2469.

Johannessen, J. V. (1977). Use of paraffin material for electron microscopy. *Pathol. Annu.* **12**, 189–224.

Jørgensen, P. L. (1974). Purification and characterization of (Na$^+$-K$^+$)-ATPase. III. Purification from the outer medulla of mammalian kidney after selective removal of membrane components by sodium dodecylsulphate. *Biochim. Biophys. Acta* **356**, 36–52.

Luft, R., Ikkos, D., Palmieri, G., Ernster, L., and Afzelius, B. (1962). A case of severe hypermetabolism of nonthyroid origin with a defect in the maintenance of mitochondrial respiratory control: A correlated clinical, biochemical, and morphological study. *J. Clin. Invest.* **41**, 1776–1804.

Malmqvist, E., Ivemark, B. I., Lindsten, J., Maunsbach, A. B., and Mårtensson, E. (1971). Pathologic lysosomes and increased urinary glycosylceramide excretion in Fabry's disease. *Lab. Invest.* **25**, 1–14.

Maunsbach, A. B. (1966). Absorption of I^{125}-labeled homologous albumin by rat kidney proximal tubule cells: A study of microperfused single proximal tubules by electron microscopic autoradiography and histochemisty. *J. Ultrastr. Res.* **15**, 197–241.

Maunsbach, A. B. (1976). Cellular mechanisms of tubular protein transport. *In* "International Review of Physiology: Kidney and Urinary Tract Physiology" (K Thurau, ed.), Vol. II, pp. 145–167. University Park Press, Baltimore.

Maunsbach, A. B., Marples, D., Chin, E., Ning, G. Bondy, C., Agre, P., and Nielsen, S. (1997). Aquaporin-1 water channel expression in human kidney. *J. Am. Soc. Nephrol.* **8**, 1–14.

Maunsbach, A. B., Skriver, E., and Hebert, H. (1991). Two-dimensional crystals and three-dimensional structure of Na, K-ATPase analyzed by electron microscopy. *In* "The Sodium Pump: Structure, Mechanism, and Regulation" (J. H. Kaplan and P. De Weer, eds.), pp. 159–172. Rockefeller Univ. Press, New York.

Maunsbach, A. B., and Thomsen, K. (1994). Atomic force microscopy of crystallized Na$^+$/K$^+$-ATPase. *In* "The Sodium Mechanism, Hormonal Control and its Role in Disease" (E. Bamberg and W. Schoner, eds.), pp. 342–345. Steinkopff Verlag, Darmstadt.

Miller, A., and Maunsbach, A. B. (1996). Electron microscopic autoradiography of rabbit reticulocytes active and inactive in protein synthesis. *Science* **151**, 1000–1001.

Nass, M. M. K., and Nass, S. (1963). Intramitochondrial fibers with DNA characteristics. II. Enzymatic and other hydrolytic treatments. *J. Cell Biol.* **19**, 593–611.

Nass, M. M. K., Nass, S., and Afzelius, B. A. (1965). The general occurrence of mitochondrial DNA. *Exp. Cell Res.* **37**, 516–539.

Ning, G., Maunsbach, A. B., Lee, Y.-J., and Møller, J. V. (1993). Topology of Na, K-ATPase α subunit epitopes analyzed with oligopeptide-specific antibodies and double-labeling immunoelectron microscopy. *FEBS. Lett.* **336**, 521–524.

Olsen, T. S., Racusen, L. C., and Solez, K. (1988). Ultrastructural investigation of renal biopsies: A discussion of artifacts and special methodology. *J. Electron Microsc. Techn.* **9**, 283–291.

Palade, G. (1975). Intracellular aspects of the process of protein synthesis. *Science* **189**, 347–358.

Rosso, J., and Sommers, S. C. (eds.) (1986–1989) "Tumor Diagnosis by Electron Microscopy," Vols. 1–3. Field and Wood, Inc., New York.

Roth, T. F., and Porter, K. R. (1964). Yolk protein uptake in the oocyte of the mosquito *Aedes aegypti*. L. *J. Cell Biol.* **20**, 313–332.

Sjöstrand, F. S. (1990). "Deducing Function from Structure: A Different View of Membranes." Academic Press, San Diego.

FIXATIVES

A constant problem faced in biological ultrastructure research is the execution of the fixation procedure. It is possible to prepare frozen sections of unfixed tissue, but this is far from a routine technique. The overwhelming number of ultrastructural studies are undertaken on sections of specimens, which have been "fixed" in one way or the other. Fixation can be accomplished by a large variety of chemical treatments that stop the metabolic processes and stabilize components in the cells.

The characteristics of the fixation procedures are discussed here under three different headings: the properties of the fixative itself, the composition of the vehicle of the fixative, and the way by which the fixative solution is brought in contact with the tissue. This and the following two chapters will each deal with one of these aspects.

The choice of the fixative itself is obviously of cardinal importance for the outcome of the study. More than 20 different chemicals have been used as fixatives. Many of these are rarely applied, but a few, notably glutaraldehyde, paraformaldehyde, and osmium tetroxide, have been used extensively. Each fixative gives a characteristic preservation that the investigator has to recognize. The properties of the fixatives are undoubtedly closely dependent on their ability to react with the chemical constituents of biological matter. Glutaraldehyde, for example, reacts by cross-linking the polypeptide chains of proteins, whereas it does not react with lipids. Osmium tetroxide undergoes a very complex series of reactions with proteins, as well as with many lipids. When comparing two fixatives it is *a priori* impossible to decide which of two images is the most true to organization in the living cell. In order to decide which fixative gives the most "life-like" preservation, it is necessary to obtain additional information, e.g., by comparing data obtained with different fixatives or by correlation to biochemical or physiological information.

Osmium tetroxide was first introduced in the electron microscopical technique by Ladislaus Marton (1934), who used it both as a fixative and as an enhancer of contrast. Osmium tetroxide had been applied much earlier in cytology to make the firefly lantern or some marine protozoa send out light. Max Schultze (1865), one of the proponents of the cell theory, found that the treated cells remained preserved in a life-like condition. Another fixative, formaldehyde, was first tried in cytology by Ferdinand Blum (1893), a bacteriologist, after an observation that it hardened the skin of his fingers. The use of tannic acid also came from an observation unrelated to cytology, namely that the tannic acid-rich juice of a Japanese fruit denatured the saliva, which led Vinci Mizuhira to try it as a fixative. Lucky coincidences or serendipity thus lies behind the use in cytology of these compounds. Or as Louis Pasteur expressed in 1854: "Dans les champs de l'observation le hasard ne favorise que les esprits préparés" ("Where observation is concerned, chance favors only the prepared mind").

Blum, F. (1893). Z. Wiss. Mikrosk. **10**, 314–315.
Marton, L. (1934). Nature **133**, 911.
Mizuhira, V. (1992). Curr. Cont. **135**, 9, June 15.
Schulze, M. (1865). Arch. Mikrosk. Anat. **1**, 124–137.

1. Osmium Tetroxide and Glutaraldehyde at Low Magnification

FIGURE 2.1A A rat liver fixed for 2 hr in 1% osmium tetroxide in 0.1 M phosphate buffer (pH 7.2). The "single-fixed" tissue was embedded in Epon and section stained with uranyl acetate and lead citrate. $\times$ 30,000.

The image is characterized by well-defined membranes in mitochondria and elsewhere. The mitochondrial membranes stand out against the mitochondrial matrix and the membranes of the endoplasmic reticulum against the ground cytoplasm. The membranes of the RER show long, continuous profiles with a relatively constant width, although many minor irregularities are seen.

FIGURE 2.1B A rat liver fixed for 2 hr in 2% glutaraldehyde in 0.1 M phosphate buffer (pH 7.2), embedded in Epon, and section stained with uranyl acetate and lead citrate. $\times$ 30,000.

Membranes are outlined by their transparency relative to the surrounding ground cytoplasm or mitochondrial matrix, rather than seen as electron-dense lines. Light areas (asterisk) correspond to glycogen-rich regions.

FIGURE 2.1C A rat liver fixed for 2 hr in 2% glutaraldehyde in the same buffer solution as in Fig. 2.1B, briefly rinsed in buffer, and then postfixed in 1% osmium tetroxide in 0.1 M phosphate buffer for 2 hr. Embedding and staining as in Fig. 2.1A. $\times$ 25,000.

The mitochondrial matrix and the ground cytoplasm in this "double-fixed" liver cell seem well preserved and stained. The mitochondrial membranes are distinct, although their density sometimes only slightly exceeds that of the surrounding matrix. The nucleus shows regions of euchromatin and heterochromatin. The dark, granular clusters in the ground cytoplasm are glycogen particles.

COMMENTS

By comparing "single" and "double-fixed" tissues the impression is gained that double fixation retains more material in the mitochondrial matrix and in the cytoplasmic ground substance. This impression is supported by studies that show a greater extraction of lipids and proteins in tissues fixed in osmium tetroxide alone as compared to glutaraldehyde followed by osmium tetroxide. The stainability of the tissue is also different, as illustrated by the generally much higher electron density of glycogen granules in double-fixed tissue. The low background staining in single-fixed tissue is helpful in emphasizing the general cell architecture in survey micrographs.

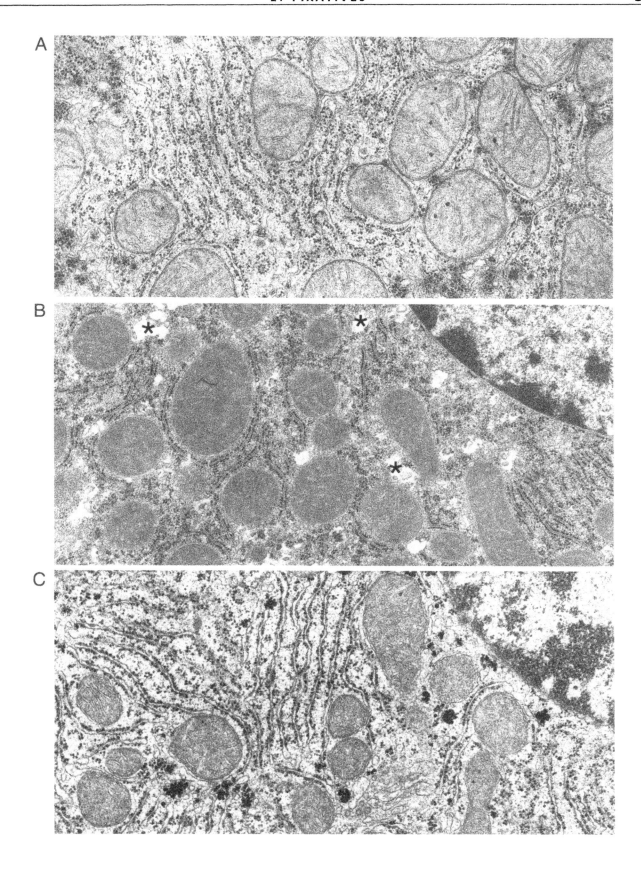

2. Osmium Tetroxide and Glutaraldehyde at High Magnification

FIGURE 2.2A parts of proximal tubule cells from a rat kidney, fixed by perfusion with 1% osmium tetroxide, block stained with uranyl acetate, embedded in Vestopal, and section stained with lead citrate. × 65,000.

Well-defined membranes stand out against a lightly stained cytoplasmic ground substance or mitochondrial matrix. The lateral intercellular spaces have a fairly constant width and appear empty.

FIGURE 2.2B Interdigitating proximal tubule cells fixed by perfusion with 1% glutaraldehyde. Embedding and section staining as in Fig. 2.2A. × 65,000.

The cytoplasmic ground substance and the mitochondrial matrix have the same high density as the outer components of the triple-layered membranes. Hence the membranes can be recognized only through their unstained middle layer, which is relatively wide. Most of the lateral intercellular spaces are completely obliterated, and a thin middle line represents the apposed outer portions of adjacent cell membranes (arrows).

FIGURE 2.2C Interdigitating proximal tubule cells fixed by perfusion with 1% glutaraldehyde and postfixed in 1% osmium tetroxide. Embedding and staining are as in Fig. 2.2A. × 65,000.

Membranes are clearly outlined and the cytoplasmic ground substance and the mitochondrial matrix are well stained. The mitochondrial membranes are not easily discerned due to the density of the mitochondrial matrix. The lateral intercellular spaces have an irregular width and show both wide dilatations and regions where they are nearly obliterated.

COMMENTS

The three fixation procedures used here result in different appearances of cellular membranes, extracellular spaces, and relationships between membranes. It is impossible to label one pattern as the true representation of the cellular organization and the others as false. Additional physiological, physical, or chemical data are necessary in order to evaluate the significance of the observed patterns. Such information may be obtained from cryosectioning, freeze-fracturing, or tracer experiments.

A

B

C

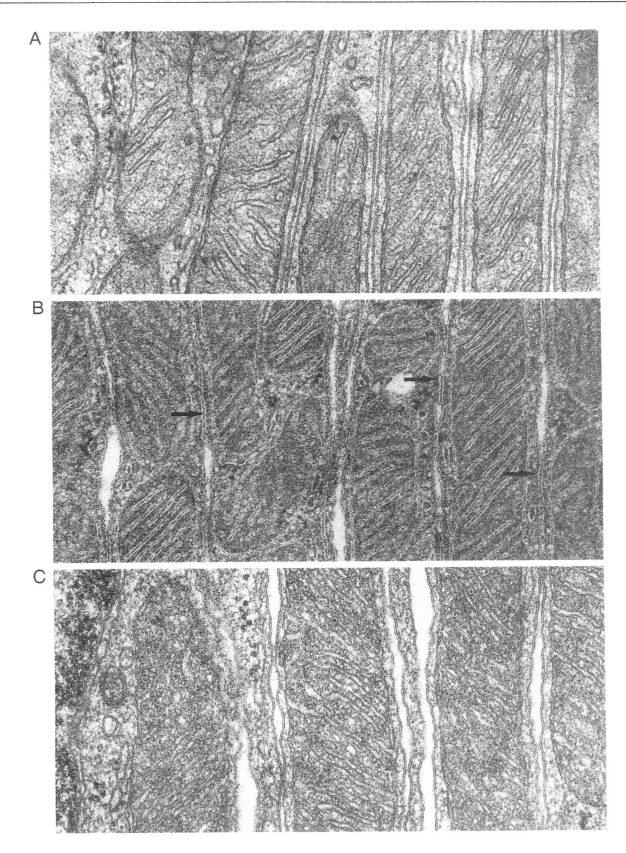

3. Glutaraldehyde Concentration: Perfusion Fixation

FIGURE 2.3A Parts of proximal tubule cells perfusion fixed with 0.2% glutaraldehyde in 0.1 M sodium cacodylate buffer, postfixed in 1% osmium tetroxide, embedded in Epon, and section stained with uranyl acetate and lead citrate. × 40,000.

Mitochondria appear swollen, and the mitochondrial matrix has a low electron density and shows some "empty" regions.

FIGURE 2.3B Parts of proximal tubule cells fixed and prepared as in Fig 2.3A except that the glutaraldehyde concentration was 2%. × 40,000.

The appearance of these cells is usually regarded as "normal" for this tissue. The matrix of the mitochondria is more stained than the cytoplasm, and the membranes of the cristae run essentially parallel.

FIGURE 2.3C Proximal tubule cells prepared as in Fig. 2.3A except that the glutaraldehyde concentration was 10%. A brief (10 sec) pre-rinse with Tyrode solution preceded the perfusion of the fixative. × 40,000.

Some of the intercellular spaces are distended, and the electron density of the cytoplasm, particularly the mitochondrial matrix, is greater than after fixation with glutaraldehyde of a lower concentration. Furthermore, the intracristal spaces are irregularly dilated.

COMMENTS

The glutaraldehyde concentration influences the appearance of the tissue in two ways: At low concentrations there may not be enough glutaraldehyde for an efficient cross-linking of the proteins and some mitochondria hence have a washed-out appearance (Fig. 2.3A), and at high concentrations the cytoplasm will show signs of osmotic shrinkage.

The concentration of the glutaraldehyde used for fixation of the tissue shown in Fig. 2.3B was sufficient for satisfactory fixation of the organelles. Osmotic effects are apparent in Fig. 2.3C, where both the entire cells and their mitochondria show evidence of shrinkage. At the macroscopic level, this shrinkage was apparent as a decrease in kidney size. Additionally, the fixed tissue was much harder to slice with a razor blade than following fixation with the less concentrated glutaraldehydes.

The osmotic effects on cells and tissues exerted by glutaraldehyde are much less than those of fixative vehicles containing corresponding concentrations of sodium chloride or sucrose. The osmotic effects on cells by 10% glutaraldehyde in 0.1 M sodium cacodylate buffer with a total osmolality of 1200 mOsm/kg H_2O (Fig. 2.3C) are thus considerably smaller than that of 1% glutaraldehyde in a Tyrode solution with twice the normal concentration of NaCl and a total osmolality of 630 mOsm/kg H_2O (compare with Figs. 3.3C, 3.4B, and 3.5C).

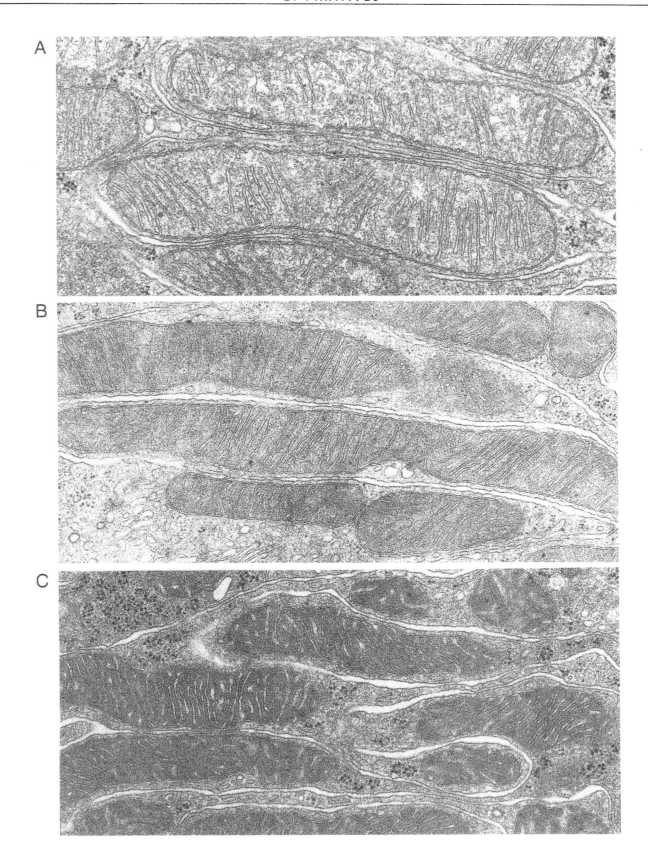

4. Glutaraldehyde Concentration: Immersion Fixation

FIGURE 2.4A Parts of proximal tubule cells immersion fixed with 0.2% glutaraldehyde in 0.1 M sodium cacodylate buffer, postfixed in 1% osmium tetroxide, embedded in Epon, and section stained with uranyl acetate and lead citrate. × 40,000.

All mitochondria appear swollen. The matrix has a low electron density in all mitochondria and shows some "empty" regions.

FIGURE 2.4B Parts of proximal tubule cells fixed and prepared as in Fig. 2.4A except that the glutaraldehyde concentration was 2%. × 40,000.

The appearance of these cells is usually regarded as "normal" for this tissue. The matrix of the mitochondria is slightly more stained than the cytoplasm, and the membranes of the cristae run essentially parallel. There is some variation in the width of the lateral intercellular spaces.

FIGURE 2.4C Proximal tubule cells prepared as in Fig. 2.4A except that the glutaraldehyde concentration was 10%. × 40,000.

The density of the mitochondrial matrix is greater than after fixation with glutaraldehyde of lower concentrations, and the intracristal spaces are irregularly dilated. A distinct feature of this preparation is that the cytoplasmic matrix varies greatly in density from cell to cell, ranging from compact to almost empty.

COMMENTS

The quality of fixation with these three aldehyde concentrations following immersion fixation resembles that obtained with the same fixatives following perfusion (Figs. 3A–3C). However, the immersion-fixed cells show a much greater variability in fixation quality than the cells fixed by perfusion. This is most apparent with the low concentration fixative where some mitochondria appear essentially washed out and with the high concentration where there is a considerable variation in the density of the cytoplasm, i.e., some cells swell and some appear to shrink.

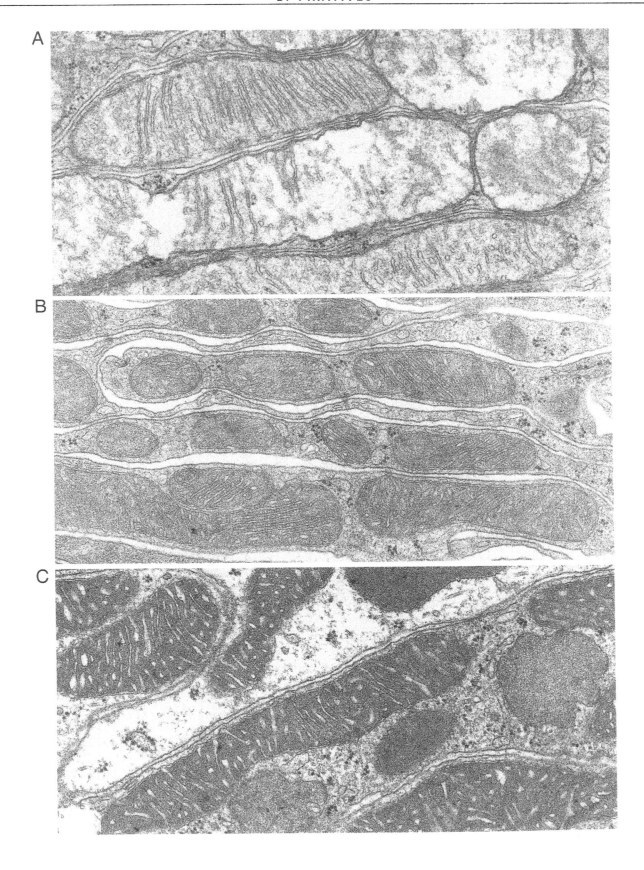

5. Long Fixation Times

FIGURE 2.5A A rat liver cell fixed with 1% osmium tetroxide in phosphate buffer for 24 hr at 4°C, embedded in Epon, and section stained with uranyl acetate and lead citrate. × 30,000.

All membranes appear continuous and well defined. The cytoplasmic ground substance is uniformly stained as is the mitochondrial matrix.

FIGURE 2.5B A rat liver cell fixed and prepared as in Fig. 2.5A except that the fixation temperature was 22°C. × 30,000.

The appearance of the cytoplasm is the same as in Fig. 2.5A except that the mitochondrial matrix and the cytoplasmic ground substance are less electron dense. Membranes therefore in general seem to stand out in high contrast.

FIGURE 2.5C A rat liver cell fixed with 2% glutaraldehyde in 0.1 *M* cacodylate buffer and stored in this solution in a cold room at 4°C for 11 years. It was then postfixed in 1% osmium tetroxide, Epon embedded, and the section double stained with uranyl acetate and lead citrate. × 30,000.

The general architecture of the cell is very well preserved. The cytoplasmic ground substance and mitochondrial matrix are densely stained. Cisternae of the endoplasmic reticulum have a rather constant width, mitochondria show well-preserved shapes, and the ribosomes are stained distinctly.

COMMENTS

Prolonged osmium fixation at room temperature results in pronounced extraction of both the nucleus and the cytoplasm, whereas the same tissue fixed at low temperature is also well preserved after 24 hr. Fixation in glutaraldehyde can be prolonged for months or even years without major structural changes. The fact that glutaraldehyde fixation is rather insensitive to variations in temperature and fixation time has the practical consequence that glutaraldehyde-fixed specimens can be shipped to other laboratories without adverse effects on fixation quality.

The temperature of glutaraldehyde fixatives has much less influence on tissue fine structure than that of osmium tetroxide fixatives, except that microtubules seem less well preserved by cold glutaraldehyde fixatives.

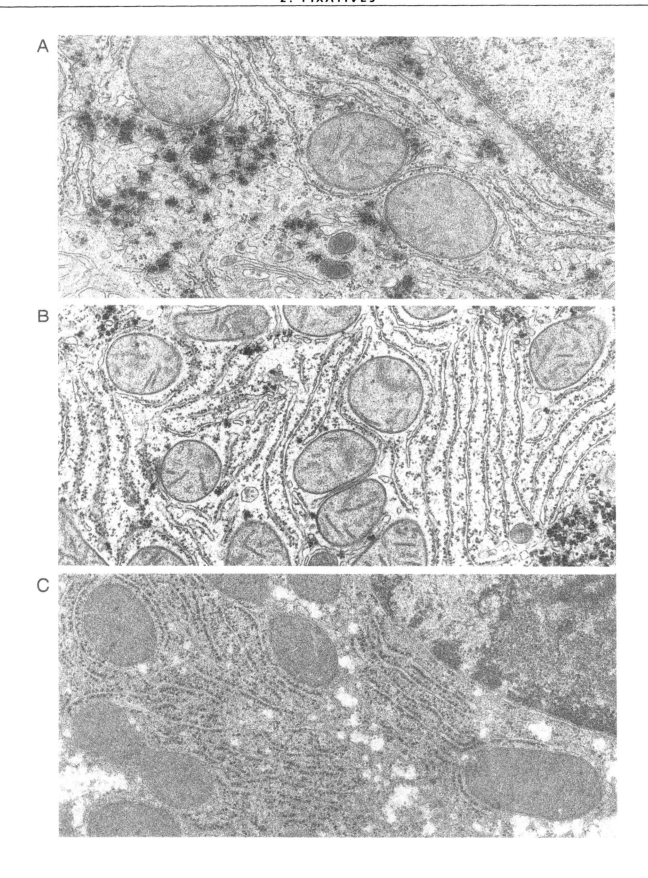

6. Formaldehyde–Glutaraldehyde Combinations

FIGURE 2.6A A rat liver perfusion fixed in a mixture of 2% formaldehyde and 2.5% glutaraldehyde in 0.1 M cacodylate buffer, pH 7.2. This mixture corresponds to half-strength "Karnovsky's fixative" (Karnovsky, 1965). The total osmolality of this fixative is about 1000 mOsm/kg H_2O. Postfixation was performed with 1% osmium tetroxide in cacodylate buffer, embedding in Epon, and staining with uranyl acetate and lead citrate. $\times$ 30,000.

The cytoplasmic ground substance is stained uniformly and densely except for the glycogen regions. The content of the endoplasmic reticulum as well as the mitochondrial matrix is well retained. Cisternae of the endoplasmic reticulum have a fairly constant width. Overall, this specimen appears to have a well-preserved cell architecture.

FIGURE 2.6B A rat liver perfusion fixed and processed as in Fig. 2.6A except that the fixative consisted of 4% formaldehyde and 0.1% glutaraldehyde. Postfixation and further processing are as in Fig. 2.6A. $\times$ 30,000.

The appearance of the tissue is similar to that in Fig. 2.6A, although the mitochondrial matrix is stained more intensely.

FIGURE 2.6C A rat liver perfusion fixed and processed as in Fig. 2.6A except that the fixative was 8% formaldehyde in 0.1 M cacodylate buffer. $\times$ 30,000.

This tissue resembles that in Fig. 2.6B, although there are expansions of the space between the outer mitochondrial membranes and more staining of glycogen and mitochondrial matrix.

COMMENTS

In principle, formaldehyde as a primary fixative gives a similar preservation as glutaraldehyde. There are, however, certain differences. Formaldehyde is a monoaldehyde, whereas glutaraldehyde is a dialdehyde and more efficient in cross-linking proteins and maintaining cell ultrastrucure. Formaldehyde, however, penetrates tissues rapidly, apparently due to its low molecular weight, whereas glutaraldehyde, contrary to common belief, penetrates the tissue rather slowly. In situations where large pieces of tissue must be fixed, e.g., certain biopsies, the use of formaldehyde or formaldehyde–glutaraldehyde mixtures should be considered rather than glutaraldehyde alone as a prefixative. The central region in tissue blocks fixed only in glutaraldehyde often shows poor tissue preservation. The favorable results generally obtained with combined formaldehyde/glutaraldehyde fixatives may be due to their dual actions: An initial rapid penetration of the small formaldehyde molecules into the cells with an arrest of metabolism and a later infiltration of glutaraldehyde that cross-links the cellular proteins more firmly. Formaldehyde is inefficient in creating osmotic gradients in the tissue. Despite its very high total osmolality (about 1000 mOsm/kg H_2O), Karnovsky's half-strength fixative has a smaller osmotic effect on the tissue than expected (see also Comments to 2.3).

For immunoelectron microscopy it is often advantageous to use mixtures of high (4–8%) formaldehyde and low (0.05–0.25%) glutaraldehyde concentrations (Fig. 2.6C). Such fixatives are more gentle to antigens and less inhibitory to enzymes than regular glutaraldehyde fixatives, yet the presence of a small amount of glutaraldehyde usually improves tissue fine structure. Formaldehyde solutions should be made up fresh from paraformaldehyde powder (Pease, 1964), as impurities present in commercial or aged formaldehyde solutions are detrimental to good tissue preservation.

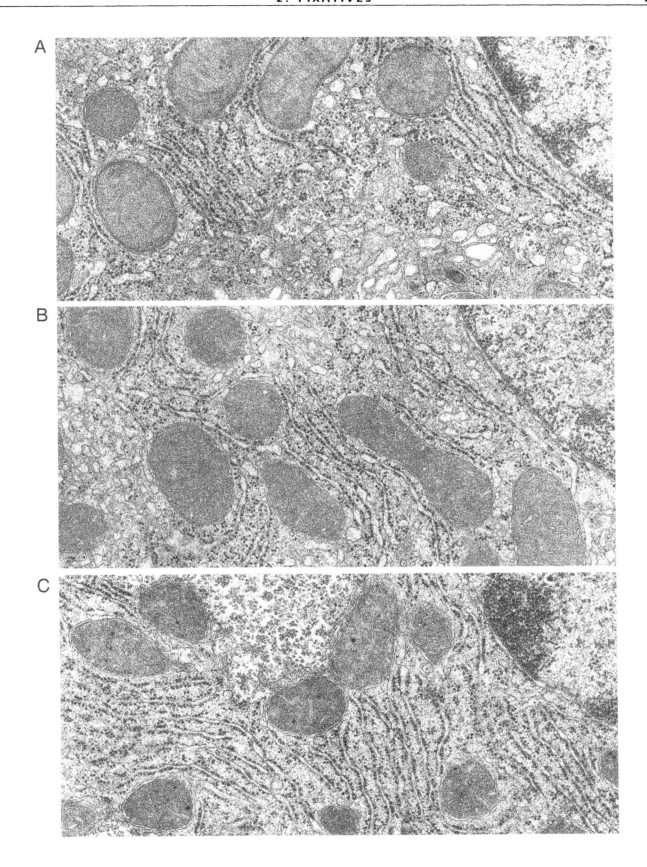

7. Potassium Permanganate, Picric Acid, and Ruthenium Red

FIGURE 2.7A A liver cell fixed in 2% potassium permanganate (Luft, 1956) in distilled water for 2 hr at 4° C. Embedding in Epon, and no section staining, × 25,000.

Electron density is largely restricted to membranes and glycogen particles. The mitochondrial matrix, as well as the cytoplasmic ground substance, has a low electron density and consists of a diffuse, flocculent material. Mitochondria appear swollen and have only a few, peripherally located cristae. The endoplasmic reticulum forms short profiles with irregular dilatations. No ribosomes can be seen.

FIGURE 2.7B Epithelial cells of thick ascending limb in rat renal medulla perfusion fixed with a mixture of 4% paraformaldehyde and 0.4% picric acid (Zamboni and De Martino, 1967) in 0.1 M cacodylate buffer (pH 7.2), followed by 1% osmium tetroxide in cacodylate buffer (pH 7.2). Embedding in Epon, and section staining with uranyl acetate and lead citrate. × 25,000.

In its general appearance, this formaldehyde–picric acid–osmium tetroxide-fixed tissue resembles that of glutaraldehyde–osmium tetroxide double-fixed matrial, but the membranes are particularly well defined and the intercellular spaces have more constant widths.

FIGURE 2.7C Connective tissue from a cock comb fixed in 2% glutaraldehyde in a cacodylate buffer to which 0.3% ruthenium red (Luft, 1971) had been added. Postfixation was performed with 1% osmium tetroxide and 0.3% ruthenium red in a Veronal–acetate buffer. Embedding in Epon, and section staining with uranyl acetate and lead citrate. × 120,000.

The collagen fiber has a distinct cross-striated appearance and is connected to several nonstriated microfibrils at the nodes of the network in the extracellular space. Electron-dense granules are associated with the collagen fiber and the microfibril network, especially at the knots of the net.

COMMENTS

Although osmium tetroxide, glutaraldehyde, and formaldehyde have been the most commonly used fixatives for years, a large number of other chemical substances have also been used as fixatives. Some of these have been abandoned gradually, whereas others are used for specific purposes. This plate illustrates three such substances that show distinctly different fixing characteristics.

The destructive, oxidative property of permanganate is likely to have caused the ribosomes to disappear and the membranes to rupture. Ruthenium red, when used in electron microscopy, can be regarded either as fixative or as a stain. When used for connective tissue, it stains the collagen fibers and reveals an otherwise invisible network of fibrils and proteoglycans in the extracellular matrix. Ruthenium red has also been used widely to demonstrate glycocalyx and other surface-associated carbohydrate components. The inclusion of picric acid in paraformaldehyde fixatives gives good structural preservation and is also useful in immunocytochemical and *in situ* hybridization studies at the light microscope level (Meister *et al.*, 1989).

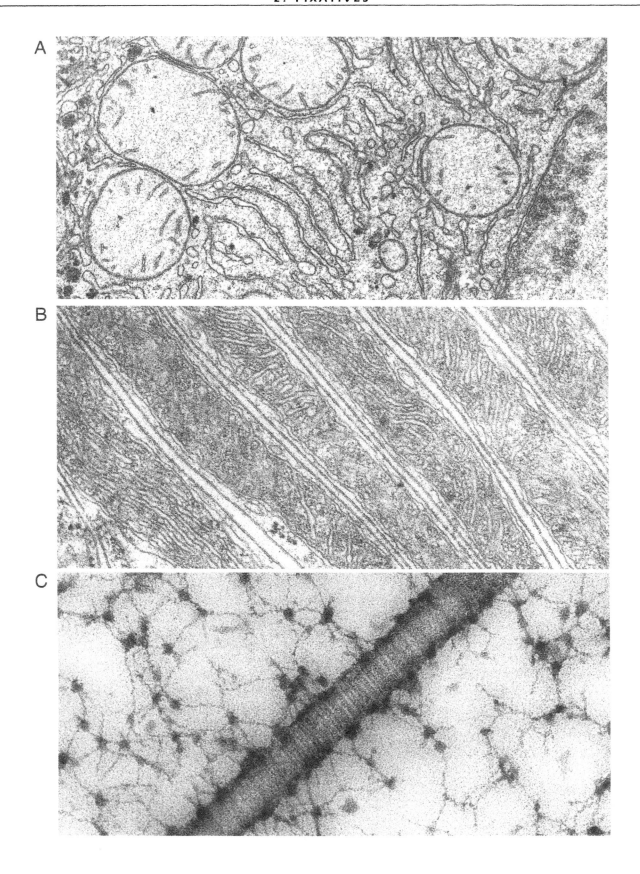

8. Lead Salts and Tannic Acid

FIGURE 2.8A Part of a proximal tubule cell from a rat kidney during endocytosis of low molecular weight dextran (Christensen and Maunsbach, 1979). The tissue was fixed with 2% glutaraldehyde, postfixed in 1% osmium tetroxide, and embedded in Epon. The sections were stained with lead citrate. × 23,000.

The lysosomes of the proximal tubule cells are distended and appear empty except for some intensely stained matrix material.

FIGURE 2.8B Preparation as in Fig. 2.8A except that the fixative contained 1.5% formaldehyde, 2.5% glutaraldehyde, 0.66% osmium tetroxide, and 0.2% lead citrate (Simionescu *et al.*, 1972) and was applied by dripping on the surface of the functioning kidney. × 21,000.

The appearance of the cell is the same as in Fig. 2.8A, although lysosomes, in addition to the electron-dense matrix material, are filled with a granular electron-dense material, representing the dextran absorbed by the cells.

FIGURE 2.8C Elastic fiber from human dermis fixed in 3% glutaraldehyde in 0.1 *M* phosphate buffer, postfixed in 2% osmium tetroxide, and embedded in Epon (Cotta-Pereira *et al.*, 1976). The section was stained with uranyl acetate and lead citrate. × 30,000.

The elastic fiber (between arrows) has an irregular outline and contains only a small amount of stained material. Collagen fibers (lower right corner) are only outlined faintly.

FIGURE 2.8D The fixation and staining procedure as in Fig. 2.8C except that 0.25% tannic acid (Mizuhira and Futaesaku, 1972) had been added to the glutaraldehyde. × 30,000.

The amorphous component of the elastic fibers is stained densely. Collagen fibrils to the left are recognizable by their banded pattern.

COMMENTS

Treatment of tissue with lead citrate during fixation enhances the contrast of certain carbohydrates, such as dextran, which otherwise would not show up. It acts both as a fixative and as an electron stain.

Tannic acid has gained wide use in ultrastructure research due to its ability to stabilize certain cellular components, including microtubules and other proteinaceous structures, as well as membrane lipids. It has been regarded as a so-called mordant, which enhances subsequent staining with lead salts (Simionescu and Simionescu, 1976). Tannic acid stabilizes many tissue components against extraction and structural changes during dehydration and embedding. High molecular mass (approximately 1400 Da) tannic acids do not penetrate cell membranes readily, whereas low molecular weight tannic acids such as penta- and hexagalloylglucoses may penetrate fixed cell membranes more readily.

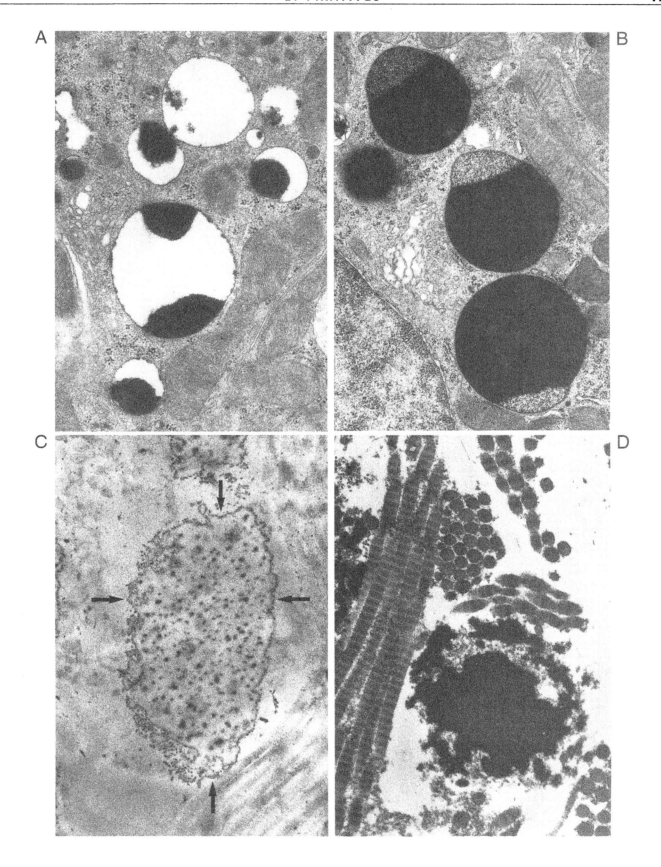

9. Uranyl Acetate Postfixation

FIGURE 2.9A A glomerular capillary wall from a rat kidney perfusion fixed with 1% glutaraldehyde in Tyrode solution and postfixed in osmium tetroxide. The tissue was rinsed in Veronal–acetate buffer at pH 7.2, and small pieces were immersed for 2 hr in 0.5% uranyl acetate in Veronal–acetate buffer, pH 5.2. The tissue was again rinsed in Veronal–acetate buffer, pH 7.2, dehydrated in acetone, and embedded in Vestopal. Thin sections were stained with lead citrate only. × 115,000.

The membranes are distinctly triple layered, and the plasma membrane has a greater width than the membranes of mitochondria. The endothelial fenestrae show a thin diaphragm, often with a central knob (single arrowhead). Some fenestrae show a double diaphragm (opposing arrowheads). Slit membranes are seen between the foot processes (arrows). Microtubules and microfilaments are prominent. Electron-dense strands connect the lamina densa of the basement membrane to the plasma membrane of the epithelial foot processes.

FIGURE 2.9B Same kind of specimen as in Fig. 2.9A perfusion fixed with a mixture of 2% glutaraldehyde, 1% tannic acid, and 1.8% sucrose in 0.1 M phosphate buffer and then postfixed in 1% uranyl acetate in distilled water, epoxy embedded, sectioned, and section stained with uranyl acetate and lead citrate. × 85,000.

Each foot process is delimited by a cell membrane with a triple-layered appearance. Diaphragms are seen between the foot processes and in some endothelial pores.

COMMENTS

Block staining with uranyl acetate reveals cytological details that may be difficult to visualize by conventional section staining with uranyl and lead salts. In addition, the image often appears more crisp. Block staining was originally introduced in the cytological technique as a secondary fixative to preserve bacterial DNA (Ryter and Kellenberger, 1958). A useful modification for general block staining of tissue is the procedure introduced by Karnovsky (1967), who used uranyl acetate in a maleate buffer (at 5.2 pH). In many laboratories, block staining is a routine procedure for preserving cytological details and for improving general contrast in micrographs recorded at both high (Fig. 2.9A) and low (see 4.1) magnifications. An additional advantage of block staining is the reduced risk of introducing contamination during section staining.

The addition of tannic acid to the primary fixative increases the general electron density of the specimen, presumably as a consequence of the mordanting action of tannic acid, allowing more heavy metal ions to be bound to cytoplasmic components. It also visualizes additional structural components, such as the protofilament architecture of the microtubules (see 1.12F).

A

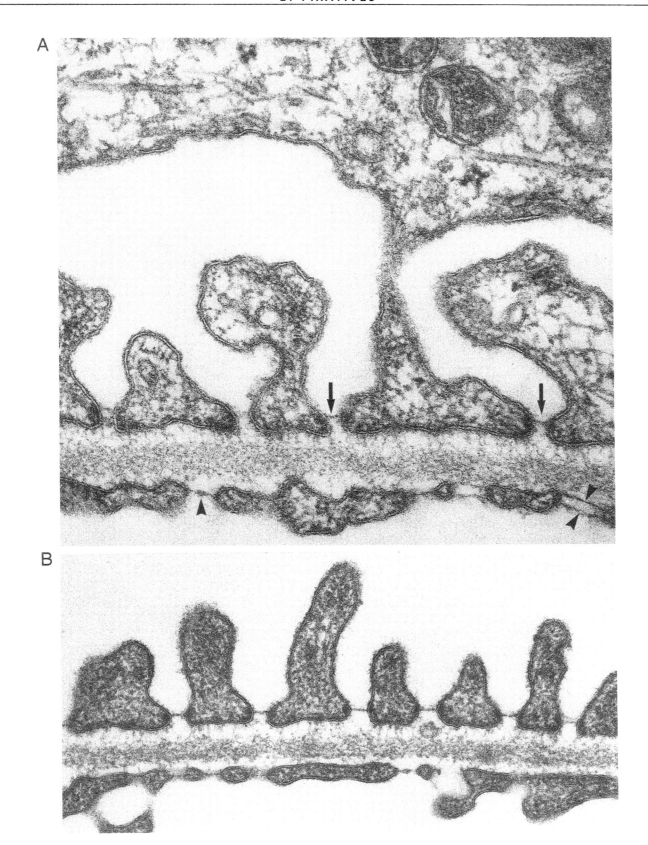

B

10. Tannic Acid–Uranyl Acetate Variations

FIGURE 2.10A A proximal tubule epithelium from a rat kidney perfusion fixed with 2% glutaraldehyde, 1% tannic acid, and 1.8 sucrose in a 0.1 M phosphate buffer. The tissue was postfixed in 1% uranyl acetate in distilled water for 3 hr and embedded in melamine (Nanoplast) according to Frösch and Westphal (1989). The thin section was analyzed unstained in an electron microscope. × 100,000.

Cell membranes are distinct, whereas mitochondrial membranes are barely visible. The mitochondrial matrix is much less stained than the cytoplasm.

FIGURE 2.10B Same preparation as in Fig. 2.10A except that the section was stained with lead citrate. × 75,000.

Section staining additionally enhances the contrast of the plasma membranes and also adds some density to the mitochondrial matrix and ribosomes, as well as membranes of the endoplasmic reticulum.

FIGURE 2.10C Same preparation as in Fig. 2.10B except that the tissue was embedded in epoxy resin (Polarbed) instead of melamine. × 75,000.

The cell structures are similar to those shown in Fig. 2.10B, although ground cytoplasm, mitochondrial matrix, and cell membranes are more intensely stained.

COMMENTS

Inclusion of tannic acid in a glutaraldehyde fixative in combination with postfixation in uranyl acetate enhances greatly the electron density of membraneous components and microtubules as first demonstrated by Mizuhira and Futaesaku (1972). These micrographs illustrate that membranes may also be distinctly imaged without osmium tetroxide, but it is likely that tannic acid and uranyl acetate only penetrate to a limited extent into mitochondria. The advantage with this fixation procedure is that a richness of cytological details in the cells is revealed, although the proteins undoubtedly are just as denatured as following osmium tetroxide fixation. There is no obvious difference in the sharpness of plasma membranes between tissue embedded in melamine and tissue in epoxy resin (compare Figs. 2.10B and 2.10C).

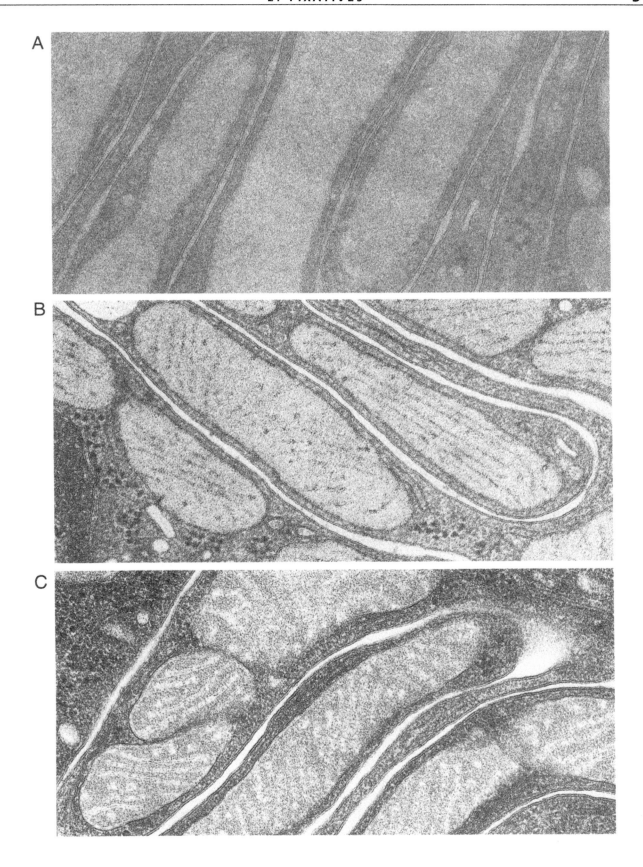

11. Osmium Tetroxide–Potassium Ferrocyanide

FIGURE 2.11A A Leydig cell from a rat testis fixed by perfusion with 5% glutaraldehyde in 0.2 M s-collidine buffer followed by postfixation in 1% osmium tetroxide with 1.5% potassium ferrocyanide. The tissue was embedded in Epon, and the sections were stained with uranyl acetate and lead citrate. × 115,000.

Both membranous structures and intermediate filaments appear distinct.

FIGURE 2.11B A proximal tubule cell from a rat kidney fixed with 2.5% glutaraldehyde and 2% formaldehyde in 0.1 M cacodylate buffer and postfixed in 1% osmium tetroxide solution containing 0.5% potassium ferrocyanide. The tissue was embedded in Epon, and the sections were stained with uranyl acetate and lead citrate. × 50,000.

Membranes stand out with a high electron density against a uniformly stained cytoplasm.

COMMENTS

Potassium ferrocyanide added to the osmium tetroxide probably forms a stable complex (Karnovsky, 1971). This method renders certain cellular components, notably membranes, microtubules, and filaments, more electron dense than after postfixation in only osmium tetroxide, and the triple-layered appearance of the cytoplasmic membranes becomes accentuated (Fig. 2.11A). The distinct but balanced contrast of membranes and other cellular components, as observed in Fig. 2.11B, facilitates the evaluation of overall cell architecture. Notice that the width of the lateral intercellular space is decreased as compared to the ordinary osmium tetroxide or aldehyde/osmium tetroxide double-fixation patterns (compare with Fig. 2.2).

A

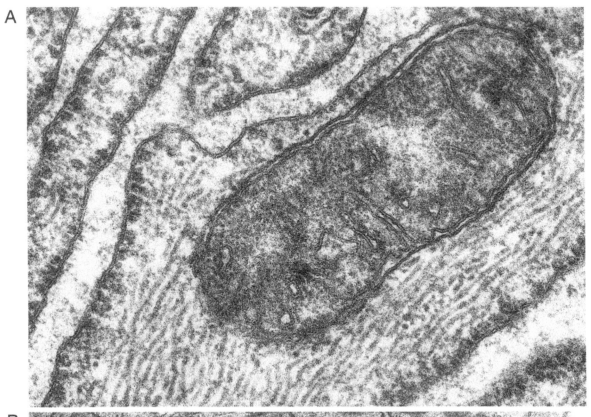

B

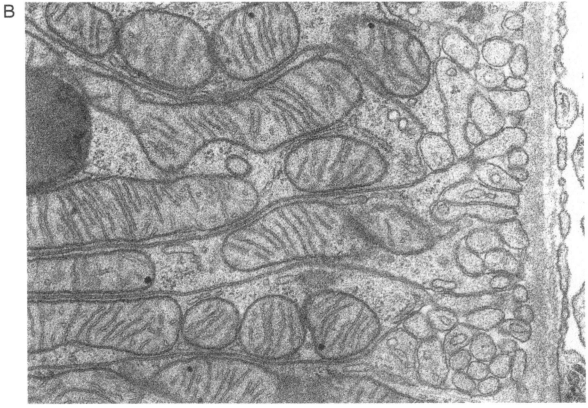

12. Osmium Tetroxide Artifacts

FIGURE 2.12A A brush border of proximal tubule cells from a rat kidney fixed in 1% osmium tetroxide in bicarbonate buffer, embedded in Vestopal, and section stained with uranyl acetate and lead citrate. × 32,000.

In many places, microvilli appear joined by one or more cytoplasmic bridges (arrows). Some microvilli appear fused at their tips.

FIGURE 2.12B Proximal tubule cells of rat kidney, fixed in 1% osmium tetroxide in phosphate buffer, embedded in Epon, and section stained with uranyl acetate and lead citrate. × 70,000.

Adjacent cells are separated by an intercellular space of fairly constant width and limited by continuous cell membranes.

FIGURE 2.12C Area from the same section as in Fig. 2.12B. × 70,000.

In places there are apparent bridges between the cells (arrows).

COMMENTS

These figures exemplify artifacts that are typical of primary osmium tetroxide fixation. In these cases, osmium tetroxide has caused membrane ruptures. In Fig. 2.12A, the ruptured membranes of the microvilli have secondarily fused with neighboring membranes, thus forming artificial bridges between the microvilli. In Fig. 2.12C, bridges have formed between separate cells and create a false impression of cytoplasmic communications.

Many authors have falsely interpreted patterns in osmium-fixed tissues to show cell fusion or endoplasmic reticulum fragmentation during certain experimental conditions. This specific osmium tetroxide artifact may be present in only part of the tissue, as illustrated in Figs. 2.12B and 2.12C, and may be difficult to identify, as the tissue otherwise may appear well preserved. With primary aldehyde fixation, these artifacts do not appear.

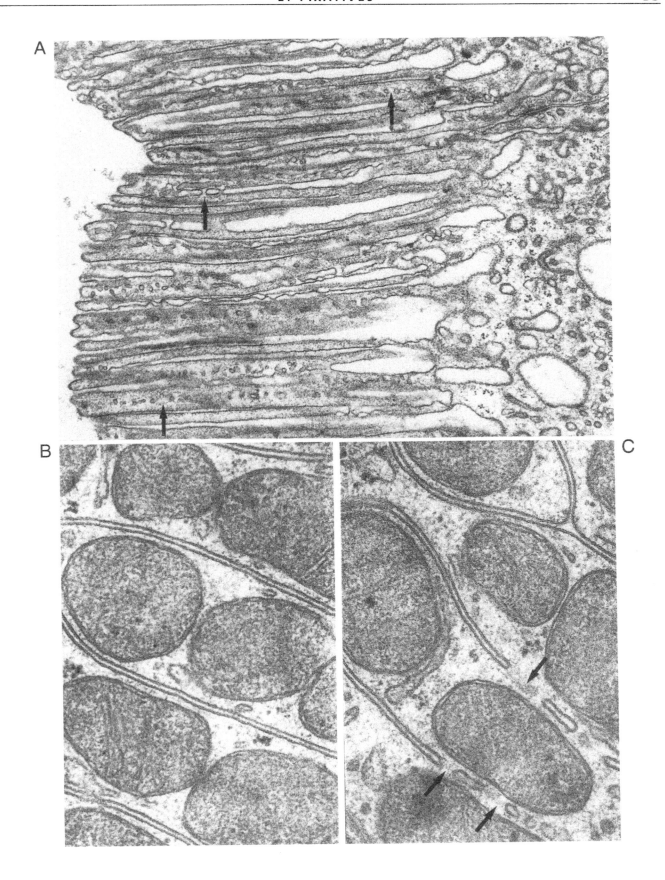

13. Glutaraldehyde Artifacts

FIGURE 2.13A Apical region of a proximal tubule cell from a rat kidney perfusion fixed with 2% glutaraldehyde and postfixed in 1% osmium tetroxide. The tissue was block stained with uranyl acetate and embedded in Epon, and the sections were stained with lead citrate. × 88,000.

In addition to endocytic vesicles surrounded by a single membrane, there are cup-shaped or doughnut-like profiles (asterisks). The cup-shaped profile has an empty center that connects with the surrounding cytoplasm. The doughnut-shaped profile probably represents a section through such a cup-shaped structure.

FIGURE 2.13B Parts of proximal tubule cells fixed with 3% glutaraldehyde and postfixed in 1% osmium tetroxide. The tissue was embedded in Epon, and the sections were stained with uranyl acetate and lead citrate. × 37,000.

One of the mitochondria has a region devoid of matrix material but contains a densely stained myelin-like material.

FIGURE 2.13C Parts of proximal tubule cells of a rat kidney perfusion fixed with 1% glutaraldehyde in modified Tyrode solution (pH 7.2), postfixed with 1% osmium tetroxide, embedded in Epon, and section stained with uranyl acetate and lead citrate. × 80,000.

In some places the cell membranes show small outbulgings connected with or consisting of a highly electron-dense material.

FIGURE 2.13D Part of intercellular space between two kidney epithelial cells. Fixation and further processing are as in Fig. 2.13A. × 70,000.

The intercellular space contains electron-dense, layered material. Some material is in contact with one of the cell membranes.

COMMENTS

The presence of myelin-like figures at the cell membrane (Fig. 2.13C) in the cytoplasm or intercellular spaces is characteristic of tissues fixed with glutaraldehyde or other aldehydes. Their presence is probably due to the inability of the aldehydes to retain lipids in their original locations. During aldehyde fixation, lipids may become dislocated and subsequently precipitate during osmium postfixation. Although prominent myelin-like materials usually are recognized as artifacts, a small amount of such material may easily be misinterpreted as being a true cell structure. Mitochondria and cell membranes are particularly vulnerable to this kind of artifact. (Fig. 2.13B).

The vesicle-containing outbulgings observed after aldehyde fixation are presumably caused by aldehyde fixation. They appear more often in embryological than in adult tissues and can easily be mistaken for real cellular structures involved in a secretory process. The cup-shaped structures observed following glutaraldehyde fixation (Fig. 2.13A) are interpreted as collapsed endocytic vacuoles, but the mechanism of their formation is unclear.

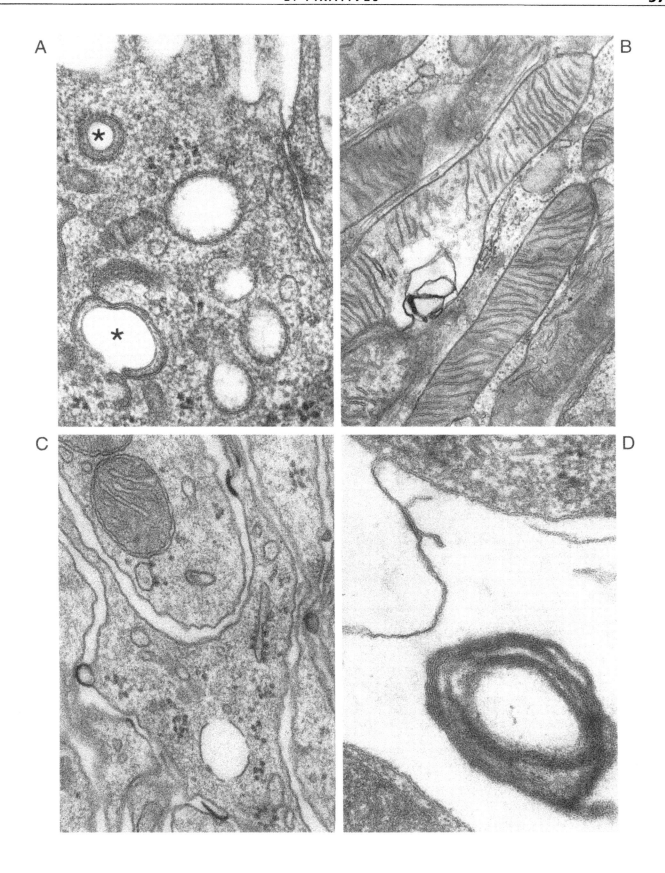

References

Afzelius, B. (1959). Electron microscopy of the sperm tail: Results obtained with a new fixative. *J. Biophys. Biochem. Cytol.* **5,** 269–278.

Bahr, G. (1954). Osmium tetroxide and ruthenium tetroxide and their reactions with biologically important substances. *Exp. Cell Res.* **7,** 457–479.

Christensen, E. I., and Maunsbach, A. B. (1979). Effects of dextran on lysosomal ultrastructure and protein digestion in renal proximal tubule. *Kidney Int.* **16,** 301–311.

Cotta-Pereira, G., Rodrigo, F. G., and David-Ferreira, J. F. (1976). The use of tannic acid-glutaraldehyde in the study of elastic and elastic-related fibers. *Stain Technol.* **51,** 7–11.

Dallai, R., and Afzelius, B. A. (1990). Microtubular diversity in insect spematozoa: Results obtained with a new fixative. *J. Struct. Biol.* **103,** 164–179.

Ebersold, H. R., Cordier, J.-L., and Luthy, P. (1981). Bacterial mesosomes: Method dependent artifacts. *Arch. Microbiol.* **130,** 19–22.

Ellis, E. A., and Anthony, D. W. (1979). A method for removing precipitate from ultrathin sections resulting from glutaraldehyde-osmium tetroxide fixation. *Stain Technol.* **54,** 282–285.

Ericsson, J. L. E., and Biberfeld, P. (1967). Studies on aldehyde fixation: Fixation rates and their relation to fine structure and some histochemical reactions in the liver. *Lab. Invest.* **17,** 281–298.

Fox, C. H., Johnson, F. B., Whiting, J., and Roller, P. P. (1985). Formaldehyde fixation. *J. Histochem. Cytochem.* **33,** 845–853.

Frösch, D., and Westphal, C. (1989). Melamine resins and their application in electron microscopy. *Electron Microsc. Rev.* **2,** 231–255.

Glauert, A. M. (1974). Fixation, dehydration and embedding of biological specimens. *In* "Practical Methods in Electron Microscopy" (A. M. Glauert, ed.), Vol. 3. North-Holland, Amsterdam.

Goldfischer, S., Kress, Y., Coltoff-Schiller, B., and Berman, J. (1981). Primary fixation in osmium-potassium ferrocyanide: The staining of glycogen, glycoproteins, elastin, and reticular structure, and intercisternal trabeculae. *J. Histochem. Cytochem.* **29,** 1105–1111.

Griffiths, G. (1993). "Fine Structure Immunocytochemistry." Springer-Verlag, Berlin.

Hopwood, D. (1985). Cell and tissue fixation, 1972–1982. *Histochem. J.* **17,** 389–442.

Karnovsky, M. J. (1965). A formaldehyde-glutaraldehyde fixative of high osmolarity for use in electron microscopy. *J. Cell Biol.* **27,** 137A–138A.

Karnovsky, M. J. (1967). The ultrastructural basis of capillary permeability studied with peroxidase as a tracer. *J. Cell Biol.* **35,** 213–236.

Karnovsky, M. J. (1971). Use of ferrocyanide-reduced osmium tetroxide in electron microscopy. *In* "Proceedings of the Fourteenth Annual Meeting of the American Society for Cell Biology," p. 146.

Kellenberger, E., and Ryter, A. (1956). Fixation et inclusion de materiel nucléaire de *Escherichia coli. Experientia* **12,** 420–421.

Kellenberger, E., Johansen, R., Maeder, M., Bohrmann, B. Stauffer, E., and Villiger, W. (1992). Artefacts and morphological changes during chemical fixation. *J. Microsc. (Oxford)* **168,** 181–201.

Langenberg, W. G. (1978). Relative speed of fixation of glutaraldehyde and osmic acid in plant cells measured by grana appearance in chloroplasts. *Protoplasma* **94,** 167–173.

Larsson, L. (1975). Effects of different fixatives on the ultrastructure of the developing proximal tubule in the rat kidney. *J. Ultrastruct. Res.* **51,** 140–151.

Lee, R. M. K. W., Garfield, R. E., Forrest, J. B., and Daniel, E. E. (1980). Dimensional changes in cultured smooth muscle cells due to preparatory processes for transmission electron microscopy. *J. Microsc. (Oxford)* **120,** 85–91.

Luft, J. H. (1956). Permanganate: A new fixative for electron microscopy. *J. Biophys. Biochem. Cytol.* **2,** 799–801.

Luft, J. H. (1971). Ruthenium red and violet. I. Chemistry, purification, methods of use for electron microscopy, and mechanisms of action. *Anat. Rec.* **171,** 347–368.

Maunsbach, A. B. (1966a). The influence of different fixatives and fixation methods on the ultrastructure of rat kidney proximal tubule cells. I. Comparison of different perfusion fixation methods and of glutaraldehyde, formaldehyde and osmium tetroxide fixatives. *J. Ultrastruct. Res.* **15,** 242–282.

Maunsbach, A. B. (1966b). The influence of different fixatives and fixation methods on the ultrastructure of rat kidney proximal tubule cells. II. Effects of varying osmolality, ionic strength, buffer system and fixative concentration of glutaraldehyde solutions. *J. Ultrastruct. Res.* **15,** 283–309.

Maunsbach, A. B. (1998). Fixation of cells and tissues for transmission electron microscopy. *In* "Cell Biology: A Laboratory Handbook" (J. E. Celis, ed.), 2nd Ed., Vol. 3, pp. 249–259. Academic Press, San Diego.

McDonald, K. (1984). Osmium ferricyanide fixation improves microfilament preservation and membrane visualization in a variety of animal cell types. *J. Ultrastruct. Res.* **86,** 107–118.

Meister, B., Fryckstedt, J., Schalling, M., Cortés, R., Hökfelt, T., Aperia, A., Hemmings, H. C., Nairn, A. C., Ehrlich, M., and Greengard, P. (1989). Dopamine- and cAMP-regulated phosphoprotein (DARPP-32) and dopamine DA$_1$-agonist-sensitive Na$^+$,K$^+$-ATPase in renal tubule cells. *Proc. Natl. Acad. Sci. USA* **86,** 8068–8072.

Mizuhira, V., and Futaesaku, Y. (1972). New fixation for biological membranes using tannic acid. *Acta Histochem. Cytochem.* **5,** 233–236.

Pease, D. C. (1964). "Histological Techniques for Electron Microscopy," 2nd Ed. Academic Press, New York.

Sabatini, D. D., Bensch, K. G., and Barrnet, R. J. (1963). Cytochemistry and electron microscopy: The preservation of cellular ultrastructure and enzymatic activity by aldehyde fixation. *J. Cell Biol.* **17,** 19–58.

Simionescu, N., and Simionescu, M. (1976). Galloylglucoses of low molecular weight as mordant in electron microscopy. I. Procedure and evidence for mordanting effect. *J. Cell Biol.* **70,** 608–621.

Simionescu, N., Simionescu, M., and Palade, G. E. (1972). Permeability of intestinal capillaries: Pathway followed by dextrans and glycogens. *J. Cell Biol.* **53,** 365–392.

Sjöstrand, F. S. (1967). "Electron Microscopy of Cells and Tissues," Vol. 1. Academic Press, New York.

Sjöstrand, F. S. (1997). The physical chemical basis for preserving cell structure for electron microscopy at the molecular level and available preparatory methods. *J. Submicrosc. Cytol. Pathol.* **29,** 157–172.

Terzakis, J. A. (1968). Uranyl acetate, a stain and a fixative. *J. Ultrastruct. Res.* **22,** 168–184.

Tormey, J. McD. (1964). Differences in membrane configuration between osmium tetroxide-fixed and glutaraldehyde-fixed ciliary epithelium. *J. Cell Biol.* **23,** 658–664.

Trump, B. F., and Bulger, R. E. (1966). New ultrastructural characteristics of cells fixed in a glutaraldehyde-osmium tetroxide mixture. *Lab. Invest.* **15,** 368–379.

Zamboni, L., and De Martino, C. (1967). Buffered picric acid formaldehyde: A new rapid fixative for electron microscopy. *J. Cell Biol.* **148,** 35.

FIXATIVE VEHICLE

In addition to the choice of the fixative itself, careful consideration must be given to the vehicle of the fixative. In the early days of electron microscopy, most investigators did not pay much attention to the characteristics of the vehicle, but now it is evident that even rather small modifications in its composition may dramatically influence the quality of tissue preservation. Osmolality, colloid osmotic pressure, and choice of buffer will influence the final result. Indeed, the composition of the fixative vehicle is usually more important than the concentration of the fixative itself (e.g., 2 or 4% glutaraldehyde, 1 or 2% osmium tetroxide).

The best preservation is often obtained when the composition of the vehicle has been adjusted to approach that of the tissue fluid, particularly with respect to osmotically active ions and molecules. Thus marine animals require a fixative solution of greater osmolality than do related freshwater species. A fixative proven successful for one tissue, such as the rat liver, is not necessarily suitable for other tissues when applied without modifications. The reason may be differences in tissue composition. As a rule, each particular biological object requires a systematic exploration of the optimal composition of the fixative vehicle. A striking example hereof is the necessity to add different amounts of sodium chloride or other osmotically active substances to the fixative when different levels of

the renal medulla are studied. In the living kidney there is a gradual increase in concentration of sodium chloride toward the tip of the medulla, and one particular fixative is not suitable for all medullary levels.

This chapter illustrates a number of modifications of the fixative vehicle that influence tissue fine structure, despite the fixative itself being the same.

Several pioneers of biological electron microscopy fixed their specimens in watery solutions of osmium tetroxide. An influential advance was that by George Palade (1952), who suggested that the osmium tetroxide fixative should be dissolved in a Veronal acetate buffer to counteract the "wave of acidity" that supposedly permeates the cell on its fixation. Osmium tetroxide is sometimes called osmic acid, although as a nonelectrolyte it does not influence the pH value of its solvent. Fritiof Sjöstrand (1953) emphasized the importance of using a vehicle that is isotonic with the tissue to be examined; this may actually be the main advantage of dissolving osmium tetroxide in a buffer.

Palade, G. E. (1952). *Anat. Rec.* **114**, 427–451.
Sjöstrand, F. S. (1953). *J. Cell. Comp. Physiol.* **42**, 15–44.

1. Absence and Presence of Buffer

FIGURE 3.1A A rat liver cell immersion fixed in 1% osmium tetroxide in distilled water, embedded in Epon, and section stained with uranyl acetate and lead citrate. × 30,000.

There is a pronounced swelling of the cell and of some organelles. The cytoplasmic ground substance has a low electron density. The endoplasmic reticulum is fragmented and shows irregular dilatations.

FIGURE 3.1B A rat liver cell fixed in 1% osmium tetroxide in 0.9% sodium chloride. Embedding and section staining as in Fig. 3.1A. × 30,000.

The cytoplasmic ground substance is more stained than in Fig. 3.1A, the fragmentation of the endoplasmic reticulum is less pronounced, and cisternae are less distended as compared to Fig. 3.1A.

FIGURE 3.1C A rat liver cell fixed in 1% osmium tetroxide in 0.1 *M* phosphate buffer. Further processing as in Fig. 3.1A. × 30,000.

A characteristic feature of cells fixed with phosphate-buffered osmium tetroxide is a higher degree of retention of cytoplasmic ground substance and mitochondrial matrix than in the unbuffered hypotonic (Fig. 3.1A) or near isotonic fixative (Fig. 3.1B).

COMMENTS

A comparison between Figs. 3.1B and 3.1C shows that a better quality of preservation is obtained when the osmium tetroxide is dissolved in a phosphate buffer than when it is dissolved in isotonic sodium chloride, yet the difference is not dramatic. A common feature of the unbuffered fixatives is a less stained appearance of the cytoplasmic ground substance, and to some degree of the mitochondrial matrix, possibly due to an acidification and extraction process that occurs in the absence of a buffer. The swelling of the endoplasmic reticulum in Fig. 3.1A is probably due to the strong hypotonicity of this fixative, as the addition of sodium chloride largely prevents these effects.

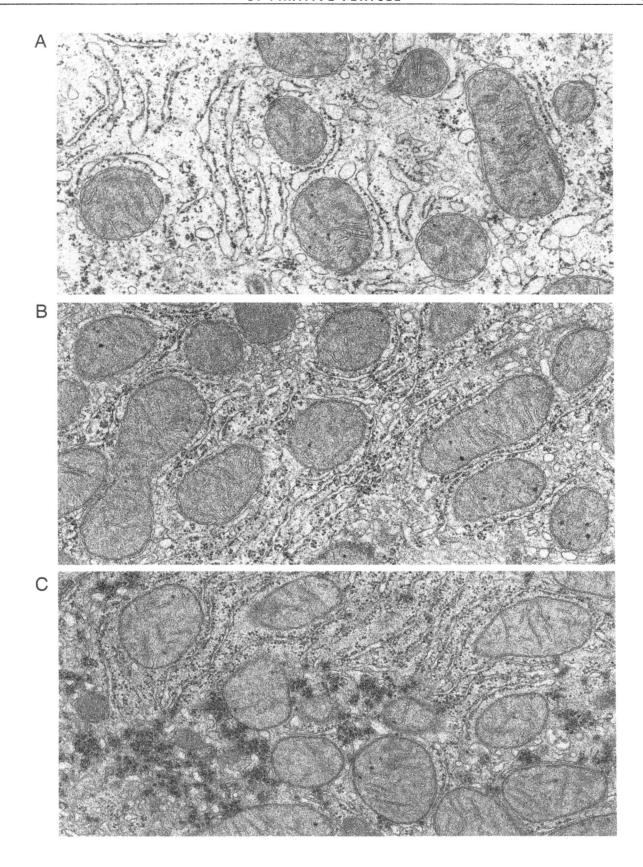

2. Comparison of Buffers

FIGURE 3.2A Cytoplasm of a rye leaf cell fixed with 3% glutaraldehyde in 0.05 M phosphate buffer and post-fixed in 1% osmium tetroxide in the same buffer. The tissue was embedded in Epon, and the section was stained with uranyl acetate and lead citrate. $\times$ 38,000.

The cell wall of the protoplast is seen at the lower left and the vacuole at the upper right. The membrane of the vacuole is distinct and intact. The thylacoid membranes stand out in good contrast.

FIGURE 3.2B Cytoplasm in a rye leaf cell, prepared as in Fig. 3.2A, except that the fixative buffer was 0.05 M cacodylate. $\times$ 38,000.

The membrane of the vacuole (lower right) is somewhat irregular and in part detached from the cytoplasm. The cytoplasm contains a large chloroplast with thylacoid membranes, which exhibit fairly low contrast.

FIGURE 3.2C AND 3.2D Enlarged portions of cytoplasm from rye leaf cells fixed in 5% glutaraldehyde in 0.05 M cacodylate buffer. The fixative used for Fig. 3.2D contained 5 mM calcium chloride, whereas that used in Fig. 3.2C was calcium free. Postfixation and further processing are as in Fig. 3.2A. $\times$ 60,000.

The appearance of the chloroplasts is quite similar in Figs. 3.2C and 3.2D. However, without calcium ions the cytoplasm outside the chloroplast adjacent to the cell membrane or vacuole membrane (tonoplast) appears wider (asterisks) than with the calcium-containing fixative. In addition, the chloroplast ribosomes are less distinct.

COMMENTS

A large number of different buffers have been used as vehicles for fixatives over the years. Phosphate and cacodylate buffers are now commonly used and with comparable results in most tissues. However, some differences exist in some tissues, and variable degrees of extraction of tissue components with different buffers have been reported. Thus Salema and Brandão (1973) demonstrated that PIPES [piperazine-N-N^1-bis (2-ethanol sulfonic acid)] buffer improved the fixation of plant cells. Coetzee and van der Merwe (1984) demonstrated that cacodylate buffer extracted more substances from plant tissue samples than did phosphate buffers. A comparison between Figs. 3.2A and 3.2B also suggests differences between these buffers, as the overall contrast in Fig. 3.2A is better than in Fig. 3.2B in this particular tissue. In other cells, only minor differences, if any, have been noted between these buffers and it appears largely inconsequential which one is used. Veronal acetate, which was used extensively in osmium tetroxide fixatives, reacts chemically with aldehydes and is thus unsuitable in such solutions. Collidine buffer has been shown to extract much cytoplasmic material and is now rarely used. In contrast, the use of imidazole buffer in osmium tetroxide solutions results in a better preservation of lipids than do other buffers.

The addition of specific ions to fixative solutions has also been advocated: Calcium ions are considered to improve fixation quality in general and magnesium ions are considered to improve the visibility of dynein arms in cilia and flagella. The different appearances of the cytoplasms and chloroplast ribosomes in Figs. 3.2C and 3.2D are consistent with an effect of calcium.

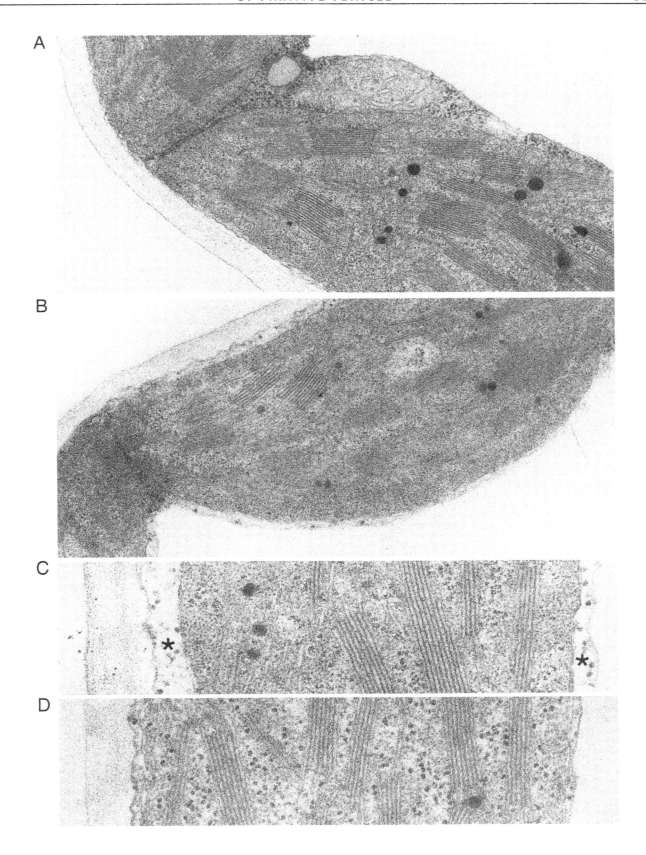

3. Osmolality of Perfusion Fixatives

FIGURE 3.3A Proximal tubules from a rat kidney cortex perfusion fixed with 1% glutaraldehyde in a modified Tyrode solution (pH 7.2). The Tyrode solution contained half the regular amount of sodium chloride (4.0 rather than 8.0 g/liter). The osmolality of the vehicle was about 180 mOsm/kg H_2O; that of the entire fixation solution was about 300 mOsm/kg H_2O. The tissue was postfixed with 1% osmium tetroxide in the same buffer and embedded in Vestopal, and the section was stained with uranyl acetate and lead citrate. $\times$ 3500.

The tubule lumen is open, but the brush border shows many irregularities. A few cells even appear disrupted (arrows). The ground cytoplasm is much less electron dense than the mitochondria. Nuclei have round and smooth contours.

FIGURE 3.3B Proximal tubules perfusion fixed with 1% glutaraldehyde in a Tyrode solution containing three-fourths (6.0 g/liter) the regular amount of sodium chloride (Maunsbach, 1966a,b). Osmolalities of the vehicle and the complete fixative solution were 235 and 354 mOsm/kg H_2O, respectively. $\times$ 3500.

The tubule lumen is round and has a smooth contour, and the brush border has a regular arrangement. The difference in electron density between the mitochondria and the ground cytoplasm is smaller than in Fig. 3.3A.

FIGURE 3.3C Proximal tubules perfusion fixed with 1% glutaraldehyde in a Tyrode solution containing twice the regular amount of sodium chloride (16.0 g/liter). Osmolalities of the vehicle and the complete fixative solution were 530 and 650 mOsm/kg H_2O, respectively. $\times$ 4500.

The tubule epithelium has a decreased height, and large intercellular spaces are evident between the epithelial cells. Ground cytoplasm, nuclei, and mitochondria have a high electron density.

COMMENTS

With a low osmolality of the vehicle, the cells tend to swell or even burst, whereas with a high vehicle osmolality, the cells shrink and the intercellular spaces widen. It is known from light microscopic observations of functioning tubules that the lumens are round and smooth, as in Fig. 3.3B. Furthermore, *in vivo* observations show an absence of large intercellular spaces such as those seen in Fig. 3.3C. It is concluded that Fig. 3.3B is more representative of functioning tubules than Figs. 3.3A and 3.3C when analyzed at this level of resolution. This conclusion does not necessarily imply that an optimal fixation is also obtained for all intracellular components.

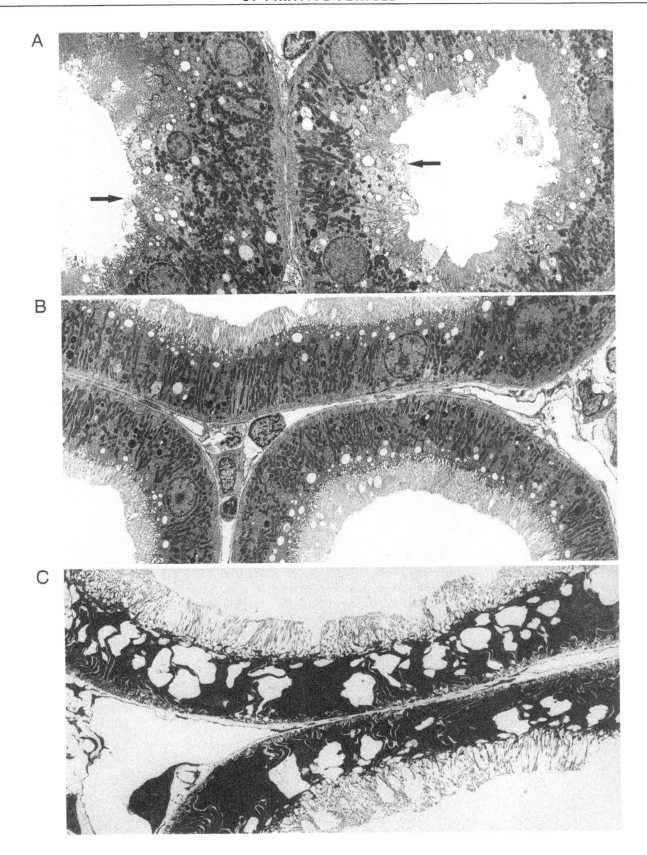

4. Effects of Osmolality on Cell Shape

FIGURE 3.4A Proximal tubule cells from a rat kidney fixed as in Figure 3.3A. × 15,000.

The ground cytoplasm has a low electron density, and mitochondrial membranes stand out with good contrast. In many places the endoplasmic reticulum is fragmented into small vesicles.

FIGURE 3.4B Part of Fig. 3.3C at a higher magnification. × 15,000.

Both ground cytoplasm and mitochondria are electron dense to an extent that the latter cannot even be distinguished. The nucleus is shrunken with an irregular outline, and the intercellular spaces are dilated greatly.

COMMENTS

These two figures illustrate that intracellular structures, as well as cell shape and size are influenced greatly by the composition of the fixative vehicle. Strikingly different results are obtained if the total osmolality is due to the fixative itself (here glutaraldehyde) or to nonfixative components in the vehicle, such as sodium chloride or sucrose. The effective osmotic force of the fixation solution depends on added salts or sucrose to a much higher degree than on glutaraldehyde or formaldehyde.

The following empirical formula may be useful in estimating the "effective osmolality" (Osm_{eff}) of the fixative:

$$Osm_{eff} = Osm_{veh} + 0.3 \times Osm_{glut} + 0.1 \times Osm_{form}$$

where the subscripts veh, glut, and form specify the osmolality of the vehicle, glutaraldehyde, and formaldehyde, respectively. The reason why formaldehyde has only one-tenth or less of the expected osmotic effect on cells is presumably that formaldehyde, and to some extent other aldehydes, penetrates readily through cellular membranes.

The sensitivity to vehicle osmolality varies considerably from cell type to cell type. For mammalian tissues with a high water content, it appears that Osm_{eff} should be approximately 300–400 mOsm/kg H_2O. The formula may be used as a guideline when designing the composition of a fixative that contains aldehydes.

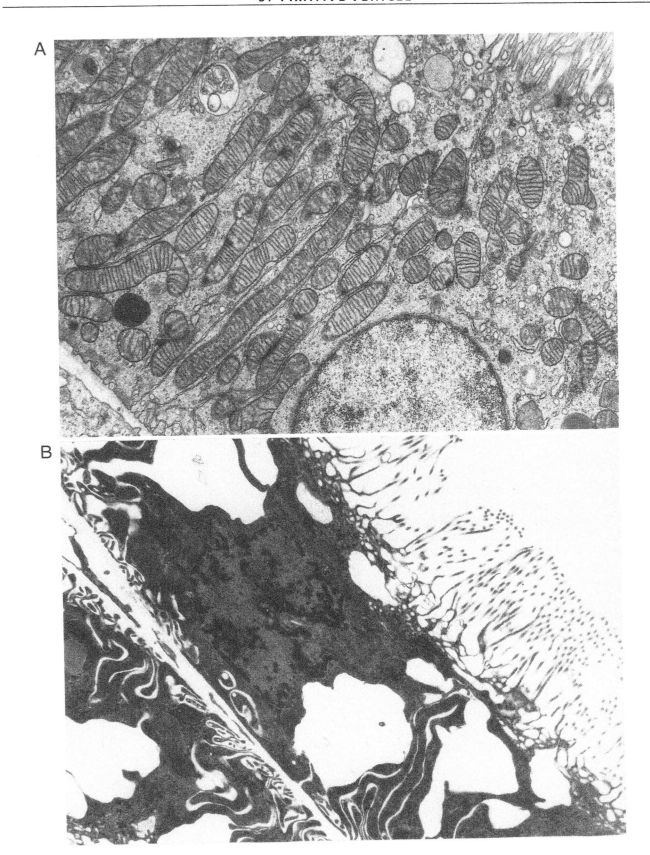

5. Effects of Osmolality on Cell Organelles

FIGURE 3.5A A proximal tubule cell fixed as in Fig. 3.3A using a hypotonic Tyrode solution containing only half (4.0 g/liter) the regular amount of sodium chloride. Postfixation and further treatment are as in Fig. 3.3A. × 90,000.

Mitochondria are swollen and their matrix and the cytoplasmic ground substance have a low electron density. Adjacent cell membranes are closely apposed at places (arrows).

FIGURE 3.5B A lysosome from the same preparation as in Fig. 3.5A. × 90,000.

The lysosomal membrane shows several ruptures, and adjacent profiles of the endoplasmic reticulum are fragmented.

FIGURE 3.5C Detail of proximal tubule cells of a rat kidney fixed as in Fig. 3.3C with 1% glutaraldehyde in a Tyrode solution made hypertonic with twice (16.0 g/liter) the regular amount of sodium chloride. Postfixation and further treatment are as in Fig. 3.3C. × 160,000.

There is a considerable shrinkage of the cell components. The mitochondrial matrix is almost as electron dense as the outer layers of the triple-layered mitochondrial membranes, which hence are visible only by their middle, electron-translucent layer. This creates what appears to be a reversed contrast of the mitochondrial structure.

COMMENTS

Different amounts of osmotically active substances in the fixative solutions not only change the general shape of the cells, but also influence the appearance of cytoplasmic details. Cell membrane contacts that resemble tight junctions may be formed following fixation in hypotonic solutions and the lysosomal membranes may rupture. Thus when observing unusual membrane relations or ruptured cell organelles in normal or pathological cells, the possible influence of the fixative composition must be taken into account. It is evident that small cytoplasmic components, such as filaments or vesicles, may be very difficult to detect in cells fixed with a fixative in a hypertonic vehicle as in Fig. 3.5C. However, they may be easy to see in cells fixed with a hypotonic vehicle, although the general cell architecture then is distorted due to cell swelling.

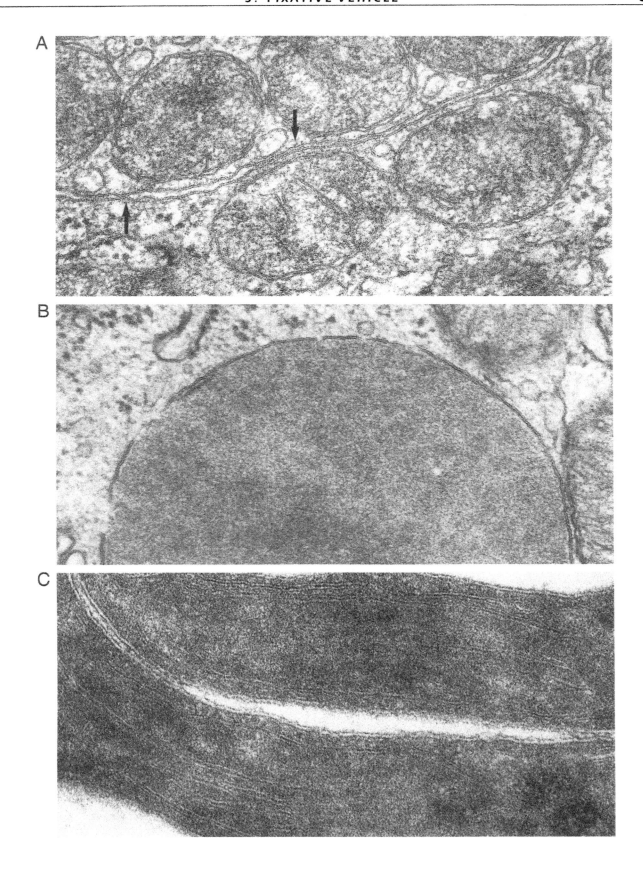

6. Adjustment of Osmolality with Sucrose

FIGURE 3.6A Cells in the thick ascending limb of the loop of Henle in the outer medulla of a rat kidney. The kidney was perfusion fixed with 0.25% glutaraldehyde and 4% formaldehyde in 0.1 M cacodylate buffer. Postfixation was performed in 1% osmium tetroxide in the same buffer, embedding in Epon, and section staining with uranyl acetate and lead citrate. $\times$ 40,000.

Mitochondria are rather well preserved, but the cytoplasmic ground substance appears largely empty. Vesicular profiles of the endoplasmic reticulum are dilated.

FIGURE 3.6B A similar preparation as in Fig. 3.6A except that the perfusate also contained 0.3 M sucrose. $\times$ 40,000.

Mitochondria and intercellular spaces are similar to those in Fig. 3.6A, but cytoplasm, including profiles of endoplasmic reticulum, is less dilated.

COMMENTS

The osmolality of the interstitium and blood in the renal medulla is unusually high due to an increased sodium chloride content in this part of the kidney. When the tissue is fixed with a solution that is suitable for other tissues, including the renal cortex, the medullary cells behave as osmometers and swell. This is illustrated in Fig. 3.6A. The swelling is at least partly a selective one, and affects the cytoplasm and endoplasmic reticulum more than mitochondria. When the high osmotic pressure of the cytoplasm is balanced by the inclusion of sucrose in the perfusate, cell swelling is diminished or absent. Because different levels of the renal medulla have different osmolalities, fixatives with different osmolalities have to be used for different levels of the kidney, as demonstrated by Bohman (1974).

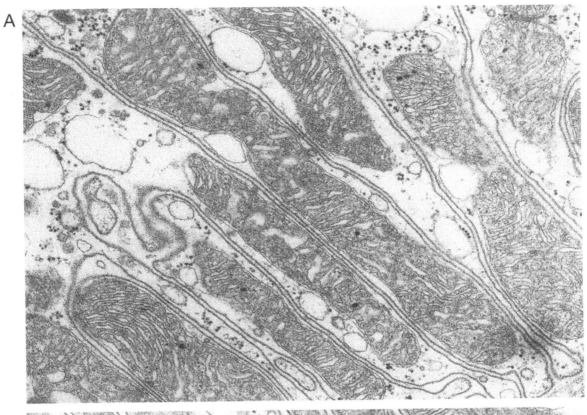

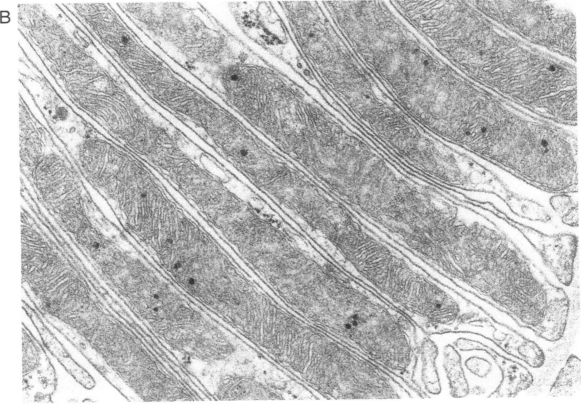

7. Colloid Osmotic Pressure: Low Magnification

FIGURE 3.7A Exocrine pancreas cells perfusion fixed with 2% glutaraldehyde in 0.1 *M* cacodylate buffer. During perfusion–fixation the pancreas was seen to swell gradually. The tissue was postfixed with 1% osmium tetroxide and embedded in Vestopal, and the section was stained with uranyl acetate and lead citrate. × 1400.

The micrograph shows several acini that are separated by wide, extravascular spaces and capillaries.

FIGURE 3.7B An exocrine pancreas after perfusion–fixation with 2% glutaraldehyde in 0.1 *M* cacodylate buffer, to which 2% dextran was added according to Bohman and Maunsbach (1970). No obvious size change of the pancreas was observed during the perfusion. Further treatment was as in Fig. 3.7A. × 1400.

The general ultrastructure of the cells is similar to that of Fig. 3.7A, but the extravascular spaces between acini are narrow.

FIGURE 3.7C Exocrine pancreas cells fixed by perfusion with 2% glutaraldehyde in 0.1 *M* cacodylate buffer, to which 5 g/liter sodium chloride was added. Further treatment was as in Fig. 3.7A. × 2000.

The cells are shrunken extensively and there are large intercellular spaces between the acini as well as between cells within the acini. × 2500.

COMMENTS

The only difference between the first two preparations is the addition of dextran to the fixature in Fig. 3.7B but not in Fig. 3.7A. This increases the colloid osmotic pressure of the solution but does not influence the osmotic pressure as measured by freezing point depression. The observed difference between the tissues fixed with dextran in the fixature (Fig. 3.7B) and those without (Fig. 3.7A) is due to retention of the capillary fluid when dextran is present. Because the gross swelling of the tissue fixed without dextran is obviously abnormal and because the only difference in the appearance of the tissue lies in the large extracellular spaces, it is concluded that these spaces are artifacts. The effects of dextran are quite different than those of salts or other low molecular weight substances, which give a general shrinkage of all cell components (Fig. 3.7C).

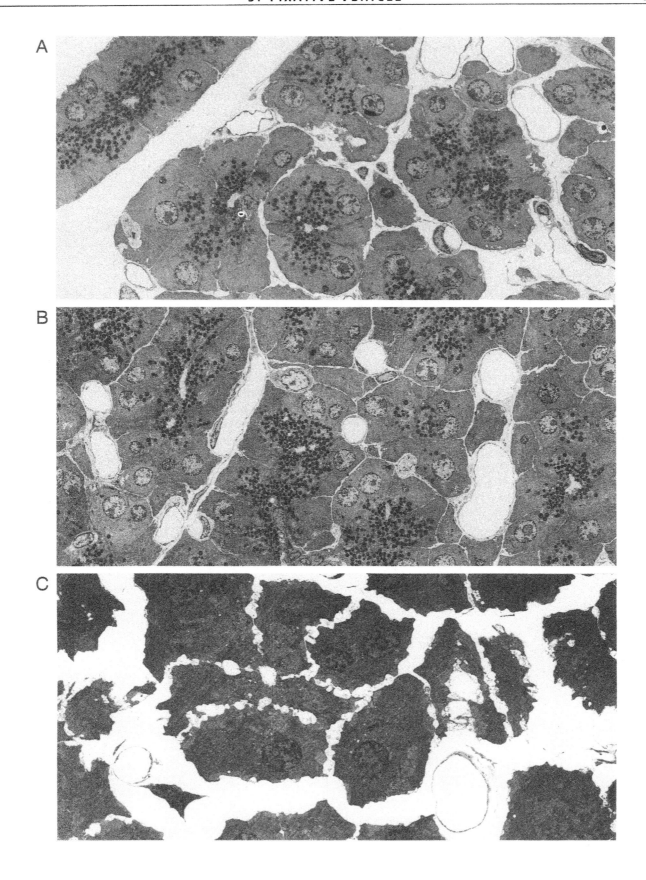

8. Colloid Osmotic Pressure: High Magnification

FIGURE 3.8A A micrograph from the same preparation as in Fig. 3.7A. × 19,000.

The capillary wall is separated from exocrine cells by wide extravascular spaces that are almost devoid of tissue components. Endothelial cells, nerve fibers, and exocrine cells all appear well preserved.

FIGURE 3.8B A micrograph from the same preparation as in Fig. 3.7B. × 17,000.

The extravascular space is narrow, of a fairly constant width, and contains various tissue components. The addition of dextran to the fixative does not influence the fine structure of the cytoplasm in exocrine pancreas cells.

COMMENTS

The dimensions of the extracellular tissue compartments evidently depend on variations in the colloidal osmotic pressure, as well as on the mode of applying the fixative and other factors. Perfusion with a fixative that does not contain a colloid osmotic substance causes swelling of some organs during perfusion, whereas immersion fixation with the same fixative will cause no swelling. Swelling occurs in organs, such as pancreas and intestine, which have no surrounding capsule, that can withstand the increase in interstitial pressure. The colloid osmotic pressure can also be increased with other high molecular weight substances such as polyvinylpyrrolidone (PVP). A drawback of adding these high molecular weight substances is the increased viscosity of the perfusate; all capillaries may not be perfused by the fixative. In practice, dextran, or PVP, is only added when it is of interest to get not only a good preservation of the cells but also of the intercellular space.

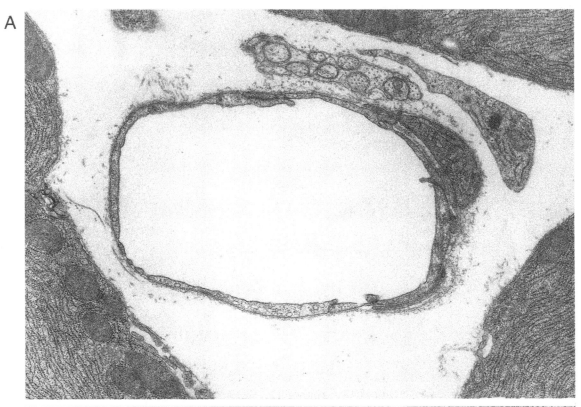

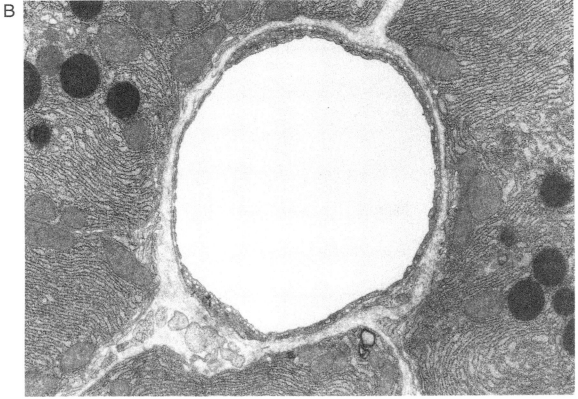

9. Phosphate Buffer Precipitate

FIGURES 3.9A AND 3.9B Proximal tubule cells of a rat kidney fixed with 1% glutaraldehyde in 0.1 M cacodylate buffer and postfixed with 1% osmium tetroxide in 0.075 M phosphate buffer (pH 7.2). The tissue was embedded in Epon and observed unstained (Fig. 3.9A) or following lead citrate staining (Fig. 3.9B). × 40,000.

The unstained section (Fig. 3.9A) has a much lower contrast than the stained section (Fig. 3.9B) but there is no precipitate in the cytoplasmic matrix or on cell organelles.

FIGURES 3.9C AND 3.9D Tissue from the same perfusion-fixed kidney as in Figs. 3.9A and 3.9B and prepared in the same fashion except that the osmium tetroxide postfixative was buffered with 0.15 M phosphate. × 48,000.

A fine precipitate is associated with cellular components, particularly with membranes of mitochondria and peroxisomes and, somewhat less so, with cell membranes and lysosomes. The precipitate is present in both unstained and stained sections, but appears with greater clarity in the latter.

COMMENTS

A precipitate similar to that shown here has been observed in many tissues fixed in glutaraldehyde and postfixed in phosphate-buffered osmium tetroxide. The mechanism for its formation is not well understood, although it apparently depends on the interaction of phosphate ions, osmium tetroxide, and lipids that have not been fixed properly by the initial aldehyde fixation (Gil and Weibel, 1968; Hendriks and Eestermans, 1982). Lipid-rich tissues are thus more vulnerable to this type of artifact than other tissues. Whatever the mechanism, the formation of precipitate depends on the concentration of phosphate ions in the postfixative, as it is absent if the phosphate concentration is low, as it is in Figs. 3.9A and 3.9B. This type of precipitate is almost exclusively found when phosphate buffers are used. A method to remove such deposits has been published by Ellis and Anthony (1979).

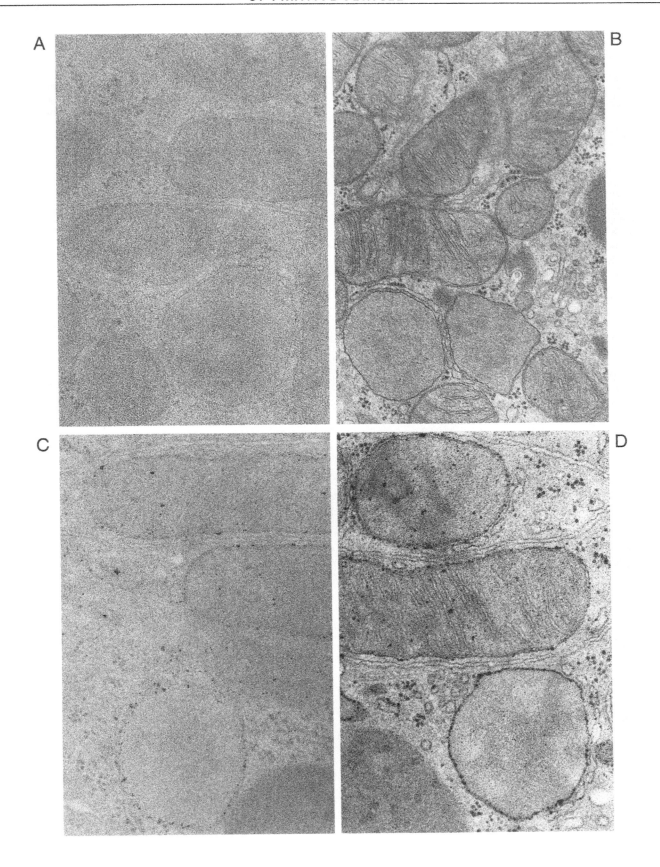

References

Angermüller, S., and Fahimi, H. D. (1982) Imidazole-buffered osmium tetroxide: An excellent stain for visualization of lipids in transmission electron microscopy. *Histochem. J.* **14**, 823–832.

Arborgh, B., Bell, P., Brunk, U., and Collins, V. P. (1976). The osmotic effect of glutaraldehyde during fixation: A transmission electron microscopy, scanning electron microscopy and cytochemical study. *J. Ultrastruct. Res.* **56**, 339–350.

Bohman, S.-O. (1974). Ultrastructure of rat renal medulla as observed after improved fixation methods. *J. Ultrastruct. Res.* **47**, 329–360.

Bohman, S.-O., and Maunsbach, A. B. (1970). Effects on tissue fine structure of variation in colloidal osmotic pressure of glutaraldehyde fixatives. *J. Ultrastruct. Res.* **30**, 195–208.

Bullock, G. R. (1984). The current status of fixation for electron microscopy: A review. *J. Micros. (Oxford)* **133**, 1–15.

Coetzee, J., and van der Merwe, C. F. (1984). Extraction of substances during glutaraldehyde fixation of plant cells. *J. Microsc. (Oxford)* **135**, 147–158.

Coetzee, J., and van der Merwe, C. F. (1987). Some characteristics of the buffer vehicle in glutaraldehyde-based fixatives. *J. Microsc. (Oxford)* **146**, 143–155.

Dvorak, A. M. (1987). Procedural guide to specimen handling for the ultrastructural pathology service laboratory. *J. Electron Microsc. Techn.* **6**, 255–301.

Ellis, E. A., and Anthony, D. W. (1979). A method for removing precipitate from ultrathin sections from glutaraldehyde-osmium tetroxide fixation. *Stain Technol.* **54**, 282–285.

Fahimi, H. D., and Drochmans, P. (1965). Essais de standardisation de la fixation au glutaraldéhyde. II. Influence des concentrations en aldéhyde et de l'osmolalité. *J. Microsc. (Paris)* **4**, 737–748.

Gil, J., and Weibel, E. R. (1968). The role of buffers in lung fixation with glutaraldehyde and osmium tetroxide. *J. Ultrastruct. Res.* **25**, 331–348.

Hayat, M. A. (1989). "Principles and Techniques of Electron Microscopy: Biological Applications." MacMillan Press, Hounsmills.

Hendriks, H. R., and Eestermans, I. L. (1982). Electron dense granules and the role of buffers: Artefacts from fixation with glutaraldehyde and osmium tetroxide. *J. Microsc. (Oxford)* **126**, 161–168.

Holt, S. J., and Hicks R. M. (1961). Use of veronal buffers in formalin fixatives. *Nature* **191**, 832.

Hopwood, D. (1985). Cell and tissue fixation, 1972–1982. *Histochem. J.* **17**, 389–442.

Karnovsky, M. J. (1965). A formaldehyde-glutaraldehyde fixative of high osmolarity for use in electron microscopy. *J. Cell Biol.* **27**, 137A–138A.

Mathieu, O., Claassen, H., and Weibel, E. R. (1978). Differential effect of glutaraldehyde and buffer osmolarity on cell dimensions: A study on lung tissue. *J. Ultrastruct. Res.* **63**, 20–34.

Maunsbach, A. B. (1966a). The influence of different fixatives and fixation methods on the ultrastructure of rat kidney proximal tubule cells. I. Comparison of different perfusion fixation methods and of glutaraldehyde, formaldehyde and osmium tetroxide fixatives. *J. Ultrastruct. Res.* **15**, 242–282.

Maunsbach, A. B. (1966b). The influence of different fixatives and fixation methods on the ultrastructure of rat kidney proximal tubule cells. II. Effects of varying osmolality, ionic strength, buffer system and fixative concentration of glutaraldehyde solutions. *J. Ultrastruct. Res.* **15**, 283–309.

Maunsbach, A. B. (1998). Fixation of cells and tissues for transmission electron microscopy. *In* "Cell Biology: A Laboratory Handbook" (J. E. Celis, ed.), 2nd Ed., Vol. 3, pp. 105–116. Academic Press, San Diego.

Maunsbach, A. B., Tripathi, S., and Boulpaep, E. L. (1987). Ultrastructural changes in isolated perfused proximal tubules during osmotic water flow. *Am. J. Physiol.* **253** (*Renal Fluid Electrolyte Physiol.* **22**), F1091–F1104.

Millonig, G. (1961). Advantage of a phosphate buffer for OsO$_4$ solutions in fixation. *J. Appl. Phys.* **32**, 1637.

Morri, S., Shikata, N., and Nakao, I. (1993). Ultracytochemistry of cytoplasmic lipid droplets. *Acta Histochem. Cytochem.* **26**, 251–259.

Rasch, R., and Holck, P. (1991). Fixation of the macula densa with fixatives of different osmolarities in normal and diabetic rats. *APMIS* **99**, 1069–1077.

Rhodin, J. (1954). "Correlation of Ultrastructural Organization and Function in Normal and Experimentally Changed Proximal Convoluted Tubule Cells of the Mouse Kidney. An Electron Microscopic Study Including an Experimental Analysis of the Conditions for Fixation of the Renal Tissue for High Resolution Electron Microscopy." Thesis, Karolinska Institutet, Stockholm.

Salema, R., and Brandão, I. (1973). The use of PIPES buffer in the fixation of plant cells for electron microscopy. *J. Submicr. Cytol.* **5**, 79–96.

Schiff, R. I., and Gennaro, J. F., Jr. (1979). The role of the buffer in the fixation of biological specimens for transmission and scanning electron microscopy. *Scanning* **2**, 135–148.

Schultz, R. L., and Karlsson, U. (1965). Fixation of the central nervous system for electron microscopy by aldehyde perfusion. II. Effect of osmolarity, pH of perfusate, and fixative concentration. *J. Ultrastruct. Res.* **12**, 187–206.

Trump, B. F., and Ericsson, J. L. E. (1965). The effect of the fixative solution on the ultrastructure of cells and tissues: A comparative analysis with particular attention to the proximal convoluted tubule of the rat kidney. *Lab. Invest.* **14**, 1245–1322.

Wood, R. L., and Luft, J. H. (1965). The influence of buffer systems on fixation with osmium tetroxide. *J. Ultrastruct. Res.* **12**, 22–45.

Zetterqvist, H. (1956). "The Ultrastructural Organization of the Columnar Absorbing Cells of the Mouse Jejunum: An Electron Microscopic Study Including Some Experiments Regarding the Problem of Fixation and an Investigation of Vitamin A Deficiency." Thesis, Karolinska Institutet, Stockholm.

FIXATIVE APPLICATION

Special consideration has to be given to the way by which the fixative is brought in contact with the tissue. When a tissue slice is excised and immersed in the fixation solution, its outer layer is fixed rapidly, whereas the interior is reached by the fixative with some delay and may undergo postmortal changes before being stabilized. When analyzed in the electron microscope, the cells in the interior therefore often have an altered ultrastructural appearance compared with the same cell types at the surface of the slice. Unless the investigator is aware of this experimental artifact, he or she may arrive at the wrong biological conclusions, in particular when the investigator happens to examine cells at different depths in blocks from different experimental conditions.

Ideally, all cells should be uniformly and simultaneously reached by the same, optimal composition of the fixative. This situation can be approached by utilizing procedures other than immersion–fixation, particularly perfusion of the fixative through the vascular system. Even after perfusion–fixation, a certain variability in the quality of fixation can be found and traced to different parameters in the perfusion technique or to differences in the handling of the tissue before or after the start of the perfusion. A special situation exists when an excised tissue block, e.g., a biopsy, is immersed in the fixative after some delay. The cells may then have undergone changes, often referred to as autolysis.

A fixation procedure, which is useful for one particular tissue, may not be suitable for another. Each particular tissue requires its own fixation procedure, including both the composition of the fixative and the method of its application. It is therefore advisable in the beginning of a research project to work out the right conditions of fixation, using the experience of previous investigators as the starting point. Sometimes much time and effort are spent on microscopy and analysis of samples in which the meaningful information was already lost at the beginning of the investigation due to inadequate fixation.

In the early days of electron microscopy the specimen to be fixed was plunged into a vial containing the fixative; it was "immersion fixed" in present-day terminology. The results were satisfactory, but were only so for the superficial layers. A remedy for the uneven fixation quality came with the method of perfusion fixation, i.e., introducing the fixative through the circulatory system. This technique was first performed successfully by Sanford Palay and colleagues in 1962. They used the fixative of that day, osmium tetroxide, which has the drawback of causing the capillaries and arterioles to contract (and of being expensive and having toxic vapors). Because vascular contraction during perfusion is deleterious to good tissue preservation, sodium nitrate was added to the fixative as a vasodilator. After the introduction of glutaraldehyde by David Sabatini, Klaus Bensch, and Russell Barrnett (1963), perfusion fixation became widely used because glutaraldehyde does not cause vascular contraction.

Palay, S., McGee-Russell, S. M. Gordon, S., and Grillo, M. A. (1962). *J. Cell Biol.* **12**, 385–410.
Sabatini, D. D., Bensch, K., and Barrnett, R. J. (1963). *J. Cell Biol.* **17**, 19–58.

1. Perfusion–Fixation versus Immersion–Fixation

FIGURE 4.1A Proximal tubules from a rat renal cortex perfusion fixed with 1% glutaraldehyde in a modified Tyrode solution (Maunsbach, 1966a). The fixative was perfused through a needle inserted retrograde in the abdominal aorta. The blood supply to the kidney was interrupted only after a sufficient perfusion pressure had been established. The tissue was postfixed in osmium tetroxide, block stained with uranyl acetate, embedded in Epon, and section stained with lead citrate. × 1900.

The tubules have open lumens and uniform brush borders. All tubules throughout the kidney cortex have the same appearance.

FIGURE 4.1B Proximal tubules in the renal cortex fixed by immersion in 1% osmium tetroxide. The tubules were located within 100 μm from the kidney surface of the excised tissue block. The tissue was embedded in Epon, and the section was stained with uranyl acetate and lead citrate. × 2000.

The tubules have closed lumens, and microvilli of the brush border are tightly packed in an irregular fashion.

COMMENTS

The functional proximal tubules of the kidney have open lumens, as can be observed with vital microscopy of the surface of the living organ. Perfusion–fixation preserves this state of the tubules and ensures an optimal preservation of cellular ultrastructure. The method of perfusion should be worked out for each particular animal species and organ as tissues vary greatly in their susceptibility to perfusion techniques. For kidney tissue it is crucial to maintain blood pressure until adequate perfusion pressure has been established, whereas this is less important for other tissues, such as the liver or pancreas.

Collapse of the tubules takes place within seconds after lowering the blood pressure in the renal artery of the animal. The absence of tubular lumens in Fig. 4.1B is the result of the excision of the kidney tissue before its immersion in the fixative. It is not an example of poor fixation per se, as the tubule was superficially located within the tissue block.

The glutaraldehyde perfusion fixative may have ice bath temperature, room temperature, or the body temperature of the animal without obvious differences in the preservation of most cell components. For some organs, a brief prewash of the vascular system has been applied, but is only necessary when the glutaraldehyde concentration exceeds 2%. Similarly, vasodilating substances may prevent vascular constriction, but are in most cases unnecessary.

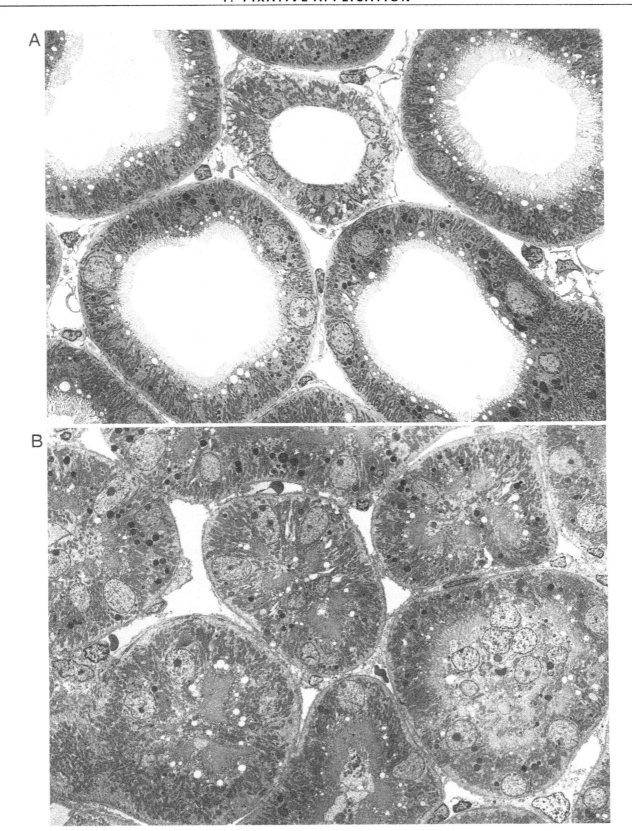

2. Perfusion–Fixation with Pressure Control

FIGURE 4.2A Cross section of a renal proximal tubule from *Necturus maculosus* that was perfusion fixed through the renal portal vein with 1% osmium tetroxide in 0.029 *M* Veronal acetate buffer. The tissue was stained *en bloc* in uranyl acetate, embedded in Vestopal W, and section stained with lead citrate. Before fixation the hydrostatic pressure inside the tubule was decreased by a microinjection of silicone in the glomerulus and the initial proximal tubule, thus creating a hydrostatic pressure gradient from capillary to lumen. During perfusion–fixation the pressure in the tubule and in a peritubular capillary was monitored through a pressure-measuring micropipette. × 3500.

Tubule cells are tall with rounded nuclei. The lateral intercellular spaces (arrows) are slightly diluted.

FIGURE 4.2B A similar preparation as in Fig. 4.2A except that in this case the intratubular hydrostatic pressure was increased by a microinjection of a silicone droplet at the very end of the proximal tubule. This maneuver increased the hydrostatic pressure in the lumen and created an elevated hydrostatic pressure gradient from the lumen to the peritubular capillaries.

Cells in this tubule are flattened as is the nucleus. The lateral intercellular space (arrow) appears decreased in width as compared to spaces in the tubule in Fig. 4.2A.

COMMENTS

Correlation of ultrastructure and function in fluid-transporting epithelia requires strict control of hydrostatic as well as osmotic pressure gradients. The complex geometry of the basolateral membranes and the intercellular space in kidney tubules has been studied extensively by electron microscopy in a variety of experimental or pathological conditions, but in most cases the hydrostatic pressures in the tissues have not been strictly controlled before and during fixation. In Fig. 4.2A the luminal hydrostatic pressure is smaller than that in the peritubular capillary, whereas in Fig. 4.2B the situation is reversed, but all other experimental parameters are the same and the perfusion pressure of the fixative is monitored and kept constant. Comparison of Figs. 4.2A and 4.2B therefore provides clear-cut evidence that the hydrostatic pressure gradients over the epithelium influence greatly the ultrastructural geometry of the cells as well as the lateral intercellular space. This example illustrates the need to control experimental parameters, such as hydrostatic pressure, before and during fixation when analyzing sensitive tissues.

A

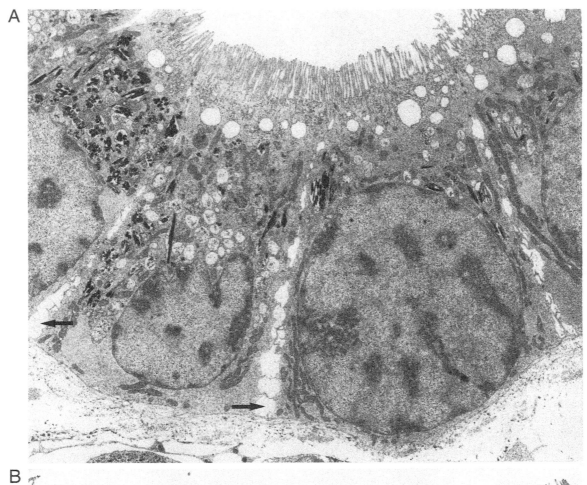

B

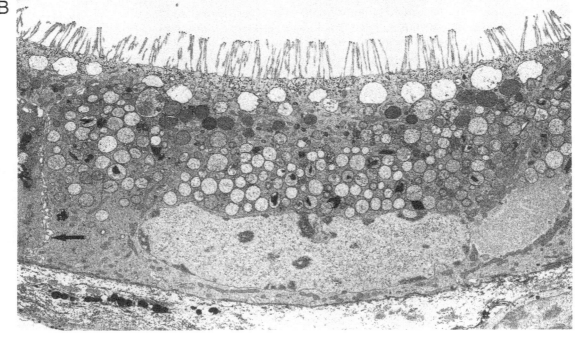

3. Fixation by Dripping *in Vivo*

FIGURE 4.3A A renal cortex fixed by dripping the fixative for 20 min onto the surface of an exposed kidney of an anesthetized rat. The fixative was 1% osmium tetroxide in a Veronal acetate buffer. A superficial slice of the kidney was then removed and immersion fixed for 1 hr. After embedding in Epon, a 1-μm-thick section was cut and examined by light microscopy. $\times$ 500.

Three different zones can be distinguished where the renal tubules differ in their gross morphology. In the outer zone, which has a depth of about 100 μm, the tubules show open lumens. In the middle zone, which has the same depth, many cells show an apical swelling that partially occludes tubule lumens. In the innermost zone the lumen is completely obliterated in most tubules.

FIGURES 4.3B–4.3D Electron micrographs of proximal tubules in the superficial, middle, and innermost zone, after the same preparation as in Fig. 4.3A. $\times$ 1500.

Tubules in the superficial zone have open, round lumens. In most tubules of the middle zone the cells protrude into the lumen; some tubules appear almost completely closed. In the innermost zone the tubules appear closed.

COMMENTS

By dripping the fixative onto the surface of an organ with an intact blood supply, a very good tissue preservation can be obtained of the superficial layers of the organ. However, a gradient in fixation quality is inevitable because the fixative reaches only the superficial layer when the microcirculation is still intact and intratubular pressure is preserved. The deep layers are penetrated by the fixative with some delay and subsequent to an interruption of the capillary circulation of the outer portions of the tissue. Swelling of the cells in the closed tubules is due to unphysiological conditions before the actual fixation. These conditions may include an impaired blood supply and detrimental effects of small amounts of the fixative reaching the cells, e.g., through the blood vessels, before the fixative concentration is sufficient to stop the metabolic events. Reliable conclusions regarding cell shape and cell volume of proximal tubule cells, which are the most sensitive cells in the renal cortex, can only be based on studies of the superficial layer, in this case the outer 100-μm-wide layer. When morphometric studies are carried out, care has to be taken to analyze only such tissue that has been fixed under optimal conditions.

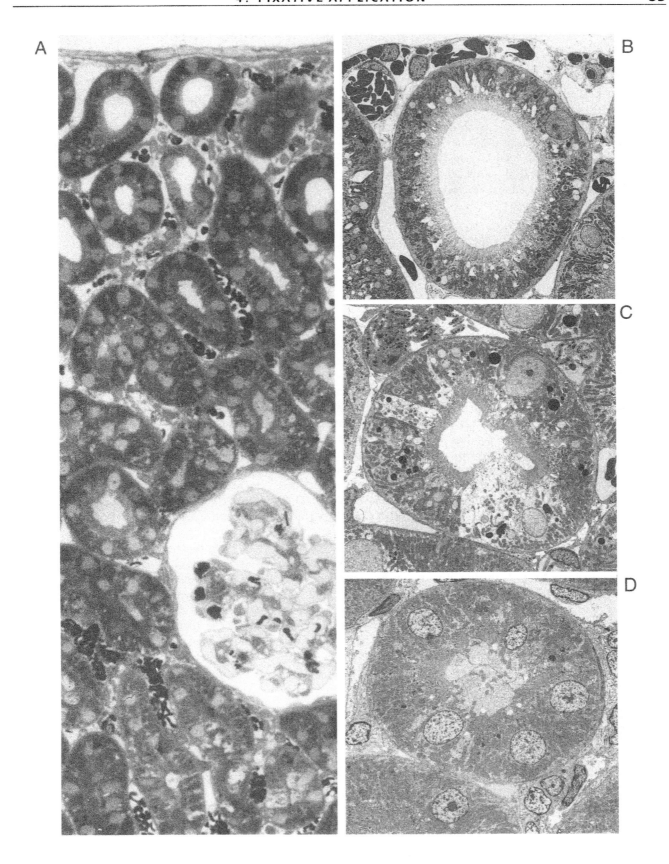

4. Immersion–Fixation

FIGURES 4.4A and 4.4B Parts of rat liver cells from the same tissue block that has been immersion fixed by immersion in 1% osmium tetroxide in 0.1 M cacodylate buffer and embedded in Epon. The sections were stained with uranyl acetate and lead citrate. The cells in Figs. 4.4A and 4.4B were located within 20 and 200 μm from the surface of the tissue block, respectively. $\times$ 15,000.

Membranes of the endoplasmic reticulum in Fig. 4.4A run approximately parallel and the mitochondrial matrix and cytoplasmic ground substance are stained uniformly. In Fig. 4.4B the smooth endoplasmic reticulum is partly fragmented and mitochondria show a rarification of the matrix in places.

FIGURES 4.4C and 4.4D Higher magnifications of the cells in Figs. 4.4A and 4.4B, respectively. $\times$ 40,000.

Figure 4.4C illustrates a uniformly fixed cytoplasmic matrix and a well-preserved endoplasmic reticulum and mitochondria. The cell in Fig. 4.4D shows an irregular distribution of cytoplasmic matrix, swelling of mitochondria, and vesiculation of the endoplasmic reticulum.

COMMENTS

These four electron micrographs from the same section illustrate the dramatic difference in tissue preparation that occurs at different depths in tissue blocks fixed by immersion. When immersion–fixation has to be resorted to, great care must be taken to evaluate optimally preserved tissue only. It may appear trivial to point out that there is a gradient in fixation quality in immersion fixed tissues, which may lead to false biological interpretations, yet such errors are not uncommonly seen in the literature. For example, the appearance of the cytoplasm seen in Fig. 4.4D could easily be—and in fact has often been—regarded as the result of an experimental procedure or a pathological disorder of the cells.

In order to avoid this basic pitfall, semithin sections for light microscopy can be cut at a right angle to the surface of the tissue block to identify zones of well-fixed tissue. A gradient of fixation quality during immersion–fixation is seen essentially in all tissues and with all types of fixative, although the zone of adequate fixation is thinner with osmium tetroxide than with aldehyde fixatives, particularly formaldehyde. Despite the inevitable gradient of fixation quality in immersion-fixed blocks, the size of the block cannot always be reduced as there is always a risk of introducing mechanical damage at the surface layers of tissues (compare with Fig. 4.7).

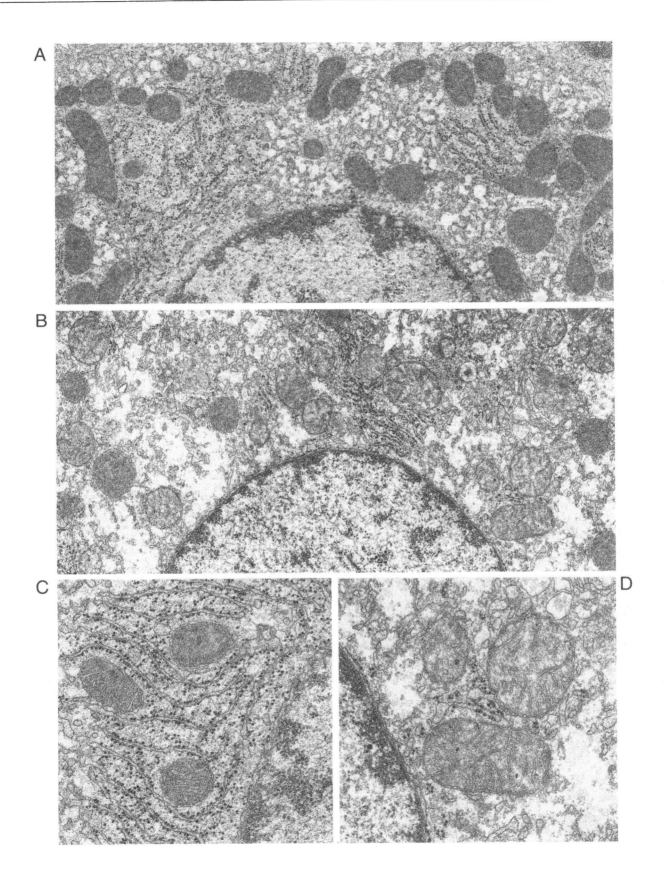

5. Variability within the Tissue

FIGURE 4.5A Basal regions of proximal tubule cells from an area corresponding to the outer zone in Fig. 4.3A. The tissue was drip fixed with 1% osmium tetroxide in a Veronal acetate buffer, embedded in Epon, and section stained with lead citrate. × 30,000.

The cytoplasmic ground substance has an even density and there is a regular distribution of stainable material. Mitochondria have a uniform appearance.

FIGURE 4.5B Basal regions of proximal tubule cells from an area corresponding to the middle zone in Fig. 4.3A. The tissue was prepared as in Fig. 4.5A. × 30,000.

There is variable density of the cytoplasmic ground substance in the processes from different cells. Some mitochondria in the cell with the light cytoplasm appear to have a larger diameter than those in the cell with a dense cytoplasm.

FIGURE 4.5C Parts of proximal tubule cells from the kidney of an aglomerular fish immersion fixed in glutaraldehyde in a phosphate buffer, postfixed in osmium tetroxide, Epon embedded, and section stained with lead citrate. × 10,000.

These tubular cells were located more than 100 μm away from the surface of the immersed tissue sample. Two of the cells project large protrusions into the tubular lumen and are largely devoid of organelles and of stainable cytoplasmic ground substance. Other portions of the cells have a rather normal appearance.

COMMENTS

Heterogeneity of the cell components is evident in Fig. 4.5B. The irregularities are still more pronounced in deeper layers, where the fixative reaches the cells even later. The uniform appearance of the cells in Fig. 4.5A is hence a more correct representation than the irregular appearance of the cells in Fig. 4.5B. When gradients in the preservation of the tissue are present, only the most superficial layer can be used for reliable ultrastructural studies. Such gradients are usually absent when the tissue is fixed by perfusion. Protrusions with an empty appearance, as in Fig. 4.5C, are clearly regarded as fixation artifacts as they are absent from tubules located superficially in immersion-fixed blocks or in cells from perfusion-fixed tissues.

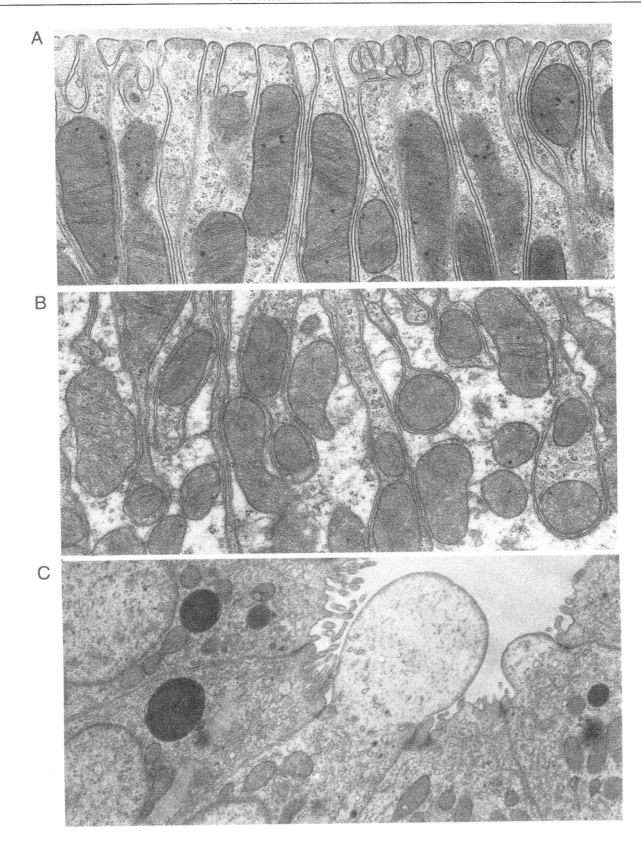

6. Unsuccessful Perfusion–Fixation

FIGURE 4.6A Proximal tubule cells from a rat kidney perfusion fixed with 1% glutaraldehyde. The vascular pressure was interrupted before perfusion was initiated. The excised tissue block was then postfixed in osmium tetroxide, embedded, sectioned, and section stained. × 4000.

The tubules are collapsed and the intercellular spaces grossly dilated.

FIGURE 4.6B Proximal tubule cells from a rat kidney cortex perfusion fixed with 1% osmium tetroxide. The perfusion pressure was above that of the blood in order to force the fixative into the kidney. The tissue was embedded, sectioned, and stained. × 8000.

Parts of the apical region of one of the cells protrude into the tubule lumen. The arrangement of microvilli and mitochondria is irregular.

COMMENTS

In the lumen of a functioning kidney tubule, the pressure is normally slightly higher than in the capillary. If the blood flow is interrupted, even for a brief period, the hydrostatic pressure conditions are changed and glomerular filtration stops. Fluid then disappears from the lumen of the proximal tubule. When such a kidney is perfused with a fixative, the hydrostatic pressure will be higher in the peritubular capillaries than in the lumen, and a dilatation of the intercellular spaces will be created. Some tubules, however, may again become distended due to regained intratubular pressure. The overall fixation pattern may hence be quite variable, which emphasizes the importance of introducing the fixative in the vascular system without interrupting the vascular pressure. The kidneys are particularly sensitive in this respect, but the same effects may also be observed in other organs.

In contrast to glutaraldehyde, osmium tetroxide causes a strong contraction of the smooth muscle cells of the blood vessels. This makes an initial perfusion with osmium tetroxide very difficult. To some extent this reaction to osmium tetroxide may be compensated for by using high hydrostatic pressure during perfusion or by using vasodilating agents, but sometimes this will cause an artifactual widening of the tubules and a modification of the normal cell structure as seen in Fig. 4.5B.

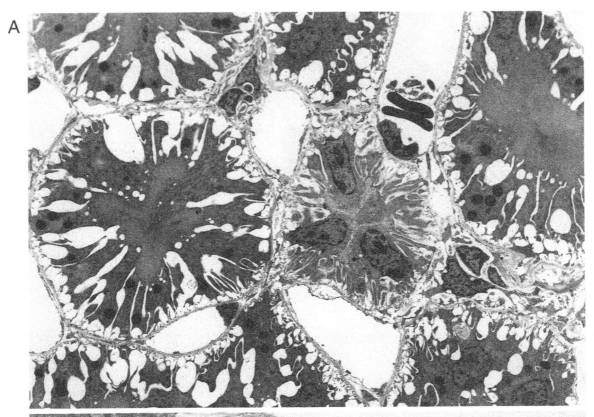

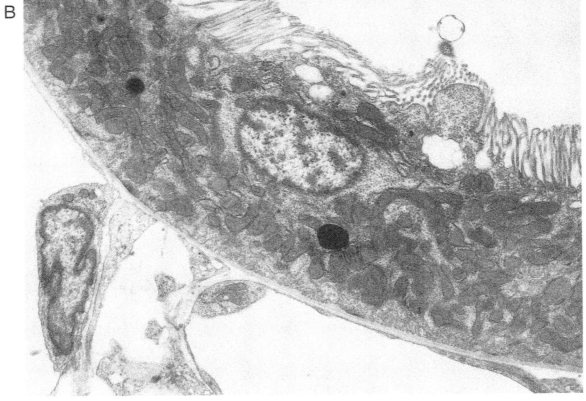

7. Superficial Tissue Damage

FIGURE 4.7A Part of a rat liver cell fixed in 1% osmium tetroxide in a phosphate buffer, embedded in Epon, and section stained with uranyl acetate and lead citrate. The cell was located at the very surface of a tissue block, which was excised with a sharp razor blade and immersion fixed. × 30,000.

Mitochondria appear well preserved, but the endoplasmic reticulum is fragmented and the cytoplasmic ground substance is virtually absent. The cytoplasm is continuous with the extracellular space (in the upper left corner) due to the absence of a plasma membrane.

FIGURES 4.7B AND 4.7C Mitochondria in superficial and severed cells with "light" cytoplasm as in Fig. 4.7A. × 26,000 and 13,000, respectively.

The conformational state of mitochondria varies. Some appear almost normal, whereas others are enlarged with either an electron-lucid matrix or with expanded intracristal spaces.

FIGURE 4.7D Parts of two intact liver cells that were close to the cut surface of the tissue slice. Preparation as in Fig. 4.7A. × 20,000.

In the cell to the lower right, the cytoplasm contains stainable material and the organelles have a normal appearance. In the cell to the upper left, the cytoplasm contains little stainable material, hence it looks like a "light cell." In addition, the endoplasmic reticulum appears fragmented and dilated in part.

COMMENTS

Excision and immersion–fixation of tissue result in mechanical damage to the superficial cell layer, presumably due to damage to the cell membrane and a subsequent loss of cytoplasmic ground substance. These cells will therefore appear lighter than the adjacent well-preserved cells. A "light cell–dark cell" pattern may also be seen inside tissue blocks of specimens immersion fixed under suboptimal condition. (Fig. 4.7D). In the early literature there are several claims of tissues containing both "light" and "dark" cell types, whereas in reality they are fixation-induced artifacts.

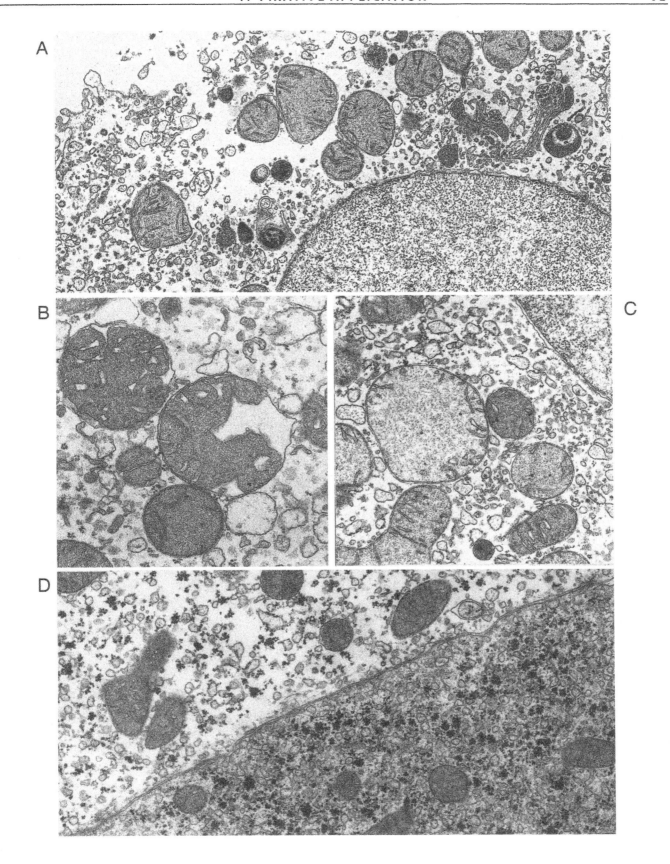

8. Early Postmortal Changes

FIGURE 4.8A Part of a rat liver cytoplasm fixed in 2% osmium tetroxide in phosphate buffer, embedded in Epon, and section stained with uranyl acetate and lead citrate. The micrograph represents the control to the autolysis experiment presented in Figs. 4.8B and 4.8C. × 25,000.

The preservation of the tissue is comparable to that normally found when single fixation with osmium tetroxide has been employed.

FIGURE 4.8B A slice of the same rat liver as in Fig. 4.8A, which was excised and then left for 6 hr at 4°C before it was immersed in the fixative. Further processing is as in Fig. 4.8A. × 25,000.

There are no pronounced alterations in the rough-surfaced endoplasmic reticulum or in the mitochondria. The smooth-surfaced endoplasmic reticulum to the left in the figure shows a higher degree of fragmentation than in the control.

FIGURE 4.8C Preparation as in Fig. 4.8B, although the tissue slice was left at 22°C for 6 hr. × 25,000.

The endoplasmic reticulum is swollen and has fragmented into vesicles. Mitochondria are also swollen.

COMMENTS

Contrary to common belief, many structural changes may not be significant within the first minutes or even hours after the removal of tissue from the living animal and before its immersion in the fixative. Many changes, which have been attributed to autolysis due to delayed fixation and consequent autolysis, in fact have been due to mechanical or chemical mistreatments of the tissue before and/or during its subsequent preparation. However, different tissues behave differently in this respect and some are more sensitive than liver tissue.

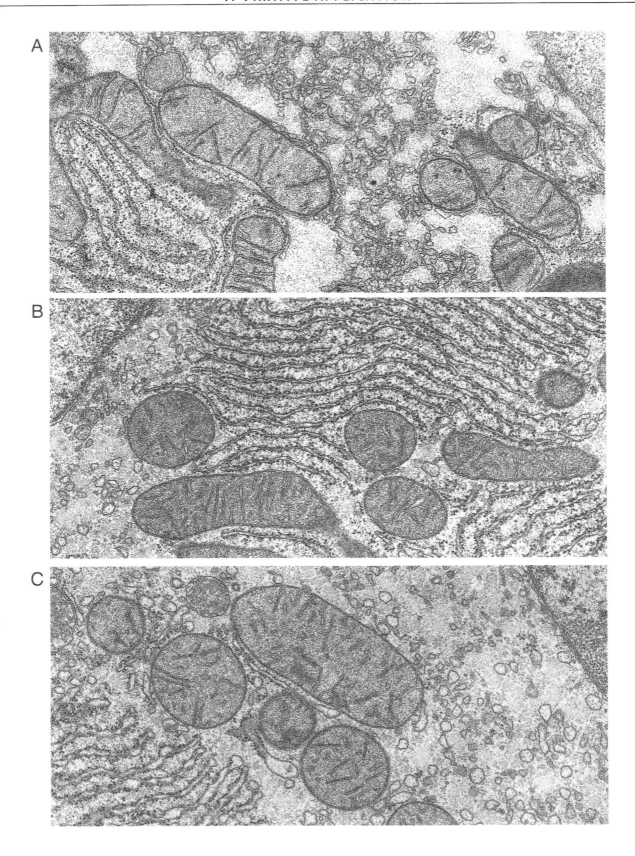

9. Late Postmortal Changes

FIGURE 4.9A A rat liver cytoplasm from tissue immersion fixed immediately after excision for 2 hr at 4°C with 1% osmium tetroxide in 0.1 M phosphate buffer, embedded in Epon, and section stained with uranyl acetate and lead citrate. $\times$ 30,000.

The hepatocyte cytoplasm has a normal appearance.

FIGURE 4.9B A rat liver cell left for 24 hr at 4°C before fixation in osmium tetroxide and further processing as in Fig. 4.9A. $\times$ 30,000.

Mitochondria are swollen and clumped together. They exhibit essentially normal shapes but the matrix is compact. The smooth endoplasmic reticulum appears fragmented or lost.

FIGURE 4.9C Part of a rat liver cell left for 24 hr at 22°C before fixation in osmium tetroxide. Further processing is as in Fig. 4.9A. $\times$ 30,000.

The endoplasmic reticulum is swollen and fragmented at places and mitochondria are enlarged, the matrix essentially gone and membranes fragmented. The cytoplasmic ground substance is flocculent or entirely absent.

COMMENTS

Structural alterations in the tissue during autolysis are highly dependent on the temperature at which the tissue is kept. An implication of this fact is the obvious recommendation to keep excised tissue at a low temperature if it cannot be fixed immediately. Autopsy material has usually undergone extensive autolytic changes. However, the structural stability of cellular components, such as cytoskeletal elements and granules in endocrine cells, still makes it meaningful in some cases to study autopsy material by electron microscopy.

Another conclusion that can be drawn from this experiment is the different sensitivity exhibited by the different organelles. Thus, the smooth-surfaced portion of the endoplasmic reticulum is more sensitive than its rough-surfaced part, the mitochondria more sensitive than the peroxisomes.

Pathologists may have to accept that autopsy specimens have undergone autolytic changes, but there is no excuse for using autolytic material in the examination of tissue from laboratory animals.

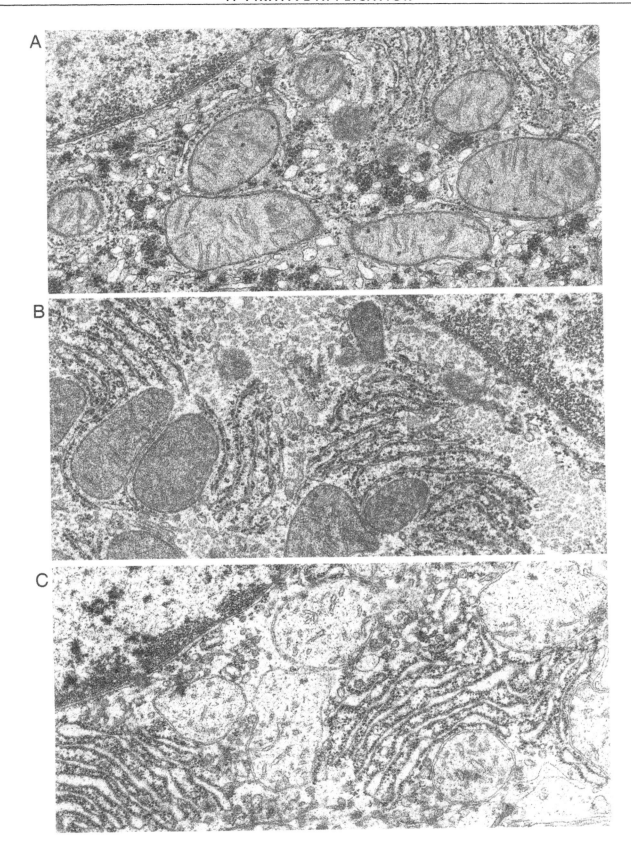

A

B

C

10. Influence of Biopsy Method

FIGURE 4.10A A proximal tubule from a biopsy that was excised during surgery from a nontumorous part of a human kidney immediately before the kidney was removed. The tissue was fixed directly in 1% osmium tetroxide and embedded in Epon, and the sections were stained with uranyl acetate and lead citrate. The tubule was located close to the surface of the tissue block. × 1900.

The tubule shows an open lumen and only moderate degrees of swelling of the luminal part of the tubule cells.

FIGURE 4.10B A proximal tubule excised from a kidney, which had just been removed surgically due to a tumor. The tissue was fixed in 3% glutaraldehyde and then treated as in Fig. 4.10A. This tubule was located in the middle of the block. × 1750.

There is extensive swelling of the apical parts of the cells and swollen cells occupy the entire tubule lumen. Notice that the magnification of Figs. 4.10A and 4.10B is almost the same.

FIGURE 4.10C Apical part of cytoplasm from tubule cells just outside the field shown in Fig. 4.10A. × 10,000.

The cells show uniform cytoplasm, and mitochondria and lysosomes have normal structures as do the apical endocytosis apparatus below the brush border.

FIGURE 4.10D Basal part of cells in the kidney tubule similar to that in Fig. 4.10B. × 15,000.

The cytoplasms of the two adjacent cells seen here are very different. The upper half of the micrograph shows a swollen cell with poorly stained cytoplasm, whereas the cell in the lower half of the micrograph has a uniformly stained cytoplasm without much evidence of cell swelling.

COMMENTS

Electron microscopy of biopsies from human tissues is sometimes obtained under conditions that are suboptimal for ideal tissue preservation. Figure 4.10A illustrates a situation where pieces of tissue have been obtained without traumatizing the biopsy. For this reason the tubule lumen is essentially retained and the epithelium appears uniform in structure. In contrast, the tissue in Fig. 4.10B was removed under suboptimal conditions, which leads to extensive cell swelling and deformation of cell architecture. The structural differences observed in Figs. 4.10A and 4.10B are maintained at the subcellular level. Figure 4.10C illustrates a well-preserved cytoplasm, whereas in Fig. 4.10D the cytoplasm of the lower cell has a uniform appearance and the upper cell is swollen, the well-known "dark cell–light cell" phenomenon, which is often encountered in tissues fixed under suboptimal conditions.

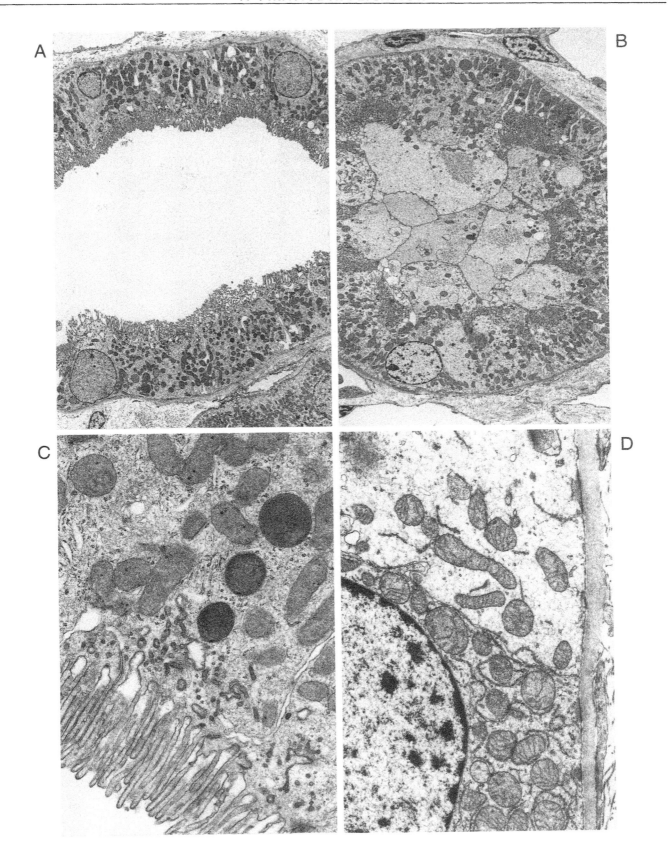

11. Microwave Treatment

FIGURES 4.11A–4.11D Part of rat liver cells in excised tissue blocks (1 × 1 × 3 mm) that were treated in microwave oven (AEG, 650W) for 10 sec in a solution consisting of 1% osmium tetroxide in 0.1 M cacodylate buffer. During microwave treatment the temperature of the solution increased from 22°C to 48°C. After microwave treatment the tissue shown in Figs. 4.11A and 4.11B was immersed in 0.1M cacodylate buffer for 2 hr, dehydrated in ethanol, and embedded in Epon. On the contrary the tissue shown in Figs. 4.11C and 4.11D was post-fixed for 2 hr in the same fixative before embedding. × 28,000.

The cell in Fig. 4.11A is located approximately 20 μm from the surface of the tissue block and shows preserved cell shape and stained ribosomes but membranes show no contrast. The cell in Fig. 4.11B, which was located approximately 100 μm from the block surface (the section was oriented at right angle to the block surface) appears completely destroyed and the cytoplasm washed out. The cells in Figs. 4.11C and 4.11D, which were post-fixed for 2 hr, and which were located approximately 20 μm and 115 μm, respectively, from the surface of the tissue block appear normal.

FIGURES 4.11E AND 4.11F Preparation as in Figs. 4.11C and 4.11D except that the tissue was not treated at all in the microwave oven. × 28,000.

The cytoplasm has a normal appearance both in cells located approximately 20 μm (Fig. 4.11E) and 115 μm (Fig. 4.11F) from the tissue block surface.

COMMENTS

Microwave irradiation of tissues during fixation in various solutions has been applied with the aim of preserving the structural organization of tissues at the electron microscopical level. The mechanism for the presumed effects of microwave treatment remains uncertain. Although some investigators refer the effect to a microwave field, others suggest that the effect, if any, is related to heating of the tissue (Leonard and Shepardson, 1994). It has also been suggested that the fixative penetrates the tissue with extraordinary speed during microwave treatment. However, a comparison between Figs. 4.11A and 4.11B suggests that this is not the case, at least not for osmium tetroxide. In fact, in Fig. 4.11A only a thin rim of the tissue block has been penetrated by osmium during the brief microwave treatment and inside this thin rim the tissue looks horrible (Fig. 4.11B). Only if the tissue blocks remained in the fixative for 2 hr subsequent to microwave treatment did the interior cells appear well preserved, but they did so also without microwave treatment.

The appearent absence of microwave effects in these experiments, and in our parallel experiments with aldehyde fixatives, may be due to differences in cell type, equipment, or procedural details. However, it should be noted that in most published studies strict attention has not been paid to the localization of cells in relation to the surface of the tissue block, i.e., whether the fine structure of cells is really uniform throughout the whole block. The possible advantage, if any, of microwave treatment for the preservation of cell ultrastructure clearly needs further stringent analysis.

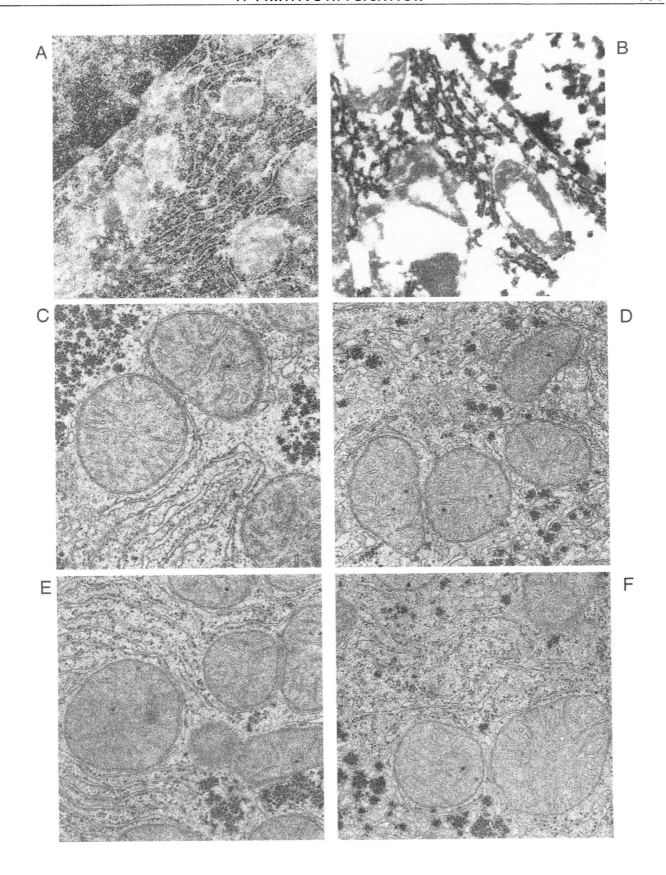

References

Benchimol, M., Goncalves, N. R., and De Souza, W. (1993). Rapid primary microwave-glutaraldehyde fixation preserves the plasma membrane and intracellular structures of the protozoan *Tritrichomonas foetus*. *Microsc. Res. Techn.* **25,** 286–290.

Bohman, S.-O. (1974). Ultrastructure of rat renal medulla as observed after improved fixation methods. *J. Ultrastruct. Res.* **47,** 329–360.

Coalson, J. J. (1983). A simple method of lung perfusion fixation. *Anat. Rec.* **205,** 233–238.

Ericsson, J. L. E., and Biberfeld, P. (1967). Studies on aldehyde fixation: Fixation rates and their relation to fine structure and some histochemical reactions in the liver. *Lab. Invest.* **17,** 281–298.

Ganote, C. E., and Moses, H. L. (1968). Light and dark cells as artifacts of liver fixation. *Lab. Invest.* **18,** 740–745.

Glaumann, B., Glaumann, H., Berezesky, I. K., and Trump, B. F. (1975). Studies on the pathogenesis of ischemic cell injury. *Virch. Arch. B* **19,** 281–302.

Hopwood, D. (1985). Cell and tissue fixation, 1972–1982. *Histochem. J.* **17,** 389–442.

Kaissling, B., and Kriz, W. (1982). Variability of intercellular spaces between macula densa cells: TEM study in rabbits and rats. *Kidney Int. Suppl.* **22,** S9–S17.

Langenberg, W. G. (1978). Relative speed of fixation of glutaraldehyde and osmic acid in plant cells measured by grana appearance in chloroplasts. *Protoplasma* **94,** 167–173.

Larsson, L. (1975). Effects of different fixatives on the ultrastructure of the developing proximal tubule in the rat kidney. *J. Ultrastruct. Res.* **51,** 140–151.

Leonard, J. B., and Shepardson, S. P. (1994). A comparison of heating modes in rapid fixation techniques for electron microscopy. *J. Histochem. Cytochem.* **42,** 383–391.

Login, G. R., and Dvorak, A. M. (1994). "The Microwave Tool Book: A Practical Guide for Microscopists." Beth Israel Hospital, Boston.

Login, G. R., Dwyer, B. K., and Dvorak, A. M. (1990). Rapid primary microwave-osmium fixation. I. Preservation of structure for electron microscopy in seconds. *J. Histochem. Cytochem.* **38,** 755–762.

Marti, R., Wild, P., Schraner, E. M., Mueller, M., and Moor, H. (1987). Parathyroid ultrastructure after aldehyde fixation, high-pressure freezing, or microwave irradiation. *J. Histochem. Cytochem.* **35,** 1415–1424.

Maunsbach, A. B. (1966). The influence of different fixatives and fixation methods on the ultrastructure of rat kidney proximal tubule cells. I. Comparison of different perfusion fixation methods and of glutaraldehyde, formaldehyde and osmium tetroxide fixatives. *J. Ultrastruct. Res.* **15,** 242–282.

Maunsbach, A. B. (1998). Fixation of cells and tissues for transmission electron microscopy. *In* "Cell Biology: A Laboratory Handbook" (J. E. Celis, ed.), 2nd Ed, Vol. 3, pp. 249–259. Academic Press, San Diego.

Maunsbach, A. B., and Boulpaep, E. L. (1980). Hydrostatic pressure changes related to paracellular shunt ultrastructure in proximal tubule. *Kidney Int.* **17,** 732–748.

Maunsbach, A. B., Madden, S. C., and Latta, H. (1962). Variations in fine structure of renal tubular epithelium under different conditions of fixation. *J. Ultrastruct. Res.* **6,** 511–530.

Mizuhira, V., Hasegawa, H., and Notoya, M. (1997). Microwave fixation and localization of calcium in synaptic terminals and muscular cells by electron probe X-ray microanalysis and electron energy-loss spectroscopy imaging. *Acta Histochem. Cytochem.* **30,** 277–301.

Møller, J. C., Skriver, E., Olsen, S., and Maunsbach, A. B. (1982). Perfusion-fixation of human kidneys for ultrastructural analysis. *Ultrastruct. Pathol.* **3,** 375–385.

Pease, D. C. (1955). Electron microscopy of the tubular cells of the kidney cortex. *Anat. Rec.* **121,** 723–743.

Rostgaard, J., Qvortrup, K., and Poulsen, S. S. (1993). Improvements in the technique of vascular perfusion-fixation employing a fluorocarbon-containing perfusate and a peristaltic pump controlled by pressure feedback. *J. Microsc.* **172,** 137–151.

Tisher, C. C., Bulger, R. E., and Trump, B. F. (1966). Human renal ultrastructure. *Lab. Invest.* **15,** 1357–1394.

Utsunomiya, H., Komatsu, N., Yoshimura, S., Tsutsumi, Y., and Watanabe, K. (1991). Exact ultrastructural localization of glutathione peroxidase in normal rat hepatocytes: Advantages of microwave fixation. *J. Histochem. Cytochem.* **39,** 1167–1174.

DEHYDRATION AND EMBEDDING

Unfixed tissues or tissues fixed by conventional methods have a consistency that is too soft for ultramicrotomy. They must therefore be infiltrated by a substance that can polymerize or otherwise harden. Alternatively, the tissue can be frozen and then sectioned.

Embedding of the tissue is usually made in compounds such as epoxy resins or polyesters, which have suitable physical properties for sectioning after polymerization. Before the tissue can be infiltrated it has to be dehydrated, as water interferes with the polymerization of most embedding media.

Dehydration can be accomplished by a number of organic compounds such as ethanol and acetone. During this process there is a significant loss of tissue components, particularly lipids. The extraction is dependent on the fixative that has been used, the nature of the dehydrating solvent, the time of dehydration, and the temperature. In addition, volume changes may occur in the tissue and cause distortion or destruction of various structures.

The choice of embedding medium should be dictated by the aim of the study as embedding media have very different characteristics. Thus, methacrylates may be difficult to polymerize with preserved tissue quality. Differences also exist in the stainability of the various media and in the tendency to sublime in the electron beam. Last but not least, some embedding media preserve antigenicity, notably Lowicryls, Unicryls, and LR White, and are thus suitable for immunoelectron microscopy (see Chapter 16).

The need of embedding objects in a material suitable for sectioning was clearly felt in the early days of biological electron microscopy, but technology did not exist. The first useful embedding medium was a methacrylate mixture developed by Sanford Newman, Emil Borysko, and Max Swerdlow (1949). This development was a spin-off effect of experiments to impregnate military boots, thus an embedding of biological material for quite another purpose. Even better embedding media were developed some years later; different epoxy resins introduced independently by Ole Maaløe and Aksel Birch-Andersen (1956), Audrey Glauert et al. (1956), and John Luft (1961) and the polyester Vestopal W by Antoinette Ryter and Edward Kellenberger (1958). A revival of the acrylic resins came with the Lowicryls and LR White resins, which are suitable for low temperature embeddings and immunoelectron microscopy.

Glauert, A. M., Rogers, G. E., and Glauert, R. H. (1956). *Nature* **178**, 803.

Maaløe, O., and Birch-Andersen, A. (1956). *In* "6th Symp. Soc. Gen. Microbiol," pp. 261–278. Cambridge Univ. Press, Cambridge, UK.

Newman, S. B., Borysko, E., and Swerdlow, M. (1949). *J. Res. Natl. Bur. Stand.* **43**, 183–199.

Luft, J. H. (1961). *J. Biophys. Biochem Cytol.* **9**, 409–414.

Ryter, A., and Kellenberger, E. (1958). *J. Ultrastruct. Res.* **2**, 200–214.

1. Stepwise versus Direct Dehydration

FIGURE 5.1A Part of a rat liver cell, perfusion fixed with 1% glutaraldehyde in 0.1 *M* cacodylate buffer, post-fixed in 1% osmium tetroxide, rinsed in buffer, and dehydrated in a graded series of ethanols: 70% (two changes, 15 min each), 90% (5 min), 95% (5 min), and 100% (two changes, 15 min each). The tissue was embedded in Epon and section stained with uranyl acetate and lead citrate. × 25,000.

There are no obvious preservation artifacts. Thus, the cell organelles show their conventional ultrastructure and the cisternae of the endoplasmic reticulum run approximately parallel and their membranes are equidistant.

FIGURE 5.1B A liver cell from the same animal as in Fig. 5.1A, prepared in the same way except that the tissue was transferred directly to 100% ethanol (two changes, 15 min each) after the buffer rinse. × 25,000.

The ultrastructural preservation is essentially the same as in Fig. 5.1A.

FIGURE 5.1C A liver cell from the same animal as in Fig. 5.1A treated in the same way except that the dehydration agent was acetone (70, 90, 95, 100%). × 25,000.

The qualitative appearance of the cell ultrastructure is essentially the same as in Fig. 5.1A.

FIGURE 5.1D A liver cell from the same animal as in Figs. 5.1A–5.1C, except transferred directly to 100% acetone (two changes, 15 min each). × 25,000.

The ultrastructural preservation is essentially the same as in Figs. 5.1A and 5.1B.

COMMENTS

The routine dehydration procedures for electron microscopy inevitably lead to protein denaturation and extraction of components of the fixed tissue. The degree of extraction depends on the fixative, the dehydrating agent, and the dehydration protocol, including time and temperature. In autoradiographic studies the removal of labeled tissue components during processing may be a serious problem, and in cytochemistry and immunoelectron microscopy the denaturation of proteins during dehydration and embedding similarly may be problematic.

Double-fixed tissues seem more resistant to modifications in the dehydration technique than those fixed in osmium tetroxide only, which is likely due to a greater degree of cross-linking of the proteins by glutaraldehyde. As illustrated here, the four different dehydration protocols give essentially the same ultrastructure of double-fixed cells.

It is well known that dehydration results in tissue shrinkage. This is a potential source of error in morphometric work and stereology; correct absolute measurements are unobtainable. In addition to an overall shrinkage, there may also be a differential shrinkage of cell components. To minimize shrinkage of the specimen, a stepwise dehydration, e.g., in ethanol, has been recommended. However, for most routine purposes it appears inconsequential whether dehydration, particularly of double-fixed tissue, is performed in ethanol or acetone and whether dehydration is gradual or takes place in a single step. The possibility of using rapid dehydration protocol is particularly advantageous in the routine processing of human biopsies (Johannessen, 1973).

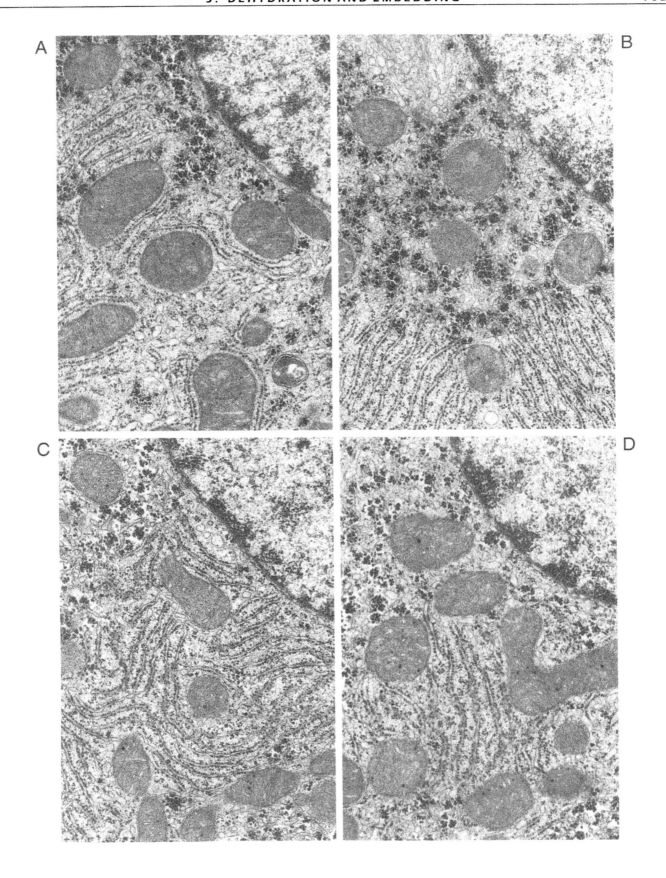

2. Prolonged Dehydration in Ethanol

FIGURES 5.2A–5.2D Cytoplasm from osmium-fixed rat liver following different schedules of dehydration in ethanol. The tissue was dehydrated in a graded series (70, 90, 95, 100%) of ethanol solutions. In Fig. 5.2A the tissue was left in 70% ethanol at 4°C for 24 hr, in Fig. 5.2B the tissue was left in 70% ethanol at 22°C for 24 hr, in Fig. 5.2C the tissue remained in 100% ethanol at 22°C for 24 hr, and in Fig. 5.2D the specimen was left in 70% ethanol at 22°C for 24 hr. All specimens were embedded in Epon and the sections stained with uranyl acetate and lead citrate. × 25,000.

In Fig. 5.2A membranes are continuous, ribosomes are distinct and the ground cytoplasm is evenly dense. There is no obvious difference between this specimen and specimens left in 70% ethanol for 1 or 2 hr only (compare with Fig. 2.1A), except that glycogen is less stained. In Fig. 5.2B only fragments of cell structures remain and the destruction is even more pronounced in Fig. 5.2D. On the contrary the cytoplasm in Fig. 5.2C, which remained in 100% ethanol, has essentially the same appearance as after standard dehydration, except that cisternae of the endoplasmic reticulum are slight dilated in places.

FIGURE 5.2E Cytoplasm from liver tissue double fixed with glutaraldehyde and osmium tetroxide and then processed as in Fig. 5.2A except that the tissue block remained in 100% ethanol for 24 hr at 22°C.

The cell ultrastructure is the same following standard preparation, including short dehydration.

COMMENTS

Tissue components may be extracted extensively during dehydration following single fixation in osmium tetroxide. Morphological evidence of extraction, as presented here, is supported by chemical studies of the extracted material found in the organic solvents used for dehydration. In particular, the temperature during dehydration is of importance (compare Figs. 5.2A and 5.2B). At room temperature, low concentrations of ethanol, e.g., 70%, cause a gradual extraction of the tissue and completely destroy cell fine structure. Such ethanol dehydration should hence be carried out in the cold. However, Fig. 5.2C illustrates that osmium-fixed specimens can remain in absolute ethanol for at least 1 day at room temperature without much change in ultrastructure. Tissues that have been double fixed in glutaraldehyde and osmium tetroxide show little change in ultrastructure even after 2 days in absolute ethanol at room temperature (Fig. 5.2E). A practical consequence is that tissue blocks, single or double fixed, may be left overnight in 100% ethanol if embedding cannot be performed the same day, but should not be left at lower concentrations. Alternatively, the fixed and dehydrated tissue may be left overnight in propylene oxide; no extraction has been detected (unpublished data). However, these conclusions may not apply to all tissues and care should be taken to adopt the dehydration protocol that gives the least adverse effects on the tissue.

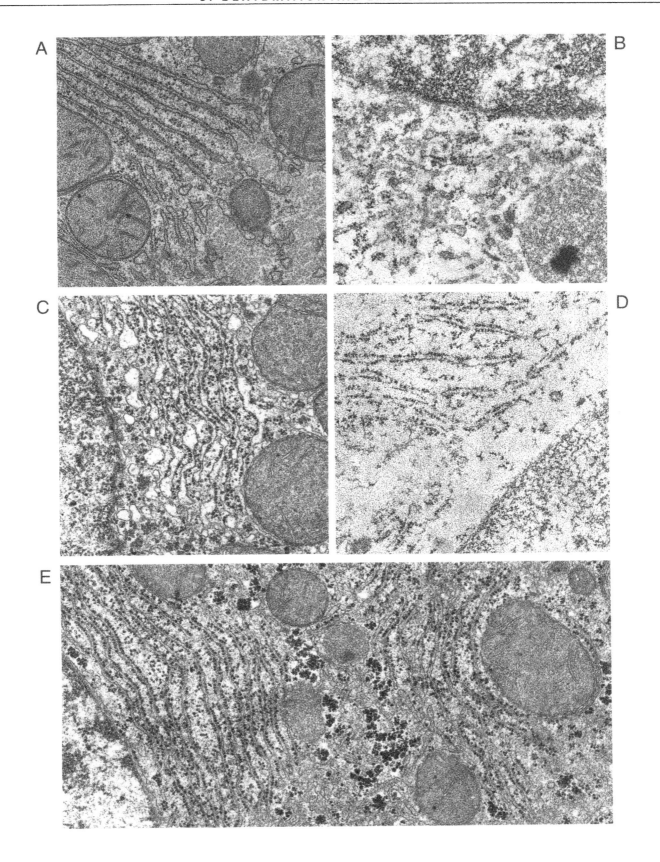

3. Prolonged Dehydration in Acetone

FIGURE 5.3A Cytoplasm from cell in rat liver osmium-fixed as in Figs. 5.2A–5.2D, but dehydrated in a graded series of acetone solutions. The specimen was left in 60% acetone for 24 hr at 4°C. Embedding was done in Epon and section staining performed with uranyl acetate and lead citrate. × 25,000.

The different types of membranes are continuous, the ribosomes are distinct, and the ground cytoplasm is distributed evenly. There is little difference between this specimen and specimens left in 60% acetone for 1 or 2 hr only (not illustrated) except that there is a slight swelling in the endoplasmic reticulum after a prolonged dehydration.

FIGURE 5.3B Specimen that also remained in 60% acetone for 24 hr, but here at 22°C. × 25,000.

The general architecture of the cytoplasmic structures is similar to that in Fig. 5.3A. However, there is less stainable material in the cytoplasmic ground substance and mitochondrial matrix.

FIGURE 5.3C Cytoplasm from a liver cell that was fixed in both glutaraldehyde and osmium tetroxide and kept in 60% acetone for 24 hr at 22°C before embedding and section staining as in Figs. 5.3A and 5.3B. × 25,000.

The fine structure of all cellular components is well preserved.

COMMENTS

These electron micrographs illustrate that acetone, similar to ethanol, may cause extraction of the tissue. However, extraction is less severe than after ethanol dehydration and affects largely the cytoplasmic ground substance and the mitochondrial matrix.

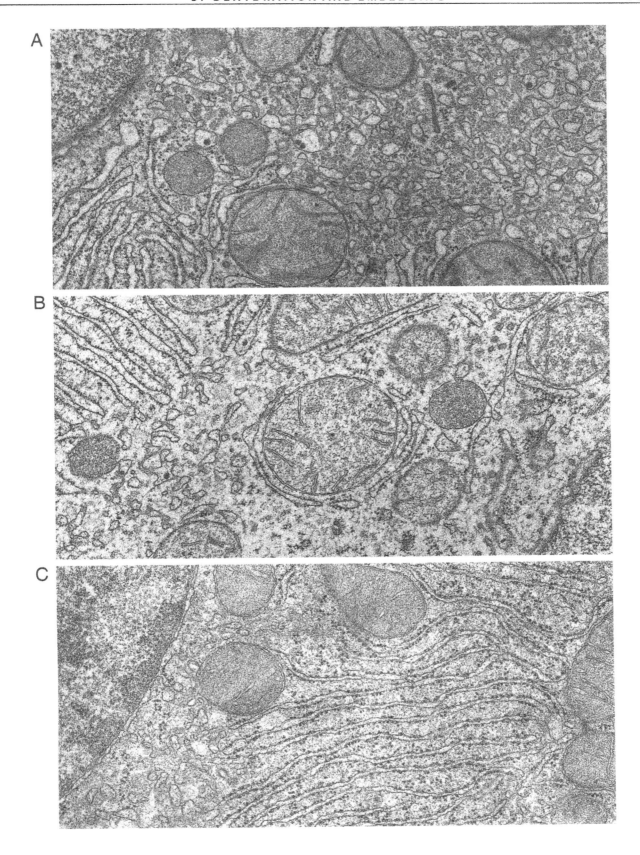

4. Inert Dehydration

FIGURES 5.4A–5.4C Renal proximal tubules, perfusion fixed with 1% glutaraldehyde in Tyrode solution, pH 7.2. The total time of fixation was 2 min. The tissues were then rinsed in Tyrode solution and cut into thin slices, 40–60 μm in thickness and then treated in three different ways.

FIGURE 5.4A After the rinse the tissue was carried through 70–100% acetone and embedded in Vestopal W starting with a Vestopal:acetone (1:3) mixture and ending with 100% Vestopal and polymerization at 60°C. × 94,000.

Mitochondrial cristae are seen as two thin electron-lucid layers and an intervening electron-dense layer with a total thickness of 160 Å (arrowheads). The lateral intercellular spaces are widened.

FIGURE 5.4B After the rinse the tissue was taken to 100% ethylene glycol where it remained for 4 min before being transferred to 100% Vestopal and polymerization. × 94,000.

Cristae have a total thickness of about 250 Å (arrowheads). Adjacent basolateral membranes are apposed with only a narrow intervening space.

FIGURE 5.4C After dehydration in 100% ethylene glycol for 3 min, the tissue was infiltrated with 100% Vestopal containing 0.6% benzoin ethyl ether and polymerized by UV light at −20°C. × 94,000.

The total thickness of mitochondrial cristae is about 350 Å (arrowheads), whereas apposed plasma membranes are considerably thinner (arrows).

COMMENTS

Pease (1996) introduced ethylene glycol as dehydrating medium because it is the weakest known protein denaturing solvent (Tanford *et al.*, 1962). He also eliminated fixation of the tissue. However, ethylene glycol denatures proteins at high concentrations and, in addition, the embedding media used by Pease denature proteins. Sjöstrand and Barajas (1968) and Sjöstrand (1977, 1997) applied a physical chemical principle when embedding the tissue. Thus, the high viscosity of Vestopal prevents extensive unfolding of the polypeptide chains and the extremely short dehydration time, making use of the high viscosity of 100% ethylene glycol at 0°C reduces the exposure of the tissue to denaturing conditions. The cristae are then 250 Å thick (Fig. 5.4B). Incomplete dehydration because of embedding too large pieces of tissue makes the cristae appear considerably thicker (Fig. 5.4C). The 250 Å thickness has been established using six different preparatory methods, including freeze fracturing (Sjöstrand, 1990).

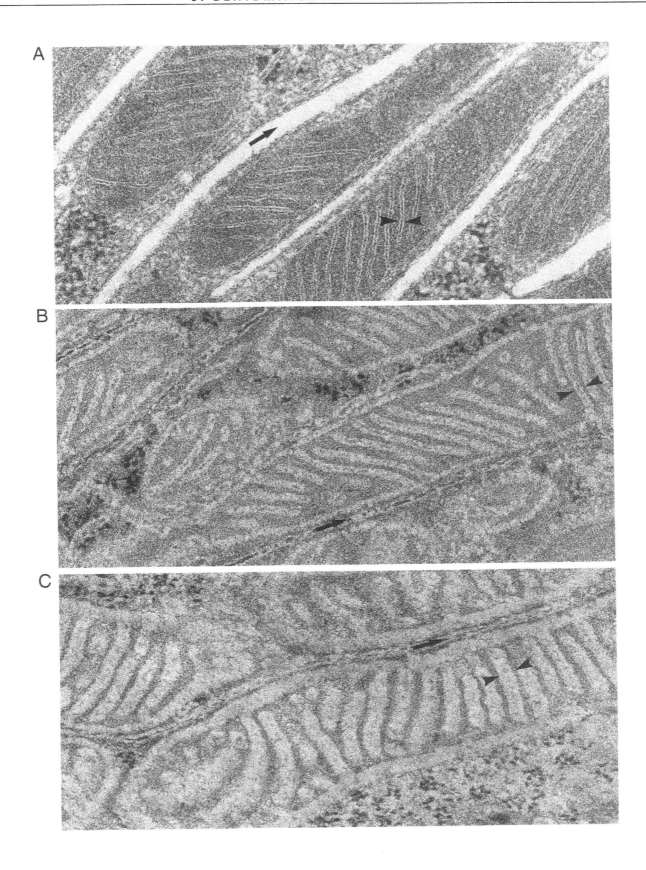

5. Choice of Intermediate Solvent

FIGURE 5.5A A rat liver tissue that has been fixed in 2.5% glutaraldehyde in a 0.1 *M* cacodylate buffer, postfixed in osmium tetroxide, dehydrated with several changes of ethanol (70, 90, and 100%), and transferred to propylene oxide for 2 hr, then to a 1:1 mixture of propylene oxide and the epoxy resin Polarbed for 1 hr, to a 1:2 mixture of these substances for 18 hr, and to undiluted Polarbed for 4 hr before embedding in Polarbed. × 25,000.

The cytoplasm and nucleus have an appearance that is typical for cells fixed with glutaraldehyde followed by osmium tetroxide and embedded in an epoxy resin.

FIGURES 5.5B–5.5D Same as Fig. 5.5A except that propylene oxide has been exchanged for other intermediate solvents: tetrachloroethylene (=perchloroethylene = CNP 30) in Fig. 5.5B, 1,1,1-trichloroethane (= inhibisol) in Fig. 5.5C, and limonene (= dipentene) in Fig. 5.5D. × 25,000.

The ultrastructural appearance of the tissue is the same as in Fig. 5.5A.

COMMENTS

The use of an intermediate solvent between the dehydrating ethanol and the embedding resin has been recommended because the resin is not readily miscible with ethanol. Propylene oxide seems to be the most commonly used solvent in electron microscopical laboratories. However, it has the disadvantage of being an alkylating agent, with which there may be a risk of carcinogenicity. For this reason various substitutes for it have been tested. The solvents tetrachloroethylene and trichloroethane have been recommended in the light microscopical technique as substitutes for the intermediate xylene (Maxwell, 1978), although they too are toxic irritants. They have also been tried as substitutes for propylene oxide in the electron microscopical technique. A more acceptable intermediate solvent is limonene (U. E. Afzelius, 1990), which has no known toxic properties and in fact is a component of orange marmelade. Mixtures of limonene and epoxy resin are more viscous than mixtures of propylene oxide and resin, but the penetration of resin into the tissues and section quality are comparable.

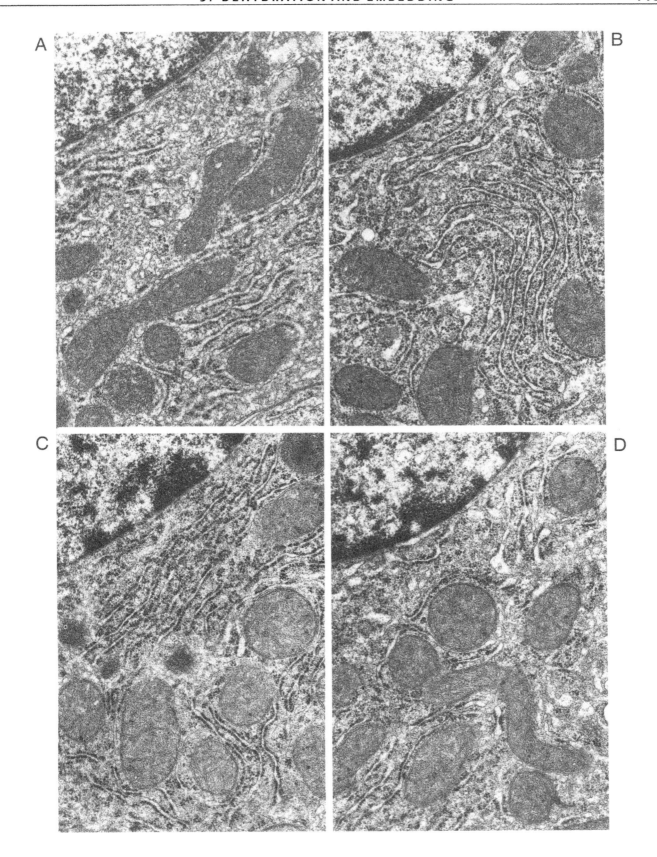

6. Epon, Araldite, and Vestopal: Unstained Sections

FIGURE 5.6A Liver cytoplasm, perfusion fixed with 1% glutaraldehyde in 0.1 M cacodylate buffer, postfixed in 1% osmium tetroxide, dehydrated in graded ethanols, and embedded in Epon. No section staining. × 25,000.

The cell ultrastructure can be discerned even though the micrograph has an overall low contrast.

FIGURE 5.6B Liver cytoplasm from the same animal as in Fig. 5.6A. The specimen has been treated in the same way except that embedding was in Araldite. × 25,000.

The cell ultrastructure is qualitatively indistinguishable from that in Fig. 5.6A but the contrast is somewhat lower.

FIGURE 5.6C Specimen from the same animal as in Fig. 5.6A and treated in the same way except that embedding was in Vestopal W. × 25,000.

The cell fine structure is largely similar to that in Fig. 5.6A, although there is noticeably less contrast.

COMMENTS

These three embedding media permit a good preservation of the cell structure in a reproducible fashion. However, they appear to have slightly different densities in the electron microscope. These differences are due in part to differences in their specific gravities, but undoubtedly an evaporation in the electron beam is a more important factor in this respect. Such an evaporation, called "clearing," is less pronounced in Vestopal than in Epon or, in fact, in most other embedding media (Luft, 1973); in particular, the methacrylates undergo a pronounced clearance in the electron beam. The evaporation makes the section thinner, sometimes to half its original thickness, and at the same time makes the electron-dense components of the tissue stand out more clearly, although at the expense of some deterioration of image quality.

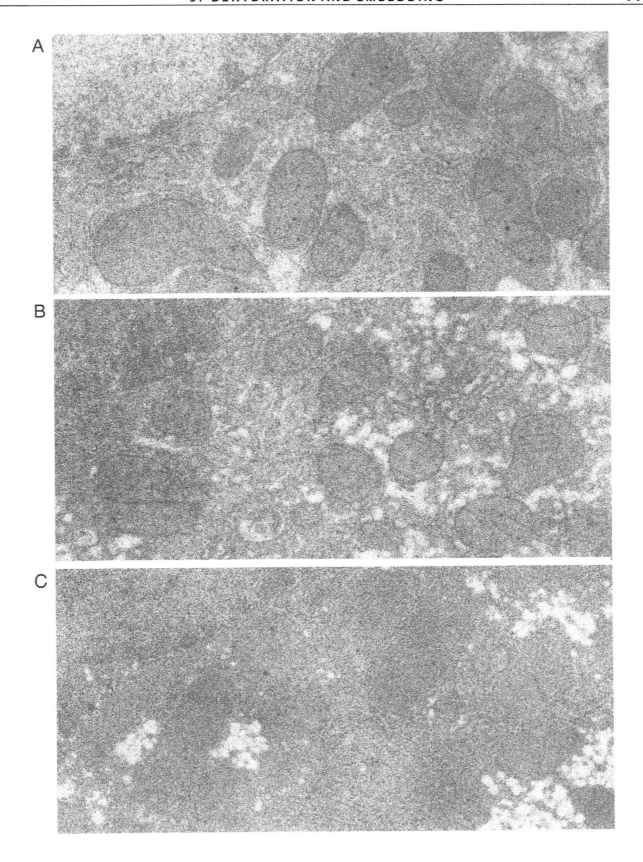

7. Epon, Araldite, and Vestopal: Stained Sections

FIGURES 5.7A–5.7C Sections from the same three embeddings shown in Figs. 5.6A–5.6C, although the sections were stained. Section staining was performed in an identical fashion and at the same occasion, with uranyl acetate and lead citrate. × 40,000.

There is no obvious difference in cell ultrastructure or even in micrograph contrast among the three embedding media.

COMMENTS

The lower contrast seen in the unstained Vestopal section seems to be largely compensated for by the greater stainability of tissue components in Vestopal. Differences may be encountered in resin viscosity (the most viscous one being Vestopal), transparency, and color; in their ability to adhere and glue together biological structures—Araldit and Epon being nearly ideal in this respect—whereas for instance the methacrylates may not always stick to fatty surfaces such as insect cuticula; in their sectionability (methacrylates being the easiest to cut); and in their behavior in the electron beam.

Quantitative measurements have shown that methacrylates show a relatively great shrinkage during polymerization, that Vestopal shrinks considerably less, and that epoxides such as Epon and Araldite show the least shrinkage of the examined embedding media (Kushida, 1962). For the latter media, shrinkage amounts to about 2% overall. A differential shrinkage within the embedding block may occur, but this effect has not been examined.

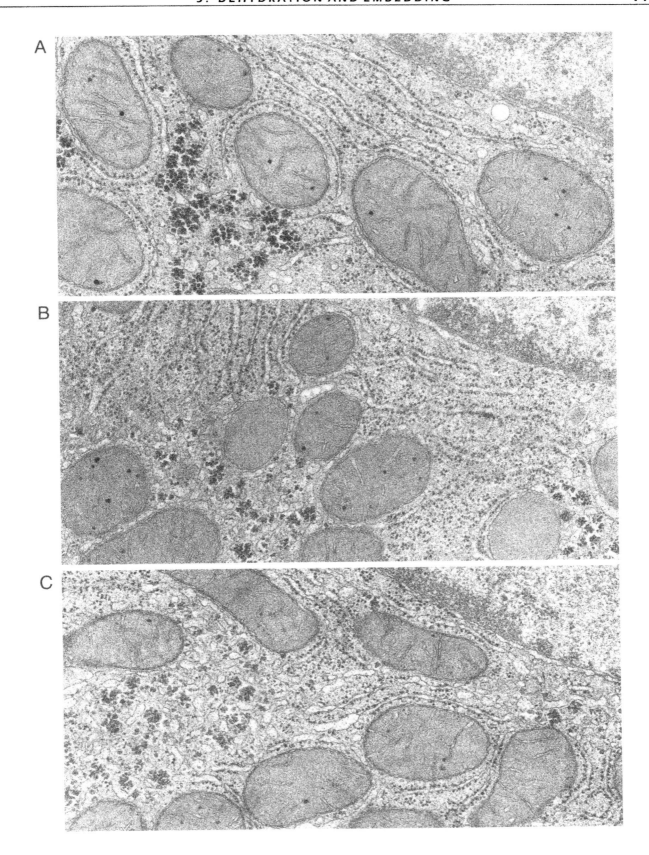

8. Different Brands of Epoxy Resins

FIGURES 5.8A–5.8C A skeletal rat muscle fixed for 2 hr at room temperature in 2.5% glutaraldehyde in 0.1 *M* cacodylate buffer, postfixed in 1% osmium tetroxide, dehydrated in ethanol, cleared in propylene oxide, and embedded in three different commercial brands of epoxy resins: Epon 812 (Electron Microscopy Sciences, Fort Washington, PA) (Fig. 5.8A), Polarbed 812 (Polaron BioRad, Cambridge, MA) (Fig. 5.8B), and EM bed (Electron Microscopy Sciences) (5.8C). The instructions for use given by the suppliers were followed. Sectioning was performed with a diamond knife, and sections were mounted on naked grids and section stained with uranyl acetate and lead citrate. × 22,000.

The three different embedding media give nearly identical results and the only differences were those that could be noted during embedding, trimming of blocks, and sectioning. In this comparison, the unpolymerized EM bed and Polarbed are less viscous than Epon 812. All three embedding resins give uniformly polymerized, good blocks. Specimens embedded in EM bed or Polarbed appeared somewhat more brittle than were Epon 812 ones; occasionally they fractured during trimming. Sections of Epon 812 were found to stretch somewhat better on the water trough during sectioning. In all cases the sections could be used either with or without supporting film.

COMMENTS

The three embedding media tested here, and undoubtedly a number of others, are suitable for work with the electron microscope and give very similar results. It is probably good advice to stick to one embedding medium that has been shown to give satisfactory results in the laboratory. When changing to another embedding medium, a new formula often has to be followed and a successful outcome is far from guaranteed.

When using these or any other embedding media, one must always keep in mind that they are toxic, mutagenic, allergenic, and perhaps carcinogenic (Causton, 1981; Ringo *et al.*, 1982). All work should be carried out in a well-ventilated hood, and it is important to use gloves that are not penetrated by the resins (Ringo *et al.*, 1984).

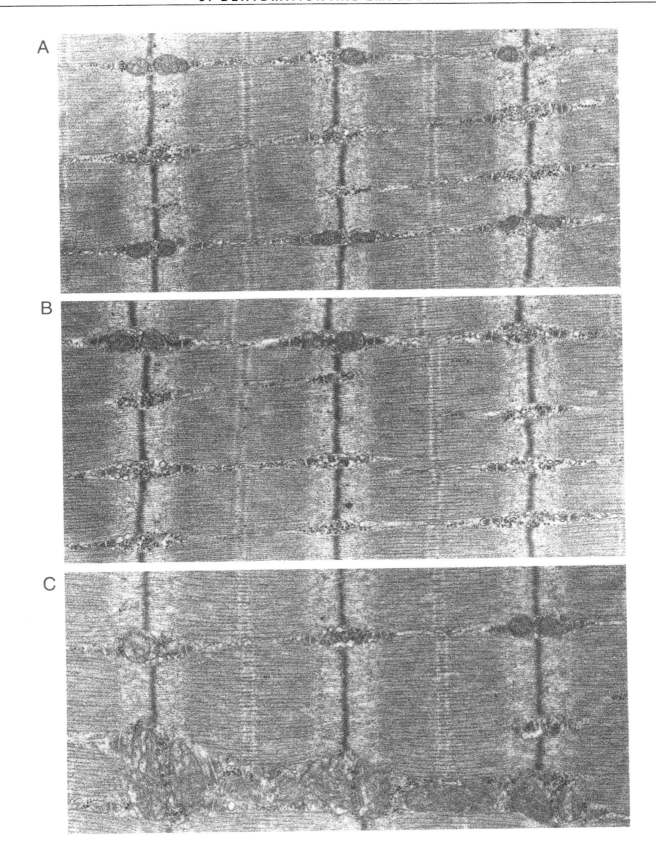

9. Spurr and LR White

FIGURE 5.9A Liver cytoplasm fixed in glutaraldehyde and osmium tetroxide and embedded in the low viscosity resin Spurr (Spurr, 1969). Section staining was done with uranyl acetate and lead citrate. × 60,000.

Mitochondria, endoplasmic reticulum, and nucleus appear as following conventional Epon embedding.

FIGURE 5.9B A liver cell fixed in glutaraldehyde (no osmium tetroxide postfixation) and embedded in LR White (Newman and Hobot, 1987, 1993). Following fixation the tissue was dehydrated in 70% ethanol, infiltrated with the LR White resin, and polymerized at 60°C. × 40,000.

Cytoplasmic components are well preserved, but all membranes appear in reverse contrast, i.e., as white lines against a surrounding stained cytoplasm or mitochondrial matrix. The white areas in the cytoplasmic ground substance represent sites of dissolved glycogen.

FIGURE 5.9C Liver cells fixed in 4% formaldehyde and 0.4% picric acid and embedded in LR White. Dehydration was in ethanol and continued to 100% before resin filtration. × 40,000.

Membranes appear in reverse contrast, as is typical for nonosmicated tissues. As compared to Fig. 5.9B, the mitochondrial matrix is stained more intensely.

COMMENTS

In addition to "classical" embedding media, methacrylate, Araldite, and Epon, a number of other embedding media have been developed and applied to more specific purposes. Thus the low viscosity embedding medium Spurr is particularly useful for embedding plant and insect materials, which are difficult to infiltrate with conventional epoxy resins. It is chemically based on vinyl cyclohexene dioxide, which contains two epoxides and is potentially carcinogenic. With respect to sectioning, staining, and contrast properties there are no major differences between Spurr- and Epon-embedded tissues. Furthermore, the fine structure of cellular components, such as those illustrated in Fig. 5.9A, do not appear distinctly different from those embedded in Epon.

LR White embedding media (London Resin White) have mainly been applied in connection with electron microscope immunocytochemistry due to their properties of preserving antigenicity. One reason for this property may be that they retain some water during polymerization. LR White resins are not used for osmium tetroxide-fixed tissues. A disadvantage with these resins is that cellular membranes are often poorly preserved and show disruptions and distortions when observed at high magnification. The intense staining of the cytoplasmic matrix often obscures finer details such as microtubules and filaments. The holes observed in the sections in Figs. 5.9B and 5.9C represent glycogen that has been extracted during LR White embedding or from the section during ultramicrotomy.

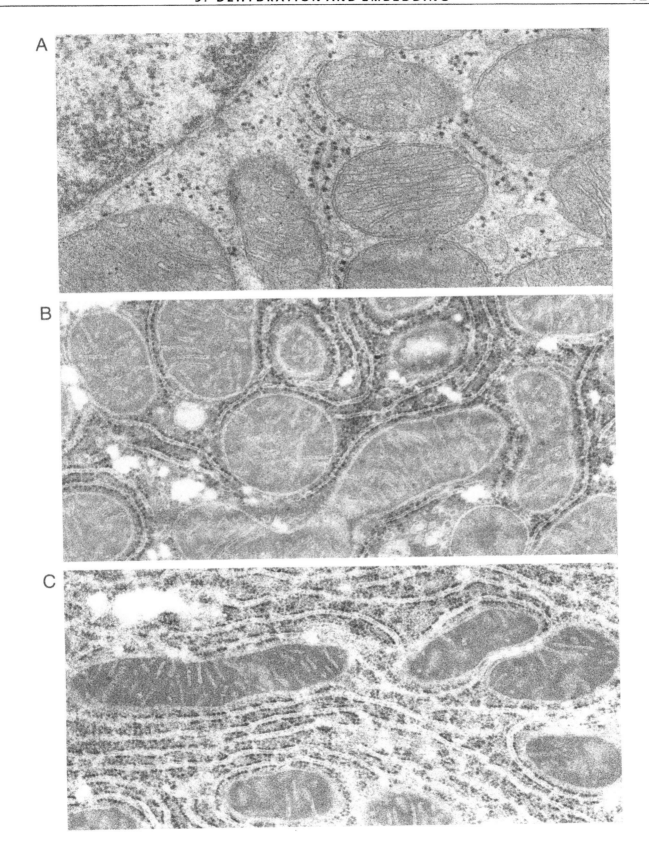

10. Embedding of Isolated Cells

FIGURE 5.10A A pellet of isolated kidney cells. The cells were centrifuged at 2000*g* in an Eppendorf tube for 10 min. A suspension of the isolated kidney cells was mixed 1:1 with 4% glutaraldehyde in cacodylate buffer and sedimented to a pellet after 2 hr in a 1-ml Eppendorf tube. The pellet was then rinsed in a buffer and dehydrated with increasing concentrations of ethanol. During the last steps of dehydration, the pellet was detached spontaneously from the bottom of the centrifuge tube and thereafter embedded in Epon as is done with a tissue block. Care was taken not to damage the pellet mechanically. × 3000.

The cells are well preserved and stand out against an empty embedding medium.

FIGURE 5.10B Isolated rye protoplasts embedded in Epon. A suspension of isolated protoplasts was fixed in glutaraldehyde, allowed to sediment, resuspended in a phosphate buffer, and concentrated by centrifugation. Droplets of 5 μl of the protoplast pellet were pipetted into the interior of droplets of a 4% agar solution kept at 40°C. After cooling, these droplets were fixed in osmium tetroxide and treated the same way as tissue specimens. × 2200.

The protoplasts are spherical, which indicates that they are intact.

FIGURE 5.10C A tissue culture cell grown on a collagen membrane with a diameter of about 5 mm. The membrane with attached cells was immersed in 2% glutaraldehyde in a cacodylate buffer, rinsed, postfixed in osmium tetroxide, and divided into smaller pieces, which were then individually Epon embedded in flat embedding molds. Sections were cut at a right angle to the collagen membrane and double stained with uranyl acetate and lead citrate. × 6000.

The cell appears well preserved and attached to the densely stained collagen membrane.

COMMENTS

Several methods have been described for the preparation of isolated and cultured cells for electron microscopy. In principle, such preparations are simple and the cells can be treated as small tissue blocks, but in practice care has to be taken not to damage or lose the cells during the repeated centrifugations and pipettings. For example, plant protoplasts are fragile and break apart upon handling; tissue culture cells tend to detach from the support membrane. The three procedures described here are examples of techniques that we have found rather easy to apply.

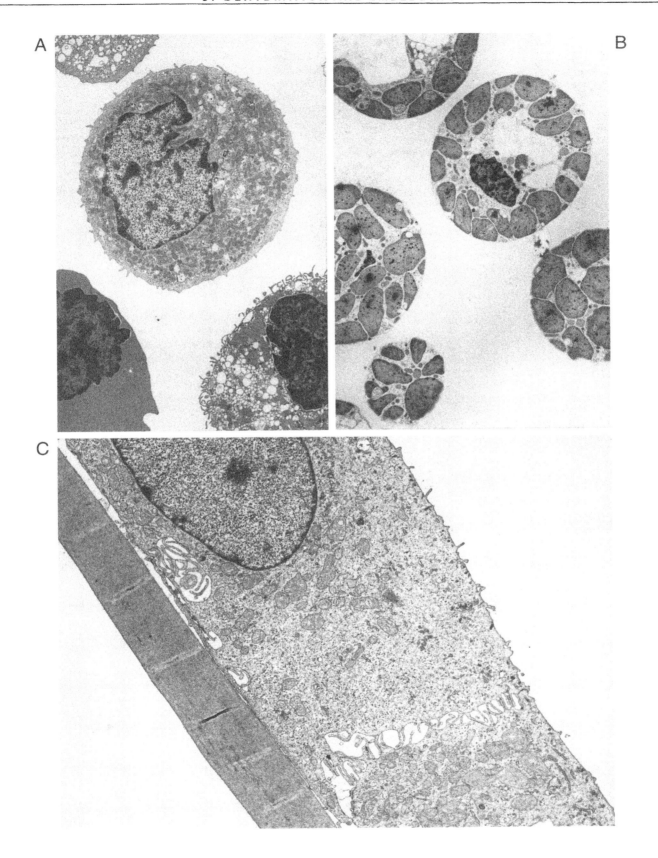

References

Acetarin, J.-D., Carlemalm, E., Kellenberger, E., and Villiger, W. (1987). Correlation of some mechanical properties of embedding resins with their behaviour in microtomy. *J. Electr. Microsc. Techn.* **6,** 63–79.

Afzelius, U. E. (1990). Limonene: A citrus fruit derived solvent as substitute for propylene oxide. *Micron Microsc. Acta* **21,** 147–148.

Ashworth, C. T., Leonard, J. S., Eigenbrodt, E. H., and Wrightsman, F. J. (1966). Hepatic intracellular osmiophilic lipid droplets: Effect of lipid solvents during tissue preparation. *J. Cell Biol.* **31,** 301–318.

Bahr, G. F., Bloom, G., and Friberg, U. (1957). Volume changes of tissues in physiological fluids during fixation in osmium tetroxide or formaldehyde and during subsequent treatment. *Exp. Cell Res.* **12,** 342–355.

Bernhard, W. (1955). Appareil de deshydratation continue. *Exp. Cell Res.* **8,** 248–249.

Boyde, A., Bailey, E., Jones, S. J., and Tamarin, A. (1977). Dimensional changes during specimen preparation for scanning electron microscopy. *Scann. Electr. Microsc.* **1977 I,** 507–518.

Causton, B. E. (1981). Resins: Toxicity, hazards, and safe handling. *Proc. Roy. Microsc. Soc.* **16,** 265–269.

Cope, G. H., and Williams, M. A. (1967). Quantitative studies on neutral lipid preservation in electron microscopy. *J. Roy. Microsc. Soc.* **88,** 259–277.

Fahimi, H. D. (1974). Effect of buffer storage on fine structure and catalase cytochemistry of peroxisomes. *J. Cell Biol.* **63,** 675–683.

Frösch, D., and Westphal C. (1989). Melamine resins and their application in electron microscopy. *Electr. Microsc. Rev.* **2,** 231–255.

Gerdes, A. M., Kriseman, J., and Bishop, S. P. (1982). Morphometric study of cardiac muscle: The problem of tissue shrinkage. *Lab. Invest.* **46,** 271–274.

Glauert, A. M. (1974). Fixation, dehydration and embedding of biological specimens. *In* "Practical Methods in Electron Microscopy" (A. M. Glauert, ed.), Vol. 3 North-Holland, Amsterdam.

Glauert, A. M. (1991). Epoxy resins: An update on their selection and use. *Eur. Microsc. Anal.* September, 13–18.

Helander, K. G. (1987). Studies on the rate of dehydration of histological specimens. *J. Microsc. (Oxford)* **145,** 351–355.

Hildebrand, C., and Müller, H. (1974). Low-angle X-ray diffraction studies on the period of central myelin sheaths during preparation for electron microscopy: A comparison between different anatomical areas. *Neurobiology* **4,** 71–81.

Johannessen, J. V. (1973). Rapid processing of kidney biopsies for electron microscopy. *Kidney Int.* **3,** 46–50.

Korn, E. D., and Weisman, R. A. I. (1966). Loss of lipids during preparation of amoebae for electron microscopy. *Biochim. Biophys. Acta* **116,** 309–316.

Kushida, H. (1962). A study of cellular swelling and shrinkage during fixation, dehydration and embedding in various standard media. *J. Electr. Microsc. (Tokyo)* **11,** 135–138.

Langenberg, W. G. (1982). Silicone additive facilitates epoxy plastic sectioning. *Stain Technol.* **57,** 79–82.

Lee, R. M. K. W., Garfield, R. E., Forrest, J. B., and Daniel, E. E. (1980). Dimensional changes in cultured smooth muscle cells due to preparatory processes for transmission electron microscopy. *J. Microsc. (Oxford)* **120,** 85–91.

Luft, J. H. (1961). Improvements in epoxy resin embedding methods. *J. Biophys. Biochem. Cytol.* **9,** 409–414.

Luft, J. H. (1973). Embedding media: Old and new. *In* "Advanced Techniques in Biological Electron Microscopy" (J. K. Koehler, ed.), pp. 1–34, Springer-Verlag, Berlin.

Maunsbach, A. B. (1998). Embedding of cells and tissues for ultrastructural and immunocytochemical analysis. *In* "Cell Biology: A Laboratory Handbook" (J. E. Celis, ed.), 2nd Ed., Vol. 3, pp. 249–259. Academic Press, San Diego.

Maxwell, M. H. (1978). Safer substitutes for xylene and propylene oxide in histology, hematology, and electron microscopy. *Med. Lab. Sci.* **35,** 401–403.

Mollenhauer, H. H. (1986). Surfactants as resin modifiers and their effect on sectioning. *J. Electr. Microsc. Tech.* **3,** 217–222.

Mollenhauer, H. H. (1993). Artifacts caused by dehydration and epoxy embedding in transmission electron microscopy. *Microsc. Res. Techn.* **26,** 496–512.

Mollenhauer, H. H., and Droleskey, R. E. (1985). Some characteristics of epoxy embedding resins and how they affect contrast, cell organelle size, and block shrinkage. *J. Electr. Microsc. Techn.* **2,** 557–562.

Morii, S., Shikata, N., and Nakao, I. (1993). Ultracytochemistry of cytoplasmic lipid droplets. *Acta Histochem. Cytochem.* **26,** 251–259.

Muller, L. L., and Jacks, T. J. (1975). Rapid chemical dehydration of samples for electron microscopic examinations. *J. Histochem. Cytochem.* **23,** 107–110.

Newman, G. R., and Hobot, J. A. (1987). Modern acrylics for post-embedding immunostaining techniques. *J. Histochem. Cytochem.* **35,** 971–981.

Newman, G. R., and Hobot, J. A. (1993). "Resin Microscopy and On-Section Immunocytochemistry." Springer-Verlag, Berlin.

Ockleford, C. D. (1975). Redundancy of washing in the preparation of biological specimens for transmission electron microscopy. *J. Microsc. (Oxford)* **105,** 193–203.

Parton, R. G. (1995). Rapid processing of filter-grown cells for Epon embedding. *J. Histochem. Cytochem.* **43,** 731–733.

Pease, D. C. (1966). The preservation of unfixed cytological detail by dehydration with "inert" agents. *J. Ultrastruct. Res.* **14,** 356–378.

Pihakaski-Maunsbach, K., Soitamo, A., and Suoranta, U.-M. (1990). Easy handling of cell suspensions for electron microscopy. *J. Electr. Microsc. Techn.* **15,** 414–415.

Ringo, D. L., Brennan, E. F., and Cota-Robles, E. H. (1982). Epoxy resins are mutagenic: Implications for electron microscopists. *J. Ultrastruct. Res.* **80,** 280–287.

Ringo, D. L., Read, D. B., and Cota-Robles, E. H. (1984). Glove materials for handling epoxy resins. *J. Electr. Microsc. Techn.* **1,** 417–418.

Rostgaard, J., and Tranum-Jensen, J. (1980). A procedure for minimizing cellular shrinkage in electron microscope preparation: A quantitative study on frog gall bladder. *J. Microsc. (Oxford)* **119,** 213–232.

Segal, A. S., Boulpaep, E. L., and Maunsbach, A. B. (1996). A novel preparation of dissociated renal proximal tubule cells that maintain epithelial polarity in suspension. *Am. J. Physiol.* **270** (*Cell Physiol.* **39**), C1843–C1863.

Sjöstrand, F. S. (1980). The structure of kidney mitochondria in low denaturation embedded and in freeze-fractured material. *J. Ultrastruct. Res.* **72,** 174–188.

Sjöstrand, F. S. (1997). The physical chemical basis for preserving cell structure for electron microscopy at the molecular level and available preparatory methods. *J. Submicrosc. Cytol. Pathol.* **29,** 157–172.

Sjöstrand, F. S., and Barajas, L. (1968). Effects of modifications in conformation of protein molecules on structure of mitochondrial membranes. *J. Ultrastruct. Res.* **25,** 121–155.

Spurr, A. R. (1969). A low-viscosity epoxy resin embedding medium for electron microscopy. *J. Ultrastruct. Res.* **26,** 31–43.

Tanford, C., Buckley, C. E., De, P. K., and Lively, E. P. (1962). Effect of ethylene glycol on the conformation of γ-globulin and β-lactoglobulin. *J. Biol. Chem.* **237,** 1168–1171.

Widéhn, S., and Kindblom, L.-G. (1988). A rapid and simple method for electron microscopy of paraffin-embedded tissue. *Ultrastruct. Pathol.* **12,** 131–136.

FREEZING AND LOW-TEMPERATURE EMBEDDING

Chemical fixation, together with subsequent specimen handling, inevitably influences the chemical composition of the tissue components and also modifies tissue fine structure in various ways, as illustrated in chapters 2–5. An alternative and, in principle, ideal method would be to freeze the biological specimen and to observe it without any chemical treatment in the electron microscope under what might be regarded as "artifact-free" conditions. In practice this is extremely difficult to achieve, although a number of low-temperature methods have been introduced, which to various degrees aim toward this goal. Usually the methods require complex and expensive instrumentation.

The first critical step in any cryoprocedure is freezing of the tissue. Sometimes this step is referred to as "cryofixation," although it actually does not involve any fixation in the common histological sense of the word. The outstanding problem in cryofixation is to avoid ice crystal formation, and great efforts have been invested into solving this problem. The direct methods of freezing include the rapid immersion of specimens in heat-conducting coolants kept at a very low temperature, freezing by rapid impact on a cooled metal surface (metal mirror freezing, "slam freezing"), or exposure to very high pressures and a low temperature (high-pressure freezing). Another approach is to fix the specimen lightly with a chemical fixative and then infiltrate the specimen with an antifreeze substance such as sucrose or glycerol before freezing.

The frozen specimen can be treated in different ways. It can be observed directly in the electron microscope at low temperature (cryoelectron microscopy; see Chapter 10), fractured and metal shadowed in a freeze-fracture apparatus (see Chapter 17), or sectioned at low temperature (see Chapter 8), particularly for immunocytochemistry (see Chapter 16). Finally, the specimen may be dehydrated at low temperature, freeze-substituted, and embedded in resins or plastics either at low temperature or after raising the temperature to room temperature.

An early attempt to introduce low-temperature procedures in biological electron microscopy was reported by Friedrich Krause (1937). He wanted to minimize heating and radiation effects on the specimen and cooled the specimen chamber with liquid air. He also cut frozen thick sections, which he dried and observed in the electron microscope.

Over the following decades the development of cryotechniques was slow but eventually resulted in a great diversification of procedures, as summarized by Patrick Echlin (1992). This field now includes plunge freezing, high-pressure freezing, metal mirror freezing, and spray freezing, which may precede freeze-substitution, freeze-drying, freeze-fracturing, cryoultramicrotomy, or microscopy of frozen hydrated specimens.

Echlin, P. (1992). "Low-Temperature Microscopy and Analysis." Plenum Press, New York.
Krause, F. (1937). *Naturwissenschaften* **51**, 817–825.

1. Plunge Freezing

FIGURE 6.1A Parts of proximal tubule cells perfusion fixed with 1% glutaraldehyde, plunge-frozen without prior treatment with a cryoprotectant, and freeze-substituted in Lowicryl HM20. The small tissue blocks were held with fine forceps and plunged rapidly into the liquid nitrogen. × 53,000.

The tissue has large, empty areas as well as fragmented and compressed cytoplasmic regions.

FIGURES 6.1B AND 6.1C Tissue cryoprotected by immersion in 2.3 *M* sucrose (Fig. 6.1B) or 30% glycerol (Fig. 6.1C) for 1 hr before plunge freezing in liquid nitrogen and further treatment as in Fig. 6.1A. × 53,000.

The tissues in Figs. 6.1B and 6.1C show well-preserved cytoplasm and clearly discernable membranes and cytoplasmic organelles.

COMMENTS

The freezing step is of outmost importance in all cryopreparation procedures. A simple method is to plunge the specimen into the coolant, either manually or with some instrument. Fixed or unfixed tissues, which are not cryoprotected and which are manually plunged into liquid nitrogen, invariably show ice crystal formation due to the low heat conductivity of the nitrogen gas formed at the interface between tissue and liquid nitrogen. Ice crystal formation can be avoided if the specimen is treated with a cryoprotectant or if it is very small, such as molecules on a grid, and plunged into a suitable coolant at high speed by means of refined instrumentation (see Robards and Sleytr, 1985; Echlin, 1992). For the best result, plunge freezing should be carried out with cooled propane or ethane, which are superior to nitrogen as coolants. However, for practical purposes, in connection with immunocytochemistry on cryosections of aldehyde-fixed tissues, cryoprotection with 2.3 *M* sucrose and subsequent manual plunge cooling into liquid nitrogen is usually quite adequate.

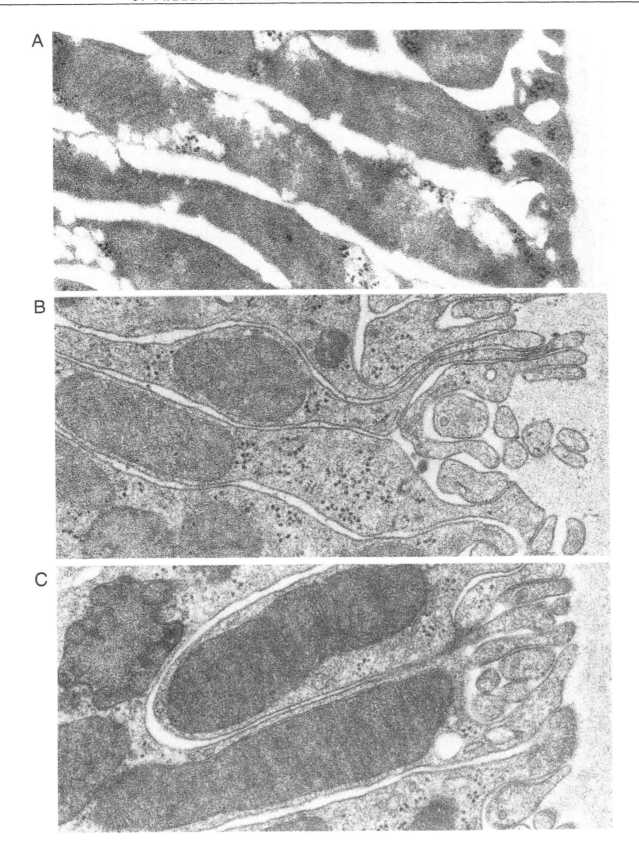

2. Contact Freezing of Unfixed Tissue

FIGURE 6.2A Kidney tissue frozen by slamming it against a polished and gold-coated copperblock cooled to liquid nitrogen temperature, so-called contact freezing, impact cooling, or "slam freezing." One kidney in an anesthetized rat was exposed, and a thin slice was excised rapidly with a thin razor blade. It was mounted on a foam rubber stud placed on the piston of a contact-freezing apparatus (CF-100, Life Cell Corporation). Within 3 min after excision of the tissue from the functioning kidney the piston was released and the tissue was frozen. Thereafter it was freeze-substituted in methanol containing 0.5% uranyl acetate, and low-temperature embedded in Lowicryl HM20. × 20,000.

Upper part of a proximal tubule that came in direct contact with the copperblock. The direction of impact is indicated by arrowheads. The cell has a normal architecture and the cytoplasmic matrix a uniform appearance.

FIGURE 6.2B Part of glomerulus located close to the kidney surface in the same preparation as in Fig. 6.2A. × 75,000.

Cross-sectioned foot processes from glomerular epithelial cells have more rounded profiles than after aldehyde fixation (compare Fig. 2.9B). Slit membranes (arrows) are preserved.

FIGURE 6.2C Cells in the same section as Fig. 6.2A but located more than 10 μm from the kidney surface that hit the copperblock. × 20,000.

The cytoplasm has a reticular appearance and mitochondria and other cell organelles are deformed.

COMMENTS

Freezing of unfixed tissue without ice crystal formation requires a freezing speed in excess of 10,000°C per second (Moor, 1987). This can be obtained by rapidly slamming the tissue against a cold, polished metal surface with high heat conductivity. As biological tissues have a low heat conductivity, this high freezing speed is achieved only in the most superficial layer of the tissue. In Fig. 6.2A the tissue was in direct contact with the cooling metal surface and the superficial tissue appears well preserved. Also, a few micrometers below the surface there is no disturbing ice crystal formation (Fig. 6.2B), whereas further from the surface (Fig. 6.2C) the cytoplasm has a net-like appearance, due to the formation of small ice crystals. In other tissues, the depth of acceptable preservation has been reported to vary between 5 and 25 μm. Complications of slam freezing include difficulties in adjusting the impact force and keeping the surface of the metal block free from ice. With too high a force of the piston the cells become flattened. Ice formation on the metal surface reduces cooling speed and leads to inadequate freezing. A great strength of slam freezing is the possibility of capturing very rapid biological events. Thus in studies of the release of synaptic vesicles, Heuser et al. (1979) were able to achieve a time resolution around 2–3 msec.

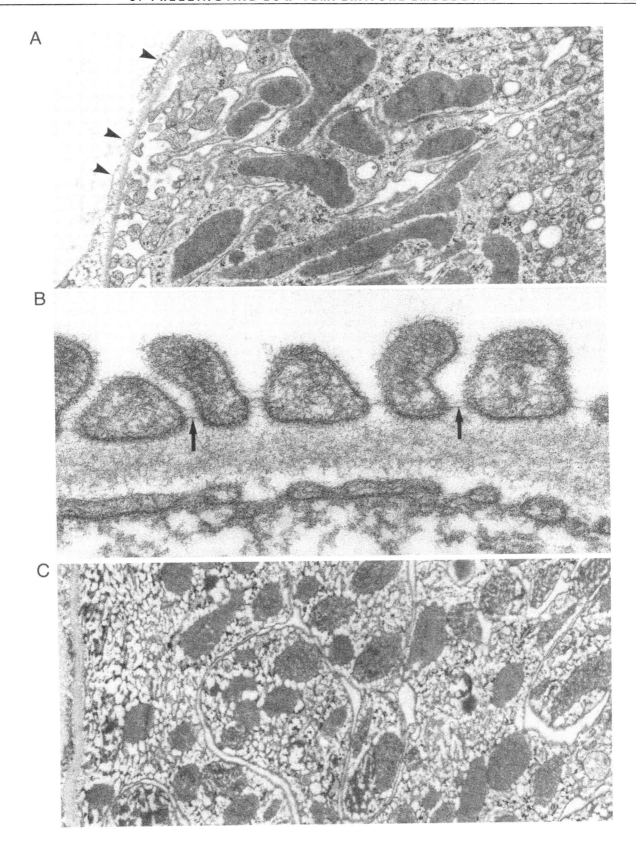

3. Contact Freezing of Fixed Tissue

FIGURE 6.3A A proximal tubule epithelium from a rat kidney perfusion fixed with 1% glutaraldehyde and contact frozen without cryoprotectant against a copperblock cooled with liquid nitrogen. The tissue was then freeze-substituted in methanol containing 0.5% uranyl acetate starting at −85°C and embedded in Lowicryl HM20 at −45° and UV polymerized. The section was stained with uranyl acetate and lead citrate. Arrowheads indicate the cell surface facing the copperblock. × 13,000.

The cytoplasm in the basal part of the cell has a uniform appearance, but the apical part of the cell already shows a slight reticular pattern.

FIGURE 6.3B A proximal tubule cytoplasm from a rat kidney perfusion fixed with 2% paraformaldehyde and 0.1% glutaraldehyde. The tissue was contact frozen without cryoprotectant and low-temperature embedded as in Fig. 6.3A. The center of this micrograph is located 2.0 μm from the tissue surface facing the metal block. × 80,000.

The cytoplasm has a uniform appearance and mitochondria are free from structural defects.

FIGURE 6.3C A proximal tubule cytoplasm from the same section as in Fig. 6.3A, although located more than 20 μm from the surface of the tissue block. × 40,000.

The cytoplasm has a reticular appearance with empty areas and mitochondria show irregular defects.

COMMENTS

It is evident that tissue fixed with aldehydes but without cryoprotection is as sensitive to ice crystal formation as is unfixed tissue. Only a very thin layer of superficial tissue has a useful preservation quality after contact freezing. In some aspects the aldehyde-fixed tissue differs from unfixed ones: It is much easier to cut a thin slice of the tissue before freezing, and the resistance against compression during slamming is somewhat greater. Structural deformation and extraction during tissue processing are also less pronounced. In fact, the fine structure of the best preserved tissue (Fig. 6.3B) closely resembles nonfrozen aldehyde-fixed tissue, which have been conventionally embedded in epoxy resin. However, the freeze-substituted and Lowicryl-embedded specimens are more useful for immunoelectron microscopical studies than tissue embedded in epoxy resin.

A

B

C

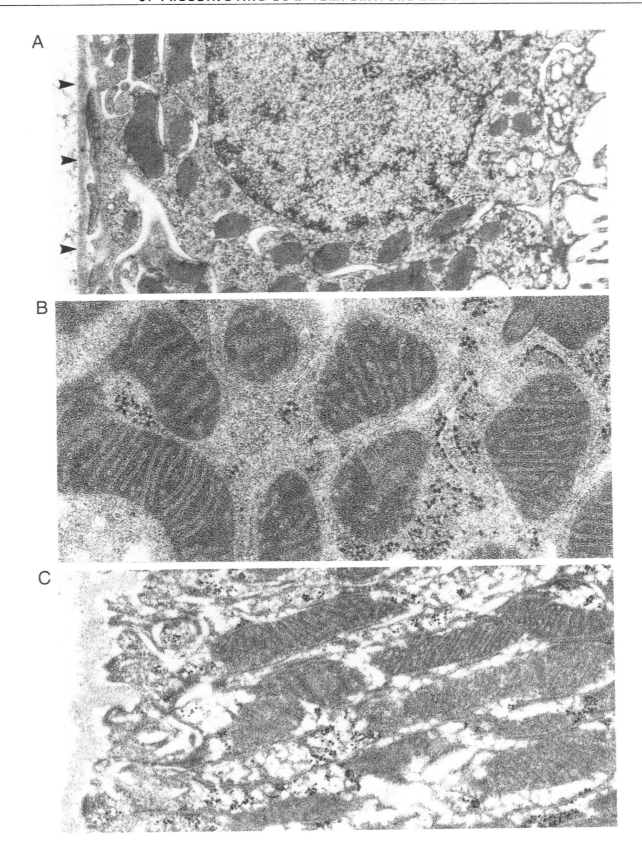

4. High-Pressure Freezing

FIGURES 6.4A AND 6.4B A kidney tissue high-pressure frozen in a Balzers high-pressure instrument (HPM 010). A cylindrical (about 1 mm) biopsy was taken from the exposed kidney of an anesthetized rat. A thin slice of the biopsy was placed in the cavity between the two metal specimen holders and frozen in the freezing chamber at a pressure in excess of 2100 bar. The specimen was frozen within 2 min after removal from the living animal. The frozen specimens were then freeze-substituted in methanol/uranyl acetate and embedded in Lowicryl HM20. The sections were stained with uranyl acetate and lead citrate. Fig. 6.4A $\times$ 12,500 and Fig. 6.4B $\times$ 32,000.

Figure 6.4A shows a well-preserved architecture of proximal tubule cells, and Fig. 6.4B illustrates the well-maintained brush border and homogeneous cytoplasm. There are narrow empty spaces around some of the lysosomes.

COMMENTS

Exposure of tissues to high pressure will decrease its freezing point, which will reduce the cooling rate that is required to prevent ice crystal growth. As a result it is unnecessary to apply extremely high cooling rates. The salient advantage of high-pressure freezing is that the thickness of adequately frozen tissue may be up to 500 μm, thus some 20–30 times greater than in other types of freezing (Studer *et al.*, 1989). The drawback of the technique is the high cost and complexity of the instrumentation. The final appearance of high-pressure frozen tissue depends not only on the freezing process, but also on subsequent freeze-substitution and embedding procedures. The empty spaces around lysosomes are not necessarily related to the freezing but also to the subsequent freeze-substitution, which was carried out without the application of glutaraldehyde or osmium tetroxide. If the tissue is freeze-substituted in acetone containing osmium tetroxide, such empty spaces around mitochondria are absent, but the tissue would then be unsuitable for immunocytochemistry.

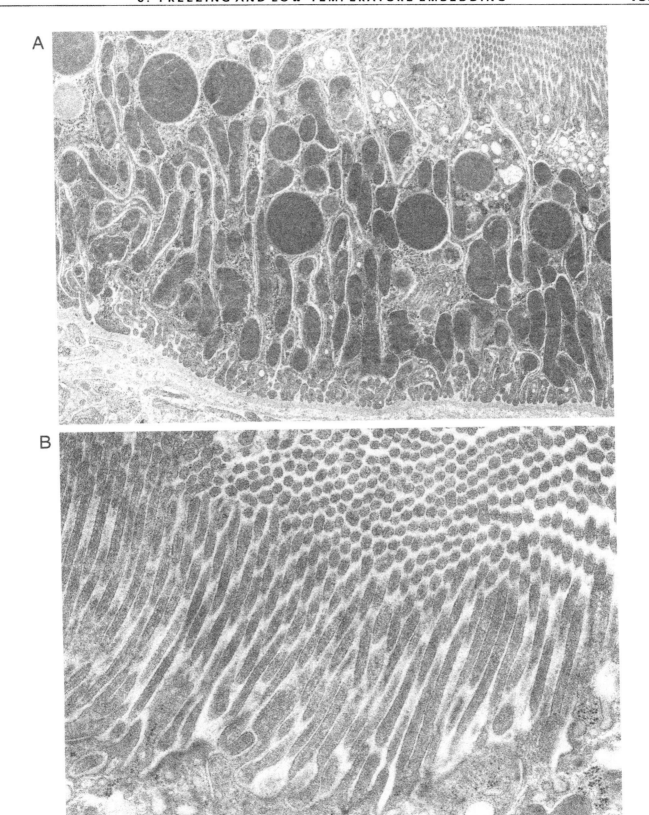

5. Freeze-Substitution in Methanol/Uranyl Acetate

FIGURE 6.5A The thick ascending limb of the loop of Henle in a rat kidney perfusion fixed with 4% formaldehyde and 0.4% picric acid in 0.1 M cacodylate buffer. Small tissue blocks (<0.5 mm in all directions) were excised, rinsed in buffer, and infiltrated with 2.3 M sucrose containing 2% paraformaldehyde. Each tissue block was held with a fine forceps and quickly plunged by hand into liquid nitrogen. The blocks were then transferred to methanol containing 0.5% uranyl acetate in a Balzers freeze-substitution apparatus cooled with liquid nitrogen. Substitution was initiated at −85°C and after 4 hr was continued at −80°C for 24 hr. The specimens were then rinsed four times in pure methanol, initially at −70°C and then at −45°C, and infiltrated at −45°C with Lowicryl HM20 : methanol in proportions 1 : 2 and 1 : 1 and finally with pure resin. They were polymerized with indirect UV light, first for 48 hr at −45°C and then for 48 hr at 0°C. × 30,000.

The membranes of the interdigitating tubule cells and the adjacent endothelial cells display a high contrast. Cell ultrastructure is well preserved and resembles that of specimens processed at room temperature and embedded in Epon.

FIGURE 6.5B Specimen prepared as in Fig. 6.5A except that fixation was 4% formaldehyde and that embedding was in Lowicryl K4M, which was infiltrated at −20°C instead of −45°C. Section staining was with uranyl acetate alone. × 60,000.

The cytoplasm is well preserved and the cell membranes are stained distinctly and in places clearly triple layered. The mitochondrial matrix is stained intensely, rendering cristae almost undetectable. Note the strictly parallel course of adjacent basolateral membranes.

FIGURE 6.5C Basal part of a kidney cell prepared as in Fig. 6.5B. × 60,000.

Cell membranes are well preserved and cytoplasm uniformly stained with ribosomes clearly identifiable.

COMMENTS

Freeze-substitution of aldehyde-fixed tissues gives an ultrastructural preservation that is comparable to "standard" fixation methods with aldehydes and Epon embedding without postfixation in osmium tetroxide. In certain aspects, however, freeze-substituted tissue are different. Basolateral membranes may show a more uniform spacing (Fig. 6.5B) and the endocytic vacuoles may appear more spherical and without wrinkles of the endosome membrane as sometimes seen after conventional preparations (see Fig. 2.13A). Freeze-substitution of aldehyde-fixed tissues has the advantage that the antigenic properties of cell components are well preserved. Despite the fact that freeze-substitutions in Lowicryls are time-consuming and somewhat cumbersome, they represent one of the preferred preparatory procedures in immunoelectron microscopy.

A

B

C

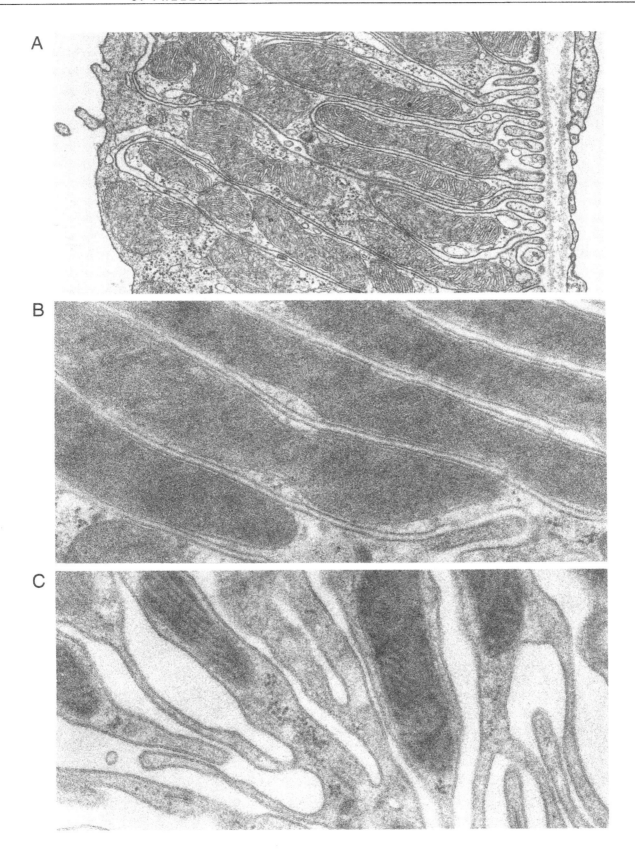

6. Freeze-Substitution in Osmium Tetroxide/Acetone

FIGURE 6.6A A kidney tissue frozen against a polished and gold-coated copperblock in a contact-freezing apparatus CF-100 (Life Cell Corporation). The procedure was the same as in Fig. 6.2 except that substitution was in 3% osmium tetroxide in pure acetone for 3 days before three rinses in acetone at $-45°C$. After raising the temperature to ambient temperature, the tissue was infiltrated with Epon and polymerized at 60°C. Ultrathin sections were cut at a right angle to the surface of the thin, embedded tissue slice and stained with uranyl acetate and lead citrate. × 15,000.

The illustrated portion of the section reaches from the surface touching the metal mirror (arrowheads) down into the tissue approximately 12 μm. A distinct gradient in preservation quality can be readily observed. The cytoplasm close to the mirror has a uniform appearance whereas the cytoplasm at the bottom is reticular in appearance. Furthermore, the nuclei are flattened.

FIGURE 6.6B The center of this field is located approximately 2 μm from the surface of the tissue block. × 80,000.

The lateral intercellular spaces are either completely closed (arrowheads) or slightly separated but filled with a stainable material (arrows). The cytoplasm has a uniform staining.

FIGURE 6.6C The center of this field is located about 5 μm from the tissue surface. × 80,000.

Mitochondria are well preserved and mitochondrial cristae have a compact five-layered appearance (arrowheads). To the left is part of the compressed nucleus that is also seen in the upper third of Fig. 6.6A.

FIGURE 6.6D Part of a cell located approximately 10 μm from the tissue surface. × 80,000.

The nucleoplasm has a reticular appearance, as has the ground cytoplasm.

COMMENTS

There is a gradual decrease in quality of tissue preservation during metal mirror freezing when going from the surface into the tissue. At a depth of 10 μm a severe ice crystal formation may distort the structure (Fig. 6.6D). Mitochondria very close to the surface are well preserved, although mitochondrial cristae deviate from the "classical" pattern in that the space between the mitochondrial membranes is absent. An unusual pattern is also observed with respect to the basolateral membranes, where adjacent membranes are either touching (arrowheads) or separated by a space apparently containing stainable material as seen in Fig. 6.6B. A comparison between this pattern and the pattern observed in various other experimental protocols (e.g., Figs. 2.2 and 6.5B) illustrates the difficulty in determining the true width of the intercellular space between epithelial cells. The flattened appearance of the nucleus (Figs. 6.6A and 6.6B) indicates that the pressure on the tissue from the metal mirror was high and deformed the cells. To what extent this has modified the fine structure of the cytoplasm remains uncertain.

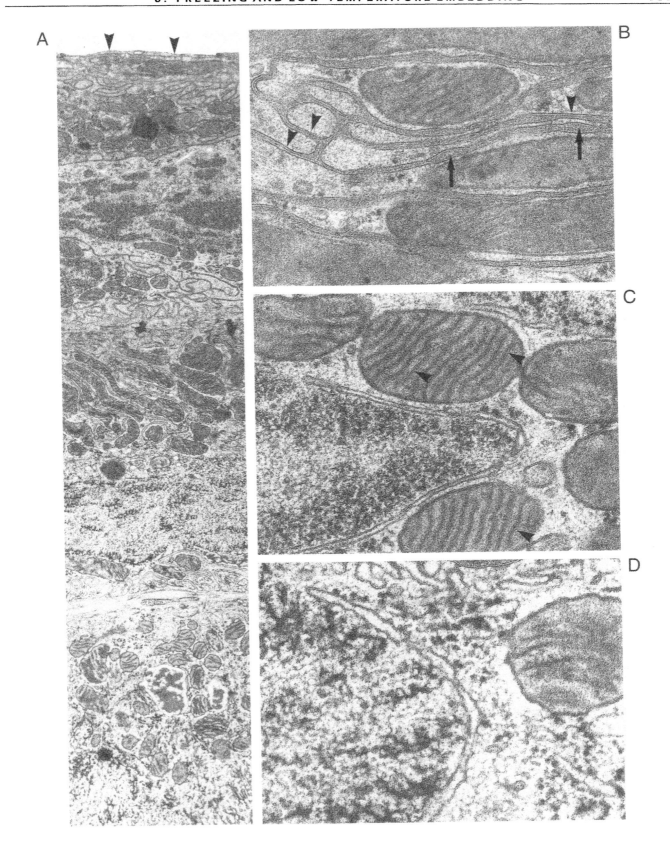

7. Progressive Lowering of Temperature Embedding in Lowicryl

FIGURE 6.7A Parts of proximal tubule cells following aldehyde fixation and low-temperature embedding in Lowicryl K4M according to Carlemalm *et al.* (1982). The kidney was perfusion fixed with 1% glutaraldehyde and excised specimens were dehydrated in increasing concentrations of ethanol without prior osmium tetroxide fixation. During dehydration in ethanol the temperature was lowered progressively to −20°C and the tissue was transferred to ethanol : Lowicryl K4M mixtures 2 : 1 and then 1 : 1, followed by pure resin. Polymerization was carried out in indirect UV light, first at −20°C for 1 day and then at room temperature for 3 days. Section staining was done with uranyl acetate and lead citrate. × 12,500.

The cell architecture is well maintained and cytoplasm and mitochondria are stained uniformly.

FIGURE 6.7B Higher magnification of the same preparation in Fig. 6.7A showing interdigitating proximal tubule cells. × 46,000.

The cytoplasm has the typical appearance of cells fixed in aldehyde but not processed through osmium tetroxide. The intercellular space has a very uniform width of approximately 400 Å.

FIGURE 6.7C Interdigitating cells of the thick ascending limb of the medullary distal tubule (mTAL) in a rat kidney in the same preparation as in Fig. 6.7A. The section was stained with 2% uranyl acetate in methylcellulose according to Roth *et al.* (1990). × 44,000.

Cell shapes are well preserved and intercellular spaces are of very constant widths, around 200 Å. The staining of mitochondria is granular and their membranes are outlined poorly. The contrast of the cell membranes is stained strongly and the outer layer of the membrane is particularly distinct. This staining pattern is characteristic of Lowicryl K4M. It is only obtained rarely with Lowicryl HM20 and not seen after Epon embedding.

COMMENTS

Embedding in acrylic resins can be carried out by freeze-substitution or by employing the "progressive lowering of temperature" technique illustrated here (Carlemalm *et al.*, 1982, 1985). This procedure provides good ultrastructural preservation and is comparatively easy to perform with the different Lowicryls and LR White (London Resin White). The Lowicryls can be sectioned at room temperature and HM20 is easier to section than K4M. Although K4M is considered a hydrophilic resin and HM20 a hydrophobic one, this difference is of little importance in immunocytochemistry.

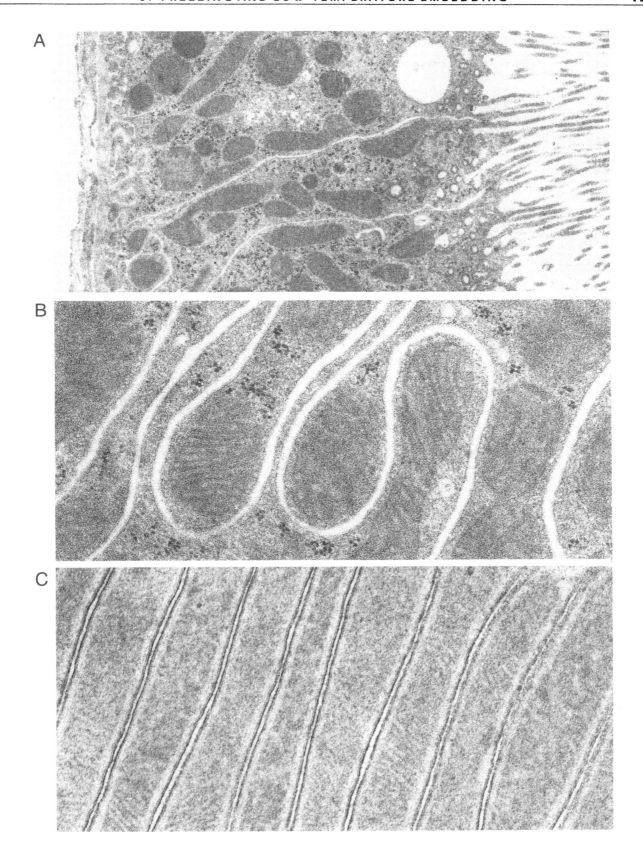

References

Ashford, A. E., Allaway, W. G., Gubler, F., Lennon, A., and Sleegers, J. (1986). Temperature control in Lowicryl K4M and glycol methacrylate during polymerization: Is there a low-temperature embedding method? *J. Microsc. (Oxford)* **144**, 107–126.

Carlemalm, E., Garavito, R. M., and Villiger, W. (1982). Resin development for electron microscopy and an analysis of embedding at low temperature. *J. Microsc. (Oxford)* **126**, 123–143.

Carlemalm, E., Villiger, W., Hobot, J. A., Acetarin, J.-D., and Kellenberger, E. (1985). Low temperature embedding with Lowicryl resins: Two new formulations and some applications. *J. Microsc. (Oxford)* **140**, 55–63.

Chiovetti, R., McGuffee, L. J., Little, S. A., Wheeler-Clark, E., and Brass-Dale, J. (1987). Combined quick freezing, freeze-drying, and embedding tissue at low temperature and in low viscosity resins. *J. Electr. Microsc. Techn.* **5**, 1–15.

Dahl, R., and Staehelin, L. A. (1989). High-pressure freezing for the preservation of biological structure: Theory and practice. *J. Electr. Microsc. Techn.* **13**, 165–174.

Dubochet, J. (1995). High-pressure freezing for cryoelectron microscopy. *Trends Cell Biol.* **5**, 366–368.

Dubochet, J., Adrian, M., Chang, J.-J., Homo, J.-C., Lepault, J., McDowall, A. W., and Schultz, P. (1988). Cryo-electron microscopy of vitrified specimens. *Quart. Rev. Biophys.* **21**, 129–228.

Echlin, P. (1992). "Low-Temperature Microscopy and Analysis." Plenum Press, New York.

Edelmann, L. (1991). Freeze-substitution and the preservation of diffusible ions. *J. Microsc. (Oxford)* **161**, 217–228.

Fujimoto, K., Noda, T., and Fujimoto, T. (1997). A simple and reliable quick-freezing/freeze-fracturing procedure. *Histochem. Cell Biol.* **107**, 81–84.

Glauert, A. M., and Young, R. D. (1989). The control of temperature during polymerization of Lowicryl K4M: There *is* a low-temperature embedding method. *J. Microsc. (Oxford)* **154**, 101–113.

Hagler, H. K. (1988). Artifacts in cryoelectron microscopy. *In* "Artifacts in Biological Electron Microscopy" (R. F. E. Crang and K. L. Klomparens, eds.), pp. 205–217. Plenum Press, New York.

Heuser, J. E., Reese, T. S., Dennis, M. J., Jan, Y., Jan, L., and Evans, L. (1979). Synaptic vesicle exocytosis captured by quick freezing and correlated with quantal transmitter release. *J. Cell Biol.* **81**, 275–300.

Hohenberg, H., Tobler, M., and Müller, M. (1996). High-pressure freezing of tissue obtained by fine-needle biopsy. *J. Microsc.* **183**, 133–139.

Humbel, B., and Müller, M. (1986). Freeze substitution and low temperature embedding. *In* "The Science of Biological Specimen Preparation for Microscopy and Microanalysis" (M. Müller, R. P. Becker, A. Boyde, and J. J. Wolosewick, eds.), pp. 175–183. Scanning Electron Microscopy, Inc., AMF O'Hare, IL.

Kaeser, W., Koyro, H.-W., and Moor, H. (1989). Cryofixation of plant tissues without pretreatment. *J. Microsc. (Oxford)* **154**, 279–288.

Maunsbach, A. B. (1998). Embedding of cells and tissues for ultrastructural and immunocytochemical analysis. *In* "Cell Biology: A Laboratory Handbook" (J. E. Celis, ed.), 2nd Ed., Vol. 3, pp. 260–267. Academic Press, San Diego.

Meissner, D. H., and Schwarz, H. (1990). Improved cryoprotection and freeze-substitution of embryonic quail retina: A TEM study on ultrastructural preservation. *J. Electr. Microsc. Techn.* **14**, 348–356.

Michel, M., Hillmann, T., and Müller, M. (1991). Cryosectioning of plant material frozen at high pressure. *J. Microsc. (Oxford)* **163**, 3–18.

Moor, H. (1987). Theory and practice of high pressure freezing. *In* "Cryotechniques in Biological Electron Microscopy" (R. A.

Steinbrecht, and K. Zierold, eds.), pp. 175–191. Springer-Verlag, Berlin.

Müller, M., Meister, N., and Moor, H. (1980). Freezing in a propane jet and its application in freezing-fracturing. *Mikroskopie (Wein)* **36**, 129–140.

Müller, T., Moser, S., Vogt, M., Daugherty, C., and Parthasarathy, M. W. (1993). Optimization and application of jet-freezing. *Scan Microsc.* **7**, 1295–1310.

Ohno, S., Baba, T., Terada, N., Fuji, Y., and Ueda, H. (1996). Cell biology of kidney glomerulus. *Intl. Rev. Cytol.* **166**, 181–230.

Plattner, H., and Knoll, G. (1984). Cryofixation of biological materials for electron microscopy by the methods of spray-, sandwich-, cryogen-jet- and sandwich-jet-freezing: A comparison of techniques. *In* "The Science of Biological Specimen Preparation for Microscopy and Microanalysis" (J.-P. Revel, T. Barnard, G. H. Haggis, and S. A. Bhatt, eds.), pp. 139–146. Scanning Electron Microscopy, Inc., AMF O'Hare, IL.

Quintana, C. (1994). Cryofixation, cryosubstitution, cryoembedding for ultrastructural, immunocytochemical and microanalytical studies. *Micron* **25**, 63–99.

Robards, A. W., and Sleytr, U. B. (1985). Low temperature methods in biological electron microscopy. *In* "Practical Methods in Electron Microscopy" (A. M. Glauert, ed.), Vol. 10, pp. 1–551. Elsevier, Amsterdam.

Roth, J., Taatjes, D. J., and Tokuyasu, K. T. (1990). Contrasting of Lowicryl K4M thin sections. *Histochemistry* **95**, 123–136.

Ryan, K. P. (1992). Cryofixation of tissues for electron microscopy: A review of plunge cooling methods. *Scan. Microsc.* **6**, 715–743.

Ryan, K. P., and Knoll, G. (1994). Time-resolved cryofixation methods for the study of dynamic cellular events by electron micrscopy: A review: *Scan. Microsc.* **2**, 259–288.

Sitte, H., Neumann, K., and Edelmann, L. (1987). Safety rules for cryopreparation. *In* "Cryotechniques in Biological Electron Microscopy" (R. Steinbrecht and K. Zierold, eds.), pp. 285–290. Springer-Verlag, Berlin/Heidelberg.

Sitte, H., Neumann, K., and Edelmann, L. (1989). Cryosectioning according to Tokuyasu vs. rapid-freezing, freeze-substitution and resin embedding. *In* "Immuno-Gold Labeling in Cell Biology" (A. J. Verkleij and J. L. M. Leunissen, eds.), pp. 63–93. CRC Press, Boca Raton, FL.

Sjöstrand, F. S. (1982). Low temperature techniques applied for CTEM and STEM analysis of cellular components at a molecular level. *J. Microsc. (Oxford)* **128**, 279–286.

Sjöstrand, F. S. (1990). Common sense in electron microscopy: About cryofixation, freeze-substitution, low temperature embedding, and low denaturation embedding. *J. Struct. Biol.* **103**, 135–139.

Sjöstrand, F. S., and Kretzer, F. (1975). A new freeze-drying technique applied to the analysis of the molecular structure of mitochondrial and chloroplast membranes. *J. Ultrastruct. Res.* **53**, 1–28.

Steinbrecht, R. A., and Zierold, K. (eds.) (1987). "Cryotechniques in Biological Electron Microscopy." Springer-Verlag, Berlin.

Studer, D., Michel, M., and Müller, M. (1989). High pressure freezing comes of age. *Scan. Microsc.* **3**, 253–269.

Van Harreveld, A., and Malhotra, S. K. (1967). Extracellular space in the cerebral cortex of the mouse. *J. Anat.* **101**, 197–207.

Weibull, C., and Christiansson, A. (1986). Extraction of proteins and membrane lipids during low temperature embedding of biological material for electron microscopy. *J. Microsc. (Oxford)* **142**, 79–86.

Zierold, K. (1990). Low-temperature techniques. *In* "Biophysical Electron Microscopy: Basic Concepts and Modern Techniques." (P. W. Hawkes and U. Valdrè, eds.), pp. 309–346. Academic Press, London.

SUPPORT FILMS

Most biological specimens studied in the electron microscope are placed on a specimen grid covered with a thin support film made of Formvar, collodium, or carbon. Ultrathin sections may also be placed on naked grids without a film, provided the sections are large enough to cover an opening in the grid and are free from large holes and deep knife scratches. For some purposes, particularly high-resolution work, the ultrathin sections are placed on a micronet ("holey" film), i.e., a film of Formvar (polyvinyl formal) or other material that is perforated by a large number of small holes and that functions as a supporting net. Those parts of the specimen that overlie the small holes can be viewed without interference of a support film.

The ideal support film should be thin, uniform, structureless, clean, and strong. As all these properties are difficult to combine, it is usually necessary to select the support film according to the aim of the investigation. At present, Formvar films and collodium films strengthened by a thin layer of evaporated carbon are in common use for sectioned material. The evaporated thin layer of carbon serves to minimize thermal specimen drift during microscopy. Support films are relatively easy to prepare, but they often exhibit defects, which may interfere with specimen interpretation. Special consideration should be given to the preparation of support films for the large openings of the so-called one-hole grids, where film strength is essential.

This chapter illustrates some of the characteristics of Formvar, collodium, and carbon films, including difficulties encountered in their production and handling.

The first biological samples to be examined by electron microscopy by necessity had to be very thin: cotton fibers, viruses, bacteria, mosquito wings, spores from molds, etc. One of the first problems encountered was to find a suitable substrate onto which to deposit the samples; it had to be sufficiently thin to be penetrated by the electrons, yet strong enough to support the specimen. Several soluble polymers were tried but the most useful one was Formvar (polyvinyl formal), which was introduced in the electron microscopical technique by Lester Germer (1939). Specimen grids were not yet commercially available but could be punched out from fine-mesh sieve nets, such as used for sifting flour. Alternatively, a small hole was drilled in a platinum disk. The Swedish botanist Ivar Elvers (1943) even manufactured microgrids from microtome sections of pinewood, which were then impregnated with osmium tetroxide.

Elvers, I. (1943) *Acta Horti Bergiani* **13**, 149–243.
Germer, L. (1939) *Phys. Rev.* **66**, 58–71.

1. Surface Topography

FIGURE 7.1A Surface of a Formvar film that has been shadowed with platinum at an angle of 30°. The Formvar film was produced by dipping a glass slide into a Formvar solution and floating the film off the slide onto a water surface. This side of the film faced the glass ("glass-facing surface"). × 30,000.

On the surface of the film there are many small spots of variable size and shape that are often linearly arrayed.

FIGURE 7.1B Surface of a Formvar film prepared as in Fig. 7.1.A, but showing that side of the film, which faced the air during manufacture of the film ("air-facing surface"). × 30,000.

The surface shows some large crater-like formations.

FIGURE 7.1C Surface of a collodium film shadowed with platinum at an angle of 30°. The film was produced by spreading a drop of the 0.5% collodium in amyl acetate on a water surface. This micrograph shows the water-facing surface. × 30,000.

The water-facing surface is essentially devoid of irregularities.

FIGURE 7.1D Air-facing surface of a collodium film prepared as in Fig. 7.1.C. × 30,000.

In contrast to the water-facing surface, the air-facing surface shows a few small craters or elevations.

COMMENTS

The glass-facing surface of the Formvar film is essentially a replication of the glass slide and reflects irregularities of its surface. It also replicates contamination on the slide, which has a tendency to become linearly arranged during the cleaning of the glass slide. The air-facing surface of the Formvar film is essentially devoid of linearly arranged defects, but may show single crater-like structures, which may have been created by small air bubbles trapped in the Formvar solution. These defects probably correspond to the so-called "pseudo holes" in the Formvar film. The water-facing surface of the collodium film is devoid of structures at this magnification. The small defects in the air-facing surface may represent small air bubbles in the collodium solution.

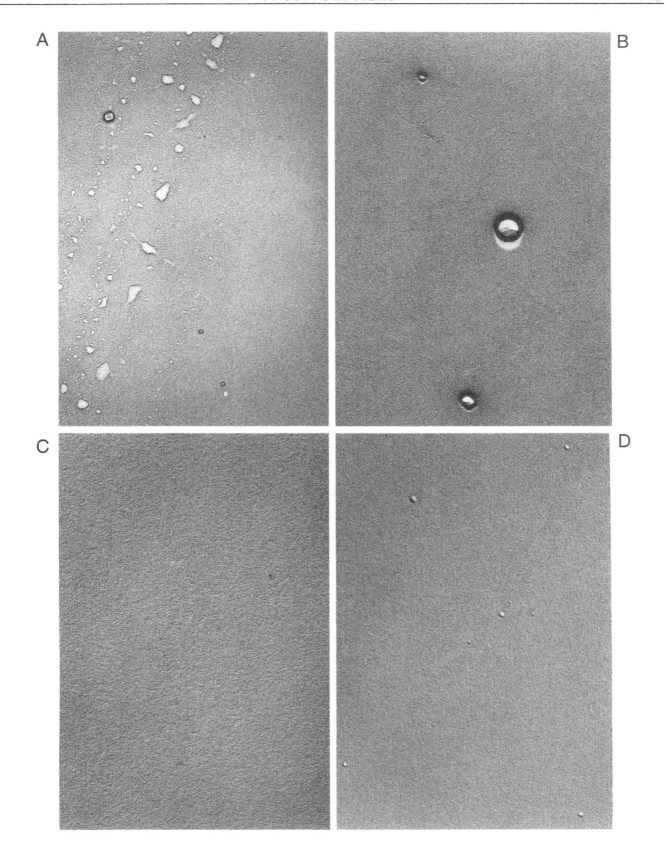

2. Stability of Film or Section

FIGURE 7.2A Latex particles on a Formvar film that was not carbon stabilized. Three exposures were made on the same photographic plate with 1-min intervals. The beam current was kept constant. The first exposure was slightly longer than the second and third ones. Grid bars are seen at the bottom and to the right. × 35,000.

The three exposures give rise to a triple image of each latex particle. Those particle triplets, which are far from the grid bars, are more extended than those closer to the grid bar. Within each triplet the longest exposure (the darkest image) is located away from the grid corner.

FIGURE 7.2B Latex particles on a carbon-coated collodion film. The same triple exposure technique as in Fig. 7.2A was used. The dark structures at the left and at the lower side are portions of the copper grid × 35,000.

The three exposures of the latex particles fall largely on top of each other and the micrograph is essentially indistinguishable from a single exposure.

FIGURE 7.2C A section of large (1.3 μm) latex particles, embedded in Epon. The section was mounted directly on a "naked" copper grid. Small (0.088 μm) latex particles were spread on the section. Three exposures were made with 1-min intervals. × 25,000.

The displacement of each structure is more pronounced at a distance away from the grid bar at the upper side.

FIGURE 7.2D An Epon section mounted on a copper grid and not coated with carbon. Small latex particles were spread on the section. Three exposures were made as in Fig. 7.2A. × 30,000.

The triple images of the peripheral particles are more extended, whereas in the center of the field the particles are almost superimposed.

COMMENTS

The multiple exposure technique gives an account of the stability and mechanical behavior of the specimen support film. It reveals whether there is movement of the film or section relative to the copper grid or movement of the specimen stage of the microscope.

Carbon films, carbon-coated collodion, or Formvar films are more stable than pure Formvar or collodion films. "Naked grids" with sections of Epon and Vestopal can be used for much routine work at low magnifications. Even without carbon coating such sections are strong enough to withstand the strain of mounting, section staining, and microscopy, provided fine mesh (e.g., 300 mesh) grids are used and the sections cover an entire opening and have been cut at about 60 nm or thicker. Without a support film, the contrast is slightly better than with support film, and dirt derived from the support film is absent. However, without carbon coating there is a shrinkage of the illuminated part of the section, which linearly may amount to 5% or more. Epon sections are not unique in shrinking when exposed to the electron beam. Organic matter in general seems to undergo a shrinkage in the beam as first demonstrated by Anderson and Richards (1942) and later by Kölbel (1972). However, carbon films seem unusual in showing a great resistance toward shrinkage.

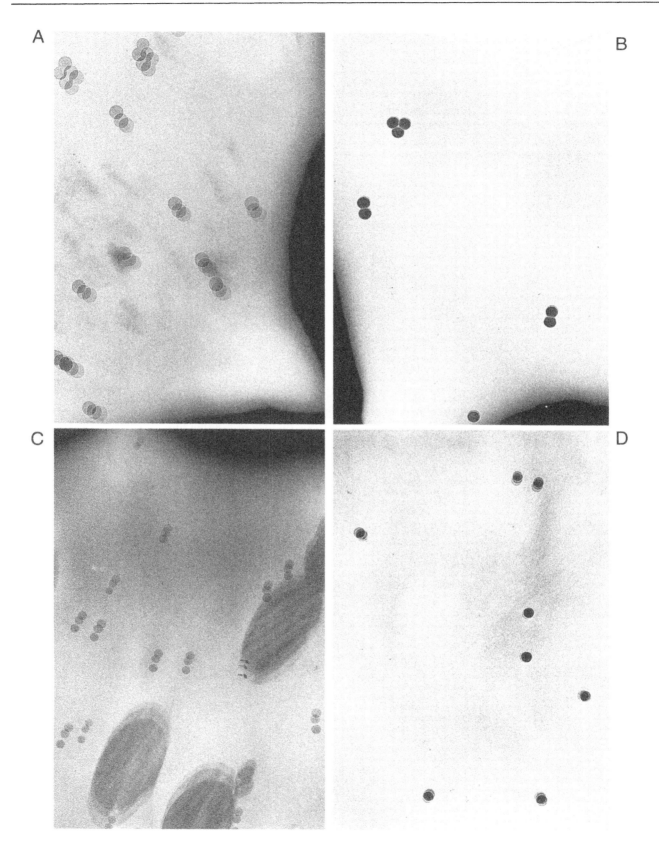

3. Holey Films

FIGURES 7.3A–7.3C Grasshopper spermatozoa fixed in glutaraldehyde, post-fixed in osmium tetroxide, and embedded in Epon. Ultrathin sections were mounted on a carbon-coated Formvar film with holes. Figure 7.3A × 42,000; Figs. 7.3B and 7.3C × 170,000.

Part of the section lies over a hole in the support film. Two indicated areas (arrows) are shown at higher magnification in Figs. 7.3B and 7.3C; they are enlargements from the same negative. Photographic paper of the same grade was used, although that in Fig. 7.3C had to be given a slightly longer exposure to give the print the same density as in Fig. 7.3B. The hole is circular and the support film has greater density (e.g., is thicker) at the rim of the hole. The slight, but noticeable, electron density displayed by the support film makes the section less electron translucent over the support film than over the hole. Over the hole (Fig. 7.3C) the section stands out with somewhat greater contrast than over the support film (Fig. 7.3B).

COMMENTS

Most support films have a thickness of 10–30 nm and will scatter electrons to a certain degree. For most biological specimens the scattering effect of the support film is without practical consequences, but for low-contrast sections the use of a holey film may allow finer details to be observed.

A

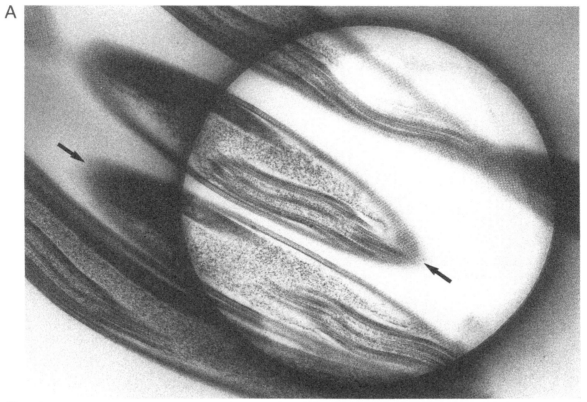

B 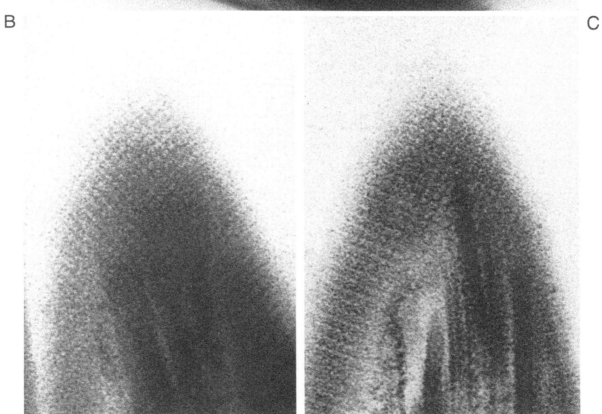 C

4. Thick and Thin Support Films

FIGURE 7.4A Ferritin molecules on a thin carbon film that was supported by a butyrate acetate micronet. × 350,000.

A micronet by definition has numerous small holes, and part of a hole is shown on the right side of the micrograph. The carbon film is thin (estimated to be less than 4 nm). The dark spots represent the iron core of the ferritin molecules and have a diameter of about 5.5 nm. Ferritin molecules on the thin carbon film appear both over the hole in the micronet and where the film is supported by the micronet.

FIGURES 7.4B AND 7.4C A higher magnification of areas in Fig. 7.4A. × 700,000.

Figure 7.4.B shows an area of Fig. 7.4A where the ferritin molecules are supported both by the micronet and by the carbon film. Figure 7.4C shows an area where the ferritin molecules are supported only by the carbon film. The ferritin molecules in Fig. 7.4C are shown with a higher contrast than those in Fig. 7.4B. This is further amplified at high magnification.

COMMENTS

In high-resolution electron microscopy of molecules and low contrast objects the electron scattering in the support film may be a limiting factor. Specimen detail may be lost and specimen resolution decreased. The general contrast in the micrograph is good when the specimen is supported by a carbon film that is very thin. Even thin carbon films, however, have an inherent granularity. Practically structureless support films can be produced from graphitized carbon, and some other materials, although these films are difficult to produce.

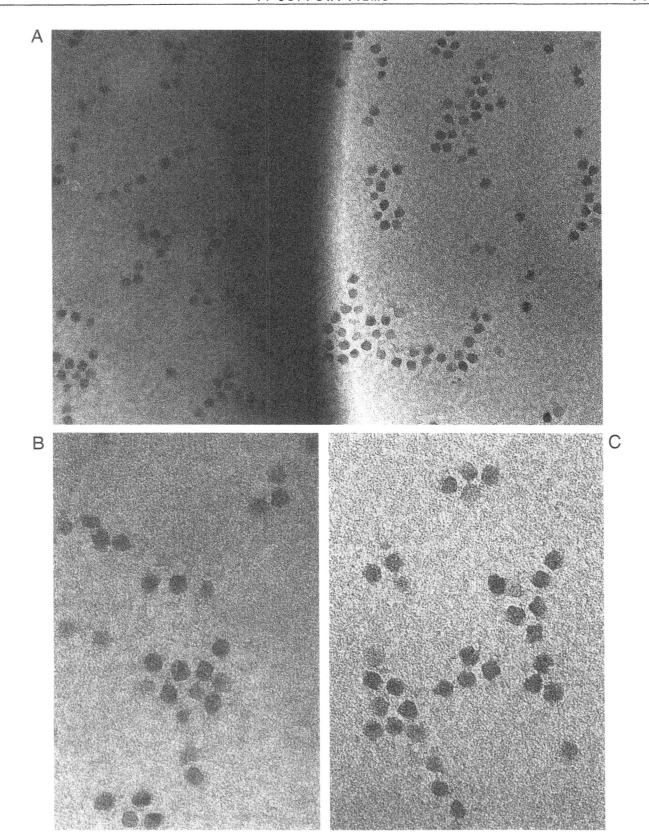

5. Folds in Support Films

FIGURE 7.5A Parts of two Epon sections on a Formvar support film. × 20,000.

A dark band runs through the entire viewing field, including the empty space between the two sections.

FIGURES 7.5B AND 7.5C Electron micrograph of kidney epithelial cells fixed and embedded by standard procedures. The sections were picked up on a Formvar film. In Fig. 7.6C the photographic enlargement was made with a much reduced exposure time in order to make the contents of the dark band visible. × 76,000.

The membranes of the round vesicles crossed by the streak and the two plasma membranes are unaffected in their course by the dense streak. Figure 7.5C confirms that the biological structures continue uninterrupted under the streak. Furthermore, a dark middle line is observed in the dark band (arrows).

FIGURE 7.5D Kidney epithelial cells prepared by conventional techniques and sections picked up on a Formvar film. × 45,000.

A wide dark band crosses the section without, however, distorting the cellular architecture.

COMMENTS

Folds in the support film appear as dark bands traversing the viewing field over relatively large distances. Their course is not interrupted at the edges of the section as seen in Fig. 7.5A. Furthermore, the biological structures are not deflected by these folds, as illustrated in Figs. 7.5C–7.5D. These characteristics distinguish folds in the support film from folds in the section (see Chapter 8, Figs. 8.6A–8.6D). The width of the thinnest folds is equal to twice the thickness of the support film. The thickness of the support film is therefore around 25 nm in Fig. 7.5C.

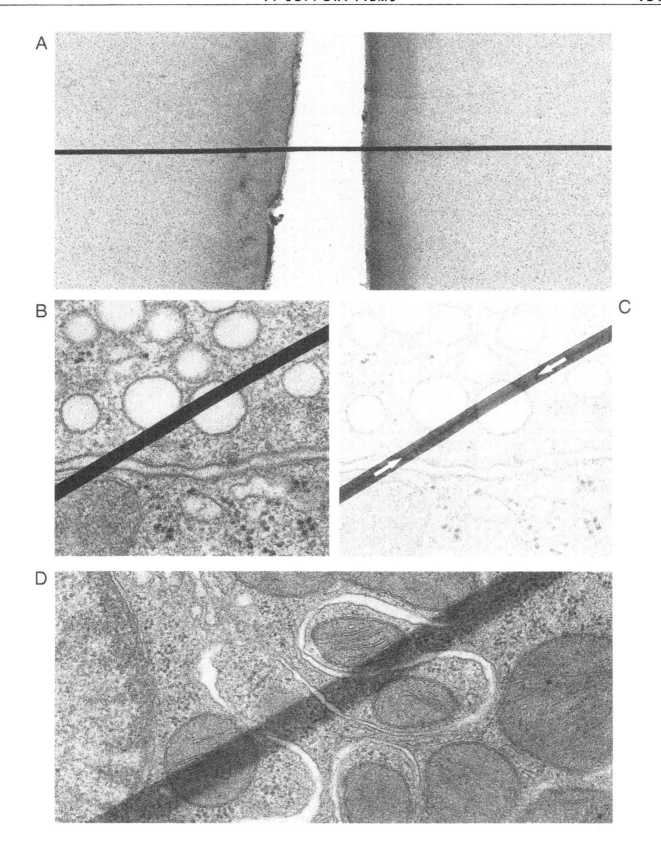

6. Defects in Formvar Films

FIGURE 7.6A A Formvar film produced in the conventional manner on glass slide and floated off on a water surface. × 10,000.

The film shows light bands in which there are some holes.

FIGURE 7.6B A Formvar film produced in the conventional manner on a glass slide. The film did not float off easily but seemed to stick to the glass surface. × 30,000.

Exotic patterns are observed in the Formvar film.

FIGURE 7.6C A Formvar film on a grid handled on purpose several times with a fine forceps. × 2000.

In some places the film is completely broken, whereas in other areas there are chevron-like regions, which are more electron dense than the surrounding Formvar film and located along a straight line. Along this line and between the chevrons there are irregular light streaks.

COMMENTS

The light bands in Fig. 7.6A represent thin regions in the Formvar film. The appearance of thin regions in Formvar films may be caused by stretching of the film when it is floated off the glass slide. Holes are more frequent in films exposed to moist air when drying on the glass (Sjöstrand, 1956). The "monsters" in Fig. 7.6B presumably represent areas of the Formvar film that formed focal attachment to the glass when the film was floated off onto water. The extended "legs" are attenuations in the film. The dense regions in Fig. 7.6C are wrinkles where the Formvar film folded upon itself. The light streaks apparently are areas where the film has been stretched and attenuated.

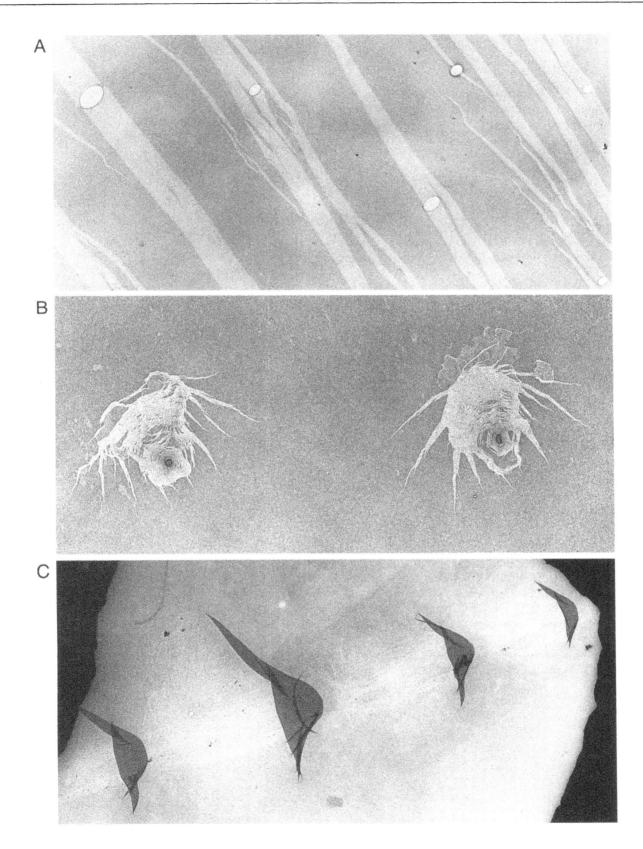

7. Common Contaminants

FIGURE 7.7A A Formvar film prepared in the conventional manner and examined in the electron microscope. × 30,000.

A faint reticular pattern is seen on part of this support film. It was not possible to decide whether this type of structure was present on the film already before microscopy or whether it was formed (or at least made more distinctly visible) during the first moments of observation.

FIGURE 7.7B A Formvar film prepared in the conventional manner and used for collection of sections. × 30,000.

An elongated structure with the general appearance of a bacterium is seen together with various forms of surrounding contamination.

FIGURE 7.7C A negative staining preparation of cell membranes utilizing uranyl acetate as negative stain and carbon as support film. × 40,000.

Contaminating bacteria are well-defined in negative stain preparations and they often exhibit a flagellum.

FIGURE 7.7D A high magnification of a structure in a preparation similar to that in Fig. 7.7C. × 200,000.

The bacterial flagellum shows a characteristic substructure.

COMMENTS

These four electron micrographs illustrate different structures that are commonly seen to contaminate support films. In many cases the origin of the contamination is difficult to trace and its nature is unknown. This is the case with the contamination shown in Fig. 7.7A and nicknamed "oil smoke." In other cases (Figs. 7.7B–7.7D) the contaminant can be identified as a bacterium or part of a bacterium, but it is usually impossible to locate the step in the preparatory procedure where bacteria appeared. The only way to reduce the amount of external contaminants is to use meticulously clean materials and tools. Stale water in distilled water bottles, dirty Pasteur pipettes, and forceps used to handle grids are common sources of contamination.

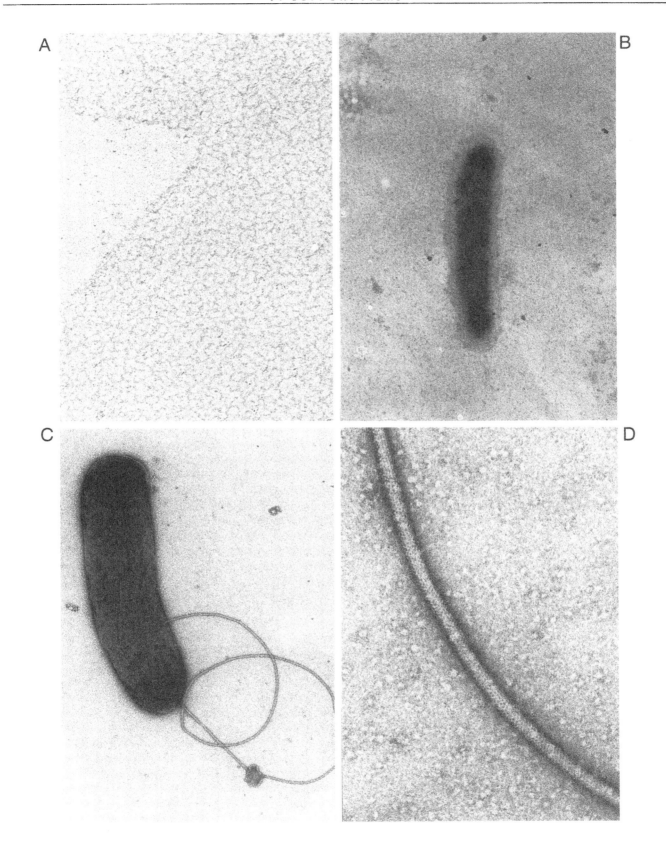

8. Volatile Contamination

FIGURE 7.8A Carbon-enforced Formvar film adjacent to a row of thin sections. The grid was double stained with uranyl acetate and lead citrate and rinsed with deionized water. This micrograph was taken within seconds after being introduced into the electron beam. × 85,000.

Objects of irregular appearance are scattered over the entire support film. A few small holes in the film are seen as small white spots to the left on the micrograph.

FIGURE 7.8B The same area recorded about half a minute later, without changes of the microscope settings. × 85,000.

The irregular objects seen in Fig. 7.8A have lost most of their contrast and structural detail, whereas the details of the support film, including the small holes, appear unchanged.

FIGURE 7.8C–7.8E A sodium chloride crystal on Formvar film exposed to increasing beam illumination. The micrograph in Fig. 7.8C was taken immediately after introducing the crystal into the electron beam. × 75,000.

In Fig. 7.8C the crystal is essentially unchanged, in Fig. 7.8D it has started to disintegrate, and in Fig. 7.8E it is largely evaporated and surrounded by a circle of contamination.

COMMENTS

The type of contaminant illustrated in Fig. 7.8A occurs sometimes on some batches of grids. Its origin is unknown. Obviously it is a volatile component that rapidly evaporates in the electron beam. It may originate from the step of Formvar film production when the film was first formed on the glass slide and then floated off onto a water surface. It may also have formed during subsequent steps of grid handling. The fact that these irregular contaminants lose most of the contrast within seconds in the electron beam means that they often remain undetected, particularly when they are present over heterogeneous and intensely stained tissue sections. Figure 7.8C–7.8E demonstrate that a focused electron beam can heat the specimen to a temperature that is sufficient to evaporate a sodium chloride crystal (melting point in vacuum around 700° C).

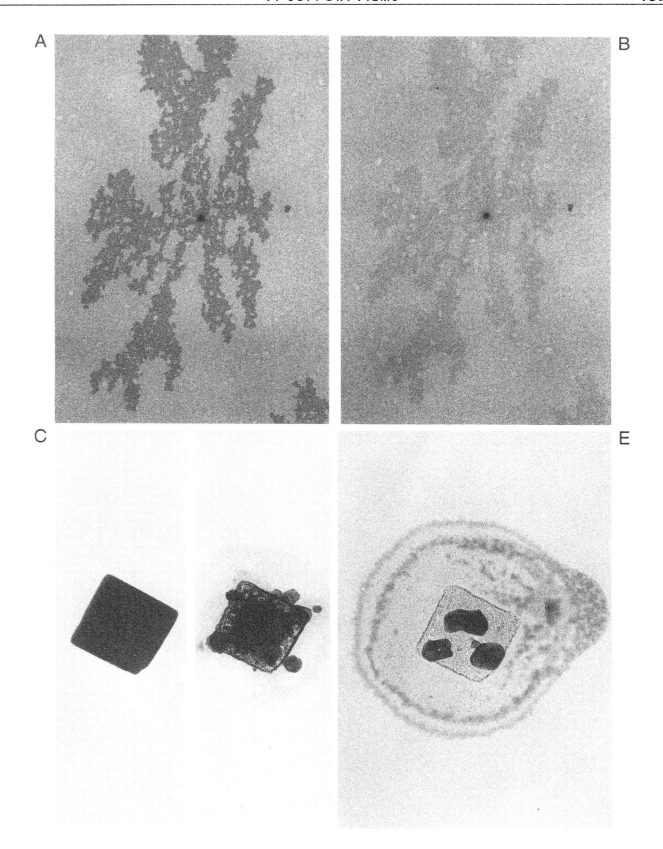

References

Anderson, T. F., and Richards, A. G. (1942). An electron microscope study of some structural colors of insects. *J. Appl. Phys.* **13**, 748–758.

Baumeister, W., and Hahn, M. (1978). Specimen supports. *In* "Principles and Techniques of Electron Microscopy: Biological Applications" (M. A. Hayat, ed.), Vol. 8, pp. 1–112. Van Nostrand-Reinhold, New York.

Baumeister, W., and Seredynski, J. (1976). Preparation of perforated films with predeterminable hole size distributions. *Micron* **7**, 49–54.

De Boer, J., and Brakenhoff, G. J. (1974). A simple method for carbon film thickness determination. *J. Ultrastruct. Res.* **49**, 224–227.

Drahoš, V., and Delong A. (1960). A simple method for obtaining perforated supporting membranes for electron microscopy. *Nature* **186**, 104.

Fukami, A., Adachi, K., and Katoh, M. (1972). Microgrid techniques (continued) and their contribution to specimen preparation techniques for high resolution work. *J. Electr. Microsc. (Tokyo)* **21**, 99–108.

Handley, D. A., and Olsen B. R. (1979). Butvar B-98 as a thin support film. *Ultramicrosc.* **4**, 479–480.

Harris, W. W. (1970). Reducing the effect of substrate noise in electron images of biological objects. *In* "Some Biological Techniques in Electron Microscopy" (D. F. Parsons, ed.), pp. 147–163. Academic Press, London.

Jahn, W. (1995). Easily prepared holey films for use in cryo-electron microscopy. *J. Micros.* **179**, 333–334.

Johansen, B. V. (1974). Bright field electron microscopy of biological specimens. II. Preparation of ultra-thin carbon support films. *Micron* **5**, 209–221.

Kölbel, H. K. (1972). Influence of various support films on image size and contrast of thin-sectioned biological objects in electron microscopy. *Mikroskopie (Wien)* **28**, 202–221.

Lünsdorf, H., and Spiess, E. (1986). A rapid method of preparing perforated supporting foils for the thin carbon films used in high resolution transmission electron microscopy. *J. Microsc. (Oxford)* **144**, 211–213.

Mihama, K., and Tanaka, N. (1976). Beryllium oxide specimen supporting film for high-resolution electron microscopy and their application to observation of fine gold particles. *J. Electr. Microsc. (Tokyo)* **25**, 64–74.

Ohtsuki, M., Isaacson, M. S., and Crewe, A. V. (1979). Preparation and observation of very thin, very clean substrates for scanning transmission electron microscopy. *Scan. Electr. Microsc.* **1979 II**, 375–382.

Ohtsuki, M., Isaacson, M. S., and Crewe, A. V. (1979). Dark field imaging of biological macromolecules with the scanning transmission electron microscope. *Proc. Natl. Acad. Sci. U.S.A.* **76**, 1228–1232.

Schmutz, M., Lang, J., Graff, S., and Brisson, A. (1994). Defects of planarity of carbon films supported on electron microscope grids revealed by reflected light microscopy. *J. Struct. Biol.* **11**, 252–258.

Sjöstrand, F. S. (1956). A method to improve contrast in high resolution electron microscopy of ultrathin tissue sections. *Exp. Cell Res.* **10**, 657–664.

Williams, R., and Glaeser, R. M. (1972). Ultrathin carbon support films for electron microscopy. *Science* **175**, 1000–1001.

ULTRAMICROTOMY

At present, most ultrathin sections are cut from tissues embedded in some type of polymerized resin or polyester, but it is also possible to cut ultrathin sections of frozen fixed, or even unfixed, tissues. Cryoultramicrotomy was introduced several decades ago, but this technique has experienced increased applications in connection with immunocytochemical studies.

An "ideal section" is thin, of uniform thickness, and free from physical distortion such as compression, scratches, vibrations, chatter, and wrinkles. An ideal section cannot be obtained despite considerable refinements of ultramicrotomy techniques over the years. Thus all sections have a finite thickness, which is large relative to the electron microscopic resolution aimed at, and sections usually show one or several of the mentioned defects. If only minor defects are present, they may not interfere appreciably with the structural analysis, whereas more pronounced defects may cause misinterpretations. A particularly demanding form of ultramicrotomy is serial sectioning, which requires uniformity of the sections, absence of section defects, and particular care when collecting the sections on grids.

This chapter illustrates characteristics of different types of sections, including cryosections, and common problems in their production. The use of cryosections in immunocytochemistry is treated in Chapter 16.

Various interesting ultramicrotome designs were tried in the early days of electron microscopy. Fritiof Sjöstrand (1943) developed a sectioning method using a sliding wedge. Albert Gessler and Ernest Fullam (1949) built an instrument, which was a modified ultracentrifuge with a cutting speed of more than 300 m/sec (as compared to 0.1–2 mm/sec in present-day ultramicrotomes)! Very successful models were built by Fritiof Sjöstrand (1953) and by Keith Porter and Joe Blum (1953). Other advances in ultramicrotomy included the development of a suitable embedding medium (see Chapter 5), the invention of glass knives by Harrison Latta and Francis Hartman (1950), the addition of a water trough to the knife by James Hillier and Mark Gettner, (1950), and of diamond knives by Humberto Fernández-Morán (1953).

Fernández-Morán, H. (1953). *Exp. Cell Res.* **5**, 255–256.
Gessler, A., and Fullam, E. (1946). *Am. J. Anat.* **78**, 245–247.
Hillier, J., and Gettner, M. E. (1950). *Science* **112**, 520–523.
Latta, H., and Hartmann, J. F. (1950). *Proc. Soc. Exp. Biol. Med.* **74**, 436–439.
Porter, K. R., and Blum, J. (1953). *Anat. Rec.* **117**, 685–712.
Sjöstrand, F. S. (1943). *Nature* **151**, 725–726.
Sjöstrand, F. S. (1953). *Experientia* **9**, 114–115.

1. Correlation of Light and Electron Microscopy

FIGURES 8.1A AND 8.1B Light micrographs of a rat renal cortex perfusion fixed with glutaraldehyde and postfixed with osmium tetroxide. The tissue was block stained with uranyl acetate and embedded in Epon. The embedding was then sectioned at 1 μm and section stained with toluidine blue. Figure 8.1A × 270, Fig. 8.1B × 1100.

Figure 8.1A shows a large area of the renal cortex with many renal tubules. The boxed area in Fig. 8.1A is shown at higher magnification in Fig. 8.1B.

FIGURE 8.1C An electron micrograph from an ultrathin section cut from the 1-μm section shown in Figs. 8.1A and 8.1B. The 1-μm section was reembedded in Vestopal (Maunsbach, 1978) and the area shown in Fig. 8.1B was thin sectioned. × 1100.

A careful comparison of Figs. 8.1B and 8.1C shows that most structural components recognized in the light micrograph are also seen in the electron micrograph. This applies to large structures, such as nuclei, and to organelles, including mitochondria, lysosomes, and endocytic vacuoles.

FIGURE 8.1D An electron micrograph showing the boxed area in Fig. 8.1C at higher magnification. × 9000.

This micrograph shows the apical region of proximal tubule cells. The ultrastructural details are well maintained in this twice-embedded and sectioned tissue. Several endocytic vacuoles (E) can also be identified in the low magnification electron micrograph (Fig. 8.1C) and in the light micrograph of the 1-μ section (Fig. 8.1B).

COMMENTS

These micrographs illustrate that it is possible to carry out a close correlation between light micrographs and electron micrographs using the present "ultrathin section of semithin section" method. Large areas can be surveyed by light microscopy and particular objects can be selected for further analysis by electron microscopy. This is advantageous in pathology and parasitology, where rare lesions and objects may be hard to find by electron microscopy alone.

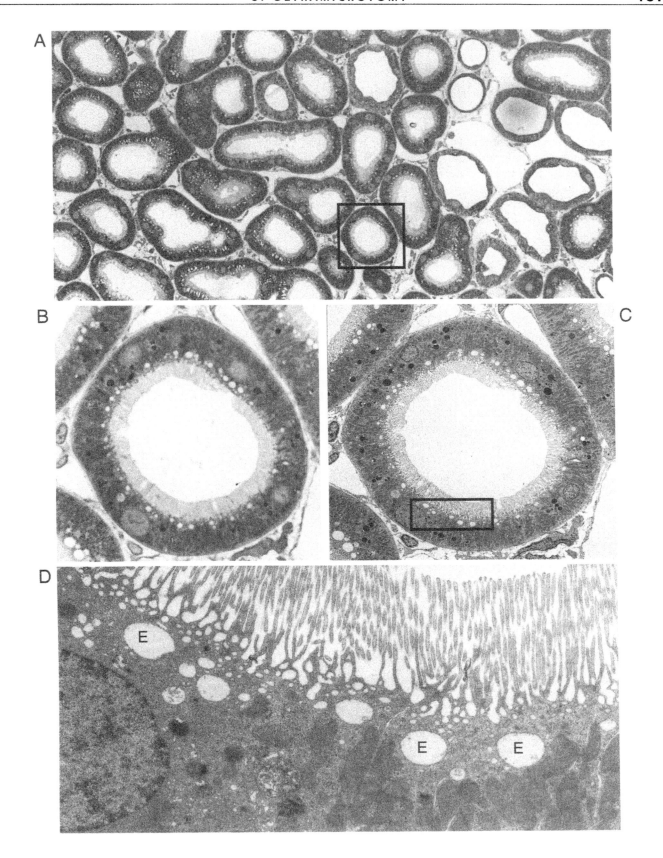

2. Section Thickness: Low Magnification

FIGURES 8.2A–8.2C Rat liver cells fixed with glutaraldehyde and postfixed with osmium tetroxide. The tissue was embedded in Epon and sectioned with a diamond knife on an LKB Ultrotome III. The sections were cut from the same tissue block and at the same occasion but with the cutting feed set at different values: in Fig. 8.2A at 30 nm, in Fig. 8.2B at 60 nm, and in Fig. 8.2C at 120 nm. The actual values of the section thickness were not determined, but the interference colors of the sections, when floating in the trough, were consistent with the mentioned values. The sections were stained with uranyl acetate and lead citrate. × 20,000.

Membranes are best defined in the thin section, where they appear as distinct lines in most places. The number of ribosomes per unit area is much higher in the thick than in the thin sections. The difference in electron density between the mitochondrial matrix and the cytoplasmic ground substance is more pronounced in the thick sections.

COMMENTS

The three micrographs illustrate the influence of section thickness on the appearance of the cells. Superposition effects are evident in thick sections as illustrated by the comparatively large number of ribosomes per unit area. However, individual ribosomes are defined more clearly in the thin sections. The number of observable mitochondrial inner membranes is higher in thin than in thick sections. It is also apparent that the range of contrast is much greater in thick sections. This is often an advantage at survey magnifications, where the associated loss in resolution is of minor importance.

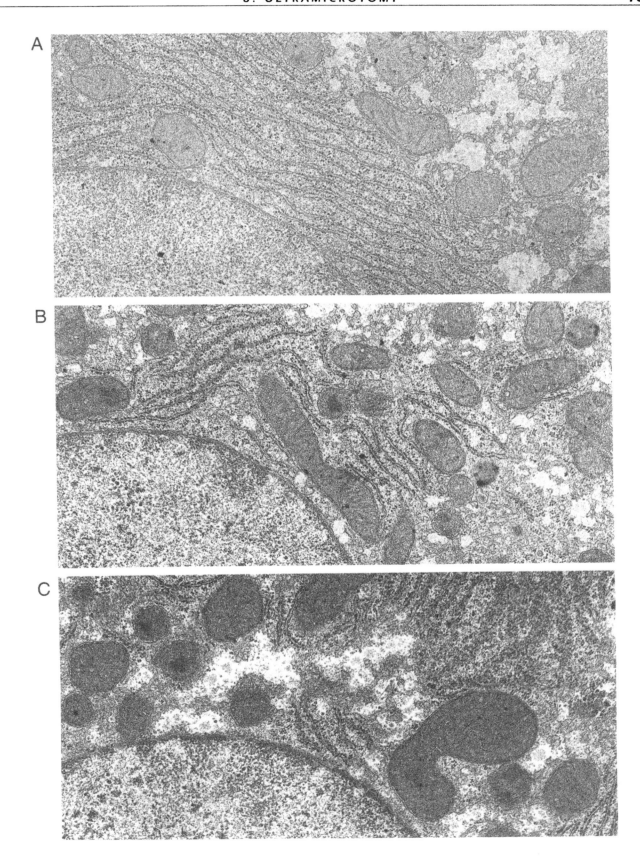

3. Section Thickness: High Magnification

FIGURES 8.3A–8.3C A higher magnification of the same three sections in Figs. 8.2A–8.2C. × 60,000.

A sharp definition and a small number of ribosomes in the thin section are evident as opposed to a lesser definition but a larger number of ribosomes in the thick sections. In the thick section only short lengths of membranes appear reasonably well defined, whereas in the thin section shown in Fig. 8.3A long regions of membranes are well defined.

COMMENTS

Precise ultrastructural information on the interrelationships of small structural details in the cytoplasm must rest on the analysis of sufficiently thin sections. If the section is too thick, superposition effects, such as observed in the center of Fig. 8.3C, may make the interpretation hazardous.

The poor definition of membranes in thick sections may be interpreted as a sign of some other deficiency in the preparation procedure, such as poor fixation quality or an unsuccessful section staining, but comparison with thinner sections provides a means to tell the difference.

Superposition effects are common in thick sections and caution is necessary when interpreting, for instance, the relationship between different kinds of membranes. When analyzing such critical preparations it is helpful either to study the thinnest possible sections or to use a tilting stage in the electron microscope (see Chapter 20).

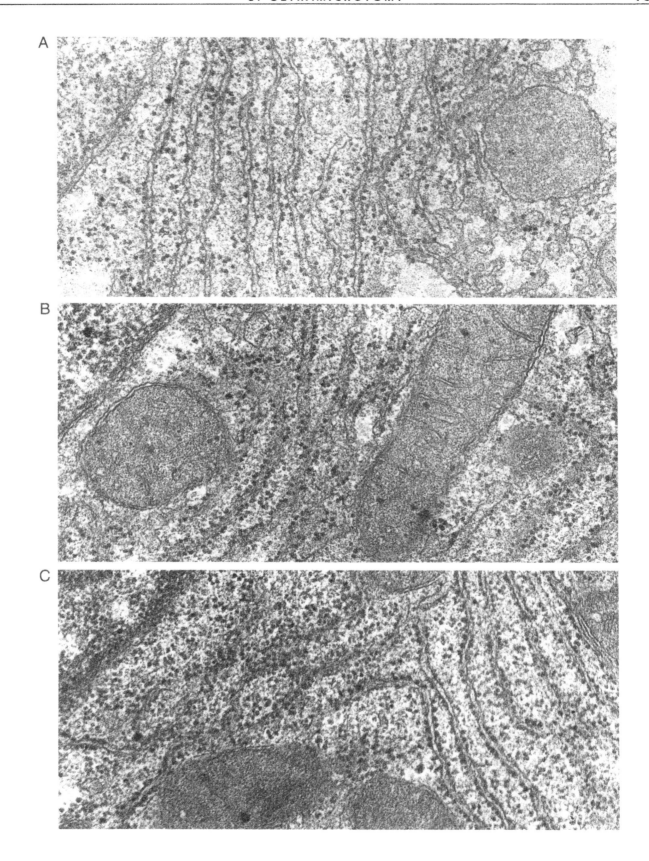

4. Section Thickness: Half-Micron Section

FIGURES 8.4A–8.4C Human spermatozoa fixed in glutaraldehyde, postfixed in osmium tetroxide, embedded in Epon, and sectioned with a diamond knife on an LKB Ultramicrotome. The sections were cut from the same block and at the same occasion, but with the cutting feed set at different values: Fig. 8.4A at 60 nm, Fig. 8.4B at 250 nm, and Fig. 8.4C at 500 nm. The sections were stained with uranyl acetate and lead citrate. Electron micrographs were taken at an accelerating voltage of 100 kV. × 25,000.

A comparison between these three micrographs shows that there is a dramatic loss in specimen resolution in the thick sections.

COMMENTS

Sections up to 1 μm thick can be observed at very low resolution in a standard electron microscope operated at 100 kV. Thick sections may occasionally be useful for survey purposes at low magnification, particularly when studying small cells that lie loosely dispersed in the medium. In most cases, however, the loss in resolution outweighs the slight gain in surveyability. In tightly packed cells or in compact tissues, too many structures will become superimposed in the micrograph to make the analysis practical. However, analysis of semithick sections may be rewarding if pairs of stereomicrographs are taken in a high voltage electron microscope.

A

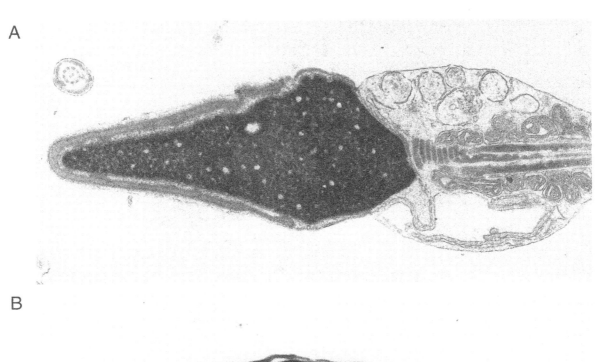

B

C

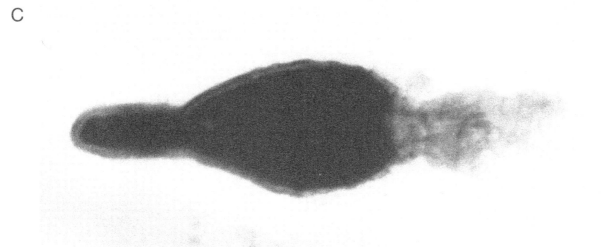

5. Determination of Section Thickness

FIGURE 8.5A An electron micrograph of a section of a kidney epithelial cell showing an electron-dense band of constant width. × 50,000.

Plasma membranes and nuclear envelope show discontinuities at the dense band (arrows).

FIGURE 8.5B Same electron micrograph, although the photographic print was made with a much shorter exposure time in order to show the pattern within the dense band. × 50,000.

In this photographic print the dense band shows a central line and some cytoplasmic structures within the bands.

FIGURE 8.5C An electron micrograph of a section of four ultrathin sections on top of each other and lying on a supporting film. The four sections and their supporting film have been reembedded and sectioned perpendicularly to the original sectioning direction in order to measure the thicknesses of sections and supporting film. Two sections are from melamine embeddings, two from Epon embeddings. They lie on a carbon-enforced Formvar film. The sections contain double-fixed rat liver cells and have been section stained with uranyl acetate and lead citrate before reembedding in Epon and resectioning. × 150,000.

The sections and supporting film and their measured thicknesses are (bottom to top): Carbon-enforced Formvar film at 30 nm, Epon 12 at nm, Epon at 60 nm, melamine at 22 nm, and melamine at 50 nm. The carbon enforcement is visible as a faint electron-dense line only. It can be seen that the sections have rather even upper and lower surfaces.

FIGURE 8.5D A similar section of a sandwich section as in Fig. 8.5C, except that the supporting film had no carbon enforcement. × 150,000.

The thin melamine section has a vertically oriented fold (between the arrows) and the width of the fold is twice that of the section thickness.

COMMENTS

Two means of measuring section thickness are illustrated here. The most direct way is to produce a cross section of the section and to record it in the electron microscope as has been done in Fig. 8.5C. It can then be seen that each ultrathin section has a uniform thickness. An easier way to measure the section thickness is to measure the width of the folds that occasionally occur during sectioning (Small, 1968). The width of a fold is twice that of the section thickness, as can be seen in Fig. 8.5D. Folds of the supporting film must not be mistaken for folds in the section (see Fig. 6.8). An easy way to discriminate between these two types of folds is to look for membrane discontinuities.

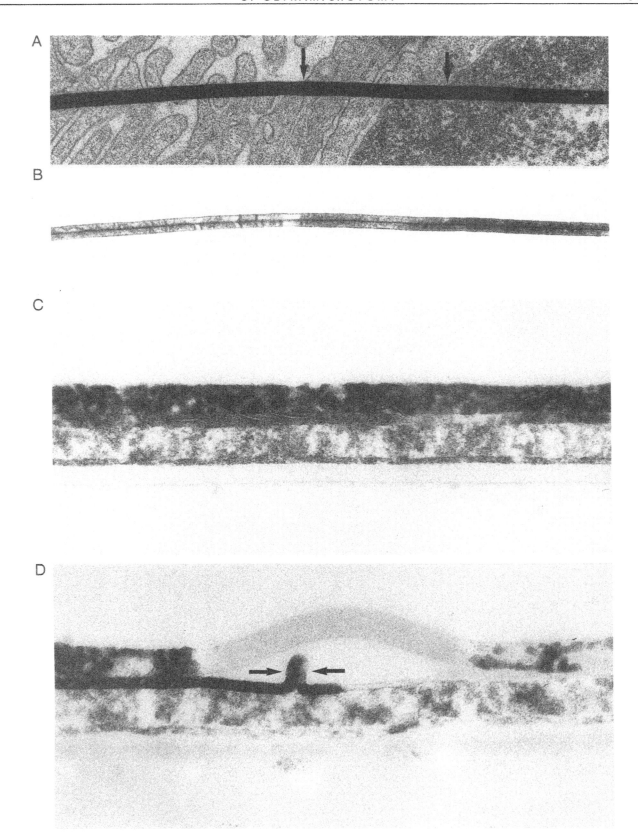

6. Folds in the Section

FIGURE 8.6A Section of a kidney tissue following fixation, Epon embedding, and sectioning with a diamond knife. × 3000.

Four dark lines with a slightly variable width can be seen. The two middle lines start and end within the section.

FIGURES 8.6B–8.6D Section of a kidney tissue prepared as in Fig. 8.6A. The edge of the section is seen at left in Fig. 8.6B. Figure 8.6B × 6500, Fig. 8.6C × 60,000, and Fig. 8.6D × 16,000.

Two long electron-dense streaks are observed in the section and end at the edge of the section, at the left. The streaks are not related to any of the tissue components. The two boxed areas are shown at higher magnification in Figs. 8.6C and 8.6D. Figure 8.6C shows a broad region of one streak that contains an N-shaped endothelial capillary wall, and Fig. 8.6D shows the forked end of another dark streak.

COMMENTS

These micrographs illustrate different types of folds in an ultrathin section. One obvious characteristic of such folds is that they cannot extend beyond the edge of the section. Several folds may run more or less parallel to each other, as shown in Fig. 8.6A. The folds may vary in width, and the broad portions of the streak are then interpreted as folds that lie flattened on the surface of nonfolded areas of the section. Cellular structures can often be seen within such folds, as in Fig. 8.6C. Not infrequently their ends appear forked and widened.

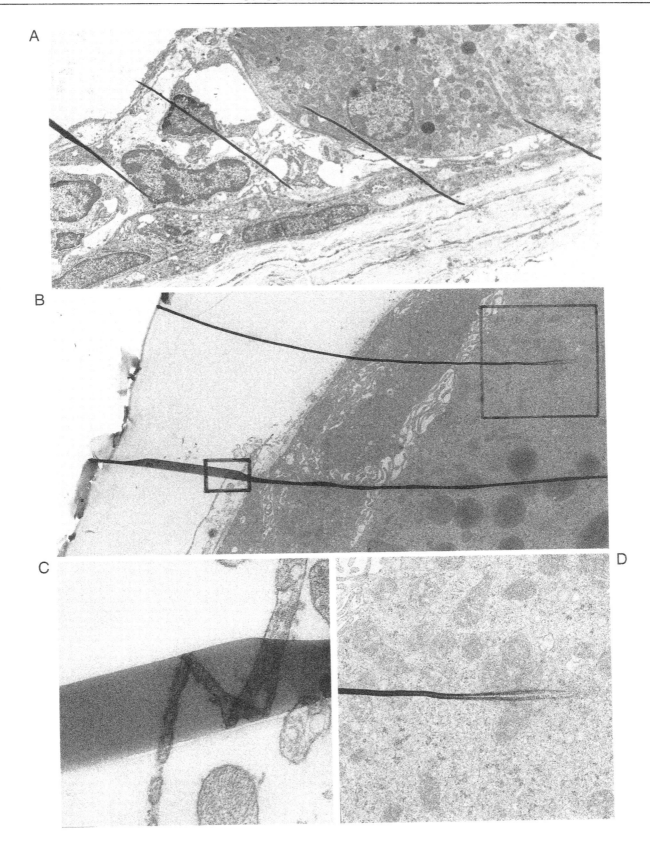

7. Collection of Sections

FIGURE 8.7A An ultrathin section of Epon-embedded tissue picked up on a Formvar-coated and carbon-enforced slot grid. The method of collection of the sections from the surface of the sectioning trough was that of Galey and Nilsson (1966), in which method the sections and a small droplet of water are lifted from the surface of the trough fluid using an empty and clean slot grid. This grid is then placed on top of a Formvar-coated slot grid. The bulk of the fluid is sucked off with a filter paper and the rest is allowed to evaporate. The sections thereby sink smoothly and unwrinkled down onto the Formvar film. × 75.

This very large section on the Formvar support film of the slot grid does not show any wrinkles.

FIGURE 8.7B Serial sections picked up with the Galey and Nilsson (1996) method and oriented onto the support film of a slot grid. × 75.

The serial sections show no wrinkles and are oriented in the original consecutive order. There is a hole in the upper part of the support film.

FIGURE 8.7C An ultrathin section picked up on the support film of a slot grid by lowering the grid from above onto the section floating on the surface of the trough fluid.

The section shows several long wrinkles, but the Formvar film seems without defects.

FIGURE 8.7D This section was collected by first immersing the slot grid with its support film into the sectioning fluid. The section was then carefully maneuvered toward the Formvar film and the slot grid slowly lifted from the trough fluid.

There are several wrinkles in the section itself as well as in the surrounding Formvar film.

COMMENTS

The collection of ultrathin sections from the surface of the sectioning fluid, which is usually pure water, is in principle a very simple procedure. However, in practice it may lead to several irritating faults. Apart from contamination of various types, wrinkles constitute a very disturbing problem. Wrinkles limit the possibility of recording large, continuous fields of a section and have a notorious tendency to appear exactly where the most interesting objects are located! In our experience, the collection of sections with the aid of an empty, clean slot grid as applied in Figs. 8.7A and 8.7B is with some practice the most reliable method for collecting wrinkle-free sections or ribbons of small sections.

In some laboratories it has been a practice to add a small amount of ethanol to the sectioning fluid. However, this is rarely necessary, and if a diamond knife is used, it may start to dissolve the material holding the diamond.

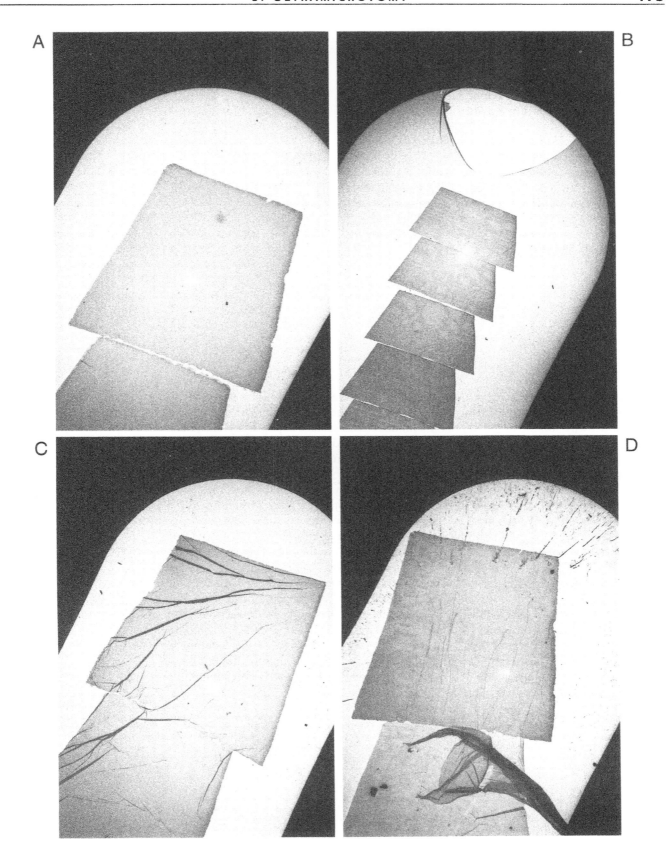

8. Surface Topography of Sections

FIGURES 8.8A–8.8D Human muscle tissue fixed in osmium tetroxide and embedded in Epon. After sectioning, one surface of each section was shadowed with platinum at an angle of 30°. Figures 8.8A and 8.8B show sections cut with a glass knife, whereas Figs. 8.8C and 8.8D sections are cut with a diamond knife. Figures 8.8A and 8.8C show those section surfaces that came in direct contact with the water in the trough during sectioning ("water–facing surfaces"), whereas Figs. 8.8B and 8.8C represent those surfaces that faced the air during sectioning ("air-facing surfaces"). × 60,000.

The sections cut with a glass knife show an irregular surface topography. On the water-facing surface (Fig. 8.8A) the irregularities are distributed randomly within the mitochondrion (M), whereas on the air-facing surface (Fig. 8.8B) the elevations tend to outline structural components within the section, particularly the inner membranes of the mitochondrion (M). The water-facing surface of the section cut with a diamond knife (Fig. 8.8C) is essentially smooth at this magnification. The air-facing surface of the section cut with a diamond knife (Fig. 8.8D) shows several elevations, which appear to correspond to structural components in the section, such as collagen fibers (lower left) and what presumably corresponds to muscle filaments (upper right).

COMMENTS

Surfaces of the sections cut with a diamond knife are smoother than surfaces cut with a glass knife, and water-facing surfaces are smoother than air-facing surfaces. It is not known why some tissue components influence the surface topography of the air-facing surface. Sections that have been reembedded and then cut perpendicularly to the first section similarly show the air-facing surface to be more coarse and to have a more flaky appearance than the water-facing surface, as shown by Favard and Carasso (1973).

Which section surface that is exposed to solutions of heavy metal stains or other solutions depends on the method of collecting the sections (compare Figs. 8.7A, 8.7C, and 8.7D). The texture of the section surface may influence the outcome of procedures, such as section staining and immunocytochemistry, although in what ways has not been studied systematically. It is conceivable that a greater number of reactive spots are exposed in the air-facing surface, particularly in sections that have been cut with a glass knife, which would be advantageous in staining and immunocytochemistry. Cryosections are likely to have an even rougher surface, especially after thawing and drying. The observation that immunogold labeling is sometimes less precise in Epon sections than in cryosections may be due to the fact that the label may redistribute on the very smooth diamond-cut resin section, whereas a more stationary distribution of the labeling in cryosections may be due to their presumably rougher surface. The finding that immunogold labeling is more intense over cryosections than over resin section is also compatible with a possible influence of surface texture.

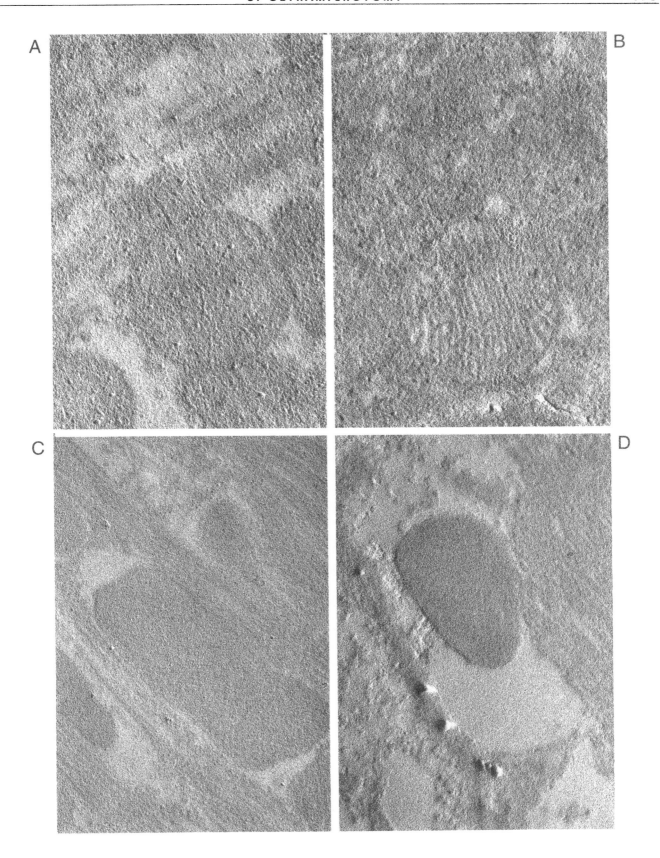

9. Knife Scratches

FIGURE 8.9A A rat liver tissue fixed with glutaraldehyde, postfixed in osmium tetroxide, embedded in Epon, and sectioned with a worn diamond knife. The upper or air-facing surface of the section was shadowed with platinum on the grid at an angle of 30°. × 40,000.

Many parallel knife marks can be seen. One of them is deeper than the others and has rough edges with crumbs of the embedding medium.

FIGURE 8.9B A pellet of isolated kidney lysosomes fixed with glutaraldehyde, postfixed with osmium tetroxide, embedded in Vestopal, and sectioned with a glass knife. × 50,000.

The electron-dense matrix of the largest lysosomes has a mottled appearance and vertical stripes (arrows).

FIGURE 8.9C A rat renal medulla fixed in glutaraldehyde, postfixed in osmium tetroxide, embedded in Epon, and sectioned with a glass knife. × 1500.

The section shows several vertical dark and light lines. Perpendicularly to those lines there are less well-defined dark and light bands with a regular periodicity. These bands are observed both over the cells and over the empty embedding medium outside the cells.

COMMENTS

Notches and contaminants of the knife edge cause scratches such as those shown in Fig. 8.9A. Whereas most of the scratches in this section are so fine that they would not have been noticed on a medium thick, nonshadowed section, some are deeper and would actually be visible in a nonshadowed section.

The vertical banding in Fig. 8.9B also represents scratches in the section caused by irregularities in the knife edge. There are no holes in the section, and the scratches have affected only the surface layer of the section. This type of artifact could be misinterpreted as a real pattern of the lysosomal matrix.

The knife edge used for cutting the section in Fig. 8.9C had many notches, which were responsible for the knife scratches that run in the vertical direction of the figure. The pronounced vibrations or chatter in the perpendicular direction differ from the knife scratches by having a fairly regular periodicity. The compression of the section, noticed by the deformation of the originally round nuclei, is parallel to the bands of the chatter.

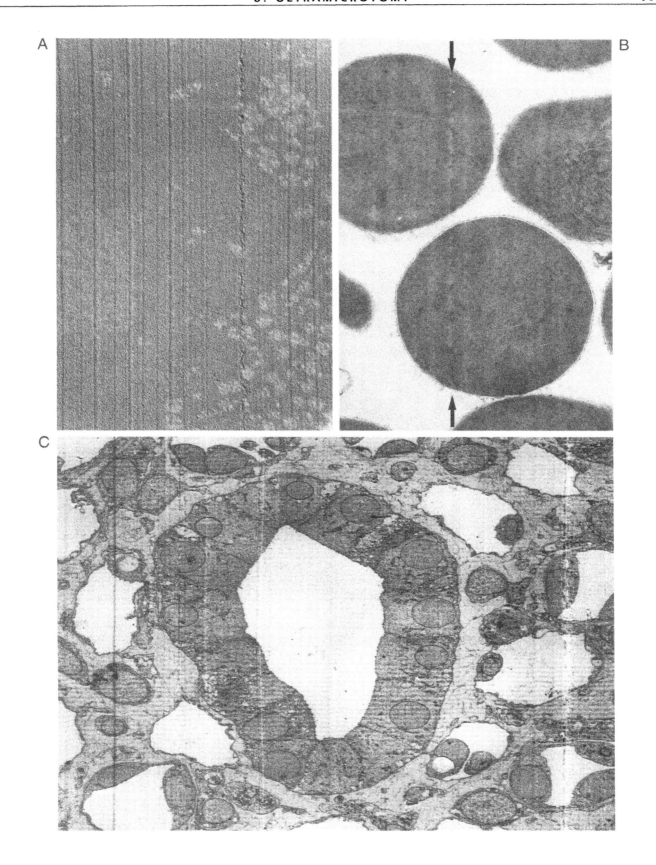

10. Mottling and Flaking

FIGURES 8.10A–8.10C Sections from human muscle tissue fixed in osmium tetroxide and embedded in Epon. The section in Fig. 8.10A was cut with a good diamond knife, whereas the sections in Figs. 8.10B and 8.10C were cut with glass knives that had been used extensively. All sections were cut with the knife edge parallel to the muscle fibers. × 50,000.

The muscle fibrils in Fig. 8.10A containing thick and thin filaments have an even density throughout. The fibrils in Fig. 8.10B show a mottled appearance with many irregularities along horizontal lines. In Fig. 8.10C there are irregularities in tissue contrast and dislocations in the filaments. This is particularly evident in the right half of the figure.

COMMENTS

Figure 8.10A has no apparent sectioning artifacts, whereas Fig. 8.10B shows what has been called mottling and Fig. 8.10C shows what can be called flaking. Mottling is probably due to defects in the knife edge, causing small shreds to be displaced from the section surface during cutting.

The flaking observed in Fig. 8.10C appears as if large areas along the knife edge have been displaced. This artifact is particularly common in relatively thick sections.

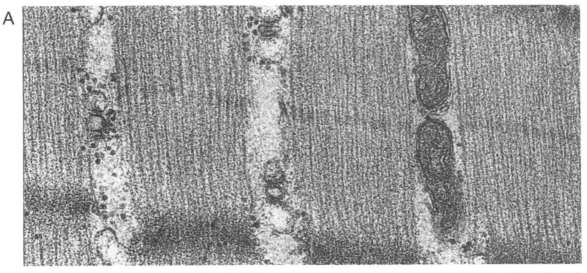

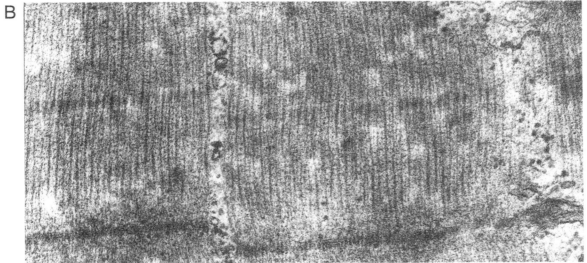

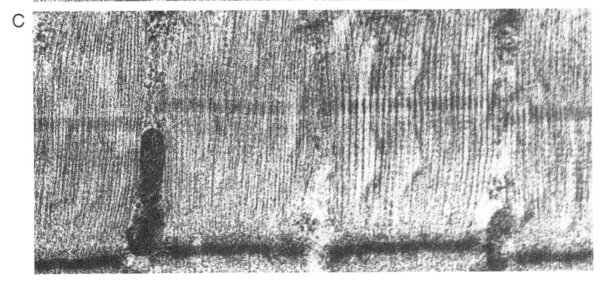

11. Worn Glass Knives

FIGURE 8.11A Rat liver cells fixed with 2% glutaraldehyde, postfixed with osmium tetroxide, embedded in Epon, sectioned with a glass knife, and section stained with uranyl acetate and lead citrate. × 3000.

There are irregular, vertical streaks, horizontal crevices, and uneven, wide bands or streaks. In the horizontal direction there is a general pattern of fine quasi-periodicities. In places, there are holes in the section close to the major dark bands.

FIGURE 8.11B Higher magnification of the central area in Fig. 8.11A. × 7000.

The section has a mottled appearance with irregular, horizontal bands and poorly defined cytoplasmic structures.

FIGURE 8.11C The central area of Figs. 8.11A and 8.11B at a considerably higher magnification. × 50,000.

Mitochondria, endoplasmic reticulum, glycogen areas, and nucleus are easily identifiable. Structural definition is good, although there is a considerable variation in electron density within each organelle, as examplified by the light and dark streaks of the mitochondria. The entire micrograph has an uneven appearance.

COMMENTS

The poor quality of this section undoubtedly is due to the use of a worn-out or otherwise dull glass knife during ultramicrotomy. The vertical streaks represent scratches caused by knife defects, whereas the horizontal streaks are variations in section mass, in part caused by compression and in part by irregularities in the section surface. Despite the poor section quality, certain structural components of the cell are recognizable and surprisingly well defined when seen at a high magnification. The remedy to these problems obviously is to change the area on the glass knife or, better still, to exchange the glass knife for a new one or, preferably, for a good diamond knife.

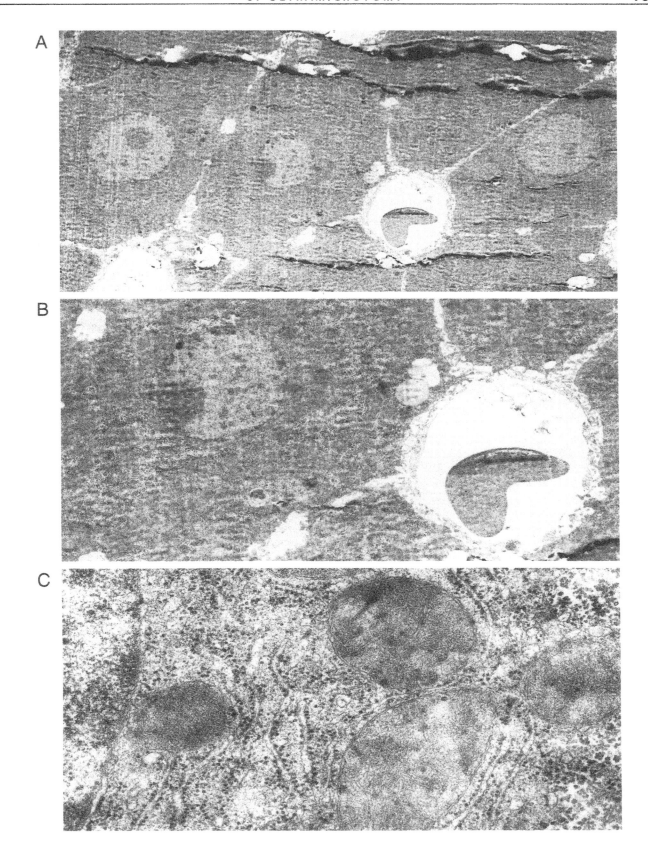

12. Transmitted Vibrations

FIGURE 8.12A Section of a double-fixed, Epon-embedded submucosa cut with an LKB Ultrotome III ultramicrotome, which was placed on a common laboratory table rather than on a vibration-damping microtome table. During cutting of this section, vibrations are transmitted to the microtome by letting a person walk on the floor close to the microtome. Ripples could be seen in the water trough. Sectioning was performed with a diamond knife, with the cutting speed set at 0.5 mm per second. The knife edge was parallel to the short sides of the figure. × 5000.

The entire section has a banded appearance with denser and lighter strands of variable widths but generally around 0.5–2.0 μm. The individual bands have a straight course and extend the entire width of the section.

FIGURE 8.12B An adjacent section from the same block, cut at the same occasion, and located in the same ribbon of sections as that in Fig. 8.12A. The person had stopped walking on the floor when this section was being cut. × 5000.

There is no banding in this section as best can be judged by the even density of the red blood cells to the left in the figure.

COMMENTS

It is sometimes difficult to trace vibrations or chatter to either a source outside the microtome or to events taking place at the knife edge. In this particular case, vibrations had been induced deliberately and could be seen as small ripples in the water trough. It is characteristic for vibrations of this kind that the widths of the bands are 0.5–2.0 μm or more and that the vibration pattern is irregular, although each dense or light band has a uniform width. As the cutting speed was 0.5 mm per second, it is evident that the transmitted vibrations had frequencies in the order of 250–1000 Hz.

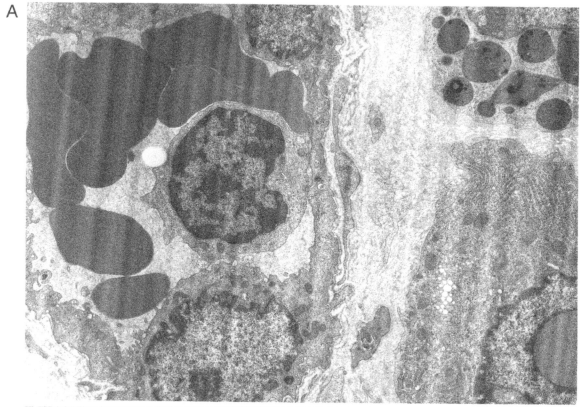

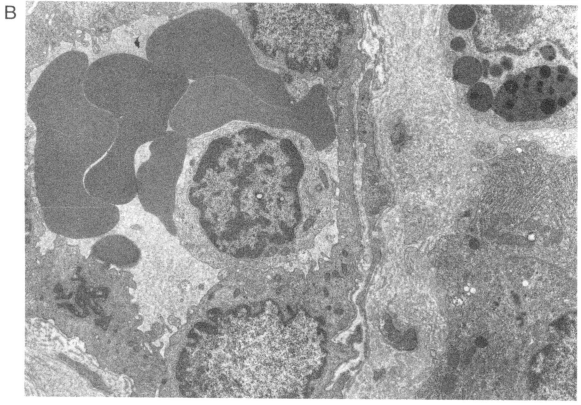

13. Vibrations and Knife Marks

FIGURE 8.13A Epon sectioned with a glass knife. The knife edge was parallel to the short sides of the figure. The cutting feed was set at 80 nm and the cutting speed at 2 mm per second. × 40,000.

Dark bands alternate with light ones with a fairly regular periodicity of about 0.23 μm. There is a gradual transition between dark and light bands. In the light bands there are two horizontal rows of holes in the section. Each hole is outlined by a dense rim.

FIGURE 8.13B Acrylic embedding of an osmium-fixed muscle tissue sectioned with a glass knife. The cutting feed was set at 80 nm and the cutting speed at 2 mm per second. The knife edge was parallel to the short sides of the figure. × 50,000.

The section shows fairly straight ribbons of an even width and thickness separated by narrow bands, which may either appear more dense or less dense than the bands. There are some faint, straight lines perpendicular to the ribbons, which represent knife marks.

FIGURES 8.13C AND 8.13D Epon sectioned with a glass knife. × 14,000 and × 20,000, respectively.

The figures show a scratch in the section extending between a hole to the left and a dense particle to the right in Fig. 8.13C and from the side of the section (not shown) to a particle to the right in Fig. 8.13D.

COMMENTS

The sectioning artifacts in Figs. 8.13A and 8.13B are due to high-frequency interactions between the knife edge and the block. With a cutting speed of 2 mm per second and a periodic pattern of 0.23 μm, the frequency of the interaction can be calculated to be about 8700 Hz. This is a high-pitched vibration, such as sometimes heard during sectioning, which is an indication that the section will exhibit this sectioning artifact. An outer source for vibrations in the sections is often looked for, but is unlikely, as most machines in the neighborhood of the microtomes will have a much lower frequency. High-frequency vibrations would furthermore be damped by the mass of the microtome and its table. Some means to reduce high-frequency vibrations are to reduce the width of the block, reduce the sectioning speed, use a knife with a low scoring angle (if using a glass knife), clean the knife edge (if using a diamond knife), and ensure that the chuck is fastened properly and that other parts of the microtome are adequately adjusted.

The knife scratches in Figs. 8.13C and 8.13D were in all likelihood caused by hard particles within the embedding. When such a particle is hit by the knife, it may be torn out, leaving a hole in the section, and attach to the knife edge, where it will cause a scratch. In Fig. 8.13C the particle has fallen off the knife edge and is seen to the right. Scratches made by the diamond knifes are usually caused by contaminations rather than by edge defects. If the contamination falls off, the section will again be free of knife scratches.

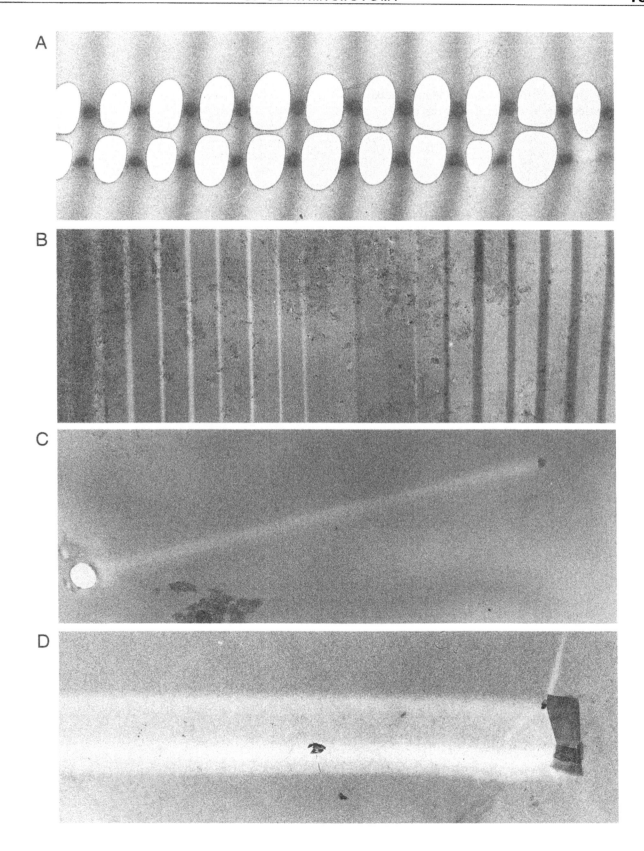

14. Selective Chatter

FIGURE 8.14A Brown adipose tissue fixed in osmium tetroxide, embedded in Epon, and sectioned with a glass knife. × 30,000.

In the round fat droplets there are alternating light and dark bands, which are oriented mainly horizontally. There are no similar bands in the surrounding cytoplasm.

FIGURE 8.14B A kidney tissue fixed in osmium tetroxide, embedded in Epon, and sectioned with a glass knife. × 20,000.

The matrix material in the several lysosomes is not uniform: each lysosome has one or two electron-dense and electron-lucid regions with their longest extension approximately in the horizontal direction of the figure.

FIGURE 8.14C A proximal tubule cell of a rat kidney fixed in osmium tetroxide, embedded in Epon, and sectioned with a glass knife. × 4500.

Although the tubular cells have a uniform density, there are closely spaced, dense and lucid bands in the tubule lumen at the center of the figure and in the capillary lumen to the left and to the right. The bands are on the whole arranged vertically and parallel to the knife edge but show many minor undulations.

COMMENTS

These three micrographs illustrate "selective chatter" in which one or a few components of the block are affected. The term "chatter" is used here for such periodicities that are somewhat wavy and sometimes branching. This type of chatter is probably due to a difference in the physical properties (hardness, elasticity, etc.) of the different regions. Variations in hardness, caused by an uneven polymerization of the embedding medium or of the specimen, may cause such a heterogeneity. Thus, it is noted frequently that fat droplets (as in Fig. 8.14A) or an empty lumen (as in Fig. 8.14C) may be softer than the rest of the tissue.

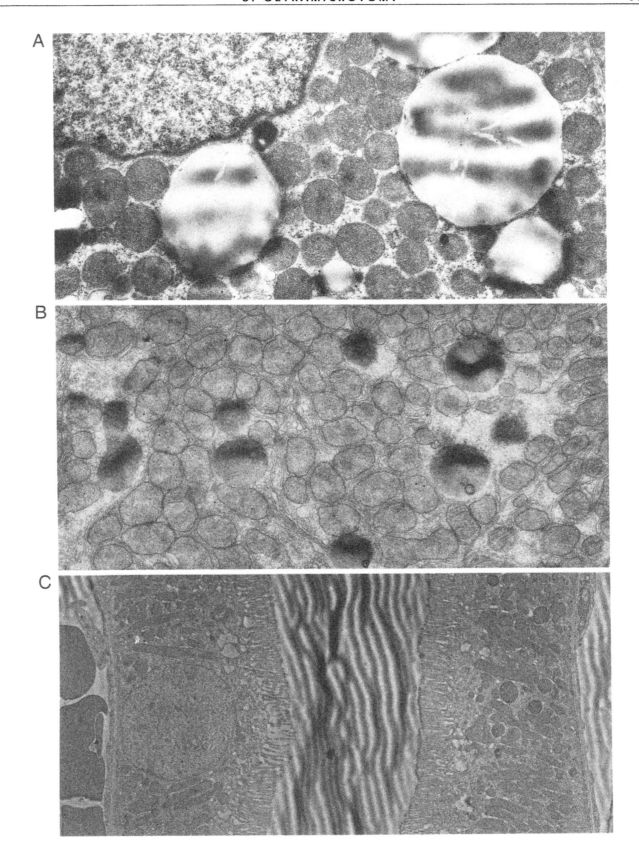

15. Compression

FIGURES 8.15A–8.15C Polystyrene latex particles with a diameter of 1.3 μm, fixed with osmium tetroxide, embedded in Epon, and sectioned with a diamond knife. Figure 8.15A comes from a section that was cut as thin as possible; a particularly thin part of the section was selected in the microscope. The section thickness is judged to be less than 30 nm. Figure 8.15B is from a section that was cut with the cutting feed set to 60 nm, whereas Fig. 8.15C was cut at 750 nm. × 40,000.

In Figure 8.15A the latex particles have an elliptical shape. In places the embedding medium is detached from the upper rim of the particles. In Fig. 8.15B the particles also have an elliptical shape but less so than in Fig. 8.15A. The lower edge of some particles shows an increased density. Along the short end of one particle there is an increased density of compressed embedding medium. In the thick section in Fig. 8.15C the particles are almost circular in projection but several are superimposed.

COMMENTS

Because latex particles are known to be spherical, round cross sections would be expected. The oval shapes actually found in thin sections must hence be due to compression, which is more pronounced in thin than in thick sections. In 60-nm sections the particles are compressed by about 30%, whereas compression of the entire section was found to be less than 5%. The compression of the section was measured by comparing the side lengths of the section to the side lengths of the square block face, from which the section was cut.

Compression of cellular components during microtomy is a commonly occurring phenomenon, although to a varying extent; it is more often encountered in tissues that are difficult to section, such as blocks with soft embedding medium. In most cases, a small to moderate compression of 5–10% has no consequences for interpretations but must be taken into account when the dimensions of cell components are measured. The degree of compression is dependent not only on section thickness and hardness of the embedding medium, but also on the embedded object, quality of the knife, and cutting speed.

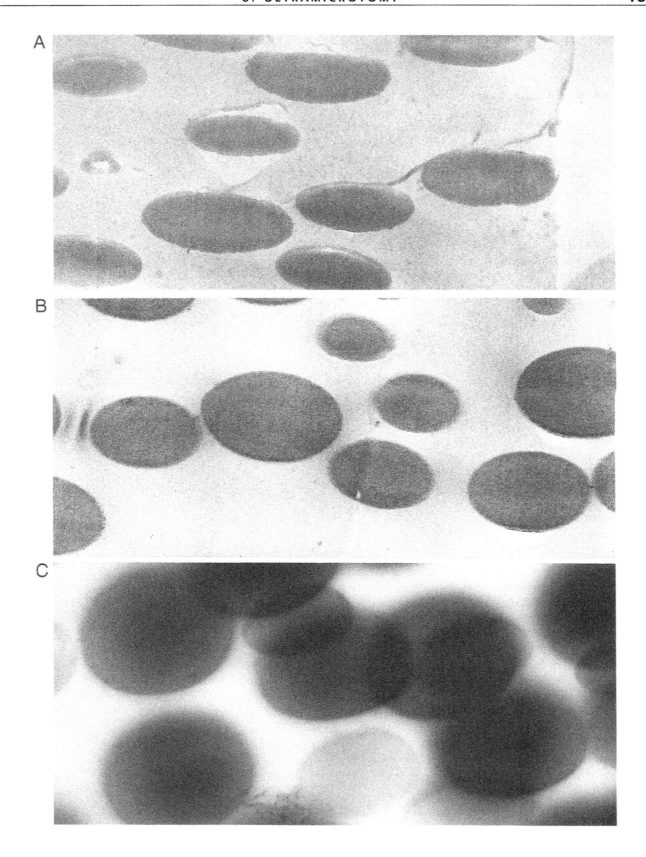

16. Holes and Deformations

FIGURE 8.16A Section of the ciliated epithelium from human nasal mucosa. The tissue was fixed with glutaraldehyde, postfixed with osmium tetroxide, embedded in Epon, and thin sectioned. The sections were mounted on a grid without supporting film and were section stained with uranyl acetate and lead citrate. × 60,000.

Several cross-sectioned cilia and microvilli are seen. In the center of the field there is a hole in the section. Ciliary cross sections adjacent to the hole are distorted, whereas all cross-cut cilia at some distance away from the hole are circular.

FIGURE 8.16B Section of osmium-fixed and Epon-embedded spermatozoa from a jellyfish. The direction of the knife through the embedded specimen was from right to left. × 60,000.

Sperm tails have been cut obliquely. The part of each sperm tail, which first encountered the knife, has a sharp and in part straight border and shows an increased electron density, whereas the embedding medium outside this part of the sperm tail shows a decreased electron density.

FIGURE 8.16C. Two sections of an osmium-fixed, methacrylate-embedded human muscle biopsy. × 1900.

The sections have a very poor quality and a highly variable thickness. In the section to the right, large areas are missing. Corresponding areas in the section to the left show an increased density. In addition, there are fine electron-dense streaks in the left section, which match light streaks in the right section.

COMMENTS

In addition to the common sectioning artifacts with a periodic appearance (chatter), ultrathin sections may show a variety of other irregular defects: Holes (Fig. 8.16A) are common in some preparations and may sometimes originate from insufficient dehydration. Unless the section is mounted on a stable supporting film, the holes can be seen to grow in the electron beam, whereupon biological structures close to the hole in the section become distorted.

Differences in consistency of the embedding medium may lead to distortions within the section. If the "empty" embedding medium is softer than the fixed, embedded structure, the knife, when going from the embedding medium to the object, has a tendency to compress the object. This will lead to a compression of the border of the object and a stretching of the embedding medium close to the border, as shown in Fig. 8.16B.

When the cutting edge is defective, when the specimen contains unusually hard-to-section objects, when the mechanical or thermal feed of the microtome is irregular, or when the tissue block is very soft, complex patterns of knife scratches or irregular patterns may appear. Regions that appear thick in one section frequently correspond to regions that are missing in the adjacent section, as seen in Fig. 8.16C.

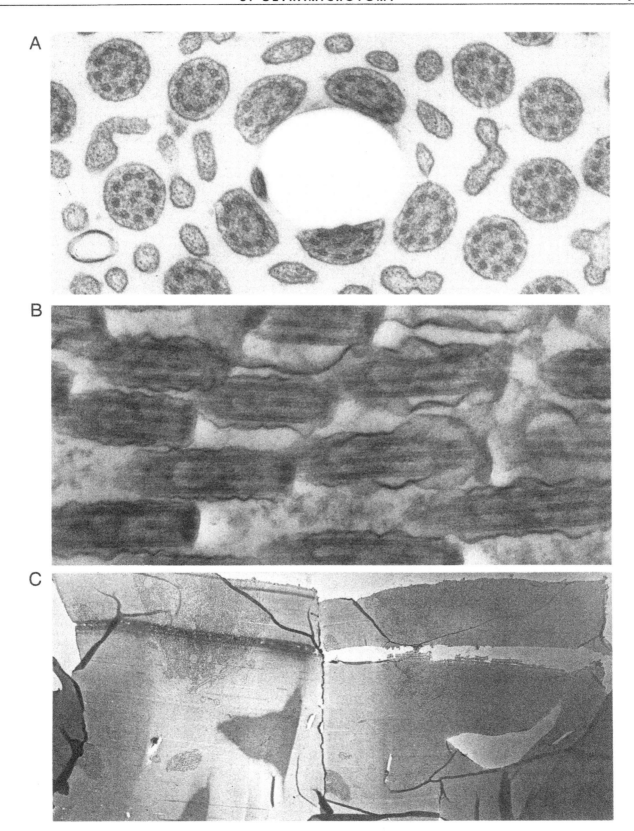

17. Contamination during Microtomy

FIGURE 8.17A Portion of an Epon section that has been cut with a diamond knife and section stained with uranyl acetate and lead citrate. × 10,000.

This part of the section did not contain embedded tissue, but the entire section exhibits a pattern of small rounded or irregular spots. The spots show their highest electron density in the center and a decreased density toward the periphery. The diameter of the spots ranges from a few nanometers to a few tenths of a micron.

FIGURE 8.17B Part of a proximal tubule cell of a rat kidney fixed with glutaraldehyde, postfixed with osmium tetroxide, and embedded in Epon. Ultrathin sections were cut with a diamond knife and section stained with uranyl acetate and lead citrate. × 30,000.

The cell has a normal fine structure, but electron-dense spots are scattered over the cell. The spots are particularly evident over the mitochondria but are also present over the large endocytic vacuoles (upper left) as well as over the tubule lumen between the microvilli (left).

FIGURE 8.17C Section of a human sperm head section stained with uranyl acetate and lead citrate. × 50,000.

The nucleus of this immature spermatid is characterized by a pattern of electron-dense dots and strands. In addition, this micrograph shows an irregular row of even more electron-dense particles in strands over the sperm head as well as outside the sperm head.

COMMENTS

These micrographs illustrate special types of contamination that may appear during ultramicrotomy. The spots over the section in Figs. 8.17A and 8.17B, in laboratory slang called "the measles," appear preferentially when the tissue has been cut with a diamond knife. It is interpreted as a contamination originating from the material that was used to cement the knife to its holder. Presumably the cement was not fully polymerized or had been dissolved by an organic solvent that had been used to clean the knife. When present in mild forms, this type of contamination may simulate cellular components, e.g., mitochondrial inclusions, but it is here present over the entire section. It is a good rule not to try to clean diamond knives with organic solvents as it may not only give rise to contamination, but also loosen the knife from the trough.

The type of contamination shown in Fig. 8.16C has not been identified, although it has some resemblance to the smoke derived from burning unsaturated organic compounds. It does not derive from the heavy metal salts used during section staining. This type of contamination happens to have some resemblance to the chromatin strands of the spermatid nucleus, but is present in both biological structures and outside the cell.

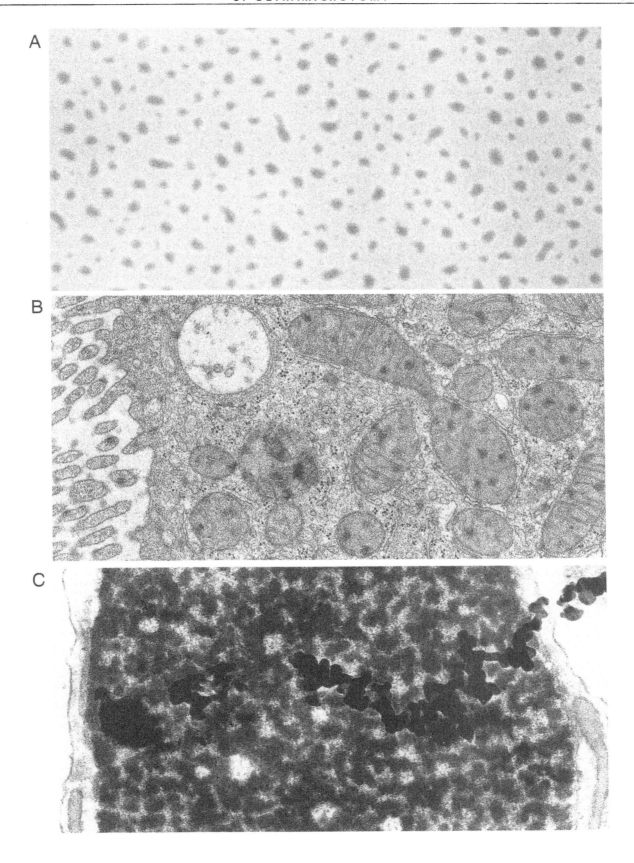

18. Extraction during Sectioning

FIGURE 8.18A Dental anlage from a rat molar. The tissue was fixed in glutaraldehyde and embedded in Epon. The section was cut at the border between the calcified and the noncalcified dental anlage. The section was cut on distilled water using a diamond knife and observed unstained. × 60,000.

Two zones are seen separated by a zig-zag border. The calcified dental anlage is to the left, the calcification zone to the right.

FIGURE 8.18B Dental anlage with dentine to the left and forming enamel to the right. This section was sectioned on a special solution containing dissolved apatite, sometimes referred to as "enamel water." No section staining was used. × 60,000.

This section shows the extensive presence of small crystallites in both the dental anlage itself and the area peripheral to the dental anlage, where small crystallites are abundant.

COMMENTS

During ultramicrotomy the sections are floated on the surface of water. If the section contains water-soluble substances, there is a risk that they are dissolved into the fluid.

Apatite has a very low solubility in water, and if the crystals have large dimensions the effect will hardly be noticeable. However, if the apatite crystals are very small, as in the case of the dental anlage, they may nevertheless dissolve if floated on pure water. This can be prevented if the sectioning fluid has been mixed well with small pieces of crushed enamel. The small amounts of dissolved enamel apatite suffice to prevent the small crystallites in the section from dissolving.

The possibility that substances in the section are dissolved during sectioning must always be kept in mind, particularly when dealing with easily water-soluble substances such as ions. In sections intended for analysis by X-ray microanalysis or electron energy loss spectroscopy, this is a particularly necessary consideration when sectioning, e.g., tissue that has been freeze-dried to prevent loss of ions. Also, enzyme reaction products formed during cytochemical reactions may dissolve during sectioning (see Fig. 15.7B). It is also possible that larger molecules, such as glycogen, may be extracted during ultramicrotomy if located at the very surface of the section.

A

B

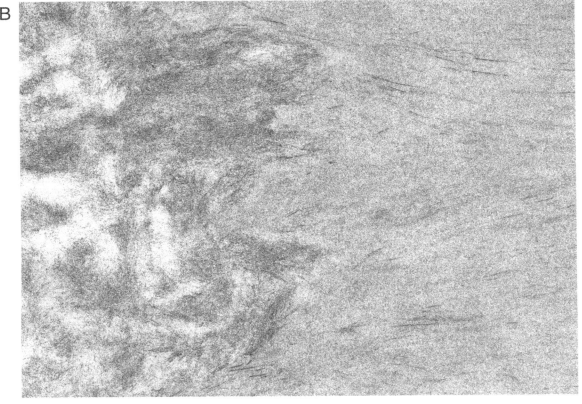

19. Cryoultramicrotomy: Survey Sections

FIGURES 8.19A AND 8.19B Parts of cells from a rat liver perfusion fixed with 2% glutaraldehyde in 0.1 M cacodylate buffer. After rinsing in cacodylate buffer, the small pieces of tissue (largest dimension 0.5–1.0 mm) were infiltrated with sucrose for 1 hr; first for half an hour with 1.15 M sucrose in 0.01 M phosphate-buffered saline (PBS), pH 7.2, containing 2% paraformaldehyde, and then for 1 hr with 2.3 M sucrose in the same solution. The tissue blocks were then placed on metal pins fitting the ultramicrotome, frozen in liquid nitrogen, and sectioned on a Reichert FCS cryoultramicrotome (Leica). After trimming, survey sections, 0.8–1.0 μm in thickness, were cut of the whole block surface at $-80°C$ using a glass knife. These sections were stained with toluidine blue for light microscopy. The block surface was then trimmed down to the preferred area, less than 0.5 mm in any direction, and ultrathin sections were cut at $-115°C$ using a 45° diamond knife (Drukker International, The Netherlands). The thin sectioning was made in the presence of a Diatome Static Line ionizer (Haug) and picked up according to the principle of Tokuyasu (1973) with a droplet in a wire loop; for this section the droplet was composed of a 1:1 mixture of 2% methylcellulose and 2.3 M sucrose (Liou *et al.*, 1996). When the droplet had melted it was placed on a dry nickel grid with a Formvar support film, which had been reinforced with carbon. The grid was then placed section side down on the droplet of PBS. After rinsing on another PBS droplet and 3 droplets of redistilled water it was placed on a droplet of 1.8% methylcellulose containing 0.3% uranyl acetate for 10 min and picked up with a wire loop. Excess methylcellulose/uranyl acetate was drawn off with dense filter paper. $\times$ 4800.

Figure 8.19A shows parts of several liver cells located around a sinusoid and separated by bile capillaries. Nuclei are essentially round, there is no evidence of wrinkles or compression, and the contrast is uniform.

The section in Fig. 8.19B shows distinct evidence of compression, which has resulted in several defects. The folds run in the direction of upper left–lower right, which is at a right angle to the direction of sectioning. Between the folds there are regions apparently without compression where section quality appears adequate (a). Ruptures (r) can be observed in places within the section.

COMMENTS

The preparation of adequately thin sections by cryoultramicrotomy is dependent on several interacting factors: consistency of the tissue, quality and angle of the diamond knife, angle between block and knife, sectioning temperature, sectioning speed, method of recovering the frozen sections, and, last but not least, skill of the technician producing the sections. The micrograph in Fig. 8.19A can be considered an optimal cryosection with a minimum of distortion and compression and with a retained contrast of cellular components. The general structure of the cells to a large extent resembles that of a conventionally fixed and Epon-embedded tissue. In contrast, the section in Fig. 8.19B, which was produced from the same tissue block, shows distortion during sectioning. The difference between the two sections may be due to minor differences in the sectioning process related to section thickness, clearing angle of the knife, the adjustment of the antistatic ionizer, temperature, or sectioning speed.

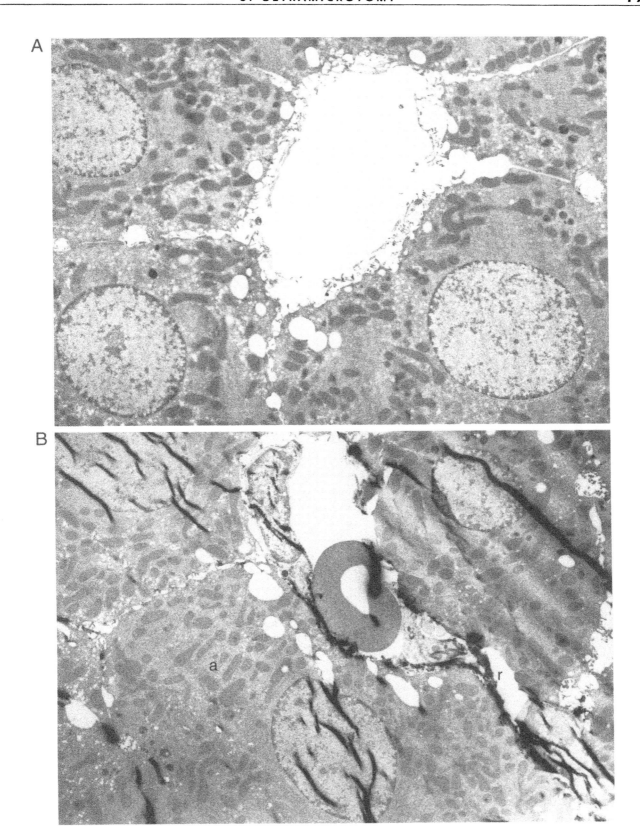

20. Collection of Cryosections

FIGURES 8.20A AND 8.20B A rat kidney was per-fusion-fixed with 4% formaldehyde in 0.1 M cacodylate buffer, processed and cryosectioned as in Fig. 8.19A. The section in Fig. 8.20A was picked up with the aid of a 2.3 M sucrose droplet in a metal loop. The loop with the droplet was carefully lowered toward the section and touched the sections immediately before it froze. After thawing the droplet the sections was touched to a Form-var-coated nickel grid, which was then placed section side down on a droplet of PBS. After a few rinses on other droplets of PBS and redistilled water, the grid was trans-ferred to a droplet of 1.8% methylcellulose containing 0.3% uranyl acetate. After 10 min the grid was picked up with a wire loop and excess methylcellulose was drawn off from the grid with a filter paper and the grid was dried. The section in Fig. 8.20B was cut from the same block but picked-up with a droplet consisting of a 1:1 mixture of 2.3 M sucrose and 2% methylcellulose ac-cording to Liou *et al.* (1996). × 50,000.

Cytoplasmic details are exceptionally well pre-served; cell membranes stand out distinctly as triple-layered structures and mitochondrial cristae are well defined when cut at right angle. The cytoplasm has a uniform appearance and there is only a moder-ate difference between the density of the cytoplasm and the mitochondrial matrix. The intercellular space has a fairly constant width.

Cell membranes appear distinctly triple layered in many, but not all, places as in Fig. 8.21A. The cytoplasm has a uniform appearance but the mito-chondrial matrix has a considerably higher density.

FIGURE 8.20C Section picked up with a droplet consisting of 2.0% uranyl acetate in 1% methylcellulose. This solution has a low viscosity and was therefore spread on the surface of a spherical Epon droplet that had been polymerized in a wire loop. After thawing the Epon loop was lowered onto the sections and then onto the grid. The loop was then lifted from the grid, and excess methylcellulose/uranyl acetate solution was immediately drawn off the grid. × 50,000.

COMMENTS

Crucial steps in cryoultramicrotomy are collecting the frozen sections and subsequent handling of the sections during the staining procedure. The current method for recov-ering ultrathin frozen sections was devised by Tokuyasu (1973, 1986) and represented a breakthrough in cryoultramictomy and also indirectly in immunoelectron microscopy by making it possible to collect sections on a grid without major deformation and drying artifacts. The technique works well with 2.3 M sucrose for most tissues. However, a mixture of sucrose and methylcellulose in the droplet further improves the integrity of the cytoplasm (Fig. 8.20B). The collection of sections with methylcellulose containing uranyl acetate (Fig. 8.20C) results in narrow intracellular spaces and a densely stained cytoplasm, but shows less cytoplasmic details.

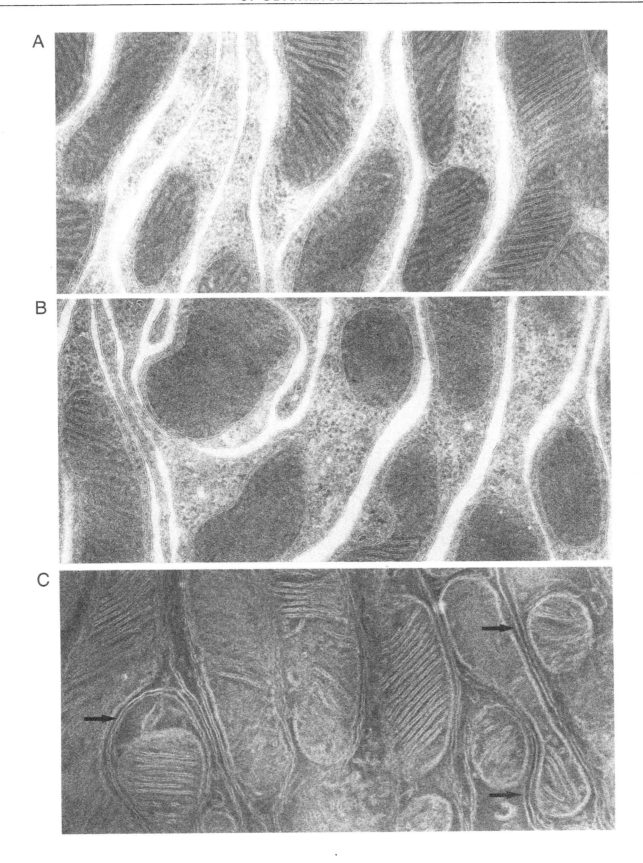

21. Thickness of Cryosections

FIGURES 8.21A AND 8.21B Cryosections of cells in the medullary thick ascending limb (mTAL) in a rat kidney perfusion fixed with 4% paraformaldehyde in 0.1 *M* cacodylate buffer, infiltrated in sucrose, and frozen as in Fig. 8.19A. The sections were cut at two different settings of the cryoultramicrotome and then collected with sucrose/methylcellulose droplets, rinsed in PBS and redistilled water, stained for 10 min with 1.8% methylcellulose containing 0.3% uranyl acetate, and picked up with a wire loop; excess fluid was drawn off with dense filter paper.

FIGURE 8.21A The cryoultramicrotome was set at a 55-nm thickness for this section and the cutting speed was about 12 mm/sec. × 50,000.

Cytoplasmic details are exceptionally well preserved; cell membranes stand out distinctly as triple-layered structures and mitochondrial cristae are well defined when cut at right angle. The cytoplasm has a uniform appearance and there is only a moderate difference between the density of the cytoplasm and the mitochondrial matrix. The intercellular space has a fairly constant width.

FIGURE 8.21B The cryoultramicrotome was set at a 120-nm thickness for this section and the cutting speed was about 1 mm/sec. × 50,000.

Cell membranes appear distinctly triple layered in many, but not all, places as in Fig. 8.21A. The cytoplasm has a uniform appearance but the mitochondrial matrix has a considerably higher density.

COMMENTS

Although the microtome setting may not always reflect the section thickness, it is evident that the section in Fig. 8.21A is much thinner than that in Fig. 8.21B. Furthermore, because the section thickness varies somewhat in most series of cryosections, it is likely that the section in Fig. 8.21A, which was the thinnest seen on the grid, is even less than 55 nm. The authors have found that thin sections, such as those in Fig. 8.21A, are most reproducibly sectioned on a 45° diamond knife and best collected with the mixture of sucrose and methylcellulose. The pattern observed in these two sections is, to a large extent, reminescent of the pattern observed in aldehyde-fixed, nonosmicated, and Epon-embedded tissue (compare with Fig. 2.2B). In the Epon-embedded tissue, however, the intercellular spaces are almost completely obliterated in places, thus illustrating the influence on the fine structure that comes from the preparatory steps subsequent to the initial fixation.

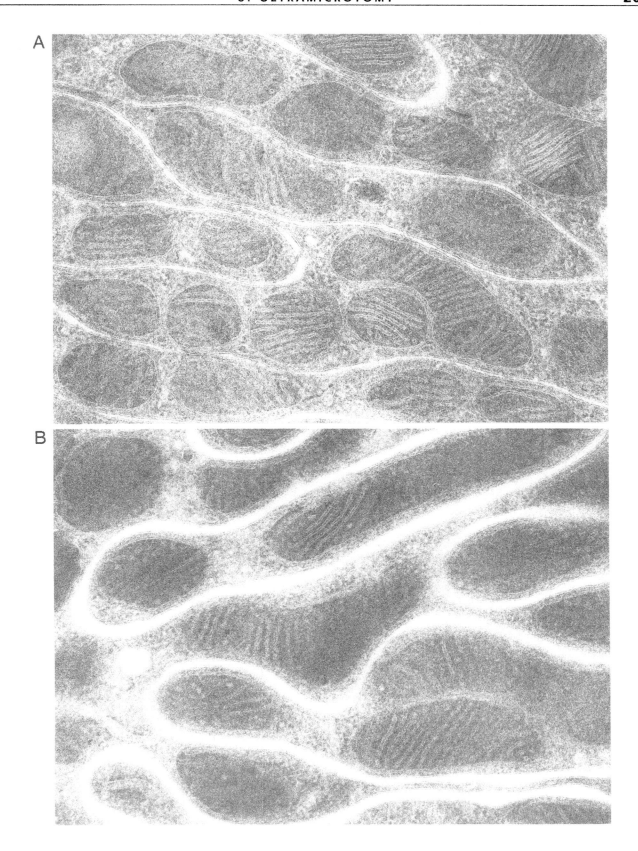

22. Staining of Cryosections

FIGURES 8.22A–8.22C These three micrographs show cryosections of kidney cells cut from the same tissue block as in Figs. 8.20 and 8.21, picked up with sucrose/methylcellulose, and rinsed in phosphate-buffered saline but treated with three different methods.

FIGURE 8.22A The section was rinsed in PBS and redistilled water and then dried without any further staining. × 50,000.

The general contrast of this section is low. The intercellular spaces are widened and in places there are holes in the section. Mitochondrial cristae appear here in weak positive contrast (arrows). In the upper left corner there is part of the nucleus, which shows a homogeneous nucleoplasm.

FIGURE 8.22B This section was stained for 10 min in the standard way with 0.3% uranyl acetate in 1.8% methylcellulose. × 50,000.

This micrograph shows a "standard" appearance of the cell cytoplasm with a good contrast of mitochondrial cristae and cell membranes, although the section thickness here is somewhat greater than that shown in Fig. 8.21A.

FIGURE 8.22C After rinsing in PBS and redistilled water, the section was stained for 5 min with a mixture of equal amounts of 4% uranyl acetate in redistilled water and 0.3 M oxalic acid and the pH was adjusted to 7.0 with 10% ammonia (Voorhout, 1988). The grid was then transferred to a droplet of 1.8% methylcellulose containing 0.3% uranyl acetate. After 5 min, excess fluid was drawn off and the grid dried. × 50,000.

Cell membranes show somewhat greater contrast than in Figs. 8.22A and 8.22B as well as in Figs. 8.21A and 8.21B. Additionally, the mitochondrial matrix has a relatively greater density.

COMMENTS

The staining procedures for cryosections are quite different than those used for resin sections and illustrated in Chapter 9. The final appearance of cryosections varies greatly, depending on several steps in the Tokuyasu procedure, e.g., collection, rinsing, staining, and drying of the sections. Without any staining and in the absence of methylcellulose, the section has a low contrast and there is a tendency for ruptures in the section, as illustrated in Fig. 8.22A. Interestingly, mitochondrial cristae are visible in positive contrast (arrows), although not as sharply defined as following uranyl acetate staining. When sections are "dried" in the presence of methylcellulose containing uranyl acetate, the cytoplasm is well preserved, with cell organelles standing out in good contrast. However, a somewhat more intense staining is sometimes obtained if the section is first stained on neutralized uranyl acetate and then stained in the presence of methylcellulose also containing uranyl acetate (Fig. 8.22C).

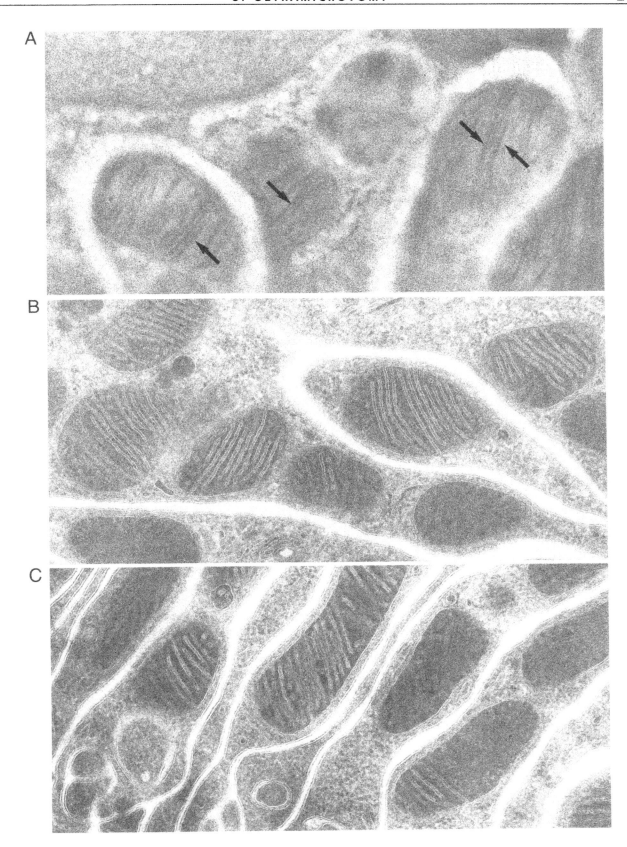

23. Defects in Cryosections

FIGURE 8.23A Part of proximal tubule cells of a rat kidney tissue infiltrated with 2.3 *M* sucrose, frozen in liquid nitrogen, and cryosectioned at 120 nm. The section was collected with a droplet of 2.3 *M* sucrose and stained with neutral uranyl acetate as in Fig. 8.22C and then in methylcellulose containing uranyl acetate. × 48,000.

The cells show numerous microvilli, but these appear tangled and located on top of each other in places (arrows).

FIGURE 8.23B Cryosection from the same block as in Fig. 8.23A sectioned at 55 nm and collected with a droplet of 2.3 *M* sucrose and stained with methylcellulose containing uranyl acetate. × 35,000.

These two kidney cells show cytoplasms with several holes or empty regions. To the right is a nucleus and to the left three mitochondria, which are very electron dense as compared to the surrounding cytoplasm.

FIGURE 8.23C Part of a kidney cell fixed with 1% glutaraldehyde, sucrose infiltrated, cryosectioned, collected with a 2.3 *M* sucrose droplet, and stained with methylcellulose containing uranyl acetate. The removal of the methylcellulose droplet from the grid was incomplete. × 60,000.

Cytoplasmic structures, such as mitochondrial cristae, show a low contrast and appear indistinct. The cells are stained intensely as compared to the light intercellular space.

COMMENTS

Cryosections may display a number of preparatory artifacts related to the various steps in their preparation. For example, cellular projections or microvilli in cryosections may appear detached from the cell body itself during the collection with the sucrose droplet or during the subsequent PBS rinses, as illustrated in Fig. 8.23A where some microvilli are superimposed. In particular, when using 2.3 *M* sucrose to collect the sections, the sections tend to expand on the surface of the sucrose droplets. This results in thinning of the cytoplasm and an apparent contraction of mitochondria, as illustrated in Fig. 8.23B. This is one of the most common artifacts in cryosectioning and section collection with pure sucrose droplets. An important precaution in order to avoid this artifact is to reduce the time that the section resides on the surface of the sucrose droplet after the droplet has melted.

Another critical step in the preparation procedure is the removal of excess methylcellulose/uranyl acetate from the section before drying. If too little methylcellulose is drawn off the section it dries with a thick layer of methylcellulose/uranyl acetate, which reduces the visibility of structural details in the section as in Fig. 8.23C. A thick layer of methylcellulose/uranyl acetate also results in increased radiation damage on the section with tendencies for "bubbling" of the stain.

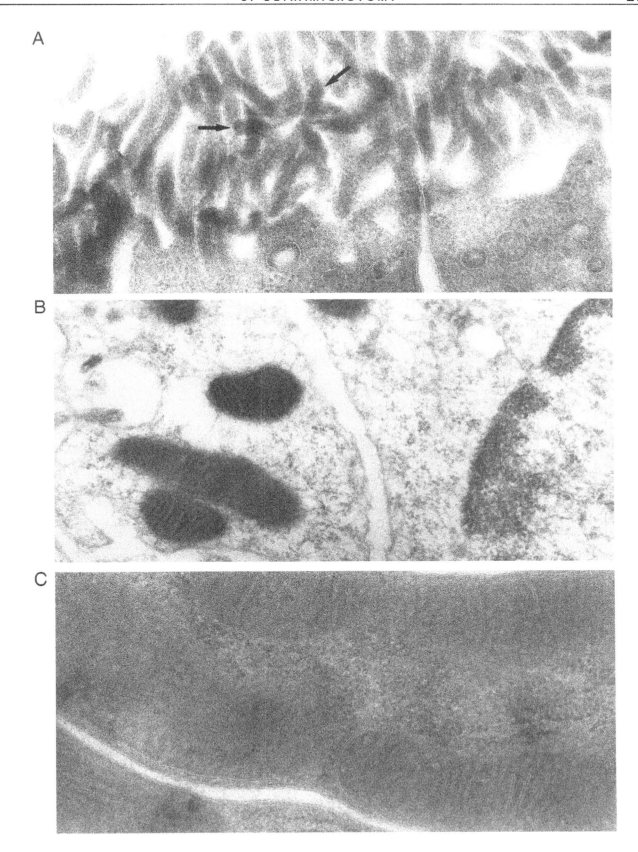

References

Abad, A. (1988). A study of section wrinkling on single-hole coated grids using TEM and SEM. *J. Electr. Microsc. Techn.* **8,** 217–222.

Acetarin, J.-D., Carlemalm, E., Kellenberger, E., and Villiger, W. (1987). Correlation of some mechanical properties of embedding resins with their behavior in microtomy. *J. Electr. Microsc. Techn.* **6,** 63–79.

Bencosme, S. A., Stone, R. S., Latta, H., and Madden, S. C. (1959). A rapid method for localization of lesions for electron microscopy. *J. Biophys. Biochem. Cytol.* **5,** 508–510.

Bénichou, J. C., Fréhel, C., and Ryter, A. (1990). Improved sectioning and ultrastructure of bacteria and animal cells embedded in Lowicryl. *J. Electr. Microsc. Techn.* **14,** 289–297.

Bernhard, W., and Leduc, E. (1967). Ultrathin frozen sections. I. Methods and ultrastructural preservation. *J. Cell Biol.* **34,** 757–771.

Bernhard, W., and Viron, A. (1971). Improved techniques for the preparation of ultrathin frozen sections. *J. Cell Biol.* **49,** 731–746.

Berthold, C.-H., Rydmark, M., and Corneliuson, O. (1982). Estimation of sectioning compression and thickness of ultrathin sections through Vestopal-W-embedded cat spinal roots. *J. Ultrastruct. Res.* **80,** 45–52.

Cecich, R. A., and Bauer, E. O. (1979). How to make your ultramicrotome vibration-free. *Stain Technol.* **54,** 103–104.

Cristensen, A. K. (1971). Frozen thin sections of fresh tissue for electron microscopy, with a description of pancreas and liver. *J. Cell Biol.* **51,** 772–804.

De Groot, D. M. G. (1988). Comparison of methods for the estimation of thickness of ultrathin tissue sections. *J. Microsc. (Oxford)* **151,** 23–42.

Favard, P., and Carasso, N. (1973). The preparation and observation of thick biological sections in high voltage electron microscope. *J. Microsc. (Oxford)* **97,** 59–81.

Galey, F. R., and Nilsson, S. E. G. (1966). A new method for transferring sections from the liquid surface of the trough through staining solutions to the supporting film of a grid. *J. Ultrastruct. Res.* **14,** 405–410.

Gillis, J.-M., and Wibo, M. (1971). Accurate measurement of the thickness of ultrathin sections by interference microscopy. *J. Cell Biol.* **49,** 947–949.

Helander, H. (1973). Some observations on knife properties and sectioning mechanics during ultramicrotomy of plastic embedding media. *J. Microsc. (Oxford)* **101,** 81–93.

Holderegger, C., and Bechter, R. (1978). Reliable method for obtaining longitudinal sections of sperm for electron microscopy. *Experientia* **34,** 1386–1387.

Jésior, J.-C. (1985). Direct determination of ultrathin section thickness using latex particles. *J. Electr. Microsc. Techn.* **2,** 161–165.

Jésior, J.-C. (1985). How to avoid compression: A model study of latex sphere grid sections. *J. Ultrastruct. Res.* **90,** 135–144.

Jésior, J.-C. (1986). How to avoid compression. II. The influence of sectioning conditions. *J. Ultrastruct. Res.* **95,** 210–217.

Jésior, J.-C. (1989). Use of low-angle diamond knives leads to improved ultrastructural preservation of ultrathin sections. *Scan. Microsc. Suppl.* **3,** 147–153.

Klomparens, K. L. (1988). Artifacts in ultrathin sectioning. *In* "Artifacts in Biological Electron Microscopy" (R. F. E. Crang and K. L. Klomparens, eds.), pp. 65–79. Plenum Press, New York.

Knobler, R. L., Stempak, J. G., and Laurencin, M. (1978). Preparation and analysis of serial sections in electron microscopy. *In* "Principles and Techniques of Electron Microscopy. Biological Applications"

(M. A. Hayat, ed.), Vol. 8, pp. 113–155. Van Nostrand-Reinhold, New York.

Langenberg, W. G. (1982). Silicone additive facilitates epoxy plastic sectioning. *Stain Technol.* **57,** 79–82.

Liou, W., Geuze, H. J., and Slot, J. W. (1996). Improving structural integrity of cryosections for immunogold labeling. *Histochem. Cell Biol.* **106,** 41–58.

Maunsbach, A. B. (1978). Electron microscopic analysis of objects in light microscopic sections. *In* "Proceedings of the Ninth International Congress on Electron Microscopy, Toronto, 1978," Vol. II, pp. 80–81.

Michel, M., Hillmann, T., and Müller, M. (1991). Cryosectioning of plant material frozen at high pressure. *J. Microsc. (Oxford)* **163,** 3–18.

Mollenhauer, H. H. (1986). Surfactants as resin modifiers and their effect on sectioning. *J. Electr. Microsc. Tech.* **3,** 217–222.

Mollenhauer, H. H., and Bradfute, O. E. (1987). Comparison of surface roughness of sections cut by diamond, sapphire, and glass knives. *J. Electr. Microsc. Techn.* **6,** 81–85.

Peachy, L. D. (1958). Thin sections: A study of section thickness and physical distortion produced during microtomy. *J. Biophys. Biochem. Cytol.* **4,** 233–242.

Pease, D. C. (1982). Unembedded aldehyde-fixed tissue, sectioned for transmission electron microscopy. *J. Ultrastruct. Res.* **79,** 250–272.

Pihakaski, K., and Suoranta, U. M. (1985). Effects of different epoxies on avoiding wrinkles in thin sections of botanical specimens. *J. Electr. Microsc. Techn.* **2,** 7–10.

Reid, N., and Beesley, J. E. (1991). Sectioning and cryosectioning for electron microscopy. *In* "Practical Methods in Electron Microscopy" (A. M. Glauert, ed.), Vol. 13. Elsevier, Amsterdam.

Richter, K., Gnägi, H., and Dubochet, J. (1991). A model for cryosectioning based on the morphology of vitrified ultrathin sections. *J. Microsc. (Oxford)* **163,** 19–28.

Sakai, T. (1980). Relation between thickness and interference colors of biological ultrathin section. *J. Electr. Microsc.* **29,** 369–375.

Sitte, H., Neumann, K., and Edelmann, L. (1989). Cryosectioning according to Tokuyasu vs. rapid-freezing, freeze-substitution and resin embedding. *In* "Immuno-Gold Labeling in Cell Biology" (A. J. Verkleij and J. L. M. Leunissen, eds.), pp. 63–93. CRC Press, Boca Raton, FL.

Sjöstrand, F. S. (1967). "Electron Microscopy of Cells and Tissues: Instrumentation and Techniques." Academic Press, New York.

Small, J. V. (1968). Measurement of section thickness. *In* "Proceedings of the Fourth European Regional Conference on Electron Microscopy, Rome," Vol. I, pp. 609–610.

Tokuyasu, K. T. (1973). A technique for ultracryotomy of cell suspensions and tissues. *J. Cell. Biol.* **57,** 551–565.

Tokuyasu, K. T. (1978). A study of positive staining of ultrathin frozen sections. *J. Ultrastruct. Res.* **63,** 287–307.

Tokuyasu, K. T. (1986). Application of cryoultramicrotomy to immunocytochemistry. *J. Microsc. (Oxford)* **143,** 139–149.

Villiger, W., and Bremer, A. (1990). Ultramicrotomy of biological objects: From the beginning to the present. *J. Struct. Biol.* **104,** 178–188.

Voorhout, W. (1988). "Possibilities and Limitations of Immuno-Electronmicroscopy." Thesis, University of Utrecht, The Netherlands.

Williams, M. A., and Meek, G. A. (1966). Studies in the thickness and topography of ultrathin sections. *J. Histochem. Cytochem.* **14,** 755–756.

SECTION STAINING

Contrast in a biological specimen that has been fixed with glutaraldehyde only or has been left unfixed is very low when studied in the electron microscope; little structural detail can be recognized and recorded. There are ways to increase the contrast and foremost among them is "staining" of the sections or of the entire tissue block with various solutions containing heavy metal compounds. The latter bind to biological structures and cause a general increase in contrast that makes it possible to identify and characterize the structures in the section. However, staining of the object is not used in cryoelectron microscopy where the biological object is embedded in vitrified ice and observed at very low temperature in the microscope (see Chapter 10.28).

The commonly used term "staining" is borrowed from light microscopy but is misleading, as the only "colors" obtained with the electron microscope are black, white, and shades grades in between. Hence "contrasting" would be a more appropriate term. The term staining may also give the impression that a certain specificity is obtained, but most section-staining procedures are rather unspecific. A number of staining solutions and formulae have been devised and some of them are entered in the reference list. In the procedures for section staining and block staining, the properties of the stains themselves, as well as the technical shortcomings of the procedures, may influence the interpretation of the biological structures profoundly.

This chapter illustrates the characteristics of some commonly used staining methods. It also shows some pitfalls in the different staining procedures, in particular different forms of precipitates, and indicates ways to avoid or remove these.

Osmium fixation and methacrylate embedding were used in much of the 1950s and in part also in the 1960s. This combination turned out to be a lucky one in that specimen contrast is adequate; osmium, as a heavy atom, gives tissue components a high electron density, whereas methacrylate has a rather low electron density. Cecil Hall was probably the first to demonstrate the usefulness of what he called electron stains in his studies of isolated paramyosin filaments treated with phosphotungstic acid. However, after section-staining procedures were developed, the life of the electron microscopist became much easier. The first report of a section-staining procedure appeared in a conference report by Ian Gibbons and J. R. G. Bradfield (1956). They used osmium tetroxide, lanthanum nitrate, or thorium nitrate as section stains of acetic acid-fixed tissues. Later, Michael Watson (1958) published two influential papers on the staining of tissue sections for electron microscopy with salts of lead and uranium. Edward Reynolds (1963) found a practical and stable formula for the lead stain based on a citrate buffer and published a short paper that was one of the most cited reports in biomedicine for several years.

Gibbons, L. R., and Bradfield, J. R. G. (1956). In "Proceedings of First European Conference on Electron Microscopy" (F. S. Sjöstrand, and J. Rhodin, eds.), pp. 121–124. Stockholm, Sweden.

Hall, C. E. (1945). J. Appl. Physics 16, 459–465.

Reynolds, E. (1963). J. Cell Biol. 17, 208–212.

Watson, M. L. (1958). J. Biophys. Biochem. Cytol. 4, 475–478 and 727–730.

1. Lead Citrate Staining

FIGURES 9.1A–9.1D A rat liver fixed with glutaraldehyde, postfixed with osmium tetroxide, and embedded in Epon that was sectioned with a diamond knife at approximately the same thickness, from the same block, and at the same occasion. The section illustrated in Fig. 9.1A was left unstained, the section in Fig. 9.1B was stained with lead citrate for 2 min at room temperature, the section in Fig. 9.1C was stained with lead citrate for 20 min at room temperature, and the section in Fig. 9.1D was stained first with lead citrate for 2 min and then rinsed intensely with distilled water. All figures were printed on photographic paper of normal contrast. × 40,000.

The stained sections show a considerable enhancement in contrast over the unstained section. Among the stained sections, the greatest contrast gain is that of prolonged lead staining, where the glycogen granules in particular have an increased electron density. Extensive rinsing after staining removes some of the lead stain, particularly from glycogen.

COMMENTS

Section staining with lead salts is the most commonly used means of improving the contrast in ultrathin sections. From a chemical point of view, lead staining is unspecific, but some cell components are more densely stained than others. Thus, membranes and glycogen granules will take up more stain than chromatin, ribosomes, or mitochondrial matrix.

Lead staining on the whole is a reliable and reproducible procedure. Variations in the outcome may depend on the type of tissue, fixative, embedding medium (see Chapter 5.7), and formula of stain solution, as well as differences in practical parameters such as duration of staining, temperature, rinsing, and drying after rinsing. Note that too much rinsing will remove some of the lead staining (Fig. 9.1D).

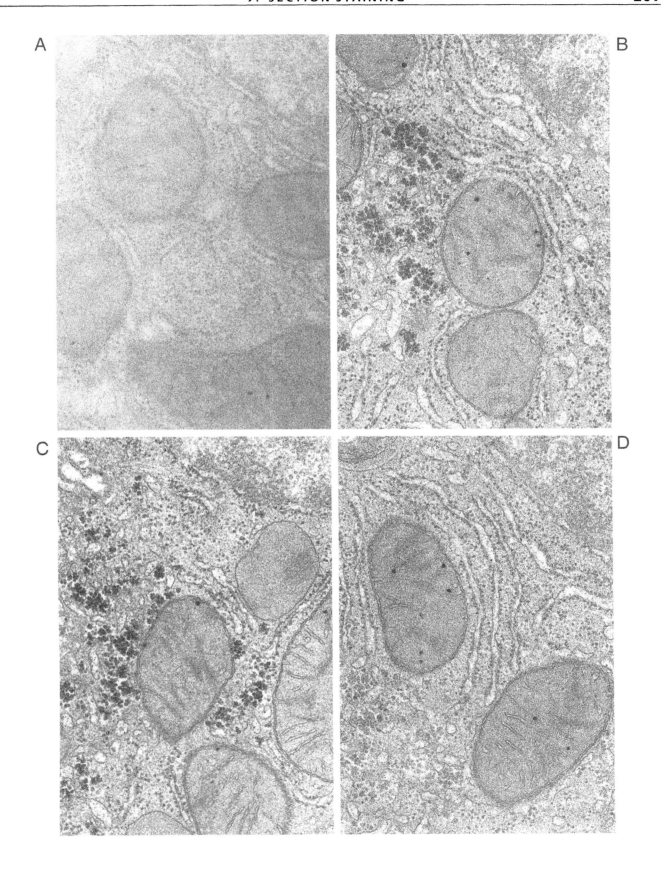

2. Uranyl Acetate Staining

FIGURES 9.2A–9.2C Sections from the same tissue block of rat liver shown in Figs. 9.1A–9.1D. In Fig. 9.2A the section is unstained, in Fig. 9.2B the section is stained for 10 min in 4% aqueous uranyl acetate at room temperature, and in Fig. 9.2C the section is stained for 60 min in 4% uranyl acetate at room temperature. × 40,000.

Compared to the unstained section in Fig. 9.2A, the section in Fig. 9.2B shows increased electron density, particularly of chromatin and ribosomes, but to some extent also of mitochondrial matrix and membranes in general. The section in Fig. 9.2C shows a slightly higher contrast than that in Fig. 9.2B.

COMMENTS

Uranyl staining, similar to lead staining, is chemically unspecific. Compared to lead staining it shows a greater affinity to cell components containing nucleic acids. Membranes, however, are only moderately contrasted and, therefore, the general cell architecture is not as distinctly revealed after uranyl staining alone as after lead staining alone.

Uranyl acetate, as well as lead citrate, has been shown to penetrate into Epon-embedded tissue. In the comparatively thin sections used for transmission electron microscopy, penetration of uranyl acetate is sufficiently rapid to be complete within 10 min and no further gain in contrast is obtained with prolonged staining. However, if the section is very thick, the duration of staining may become an influential factor.

When small cellular structures are measured, it is essential that the influence of the staining procedure is considered. Typically the dimensions of small objects, such as ribosomes or filaments, will appear larger in stained than in unstained sections. Furthermore, the dimensions of stained structures will depend on the choice of heavy metal stain and the mode of its application.

A

B

C

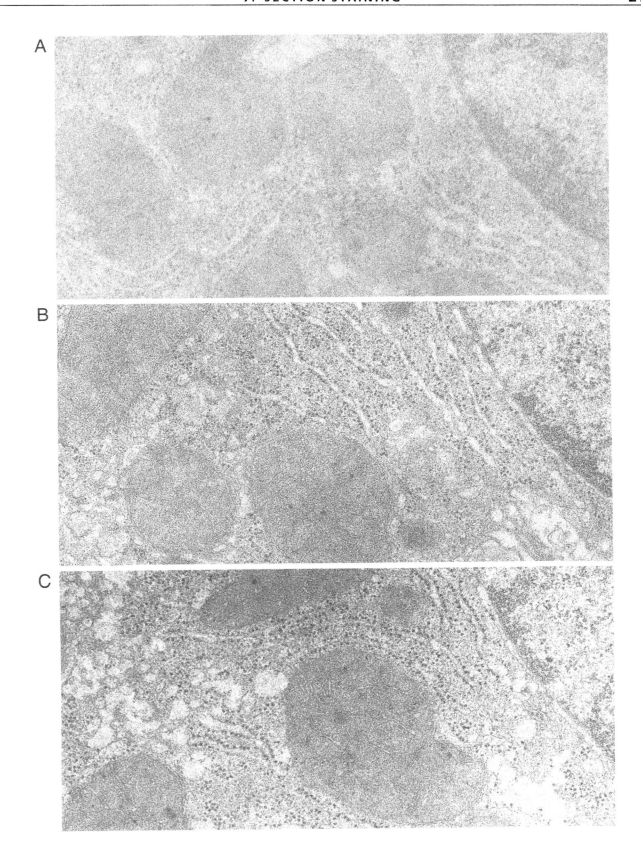

3. Enhanced Section Staining

FIGURES 9.3A–9.3C Sections from the same tissue block of rat liver as shown in Fig. 9.1A–9.1D and 9.2A–9.2C. In Fig. 9.3A the section was first stained for 10 min at room temperature in 4% uranyl acetate and then for 2 min in lead citrate. In Fig. 9.3B the section was first stained with 4% uranyl acetate for 30 min at 60°C and then with lead citrate for 2 min. The section shown in Fig. 9.3C was stained in succession at room temperature with lead citrate for 1 min, in 4% uranyl acetate for 10 min, and finally in lead citrate for 1 min according to Daddow (1983). × 40,000.

All cellular components in Fig. 9.3A show considerable electron density and their membranes stand out distinctly. The mitochondrial matrix is densely stained as is the cytoplasmic matrix. In Fig. 9.3B the staining pattern is similar to that in Fig. 9.3A, although staining intensity generally appears even greater. The triple-stained section in Fig. 9.3C exhibits mitochondria where the matrix is exceptionally densely stained to the extent that inner mitochondrial membranes are difficult to distinguish. Nuclear chromatin and ribosomes stand out with great contrast.

COMMENTS

Double staining with uranyl and lead renders essentially all cellular components considerably electron dense. However, because both the mitochondrial matrix and cristae stain intensely, the contrast differences in the sections become less pronounced than following, e.g., lead staining alone (compare Figs. 9.3A and 9.1B). From a practical point of view the high contrast of double staining is advantageous at low magnifications (less than, e.g., × 5000) and for detailed studies at high magnifications. In the middle range of magnifications, lead stains alone usually give a better overall view.

The dramatic increase in staining intensity during double staining is probably due to the great affinity for ionic lead complexes to derivatives of osmium and/or uranium in the ultrathin sections. The triple staining technique demonstrated in Fig. 9.3C is particularly useful when a very high contrast is required. Uranyl acetate staining at increased temperature as demonstrated by Brody (1959) is also useful in demonstrating cell structures that are not visualized with standard staining schedules. However, it should be kept in mind that overstaining can take place, which may obscure small details to be studied.

Staining can be sped up if performed in a microoven. It is not known if the staining is due exclusively to the increase in temperature, as in Brody's studies, or if there is also a direct effect of the microwaves. At any rate, caution is necessary, as the stains, notably uranyl acetate, may boil suddenly in the microoven.

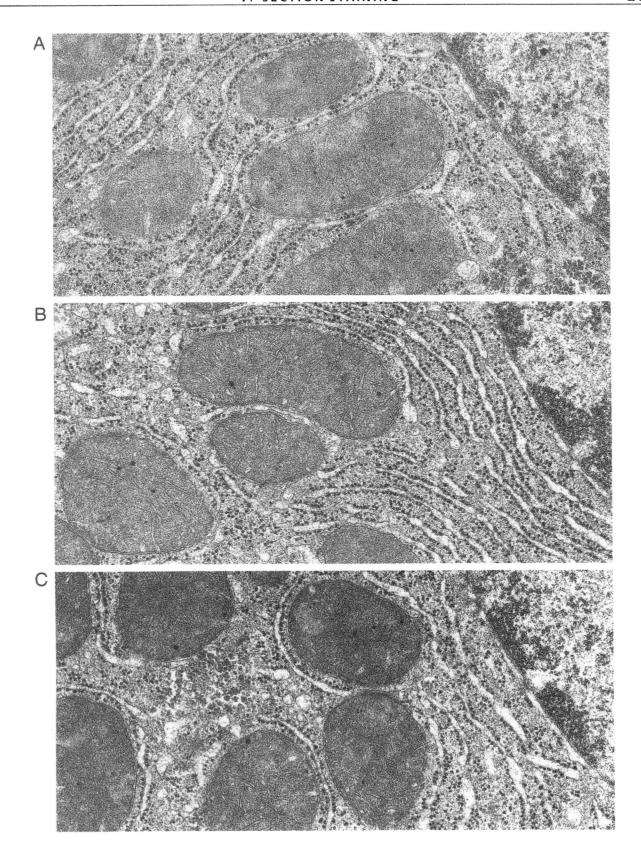

4. Effects of Grid Storage

FIGURE 9.4A An electron micrograph of a human ciliated epithelium fixed with glutaraldehyde, postfixed with osmium tetroxide, embedded in Epon, and section stained with uranyl acetate for 60 min and lead citrate for 5 min. The electron micrograph was recorded immediately after section staining. × 65,000.

The various cellular details are shown with good contrast.

FIGURE 9.4B An electron micrograph of the same section as in Fig. 9.4A but inserted in the electron microscope and recorded 12 years later. × 65,000.

The cellular structures exhibit essentially the same clarity and contrast as those seen in Fig. 9.4A.

FIGURE 9.4C An electron micrograph of a section cut from the same block as that of Fig. 9.4A. However, the section was left unstained in the gridbox for 12 years before section staining was performed and then left for another 2 years before being examined in the electron microscope. Section staining was made with the same protocol as was used for Figs. 9.4A and 9.4B, and the section was recorded in the same microscope as was used for Fig. 9.4A. × 65,000.

Although this area is different than that in Figs. 9.4A and 9.4B, contrast and cellular details are very similar to that in the other two figures.

COMMENTS

It is evident that stained sections can be stored for many years without an appreciable change in the contrast. It should be noted that the quality of section staining may sometimes become deteriorated, perhaps depending on contamination, e.g., by lipids or atmospheric moisture.

It is also evident that unstained sections remain stainable for very long periods of time and that the cytological details, when sections eventually are stained, are similar to those in newly cut and stained sections. For this reason the investigator should save a few grids with unstained sections. In this way a spare set of the specimen is obtained, should the first staining be suboptimal. An improved section-staining technique may become available at a later date.

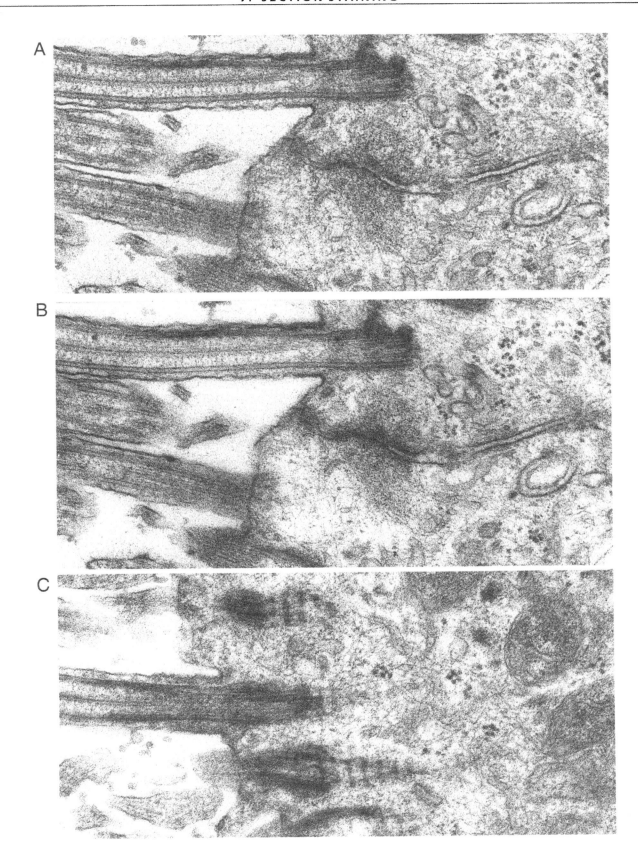

5. Section Exposed to Electron Beam

FIGURES 9.5A AND 9.5B A rat liver cell fixed with glutaraldehyde, postfixed with osmium tetroxide, embedded in Epon, sectioned with a diamond knife, and inspected in the electron microscope for some time before submitting the section to section staining with uranyl acetate and lead citrate and a second inspection in the electron microscope. During the first inspection the electron beam was focused by a double condensor to a spot with a diameter of about 8 μm. Some areas of the section were exposed to the electron beam for periods of 1 min, whereas other parts of the section were not exposed to the beam. Figure 9.5A shows one of these spots, whereas Fig. 9.5B shows the transitional zone between an exposed and an unexposed part of the section. Figure 9.5A × 10,000 and Fig. 9.6B × 40,000.

There is a great difference in contrast between the area exposed to the electron beam and the surrounding unexposed areas. During printing of the micrograph, the central exposed area has been given a longer exposure time in order to compensate for the great difference in density of the photographic negative. The beam-exposed area would otherwise appear completely empty in the print.

COMMENTS

The lack of contrast in the part of the section that has been exposed to the electron beam is due to a resistance of these parts to subsequent section staining with uranyl acetate and lead citrate. This resistance is believed to be due to a degradation of the organic substances in the section, perhaps in combination with a surface contamination of the exposed area and an evaporation of Epon in the electron beam.

A

B

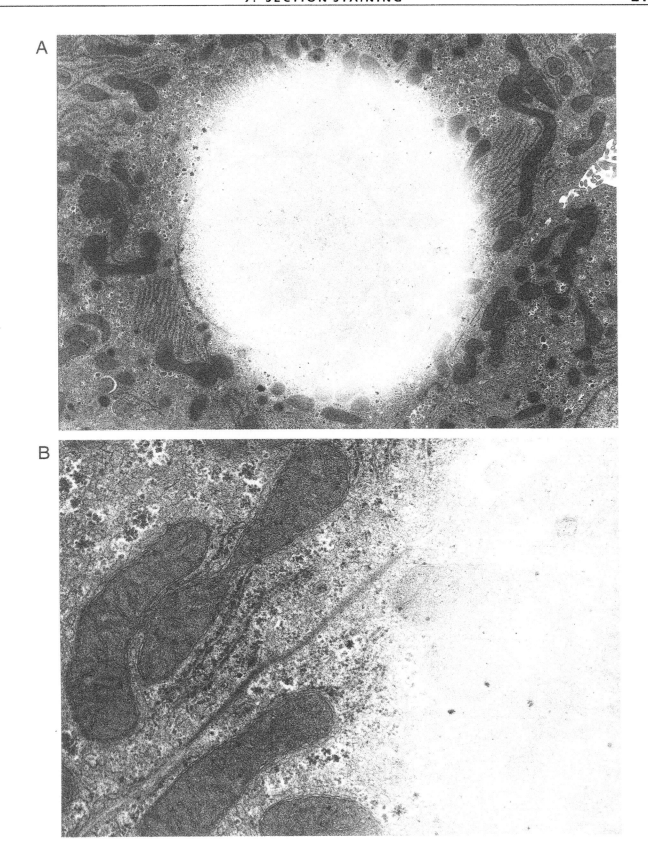

6. Effect of Electron Beam

FIGURES 9.6A–9.6C Mitochondria and cytoplasm of epithelial cells of a rat kidney fixed with 2.5% glutaraldehyde in cacodylate buffer, postfixed with osmium tetroxide, and embedded in Epon. Thin sections were stained with uranyl acetate and lead citrate. The micrograph in Fig. 9.6A was taken immediately after this part of the section had been introduced in the electron beam, which was kept at low intensity. Figure 9.6B shows the same field as in Fig. 9.6A, but the electron micrograph was taken after exposure of the section to the electron beam for approximately 2 min. The beam intensity was somewhat higher than in Fig. 9.6A. Figure 9.6C is a higher magnification of Fig. 9.6B illustrating in more detail the distribution of a precipitate. In Figs. 9.6A and 9.6B the primary magnification was 20,000 times and the final magnification was 50,000, whereas in Fig. 9.6C the final magnification was × 120,000.

The membranes in Fig. 9.6A have smooth contours and an even electron density. There is only little electron-dense precipitate. In Fig. 9.6B a fine electron-dense precipitate is sprinkled over the entire field. It is particularly associated with membranes but present also over mitochondrial matrix and ground cytoplasm. In Fig. 9.6C the membranes in places appear transformed into rows of electron-dense dots of variable size. Outside the membranes the precipitate is distributed randomly, although more precipitate is present over electron-dense structures.

COMMENTS

This series of micrographs demonstrates that a particular kind of precipitate can be induced by overexposure of the section to the electron beam. The precipitate originates, at least in part, from the section staining, as an unstained section of the same material will not develop this kind of granularity on exposure to the beam. The precipitate apparently forms by coalescence of stain deposits following heating and, presumably, by melting or softening of the section in the beam.

The formation of this type of precipitate is most commonly observed in sections of epoxy resins. However, it varies considerably from preparation to preparation, most likely due to differences in the degree of polymerization of the embedding medium. To avoid this type of artifact it is important that the resin is well polymerized. Furthermore, it must be emphasized that sections should not be exposed to strong electron beams and that micrographs should be taken shortly after the field of interest has been introduced into the beam.

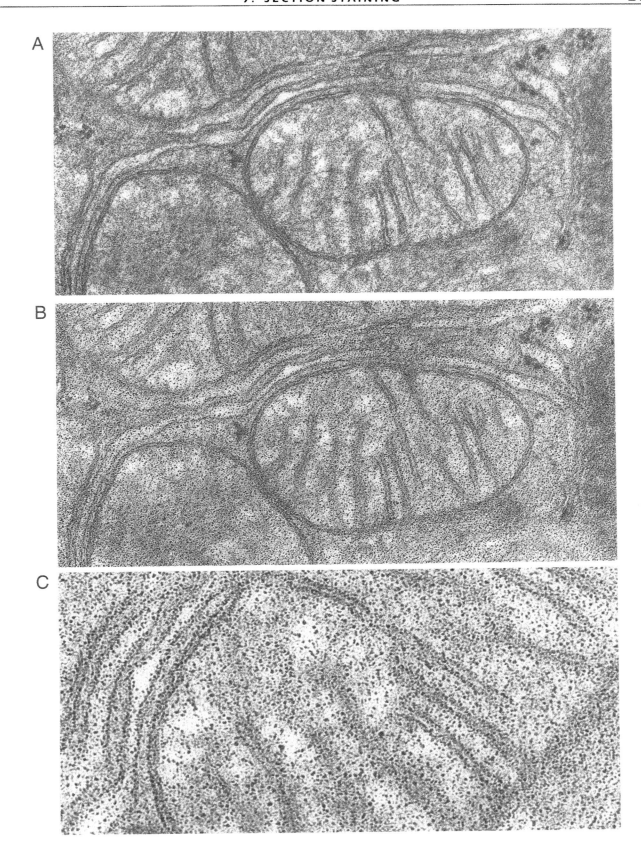

7. Lead-Staining Granularity

FIGURES 9.7A–9.7D Portions of intestinal epithelial cell (Fig. 9.7A) and of renal proximal tubule cell (Figs. 9.7B–9.7D). The tissues were osmium fixed and embedded in Epon, and the sections were stained with lead citrate. The sections were observed in the electron microscope at relatively high magnifications, and the electron beam was saturated. The micrographs were taken shortly after the sections had been introduced in the electron beam. Figure 9.7A × 30,000, Figs. 9.7B and 9.7C × 50,000, and Fig. 9.7D × 150,000.

Almost all components of these thin sections show a pronounced granularity. This granularity is severe in Fig. 9.7A, where irregular electron-dense grains outline the underlying structure to such an extent that the plasma membrane of the microvilli of the cell is totally invisible. There are no such electron-dense grains over the empty embedding medium in the upper part of the figure.

In Fig. 9.7B, mitochondrial membranes and plasma membranes are contrasted, but appear to consist of small dense grains. The cytoplasmic matrix (Fig. 9.7C) also contains numerous small grains with diameters of 20–80 Å. These cannot be confused with ribosomes, which range from 150 to 200 Å in diameter in this preparation.

In Fig. 9.7D the mitochondrial membranes at close inspection appear to contain globular subunits (arrows). Similar size granules are scattered in the cytoplasm.

COMMENTS

The granularity shown in these four figures originates from lead staining. More granules tend to appear with prolonged lead staining. Furthermore, they usually appear during the first seconds of exposure to the intensely hot electron beam. It is possible that they form during a partial melting of the section. Using an anticontamination device does not appear to affect the appearance of these grains. In cases, when this kind of stain precipitate occurs, sections may be stained with uranyl acetate only. When the granules are small and associated with membranes, they may be mistaken for real membrane subunits.

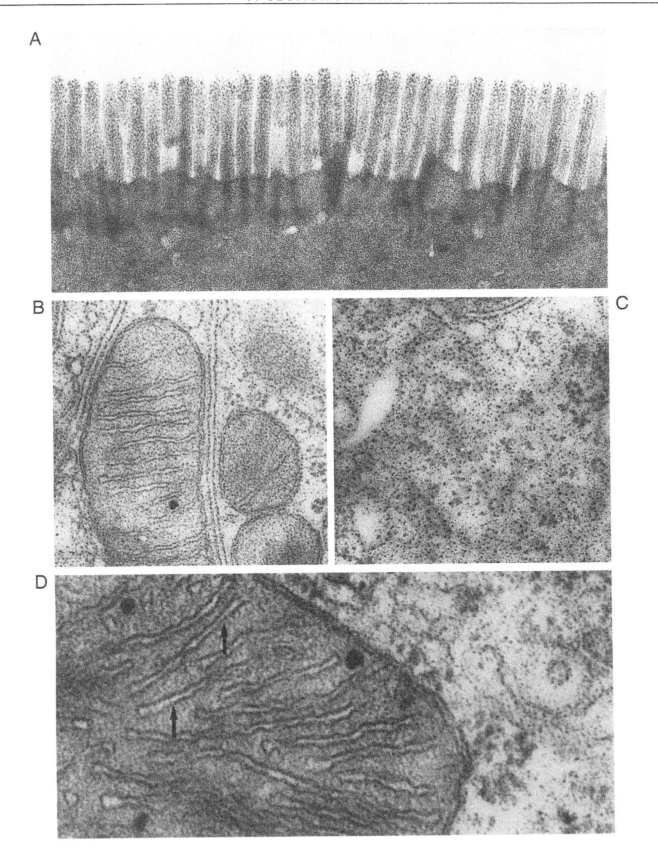

8. Contamination

FIGURE 9.8A A section of glutaraldehyde–osmium-fixed kidney tubules that have been section stained with uranyl acetate for 10 min at room temperature, rinsed briefly in distilled water, stained with lead citrate for 1 min, and again rinsed with distilled water. × 8000.

The general contrast of the cells is good and most of the section is clean. However, a few electron-dense particles are seen, varying in size from approximately 0.1 to 2 μm. The largest particles have a hexagonal shape.

FIGURE 9.8B Parts of proximal tubule cells from a rat kidney cortex fixed with 1% glutaraldehyde and postfixed with 1% osmium tetroxide. The section was stained for 2 min with lead citrate and then rinsed with distilled water. × 40,000.

The general contrast in this micrograph is adequate, but a finely granular electron-dense precipitate is observed over the heterochromatin. Nearby plasma membranes, cytoplasmic, and intercellular regions are nearly free of precipitate.

FIGURE 9.8C A section of intestinal epithelial cells stained with uranyl acetate for 30 min at 60°C, rinsed in distilled water, and then stained for 2 min with lead citrate and again rinsed in distilled water. × 8000.

Numerous needle-like objects are located in a band over the specimen. This band of contamination runs approximately parallel to one of the bars of the copper grid at the bottom of the figure.

COMMENTS

During microscopy of ultrathin sections, contamination of different kinds is commonly encountered. The source of contamination has to be identified for the defect to be eliminated. Contamination from section staining may appear in different forms. Some of these are easy to identify. Large crystals of lead salts typically have the appearance shown in Fig. 9.8A; they may be seen as solitary particles over otherwise clean areas. They are not seen in sections that have been section stained with uranyl acetate only.

In Fig. 9.8B, adjacent unstained sections were devoid of the granular electron-dense precipitation. The precipitates thus have appeared in connection with the staining procedure, either during staining itself or during the subsequent rinse with distilled water. Alternatively, it represents a contaminant that becomes stained together with the cell. The reason for this type of staining artifact is unknown but apparently it is related to the physicochemical conditions of the section surface.

Needle-like crystals, like those in Fig. 9.8C, have only been observed in preparations treated with uranyl acetate, after either block staining or section staining. Following section staining, such crystals may appear in bands running parallel to the grid bar. This contamination appears to form when the grid dries after the rinse with distilled water.

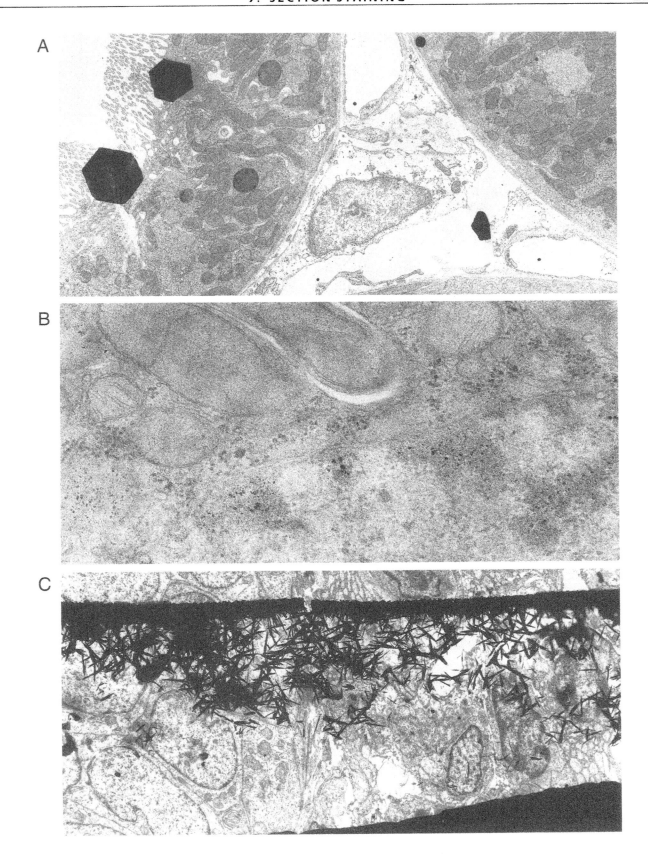

9. Block-Staining Precipitate

FIGURE 9.9A Parts of rat liver cells fixed with glutaraldehyde, postfixed with osmium tetroxide, and block stained with 1% uranyl acetate. The tissue was embedded in Epon, and the section was stained for 2 min with lead citrate. × 20,000.

Cellular structures show good contrast. A fine needle-like precipitate is, however, seen in the nucleus.

FIGURE 9.9B A section from the same block as in Fig. 9.9A but observed unstained in the electron microscope. × 20,000.

A fine needle-like precipitate is present, particularly in the capillary endothelium and on the surface of the erythrocyte (in the middle upper part).

FIGURE 9.9C Parts of two cells from a rat exocrine pancreas. The tissue was perfusion fixed with glutaraldehyde, postfixed with osmium tetroxide, and block stained with 1% uranyl acetate in Veronal–acetate buffer. Embedding was made in Epon and section staining with lead citrate. × 40,000.

Cellular details in the cell at the upper side have a considerably lower contrast than have those in the cell at the lower side. No irregular precipitate is present and the cell membranes appear intact in both cells.

COMMENTS

Block staining with uranyl acetate, which in reality is also a fixative, is an efficient way to obtain specimens with high contrast and usually without contamination. Penetration of the uranyl acetate may be uneven, however, possibly due to uneven fixation. As illustrated in Fig. 9.9C, one of the two cells has not been penetrated by the uranyl acetate. The unpenetrated cell therefore stains less intensely during the subsequent lead citrate section staining. Another problem may be the needle-like precipitate that seems to be caused by a high affinity of uranyl ions to some parts of the tissue, such as the nucleus (Fig. 9.9A). Because the needle-like precipitate is also present in unstained sections (Fig. 9.9B), it must originate from block staining.

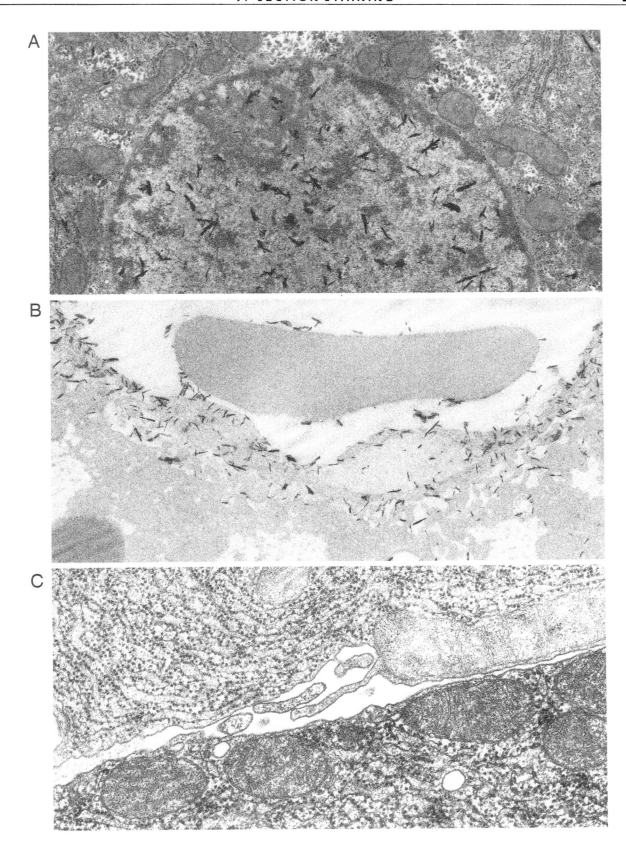

10. Removal of Contamination

FIGURE 9.10A Part of a kidney epithelial cell fixed with glutaraldehyde, postfixed with osmium tetroxide, embedded in Epon, and section stained with uranyl acetate and lead citrate. × 20,000.

In addition to well-defined cellular structures, this electron micrograph shows an irregularly distributed, electron-dense precipitation over the ground cytoplasm as well as over mitochondria and other organelles.

FIGURE 9.10B Electron micrograph of the same section as in Fig. 9.10A. Before this second electron micrograph was recorded, the grid was removed from the microscope, floated for 1 min (section side down) on a solution of 1% acetic acid in water, rinsed in distilled water, and reinserted in the microscope. × 20,000.

In this micrograph there is no irregular electron-dense precipitate sprinkled over the cytoplasm. The contrast of the section is otherwise essentially the same. Note also that the lysosome (asterisk) contains electron-dense material, which represents inclusions that occur normally in this organelle and which has remained after treatment with acetic acid.

These electron micrographs show that an unspecific precipitate, formed during section staining, may be removed, at least under some circumstances, by a treatment that has been recommended by Kuo (1980). The length of the treatment with acetic acid is somewhat critical, as too long a treatment may destain the section. If this is the case, the low-contrast sections may be exposed to another round of section staining.

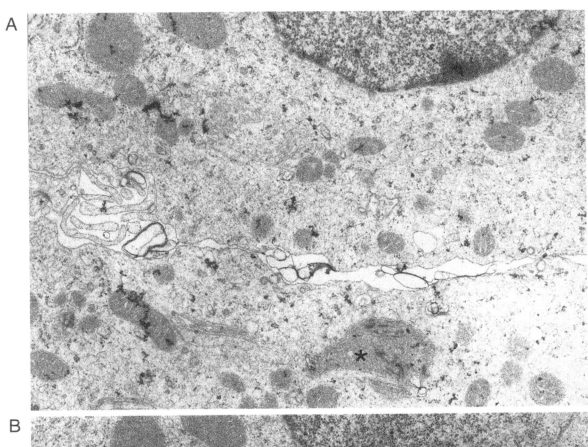

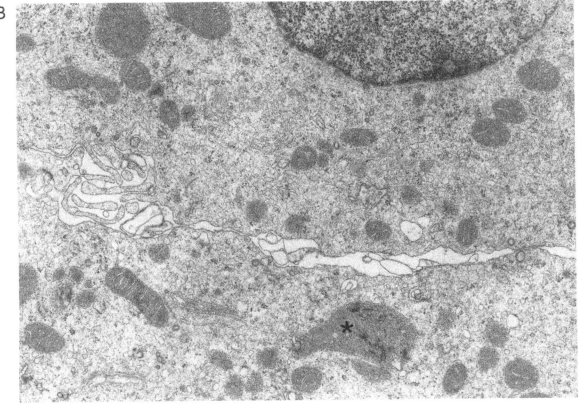

References

Afzelius, B. A. (1992). Section staining for electron microscopy using tannic acid as a mordant: A simple method for visualization of glycogen and collagen. *Microsc. Res. Tech.* **31,** 65–72.

Avery, S. W., and Ellis, E. A. (1978). Methods for removing uranyl acetate precipitate from ultrathin sections. *Stain Technol.* **53,** 137–140.

Bell, A. (1988). Artifacts in staining procedures. *In* "Artifacts in Biological Electron Microscopy" (R. F. E. Crang and K. L. Klomparens, eds.) pp. 81–106. Plenum Press, New York.

Brody, I. (1959). The keratinization of epidermal cells of normal guinea pig skin revealed by electron microscopy. *J. Ultrastruct. Res.* **2,** 482–511.

Carasso, N., Delaunay, M.-C., Favard, P., and Lechaire (1973). Obtention et coloration de coupes épaisses pour la microscopie électronique à haute tension. *J. Microsc. (Paris)* **16,** 257–268.

Colquhoun, W. R., and Rieder, C. L. (1980). Contrast enhancement based on rapid dehydration in the presence of phosphate buffer. *J. Ultrastruct. Res.* **73,** 1–8.

Daddow, L. Y. M. (1983). A double lead stain method for enhancing contrast of ultrathin sections in electron microscopy: A modified multiple staining technique. *J. Microsc. (Oxford)* **129,** 147–153.

Ellis, E. A., and Anthony, D. W. (1979). A method for removing precipitate from ultrathin sections resulting from glutaraldehyde-osmium tetroxide fixation. *Stain Technol.* **54,** 282–285.

Hanaichi, T., Sato, T., Iwamoto, T., Malavasi-Yamashiro, J., Hoshino, M., and Mizuno, N. (1986). A stable lead by modification of Sato's method. *J. Electr. Microsc.* **35,** 304–306.

Hayat, M. A. (1993). "Stains and Cytochemical Methods." Plenum, New York.

Haydon, G. B. (1969). Electron phase and amplitude images of stained biological thin sections. *J. Microsc. (Oxford)* **89,** 73.

Kuo, J. (1980). A simple method for removing stain precipitates from biological sections for transmission electron microscopy. *J. Microsc. (Oxford)* **120,** 221–224.

Lewis, P. R., and Knight, D. P. (1992). Cytochemical staining methods for electron microscopy. *In* "Practical Methods in Electron Microscopy" (A. M. Glauert, ed.) Vol. 14, pp. 1–321, Elsevier, Amsterdam.

Locke, M. (1994). Preservation and contrast without osmification or section staining. *Microsc. Res. Techn.* **29,** 1–10.

Locke, M., and Krishnan, N. (1971). Hot alcoholic phosphotungstic acid and uranyl acetate as routine stains for thick and thin sections. *J. Cell Biol.* **50,** 550–557.

Locke, M., Krishnan, N., and McMahon, J. T. (1971). A routine method for obtaining high contrast without staining sections. *J. Cell Biol.* **50,** 540–544.

Marinozzi, V. (1968). Phosphotungstic acid (PTA) as a stain for polysaccharides and glycoproteins in electron microscopy. *In* "Fourth European Regional Conference of Electron Microscopy" (S. D. Bocciarelli, ed.), Vol. 2, pp. 55–56. Tipographia Poliglotta Vaticana, Rome.

Maunsbach, A. B. (1998). Immunolabeling and staining of ultrathin sections in biological electron microscopy. *In* "Cell Biology: A Laboratory Handbook" (J. E. Celis, ed.), 2nd Ed. Vol. 3, pp. 268–276. Academic Press, New York.

Mollenhauer, H. H. (1987). Contamination of thin sections: Some observations on the cause and elimination of "embedding pepper." *J. Electr. Microsc. Techn.* **5,** 59–63.

Mollenhauer, H. H., and Droleskey, R. E. (1985). Some characteristics of epoxy embedding resins and how they affect contrast, cell organelle size, and block shrinkage. *J. Electro. Microsc. Techn.* **2,** 557–562.

Neiss, W. F. (1984). A coat of glycoconjugates on the inner surface of the lysosomal membrane in the rat kidney. *Histochemistry* **80,** 603–608.

Ottensmeyer, F. P., and Pear, M. (1975). Contrast in unstained sections: A comparison of bright and dark field electron microscopy. *J. Ultrastruct. Res.* **51,** 253–260.

Peters, A., Hinds, P. L., and Vaughn, J. E. (1971). Extent of stain penetration in sections prepared for electron microscopy. *J. Ultrastruct. Res.* **36,** 37–45.

Rambourg, A. (1971). Morphological and histochemical aspects of glycoproteins at the surfaces of animal cells. *Int. Rev. Cytol.* **31,** 57–114.

Revel, J. P., and Karnovsky, M. J. (1967). Hexagonal array of subunits in intercellular junctions of the mouse heart and liver. *J. Cell Biol.* **33,** C7–C12.

Reynolds, E. S. (1963). The use of lead citrate at high pH as an electron-opaque stain in electron microscopy. *J. Cell Biol.* **17,** 208–212.

Roth, J., Taatjes, D. J., and Tokuyasu, K. T. (1990). Contrasting of Lowicryl K4M thin sections. *Histochemistry* **95,** 123–136.

Ryter, A., and Kellenberger, E. (1958). Etude au microscope électronique de plasma contenant de l'acid desoxyribonucleique. I. Les nucleotides des bacteries en croissance active. *Z. Naturforschung. B.* **13,** 597–605.

Scott, J. E. (1971). Phosphotungstate: A "universal" (nonspecific) precipitant for polar polymers in acid solution. *J. Histochem. Cytochem.* **19,** 689–691.

Spurr, A. R. (1969). A low-viscosity epoxy resin embedding medium for electron microscopy. *J. Ultrastruct. Res.* **26,** 31–43.

Thomopoulos, G. N., Schulte, B. A., and Spicer, S. S. (1987). Postembedment staining of complex carbohydrates: Influence of fixation and embedding procedures. *J. Electr. Microsc. Techn.* **5,** 17–44.

Ting-Beall, H. P. (1979). Interactions of uranyl ions with lipid bilayer membranes. *J. Microsc. (Oxford)* **118,** 221–227.

Watson, M. L. (1958a). Staining of tissue sections for electron microscopy with heavy metals. *J. Biophys. Biochem. Cytol.* **4,** 475–478.

Watson, M. L. (1958). Staining of tissue sections for electron microscopy with heavy metals. II. Application of solutions containing lead and barium. *J. Biophys. Biochem. Cytol.* **4,** 727–730.

MICROSCOPY

Most modern transmission electron microscopes used in biological research have a resolving power on the order of 0.2–0.5 nm (2–5 Å). These values are far better than those obtainable in biological specimens, whether sectioned, replicated, or negatively stained. The resolution in a biological specimen observed under normal conditions is rarely better than 2 nm and, in most cases, is far worse. Only in very regular specimens, e.g., highly ordered protein crystals, in particular if observed at low temperatures with cryoelectron microscopy, is it possible to achieve better resolutions. A distinction should therefore be made between the resolving power of the instrument and the image resolution, which defines the finest detail that can be resolved in any given image. The high resolving power of a transmission electron microscope used in biological research is often irrelevant for the biologists performing studies of sectioned material.

The complexity of the electron microscope leaves open the possibility of erroneous or suboptimal use of the instrument. Most important are errors in focusing, alignment of the electron lens system, and correction of astigmatism. The image quality is also dependent on the contamination level and radiation damage in the microscope and can, at least in part, be influenced by the operator.

The birthday of electron microscopy is usually given with great accuracy: April 7th, 1931. On that date the 25-year-old Ernst Ruska at the Technische Hochschule in Berlin made a recording of a metal net with the two-lens instrument built by himself and his supervisor Max Knoll. The magnification was a modest 16 times, but it was unquestionably an electron microscopic enlargement. Some 8 weeks later, on June 4th, he gave a seminar on the findings. However, before the seminar, Max Knoll had discussed the results with the assistant of a man who was employed by the Siemens factory in Berlin, who immediately applied for a patent for the principle of an electron microscope on May 30th. Without ever having participated in the design and construction of the instrument, a Siemens employee became the inventor of the electron microscope in U.S. patent law terms. However, in the scientific world Ernst Ruska is recognized as the inventor and he received the Nobel Prize in physics in 1986.

Knoll, M., and Ruska, E. (1932). *Z. Phys.* **78,** 318–339.
Ruska, E. (1934). *Z. Phys.* **102,** 417.

1. Resolving Power

FIGURE 10.1A Graphitized carbon black on a thin carbon film recorded in a Philips CM 100 electron microscope at a direct magnification of 250,000 times and enlarged photographically to a final magnification of 2,000,000.

The characteristic 3.4-Å periodicity of graphite is distinct in directions at right angles to each other.

FIGURES 10.1B AND 10.1C Two recordings of a thin carbon film with deposited thin gold crystals (test specimen S142, Agar Scientific, Cambridge, England). The micrographs were taken with a JEOL 3010 microscope operated at 300 kV. Primary magnification was 300,000 times and final magnification 4,500,000.

The carbon support film has small point images (above the crescent) that can be used to estimate the point-to-point resolution. Distances between two adjacent points are from 4 Å and up. In order to be valid, the same pairs of point images have to be present in both recordings and the pairs oriented in different directions.

FIGURE 10.1D Same test specimen as in Figs. 10.1B and 10.1C recorded at a direct magnification of 400,000 times and enlarged photographically 15 times to a final magnification of 6,000,000.

The characteristic 2.44-Å periodicity can be resolved in the gold crystals to the left, and the 2.04-Å spacings are indicated as horizontal lines in the right gold crystal.

FIGURE 10.1E Test specimen consisting of crocidolite (asbestos) crystal recorded in a Philips 208 electron microscope at a direct magnification of 40,000. Final magnification × 150,000.

The vertical lines represent the 9.03-Å spacings of the crystals. Oblique lines representing the 4.53-Å spacings are indicated at about 60° from these.

COMMENTS

A characteristic of an electron microscope is its resolving power, which can be determined in different ways. One way is to measure the width of the overfocused Fresnel fringe. The most common test, however, for microscopes used in biological studies is the ability to resolve the 3.4-Å periodicity of graphite as in Fig. 10.1A. Other test objects are various crystals with different plane spacings, such as the thin gold particles with characteristic 2.44- and 2.04-Å lattices (Fig. 10.1D). For biological purposes the point-to-point resolution test may be the most meaningful one, as there are normally no regular periodic structures of such small dimensions in biological objects.

The resolutions recorded here, 2 Å and up, represent the resolving power of the instrument. This is quite different from specimen resolution, which depends on the characteristics of the object. A resolution better than 20 Å is thus achieved very rarely in an ultrathin section of a cell due to the section thickness and staining conditions. Resolution in negatively stained preparations is also limited by the granularity of the electron stain. In most aspects of biomedical electron microscopy, the limiting factor of resolution therefore resides in the specimen rather than in the microscope. The advantage of a high instrumental resolving power is that it is a sign of high mechanical and electrical stability.

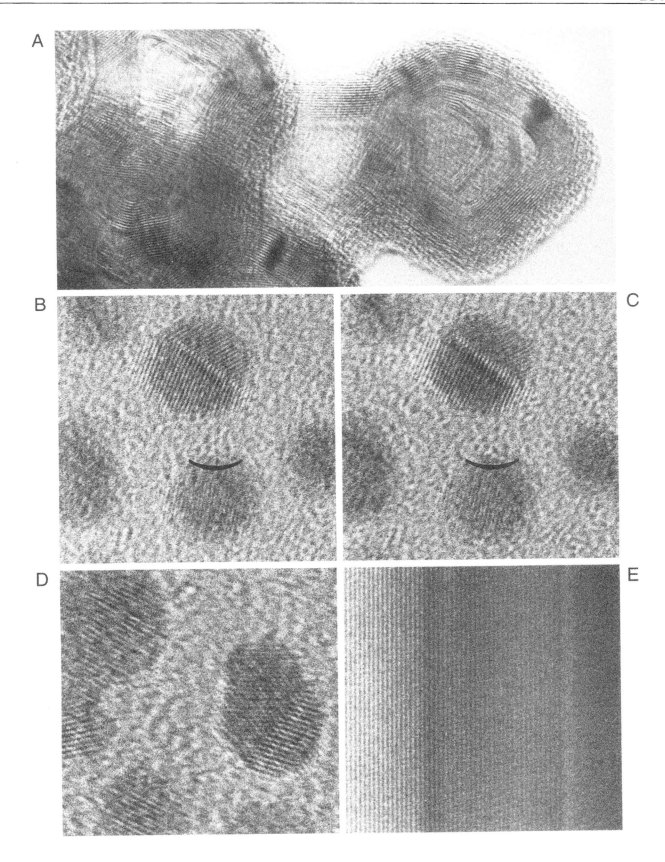

2. Through-Focus Series: Hole and Latex Particle

FIGURES 10.2A–10.2E Through-focus series of a hole in a thin, carbon support film at five different focus levels: underfocused (Fig. 10.2A), slightly underfocused (Fig. 10.2B), close to focus (Fig. 10.2C), slightly overfocused (Fig. 10.2D), and overfocused (Fig. 10.2E). The focus increments between two adjacent micrographs are the same and correspond to two fine focus steps (approximately 26 nm) in the JEOL 100 CX electron microscope. × 260,000.

FIGURES 10.2F–10.2J A corresponding through-focus series of a latex particle located on a thin supporting film and recorded with the same increment steps between each exposure. Figure 10.2H is close to focus. × 260,000.

A typical underfocus fringe is seen in Figs. 10.2A and 10.2F and also, although it is less prominent, in Figs. 10.2B and 10.2G. It is seen as a bright rim inside the hole or outside the particle. The even more characteristic overfocus fringe is visible in the overfocused micrographs: Figs. 10.2E and 10.2J and Figs. 10.2D and 10.2I. In the most underfocused micrographs (Figs. 10.2A and 10.2F) and the most overfocused ones (Figs. 10.2E and 10.2J) the carbon film shows a distinct granularity, which is of the same magnitude on both sides of focus. In the only slightly underfocused or slightly overfocused micrographs (Figs. 10.2B, 10.2G, 10.2D, and 10.2I, respectively) the granularity is less pronounced while still visible, whereas in micrographs close to focus (Figs. 10.2C and 10.2H) the supporting film is practically devoid of a granularity.

COMMENTS

When electrons pass the edge of a hole in the supporting film or the edge of an electron-dense particle, a complicated set of diffraction phenomena takes place, which gives rise to either a bright underfocus fringe or a dense overfocus fringe. They are also called Fresnel fringes, and by observing them one can determine whether the object is over- or underfocused. If the Fresnel fringes have exactly the same width around the hole in a planar film or particle, the objective lens is free of astigmatism. The observation of round holes or round electron-dense particles is thus useful both for determining the focus level and for correcting the astigmatism of the objective lens.

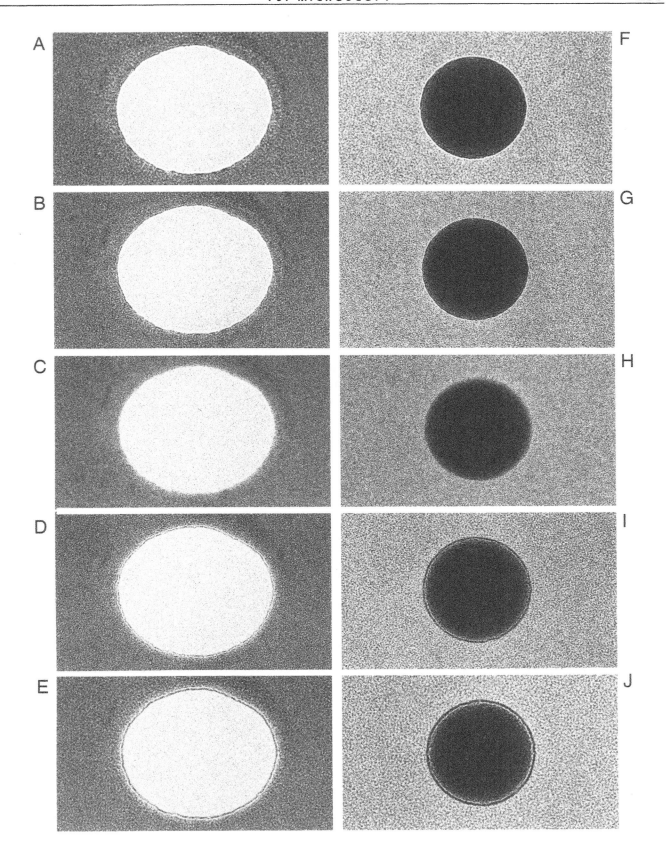

3. Through-Focus Series: Myelin Sheath

FIGURES 10.3A–10.3F A myelin sheath from a human nerve fiber fixed with glutaraldehyde, postfixed with osmium tetroxide, and embedded in Epon. Thin sections were stained with uranyl acetate and lead citrate. Six electron micrographs were taken with equal increments in the objective lens current from a very underfocused image (Fig. 10.3A) to a grossly overfocused image (Fig. 10.3F). Of the intermediate figures, Fig. 10.3B is underfocused, Fig. 10.3C is also underfocused but not far from focus, Fig. 10.3D is close to focus, and Fig. 10.3E overfocused. × 120,000.

In the most underfocused micrographs the major dense layers are smooth and comparatively thick. The intraperiod space has approximately the same thickness but shows no intraperiod layer. In Fig. 10.3E and even more so in Fig. 10.3F the major dense periods are triple layered. In Fig. 10.3F the doubling of the main period has resulted in a pattern with twice as many dense lines as in Fig. 10.3A.

COMMENTS

The focus level influences the electron microscopic appearance of biological structures dramatically. In nonordered objects, out-of-focus images usually show a characteristic focus granularity but no linear pattern. Out-of-focus images of periodic structures may show distinct patterns, which may be misinterpreted as true features of the object. The doubling of the major dense layers of the myelin sheath in overfocused images is an example of such a false periodicity.

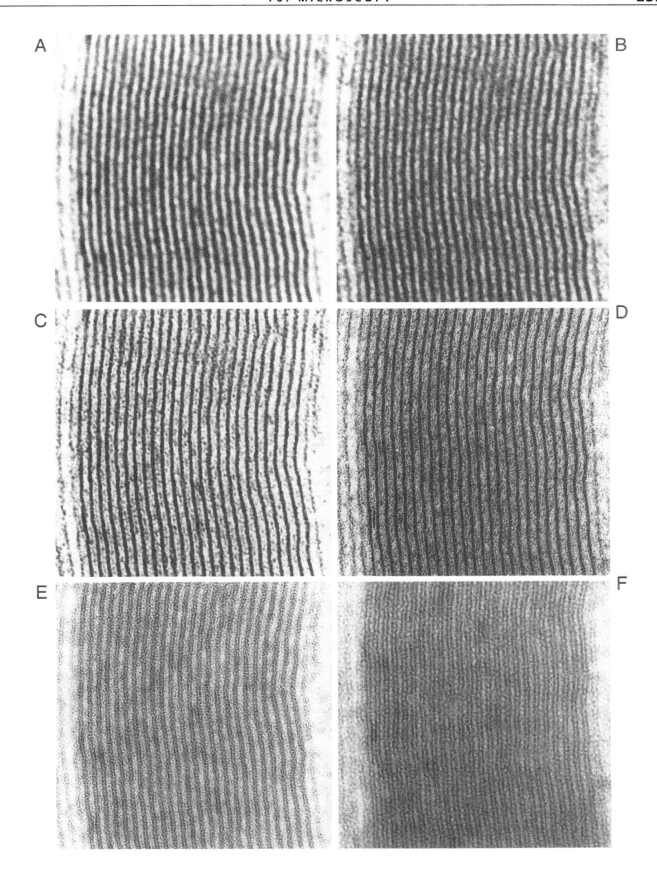

4. Through-Focus Series: Cells

FIGURES 10.4A–10.4E Parts of distal tubule cells from a rat kidney cortex fixed with 2.5% glutaraldehyde and 2% formaldehyde and postfixed in a solution containing 1% osmium tetroxide and 1% potassium ferrocyanide. The tissue was embedded in Epon, and the sections were stained with uranyl acetate and lead citrate. Different focus levels were used for this micrograph series, where Fig. 10.4A is underfocused, Fig. 10.4C is close to focus, and Fig. 10.4E is overfocused. × 60,000.

FIGURES 10.4F–10.4J A higher magnification of details from the same negatives used for Figs. 10.4A–10.4E. × 130,000.

The contrast in the underfocused micrographs is considerably greater than in the overfocused ones, where membranes appear blurred. Although the underfocused micrographs (Figs. 10.4A and 10.4B) appear acceptable at a comparatively low magnification, the overfocused ones are much less, if at all, acceptable. At a higher magnification the underfocused micrograph in Fig. 10.4G has a good contrast, although a coarse granularity obscures fine details such as the triple-layered membranes. The granularity in Fig. 10.4G is less coarse than in Fig. 10.4F, but some fine structural details are still lost, as can be seen in comparison with the only minimally underfocused micrograph in Fig. 10.4H. In Fig. 10.4F the electron-dense regions consist of small dark dots, whereas in the overfocused micrographs (Fig. 10.4I and particularly Fig. 10.4J), the dense material appears as a dark network that encloses white dots. Furthermore, in Fig. 10.4J the plasma membrane shows a five-layered appearance (dark–light–dark–light–dark) due to overfocusing, whereas in Fig. 10.4H it has a triple-layered appearance (dark–light–dark).

COMMENTS

The overall contrast is accentuated at the underfocus side, although with a simultaneous slight decrease in the high resolution contribution. However, the slightly underfocused micrographs are usually preferred in studies at low or moderate magnifications. In fact, "optimum underfocus" is a term used to signify a degree of underfocus, where contrast is higher than in the in-focus position but visibility of some details is not or just barely decreased. Most microscopes today can be preset on Schertzer focus or any other focus level desired.

It is tempting to imagine that different types of subunits exist in the membranes when these are either underfocused or overfocused. Such patterns include small stained specks in underfocused micrographs or ring-like structures in overfocused ones. In Fig. 10.4H, which in comparison is only slightly underfocused, such "substructures" are largely absent. Instead there is a sprinkling of minute dots, which represent a precipitate from a suboptimal section staining. The smallest of the dense stain specks are approximately 2 nm in diameter, but the point-to-point resolution of this particular micrograph is probably not better than 3 nm.

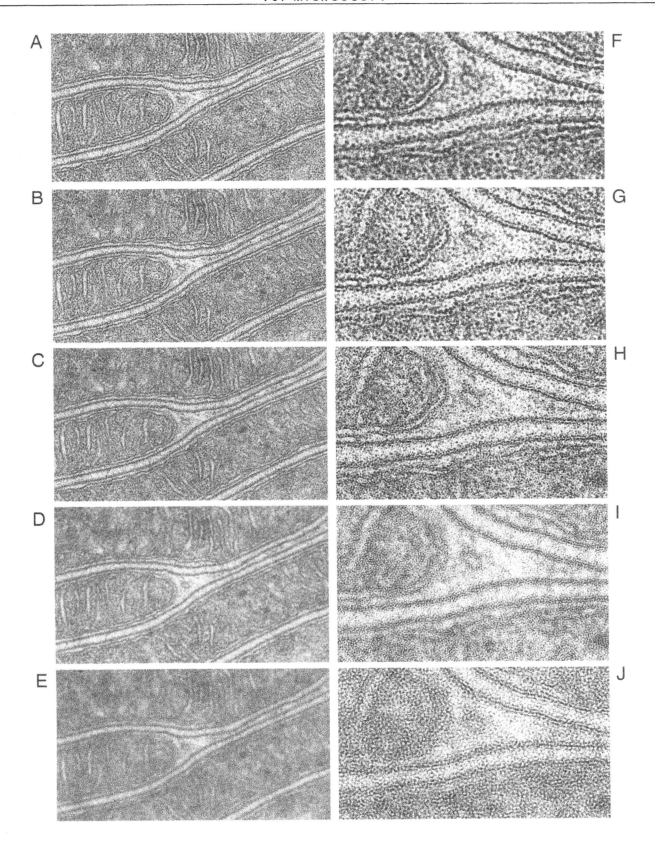

5. Minimum Contrast Focusing

FIGURES 10.5A–10.5C Parts of a renal proximal tubule fixed with glutaraldehyde, postfixed with osmium tetroxide, embedded in Epon, and section stained with uranyl acetate and lead citrate. The section was recorded in the microscope at an original magnification of × 2000 at three different focus levels from underfocused (Fig. 10.5A) to overfocused (Fig. 10.5C) and with a large (100 μm) objective aperture. × 6000.

FIGURES 10.5D–10.5F Small details in the micrographs shown in Figs. 10.5A–10.5C enlarged photographically 15 times. × 30,000.

The proper focus in Fig. 10.5B was, in this case, found by a dramatic loss in contrast that is seen at the in-focus value when changing the focus settings from underfocused to overfocused or from overfocused to underfocused. The correctness of the focusing could later be verified in the print by checking, at a higher magnification, the absence of Fresnel fringes at the rim of a hole (Fig. 10.5E).

In the greatly enlarged print of Fig. 10.5D, the small hole shows a distinct underfocus fringe and the cellular membranes stand out clearly, as do the ribosomes. No Fresnel fringes are visible in the in-focus micrograph (Fig. 10.5E), where all cellular details are poorly outlined and lack contrast. The overfocused micrograph, finally, (Fig. 10.5F) has distinct overfocus fringes in the hole and cell membranes appear double because of overfocusing.

COMMENTS

These micrographs illustrate that the focus level influences contrast and visibility of cytological details greatly, even at a low magnification. This is particularly true of micrographs taken with a large or no objective aperture. Unless the micrographs are slightly underfocused, cytological details are lost and contrast is poor. Because true focus is identified as the point of minimum contrast (and in this case also verified by the absence of Fresnel fringes), this method is usually referred to as the minimum contrast method for finding true focus. The operator should, however, defocus the microscope a certain small step toward the underfocus side in order to enhance contrast and get a useful micrograph.

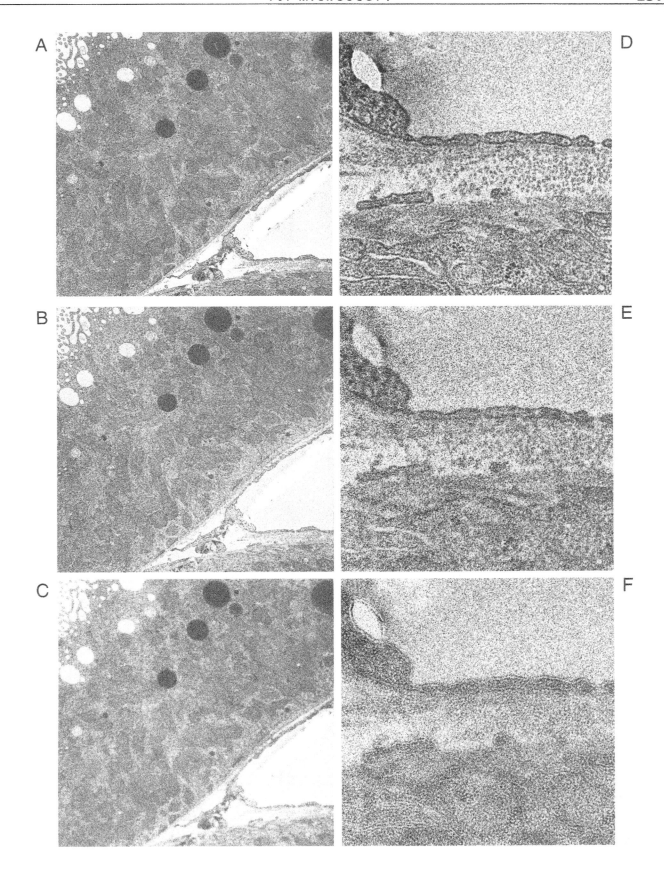

6. Wobbler Focusing

FIGURES 10.6A–10.6C Section of a liver cell focused with the aid of a wobbler. In Fig. 10.6A the wobbler was in operation and the objective lens out of focus. The image of the object was seen to wobble, i.e., it jumped back and forth. In Fig. 10.6B the wobbler was working, but the objective lens was adjusted to achieve an image where the wobbler movement was minimized. The wobbler was switched off at this position and a third micrograph recorded (Fig. 10.6C). × 90,000.

In Fig. 10.6A the image is blurred, particularly so in the vertical direction. Horizontally oriented membranes appear doubled. Such a doubling is not apparent in Fig. 10.6B, although the image is by far not as crisp as when the wobbler is switched off, as it is in Fig. 10.6C.

COMMENTS

The wobbler provides a simple, rapid, and efficient way to find a reproducible focus level in the microscope. It is particularly valuable in the low-to-medium magnification range. In some microscope models the focus level can be set at values other than true focus, e.g., to slight underfocus, which for most purposes will provide the best micrographs. In some microscopes the wobbler is automatically turned off at exposure.

A

B

C

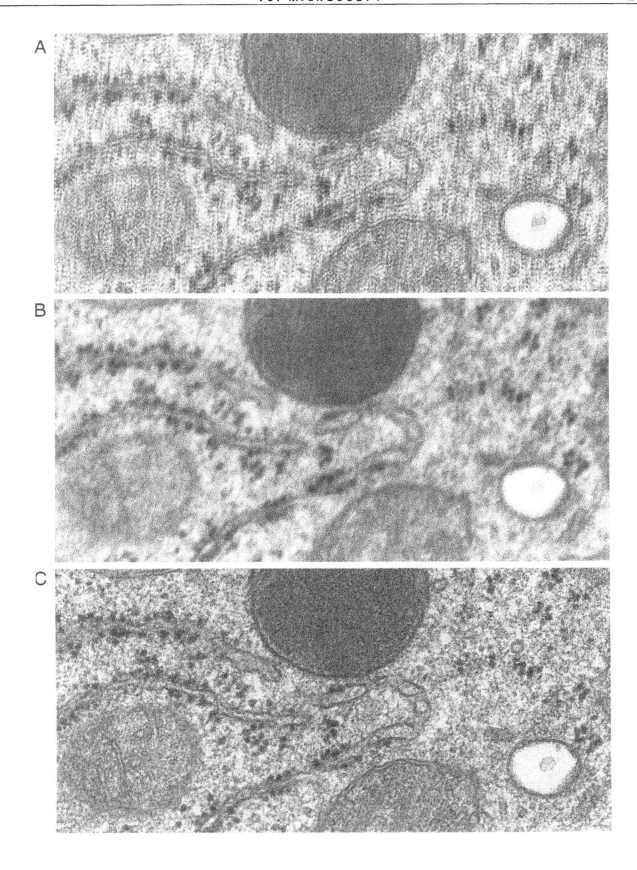

7. Accelerating Voltages 20–100 kV

FIGURES 10.7A–10.7E A thin section of a rat liver cell fixed with glutaraldehyde, postfixed with osmium tetroxide, and embedded in Epon. Sections were stained with uranyl acetate and lead citrate. Electron micrographs were taken with the same apertures at accelerating voltages of 100 kV (Fig. 10.7A), 80 kV (Fig. 10.7B), 60 kV (Fig. 10.7C), 40 kV (Fig. 10.7D), and 20 kV (Fig. 10.7E). The micrographs were taken in the mentioned order, the plates were developed together, and the micrographs were printed on the same grade photographic paper. × 12,000.

There is a gradual increase in contrast from the micrographs taken at 100 kV to the one taken at 20 kV. The contrast difference is apparent when comparing, within each micrograph, the density difference between the mitochondrial matrix and the cytoplasm, or within different regions of the nucleus. In Fig. 10.7A, and to some extent also in Fig. 10.7B, the contrast is so low that neither completely white nor completely black regions are present. However, the low voltage micrographs have black or white components but show intermediate grades less well.

COMMENTS

High resolution and low electron beam damage speak in favor of a high accelerating voltage, whereas high contrast speaks in favor of a low accelerating voltage. However, if the specimen is thick and the accelerating voltage low, chromatic aberration may degrade the image quality. In most cases, thin biological objects are therefore viewed at 60–100 kV. The loss in contrast at higher voltages can in part be compensated for by the use of hard photographic paper when reproducing the negatives or by image processing.

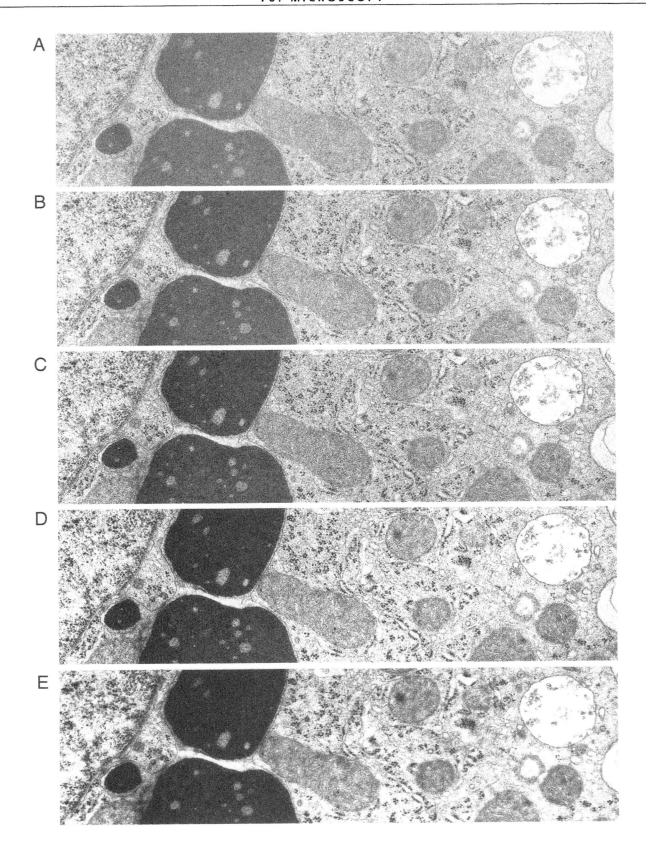

8. Accelerating Voltages 80–200 kV

FIGURES 10.8A–10.8C Electron micrographs of an approximately 0.5-μm-thick Epon section through a cat spermatozoon. The specimen was fixed with glutaraldehyde, postfixed with osmium tetroxide, embedded in Epon, and section stained with uranyl acetate and lead citrate. Micrographs were taken at accelerating voltages of 80 kV (Fig. 10.8A), 120 kV (Fig. 10.8B), and 200 kV (Fig. 10.8C), and the ensuing negatives were treated similarly and printed on the same grade photographic paper. × 75,000.

The three micrographs are rather similar, although the contrast is somewhat higher at the lower accelerating voltage than at the higher ones. There are no obvious differences with respect to fine structural details.

COMMENTS

High voltage electrons have a smaller scattering cross section in the specimen than those with lower voltages (and hence with lower speed); they can penetrate a thick specimen, which therefore will appear more transparent than the same specimen examined at lower voltages that are standard in biological electron microscopy (60–100 kV). High voltage microscopy can also give a higher resolution in thick or thin specimens, but this occurs at the cost of a decreased contrast. Another factor to be taken into consideration when working with thick specimens and high voltage electron microscopy is superposition effects; cytological details are projected on top of each other and may create a micrograph that is difficult to interpret.

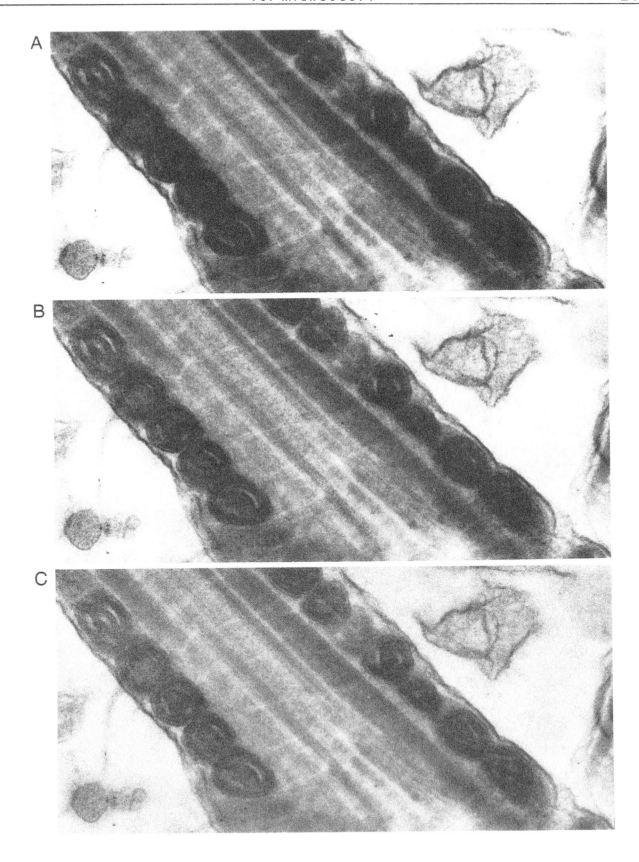

9. Unsaturated Electron Beam

FIGURE 10.9A Parts of epithelial cells from a rat kidney fixed and processed for electron microscopy by conventional methods. The micrograph was recorded with a focused double condenser but the electron beam was not fully saturated. × 90,000.

Because the electron beam is uneven the micrograph has become unevenly exposed and shows both overexposed and underexposed areas.

FIGURE 10.9B Electron micrograph of the same area as in Fig. 10.9A but exposed with a saturated and focused beam. × 90,000.

The electron beam has a small diameter and its intensity decreases from the center outward. The intensely illuminated area is symmetrical and has a diameter of slightly over 1 μm.

FIGURE 10.9C An electron micrograph of the same area after the condenser has been defocused and the illuminated area thereby increased in size. × 90,000.

The illustrated field and a large surrounding area are illuminated evenly.

COMMENTS

Modern electron microscopes are equipped with very efficient illumination systems, including an easily adjustable double condenser lens. Among the factors that influence the illumination are the current through the filament and the focus of the condenser. If the filament current is below the saturation point and the condenser lens is focused, the illuminated area of the specimen has an irregular pattern such as shown in Fig. 10.9A. Usually there is an illuminated elliptic field in which there are dark streaks and patches. With more filament current the electron beam levels off at a maximum brightness and the illuminated area becomes essentially even. It can be adjusted to a circular shape with the condenser stigmator.

The size of the illuminated area can be varied with the degree of defocus of the condenser system. It should be adjusted to a diameter at least exceeding the size of the specimen area to be recorded on the photographic material and further to the intensity required for a proper exposure time, which usually is in the order of 1–2 sec. Normally a slight undersaturation is used as it increases the lifetime of the filament and the coherence of the beam is improved. The small uneven illumination as observed with a focused beam disappears when the beam diameter is increased to cover the specimen area to be recorded.

A

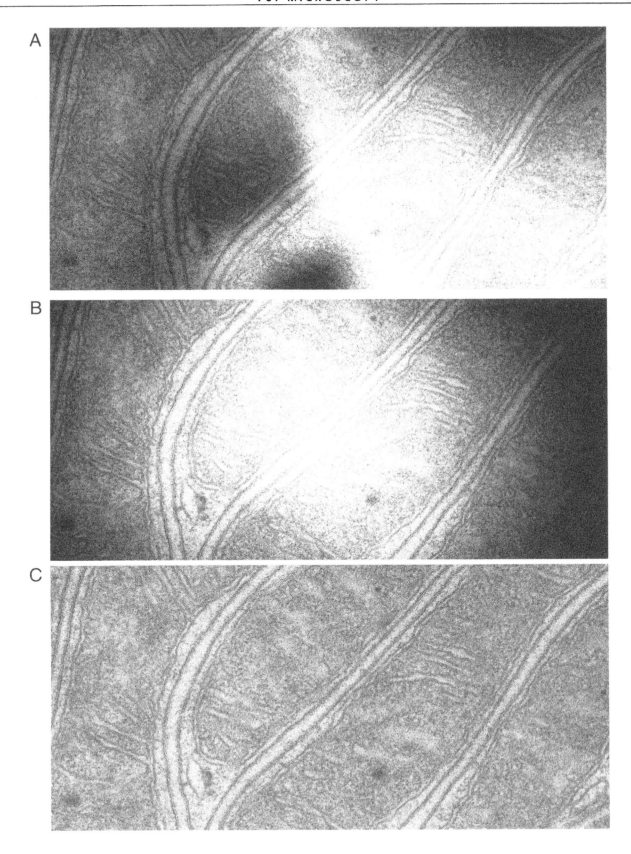

B

C

10. Condenser Apertures

FIGURES 10.10A–10.10D A rat liver cell fixed with glutaraldehyde, postfixed with osmium tetroxide, and Epon embedded, and the sections were stained with uranyl acetate and lead citrate. The micrographs were taken in an electron microscope operated at 80 kV without a condenser aperture (Fig. 10.10A), using a 400-μm condenser aperture (Fig. 10.10B), a 300-μm condenser aperture (Fig. 10.10C), and a 200-μm condenser aperture (Fig. 10.10D). The objective aperture in all cases had a diameter of 60 μm. The exposures were made with the same beam intensity, the micrographs developed together, and the enlargements printed on the same grade photographic paper. × 20,000.

> The appearance of the image is the same without an aperture or with condenser apertures of different sizes.

COMMENTS

For sectioned biological specimens at low magnifications (<× 20,000) no appreciable gain in contrast can be obtained using a small condenser aperture. The intensity of the beam is higher with a wide aperture or without an aperture, which sometimes is useful when studying electron-dense specimens. It is not recommended, however, to operate old microscope models without a condenser aperture, as it cannot be excluded that the level of parasitic X-rays may be a health hazard. At high magnifications (>× 30,000) it is necessary to use an intense illumination and a large condenser aperture, although this involves risks of thermal drift and radiation damage of the specimen. However, the condenser aperture also affects the coherence of the electron beam and the resolution is improved by selecting a small condenser aperture. This is important in high-resolution imaging and electron diffraction from crystalline specimens.

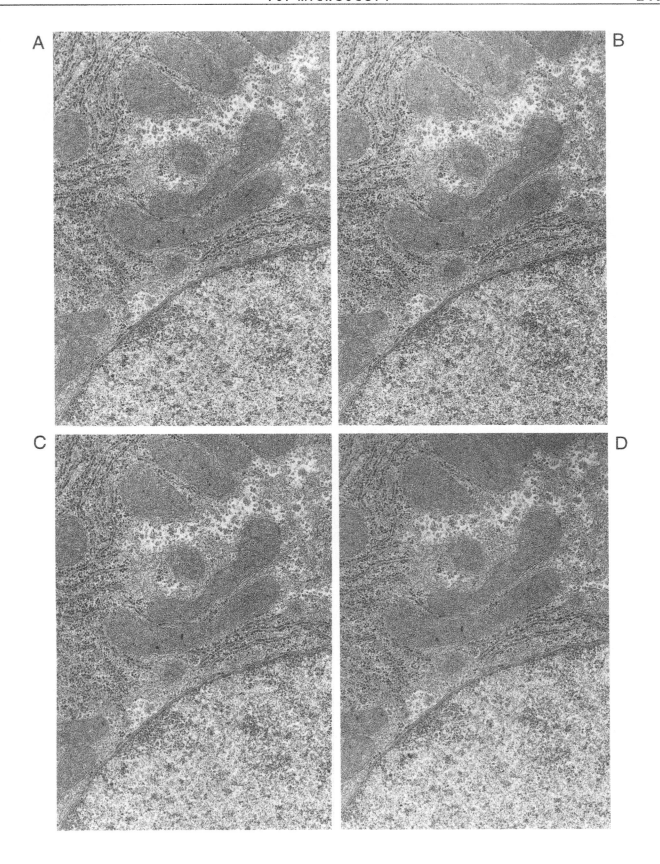

11. Objective Aperture

FIGURES 10.11A–10.11C Parts of distal tubule cells from a rat kidney cortex, fixed with 2.5% glutaraldehyde and 2% formaldehyde in 0.1 M cacodylate buffer, and postfixed with a solution containing 1% osmium tetroxide and 1% potassium ferrocyanide. The tissue was embedded in Epon and the sections were stained with uranyl acetate and lead citrate. The microscope was operated at an accelerating voltage of 80 kV without an objective aperture (Fig. 10.11A), with an objective aperture of 120 μm (Fig. 10.11B) and 20 μm (Fig. 10.11C). Other parameters were kept identical. × 75,000.

There is a considerable gain in contrast when going from no aperture to objective apertures of decreasing sizes.

COMMENTS

The use of a small objective aperture is a simple means to gain contrast in the image. However, a small aperture will reduce the field of view, give rise to diffraction phenomena, and is more likely to introduce astigmatism than a large one. The thin foil apertures used today, however, are essentially self-cleaning in order to minimize the risk of a buildup of an astigmatism-inducing layer.

For maximum resolution the objective aperture should be large or entirely removed. For normal use the size of the aperture is adjusted to the focal length and the spherical aberation of the objective lens and is usually around 40–50 μm.

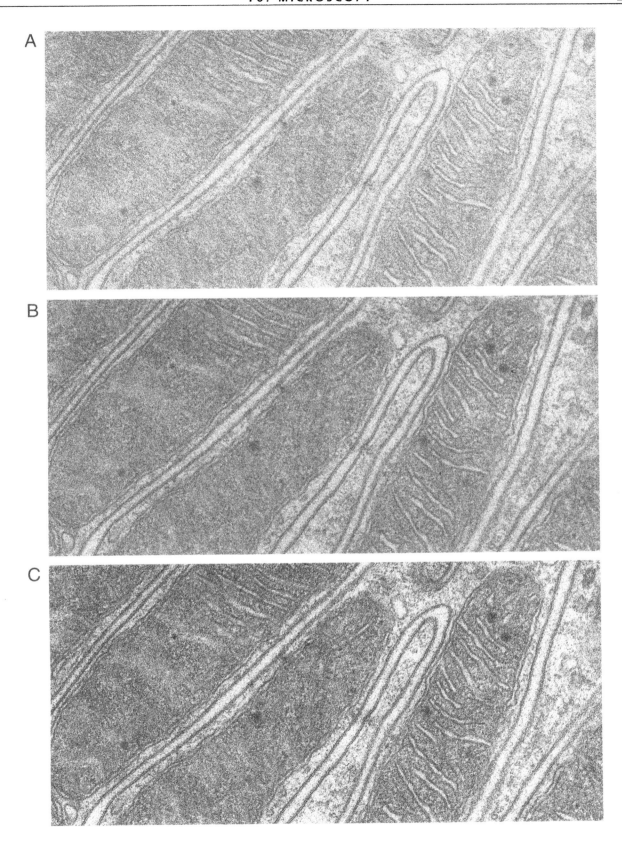

12. Through-Focus Series: Astigmatism

FIGURES 10.12A–10.12J Through-focus series of the same hole and latex particle as shown in Figs. 10.2A–10.2J, but recorded with an astigmatic objective lens. The astigmatism was introduced in the lens deliberately by changed settings of the stigmator. The focal increment steps are the same as in Fig. 10.2. × 260,000.

In all micrographs the Fresnel fringes are more prominent in one direction than in the perpendicular one. Thus in Figs. 10.12D and 10.12I the object is close to the focus in the vertical direction, whereas it is considerably overfocused in the horizontal direction. Figures 10.12C and 10.12H are simultaneously underfocused and overfocused in the two directions. The supporting carbon film shows a faint striated pattern, which changes direction at a certain level of focus and where the degree of linearity is dependent on the focus level.

COMMENTS

Fresnel fringes are visible at each focal level and differ in width depending on the distance from the closest focus level. The astigmatism of the microscope is eliminated when the width of the fringes is uniform around the circumference of the hole or electron-dense particle. Inspection of Fresnel fringes is therefore a very useful method for determining the instrumental performance, as first suggested by Haine and Mulvey (1954). For this type of test, the greatest sensitivity is obtained when the fringe is as thin as possible, although still present around the entire hole or particle. The width of the fringe, when seen all around the object, approximates the resolution of the microscope. Because stray fields, caused for instance by contamination, may build up and influence lens properties, the presence or absence of astigmatism should be checked frequently at high magnification work. It is often advantageous to correct the astigmatism directly on the supporting film. In this case the stigmator is adjusted gradually until a linear pattern is exchanged for a granular pattern.

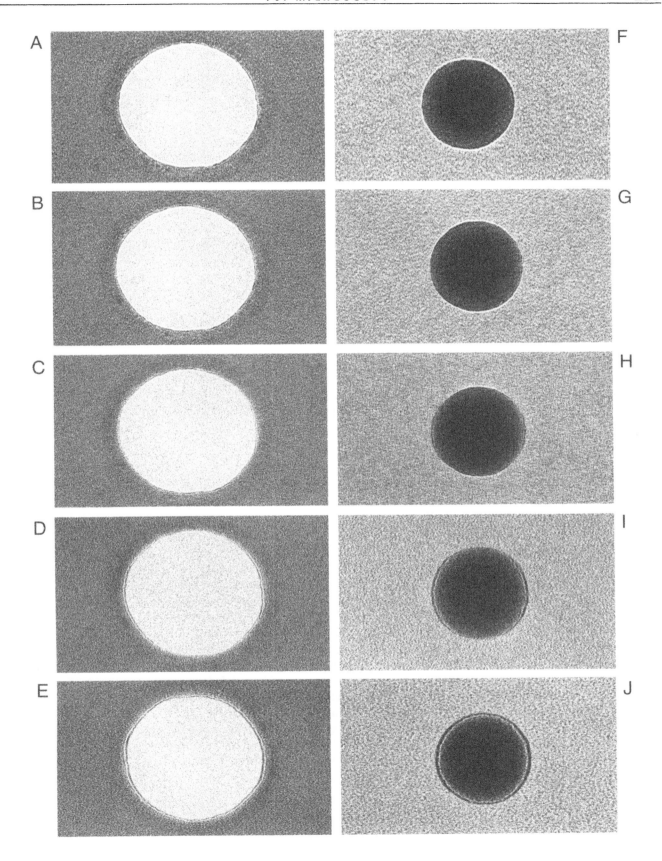

13. Image Distortion

FIGURE 10.13A Micrograph of a diffraction grating replica with two nearly perpendicular line systems having 960 lines per millimeter. The picture was taken with a microscope that is essentially distortion free at this magnification. Electron optical magnification $\times$ 1000; final magnification $\times$ 3000.

The lines of the grating replica are parallel throughout the micrograph. Two straight lines, perpendicular to each other, have been drawn on the micrograph for comparison.

FIGURE 10.13B Micrograph of the same specimen as in Fig. 10.13A. The micrograph was taken in a microscope that shows some distortion. $\times$ 3000.

The straight parallel lines of the replica curve slightly toward their left side when going from the center to the periphery. The lines are thus imaged as sigmoid. The straight lines drawn on the micrograph serve as reference lines.

FIGURES 10.13C AND 10.13D Micrographs of the same grating replica taken in a Siemens Elmiskop I with lens settings that are not recommended by the manufacturer, as they give distorted images. The micrographs were taken above and below the point of diffraction of the intermediate lens, respectively. In the center of Fig. 10.13C the magnification is about 2500 times and in the periphery about 3000 times, as calculated by the spacings of the replica. In the center of Fig. 10.13D the magnification is about 3000 times and in the periphery about 2700 times.

In Fig. 10.13C grating lines are curved in such a way that they appear concave, whereas in Fig. 10.13D they appear convex as related to the center (to the left in the figures).

COMMENTS

The distortion shown in Fig. 10.13B is usually called sigmoid distortion, sometimes called anisotropic or spiral distortion. It arises as a consequence of the spiral trajectory by electrons in electromagnetic lenses. The sigmoid distortion is more distinct at an even lower magnification than illustrated here. Some old microscopes show this distortion when operated at low magnification. If photo montages are made of micrographs showing distortion, the investigator will be frustrated by the mismatch at the edges. The imaging defects in Figs. 10.13C and 10.13D are called pincushion and barrel distortions, respectively. The squares of a grating grid will show concave or convex sides, and it is impossible to give an accurate magnification valid for all parts of the viewing field, which has obvious consequences for the direct measurement of objects. Although older types of transmission electron microscopes usually showed some degree of image distortion, particularly at low magnifications, it should be emphasized that these defects are rare in modern, well-adjusted instruments.

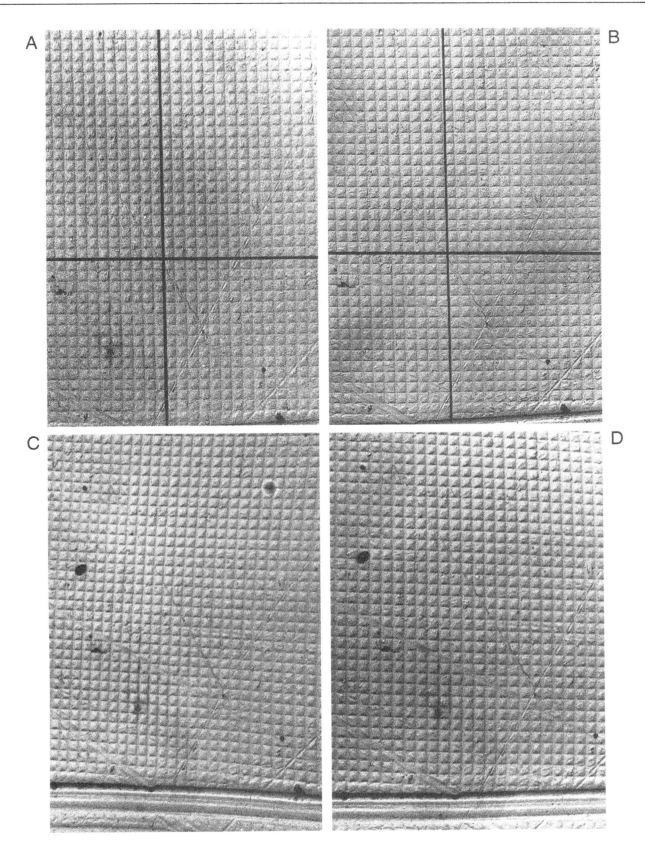

14. Chromatic Aberration

FIGURE 10.14A Part of a rat liver cell fixed with glutaraldehyde followed by osmium tetroxide and embedded in Epon. The microtome was set at a sectioning thickness of 60 nm. The section was double stained with uranyl acetate and lead citrate, and the micrograph was taken at an accelerating voltage of 40 kV. × 15,000.

At low magnification this image appears sharp.

FIGURES 10.14B–10.14D Enlargements of different areas of Fig. 10.14A. Close inspection of the negative shown in Fig. 10.14A reveals a gradient in sharpness of the image from center to periphery. × 60,000.

The image is sharp in the center of the enlarged film (Fig. 10.14D), but all structures appear blurred near the corners (Figs. 10.14B and 10.14C). The deterioration of image quality in the periphery is not an effect of the photographic enlargement as the electron micrograph negative was placed in the enlarger in such a way that the enlarged areas were close to the optical axis of the enlarger.

COMMENTS

The blurring of the image at some distance from the optical axis in the electron microscope is an electron optical defect caused by chromatic aberration. Because of the low accelerating voltage, the electrons became retarded in the section, thereby increasing their wavelength and becoming more disperse in energy. Such electrons cannot be focused accurately, and the image becomes blurred except close to the optical axis. The effect of chromatic aberration is dependent on the focal length of the objective lens and on its design and differs therefore between microscope models. Furthermore, if the microscope is used at 60 kV or lower, is poorly aligned, or if the section is thick, this aberration is accentuated. In fact, chromatic aberration is probably more often than not the source of unsharp micrographs, as routine "thin" sections in many laboratories tend to be too thick.

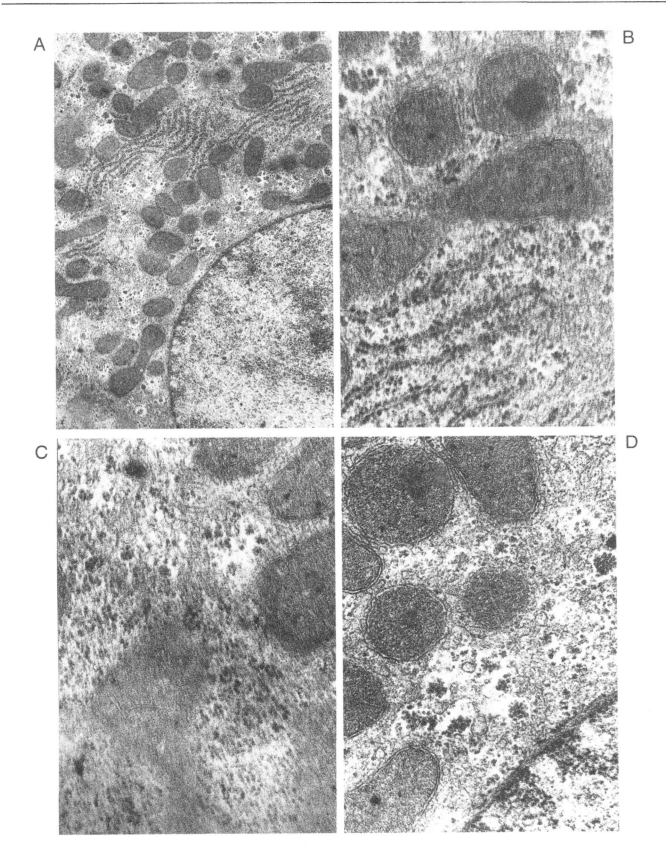

15. Mechanical Instability

FIGURE 10.15A The same field of distal tubule cells as in Fig. 10.11. The micrograph was taken in slight underfocus under optimal conditions. × 75,000.

The micrograph appears sharp, and plasma membranes limiting the lateral intercellular spaces and membranes of the mitochondria appear well defined when oriented approximately at a right angle to the section.

FIGURE 10.15B Electron micrograph of the same field as shown in Fig. 10.15A. Immediately before exposure, which lasted for 1 sec, the microscope column was tapped with a hand, which created movements of the column for a short period. The focus setting was the same as for the exposure used in Fig. 10.15A. × 75,000.

The general appearance of the image is the same as in Fig. 10.15A except that the membranes are unsharp.

FIGURE 10.15C Enlargement from the same negative as in Fig. 10.15A. × 200,000.

Even at a higher magnification the image appears sharp. The small, dense particles sprinkled over the image are attributed to section staining.

FIGURE 10.15D Enlargement from the same negative as in Fig. 10.15B. × 200,000.

At the higher magnification, the membranes almost completely lack their characteristic triple-layered appearance.

COMMENTS

These micrographs illustrate that blurred images may have a mechanical origin. In Figs. 10.15B and 10.15D the blurring was induced intentionally but the same effect can be seen with some types of microscopes, e.g., if the operator leans on the microscope table or unintentionally touches the microscope base. Another possible source of mechanical instability may be vibrations of the floor in the microscope room caused by a nearby elevator, air-conditioning equipment, or large, vibrating instruments in adjacent rooms.

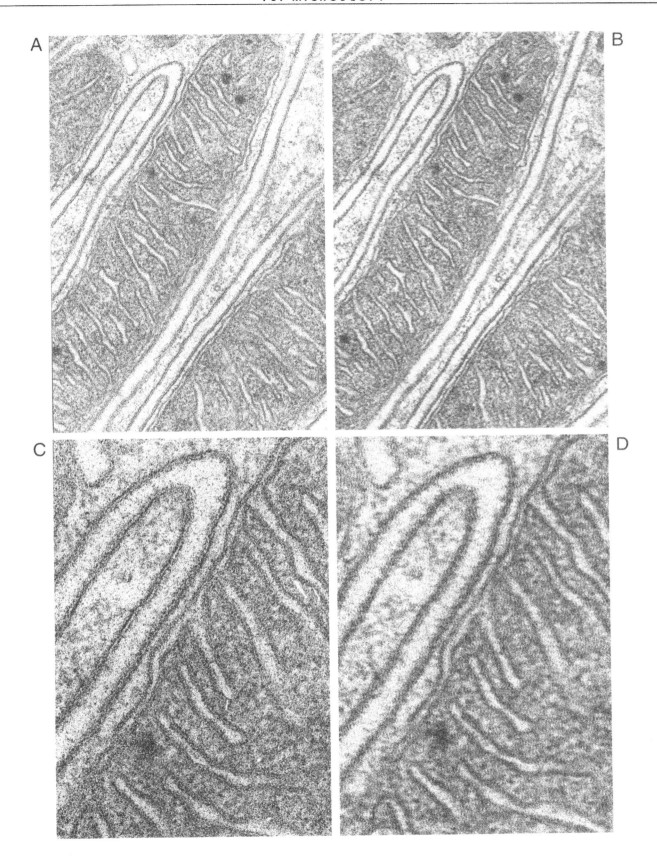

16. Specimen Drift versus Astigmatism

FIGURES 10.16A–10.16D Electron micrographs of the same distal tubule cell as in Fig. 10.11. In Fig. 10.16A the exposure lasted for 16 sec. The inset is a Fourier transform of the electron micrograph negative recorded in an analog optical diffractometer according to Johansen (1975). In Fig. 10.16B the exposure lasted for 1 sec and was made immediately after the specimen had been moved a short distance with the specimen stage controls. The inset is a Fourier transform of the electron micrograph. × 200,000.

Electron-dense regions appear to consist of fine lines, which are elongated in a direction approximately parallel to the diagonal upper left–lower right. The elongated spots in this Fourier transform are oriented at a right angle to the elongated fine lines in the micrograph. In Fig. 10.16B cell membranes appear sharp when oriented approximately parallel to the diagonal lower left–upper right, whereas inner mitochondrial membranes oriented approximately at right angles to this direction appear blurred. In the Fourier transform the elongated spots are oriented at a right angle to the small dense lines in the micrograph.

FIGURES 10.16C AND 10.16D These micrographs were obtained after intentionally changing the settings of the objective lens stigmator. The direction of the resulting astigmatism was set to be approximately parallel to either of the two diagonals of the micrographs. × 200,000

No membranes appear distinctly triple layered in these two micrographs. Linearities oriented approximately diagonally upper left–lower right in Fig. 10.16C and lower left–upper right in Fig. 10.16D are visible. The Fourier transforms show Maltesean crosses, which are typical for Fourier transforms of astigmatic micrographs.

COMMENTS

Specimen drift is one of the most common defects in electron micrographs. It results in a unidirectional blurring of the specimen. It may be mistaken for astigmatism, but a distinction is possible when observing the specimen in the electron microscope: Drift can be observed directly as a movement of the image. Astigmatism can be detected with the aid of holes, dense spheres, or background granularity. It is less easy to differentiate between these two types of defects in already recorded electron micrographs, but focus fringes, if they occur, are revealing. The best way to determine whether the blurring of the image is due to specimen movement or astigmatism is to examine the negatives in an optical diffractometer or by computed-aided image analysis.

The blurred image in Fig. 10.16A is due to a slow specimen drift during the long exposure. Consequently, all specimen points are elongated. To avoid this effect it is preferable to use short exposure times and to make sure that the specimen drift is absent or very small. The easiest way to ensure that the specimen is stable is to observe a specimen point located very close to an identifiable point on the screen, such as a small dust particle. If the specimen is stable, there will be no movement of the specimen point relative to the fix point on the screen when observed over a period of several seconds.

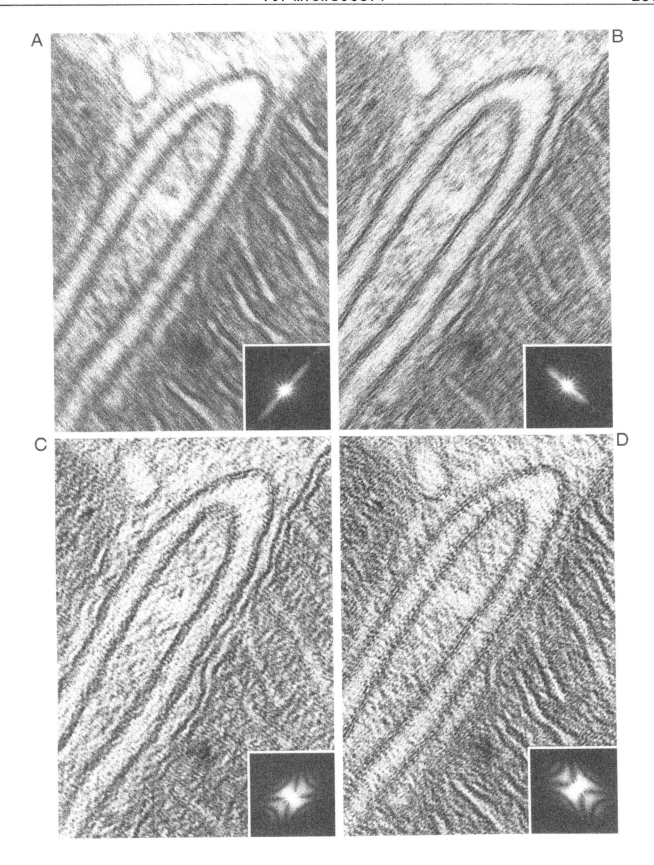

17. Focus Drift

FIGURES 10.17A–10.17E Electron micrographs of a hole in a thin carbon film recorded at an electron optical magnification of 100,000 and enlarged photographically eight times. Five consecutive micrographs were taken with the same focus settings at 1-min intervals. × 800,000.

The focus changed during the 4 min that elapsed between the first and the fifth exposure, from slightly overfocused (Fig. 10.17A) to slightly underfocused (Fig. 10.17E) with a close to focus recording in Fig. 10.17C. Note also that the granularity of the carbon film is minimal close to focus.

COMMENTS

The gradual change of the focus level illustrated here was caused by an instability of the lens current, which showed slow changes with time. Such an electrical defect makes it impossible to make well-focused micrographs at high magnifications in a reproducible way. Sometimes focus drift can be detected with the eye, but for detecting slow drift it may be necessary to take consecutive micrographs with intervals of one or several minutes. In general it is a good habit to make recordings immediately after having focused the specimen in order to avoid the effects of slow focus drift as well as radiation damage and contamination of the specimen. Using an optical diffractometer or a computer-aided image analysis, the focus drift in angstroms per second can be calculated.

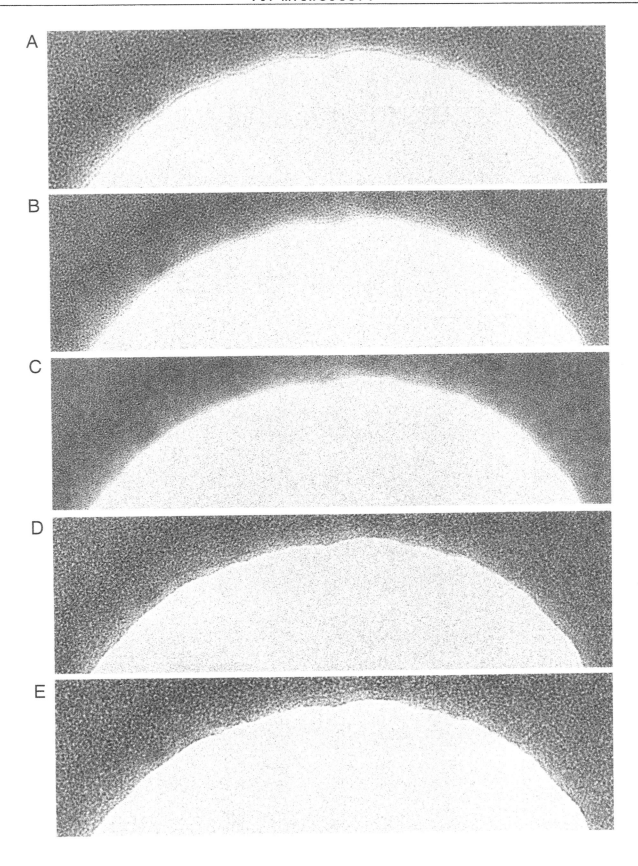

18. Electrical Instabilities

FIGURE 10.18A Part of renal epithelial cells fixed and embedded for electron microscopy by conventional methods. The ultrathin sections were supported by a carbon-coated Formvar film and were stained with lead citrate. Exposure time in the microscope was approximately 1 sec. × 20,000.

At low magnification this micrograph is somewhat blurred, although some membrane profiles appear sharp.

FIGURE 10.18B Higher magnification of part of Fig. 10.18A. × 60,000.

At high magnification vesicular profiles in the cytoplasm appear limited by a single membrane in the vertical direction, whereas all membrane structures running in the horizontal direction appear double.

FIGURE 10.18C A rat liver cell with conventional preparation. × 25,000.

Cytological details appear sharp throughout the micrograph. Two bright sectors can be seen radiating from a point near the center of the micrograph and have their brightest portions near the center. Cellular structures within the bright sectors are displayed with the same sharpness as in other areas.

FIGURE 10.18D Ultrathin section of a seminiferous tubule from a monkey testis that was conventionally fixed and embedded for electron microscopy. Exposure time in the electron microscope was 2 sec. × 4,000.

The cellular structures are well defined and appear sharp throughout. However, over the entire micrograph there is a superimposed faint pattern of radiating light and dark streaks.

COMMENTS

Brief instabilities in the image-forming system of the electron microscope may be caused by charging–discharging phenomena. It is often difficult to determine the cause of the instability. Contamination of the column may cause the image to move slowly in one direction and then to "jump" back. Repeated "jumps" were noted at the occasion when this micrograph was taken. The double image in Figs. 10.18A and 10.18B is hence interpreted as an image jump during exposure.

The precise causes of the two radiating patterns in Figs. 10.18C and 10.18D are difficult to understand. Voltage flashes were noted during the period in which the micrograph in Fig. 10.18D was taken. It is possible that the convergence center of the radial streaking is the voltage center for the microscope and that an instantaneous voltage change has caused the streaks. Another possibility is that these two patterns are caustic effects from the condenser lenses. The remedy for both disturbances is to clean the column thoroughly, as contamination may cause these high voltage discharges, and to align the microscope carefully.

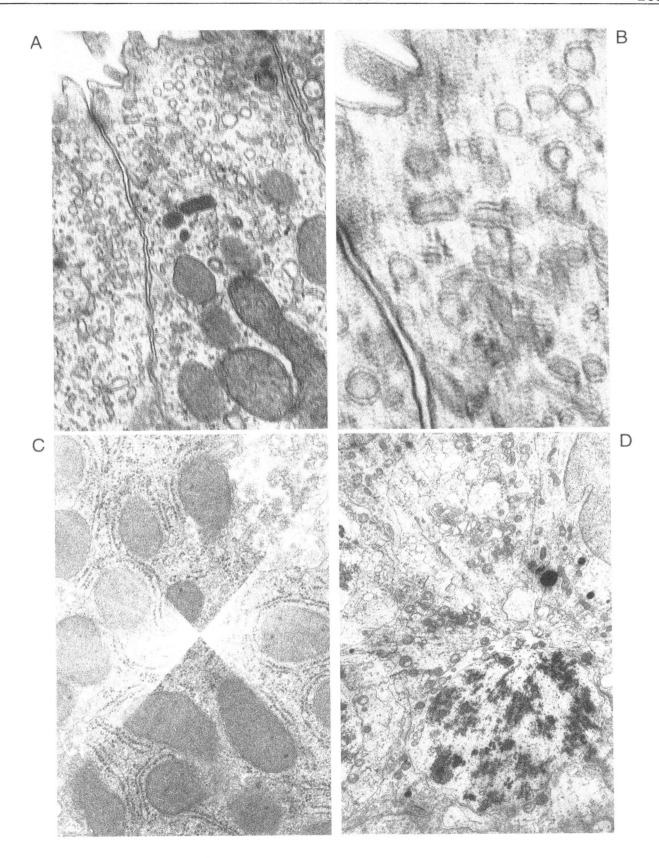

19. Contamination in the Electron Beam

FIGURES 10.19A–10.19C Electron micrographs of latex particles, 0.088 μm in diameter, lying on a Formvar support film and examined at 60 kV. The first micrograph (Fig. 10.19A) was taken shortly after this particular part of the object had been introduced into the electron beam. The same field then remained in the focused electron beam for 25 min, whereupon a new micrograph was made (Fig. 10.19B), and then for another 35 min, when a third micrograph (Fig. 10.19C) was taken. × 90,000.

It can be seen that the diameter of the latex particles has increased gradually while remaining in the electron beam. By measuring the increase in diameter of the latex particles as a function of time, the rate of contamination can be calculated. In this case contamination had a low value, about 2 Å per minute.

COMMENTS

The effect demonstated here is due to the presence of small amounts of contaminating hydrocarbon vapors in the column of the microscope. These may be derived from pump oil, photographic plates, the specimen itself, including its embedding medium, or substances associated with the inner wall of the lenses. Contamination is of particular concern when the operator aims at high-resolution studies. If an anticontamination device, commonly referred to as a "cold finger," or a microscope with very effective pumps is used, contamination is less severe and may be totally absent or even changed into its opposite effect, namely "etching" (see Figs. 10.20 and 10.22).

A

B

C

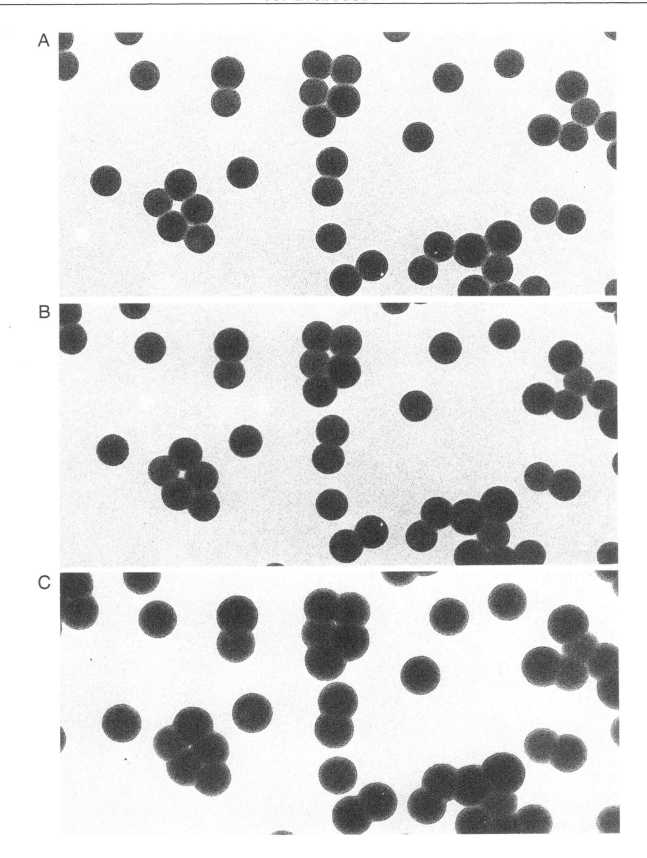

20. Radiation Damage

FIGURE 10.20A Part of an Epon section on a Formvar film imaged with a defocused condenser lense system. × 2200.

The section is evenly transparent over the entire field except for small contaminating particles and holes.

FIGURE 10.20B Same Epon section recorded with a defocused condenser but only after the operator had focused the condenser to a spot (diameter about 2 μm) and moved the spot around relative to the specimen. The photographic print was given a somewhat longer exposure than that of Fig. 10.20A in order to demonstrate the effects of the electron beam. × 2200.

The section is cleared up at sites exposed to the focused electron beam.

FIGURE 10.20C Ferritin molecules on a very thin carbon film supported by a micronet ("holey film"). The film is estimated to be less than 3 nm thick. The micrograph was taken shortly after introducing this field into the electron beam. The illumination of the specimen sufficed for direct observation at a magnification of × 100,000.

The ferritin molecules are distributed on an intact carbon film without holes.

FIGURE 10.20D Same field as in Fig. 10.20C but the micrograph was taken about 1 min later. × 100,000.

The same ferritin molecules can be identified but their relative locations are slightly changed. In addition, there are irregular holes in the supporting carbon film.

COMMENTS

The electron beam may ruin the biological objects at the molecular level, transforming at least part of them to what has been referred to by Kellenberger (1986) as "heaps of ashes." These electron micrographs illustrate that the electron beam may also interfere dramatically with the specimen embedding medium and support film. In Fig. 10.20B the electron beam has heated the specimen to the extent that parts of the Epon have undergone mass loss, thus making parts of the specimen more electron translucent. The same process occurs when ultrathin sections of biological specimens are exposed to the electron beam and seem to clear up and acquire more contrast. If the specimen consists of ordered structures there is a fading of the electron diffraction pattern and in stained specimens there may be migration of stain. The mass loss effect is most pronounced with different methacrylates. It is also visible in epoxy resins, but is somewhat smaller in Vestopal.

The very thin carbon film in Fig. 10.20D has been etched to such an extent that holes appear and ferritin molecules have been moved relative to each other. Although this effect may not be apparent when using a normal carbon film, it illustrates the damage that can be caused by high beam intensity.

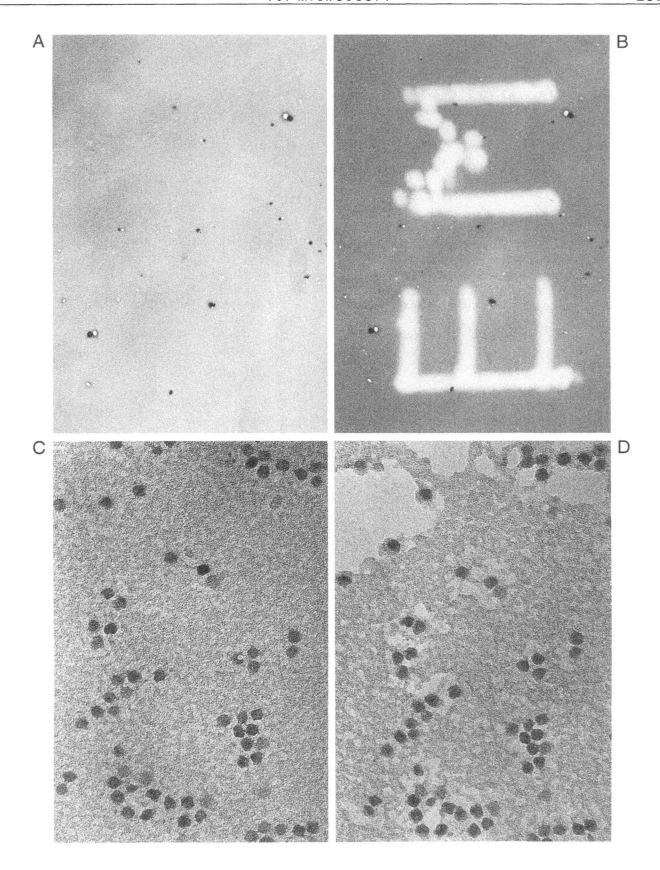

21. Radiation Damage and Contamination

FIGURES 10.21A AND 10.21C A kidney epithelium fixed with glutaraldehyde and osmium tetroxide and embedded in Epon. The section was stained with uranyl acetate and lead citrate. The micrographs were taken at 40 kV and with a defocused condenser. $\times$ 17,000 and $\times$ 60,000.

Both micrographs have been illuminated evenly and show normal contrast.

FIGURES 10.21B AND 10.21D After the recording of Figs. 10.21A and 10.21B the condenser was focused to minimum spot size and allowed to irradiate the specimen for 20 min. No anticontamination device was used. The condenser was then defocused again and Figs. 10.21B and 10.21D were recorded. $\times$ 17,000 and $\times$ 60,000.

The location of the focused beam is seen as a light spot in the specimen and is surrounded by a dark zone, the so-called doughnut effect. The remaining part of the specimen is unchanged in comparison with Fig. 10.21A. In Fig. 10.21D a physical loss of structural components can be seen in the center, whereas in the surrounding dense area the specimen contrast appears decreased.

COMMENTS

Two different defects of a prolonged irradiation of the specimen are shown here. In the center, where the beam was focused, material is removed from the specimen, causing deformation of structural details. This phenomenon is called etching, beam damage, or stripping. It may cause small holes in the irradiated part of the specimen. The radiation dose applied here is rather extreme. For more sensitive specimens, such as frozen-hydrated samples, similar effects are already seen after a few seconds exposure to electrons.

The part of the specimen surrounding the etched area is heavily contaminated with a consequent decrease in specimen contrast. If the vacuum is poor or the microscope column dirty the contamination will be increased further. Conversely, etching would have been the main or only irradiation effect had an anticontamination device ("cold finger") been used.

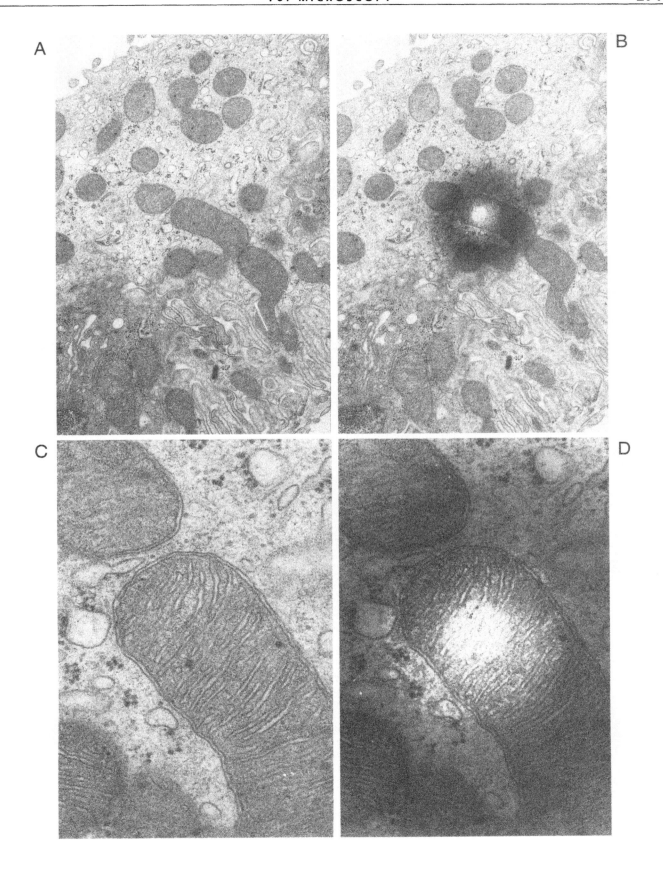

22. Low-Dose Exposure

FIGURE 10.22A A proximal tubule epithelium of a rat kidney perfusion fixed with 4% formaldehyde, cryoprotected in 2.3 *M* sucrose, and frozen in liquid nitrogen. The tissue was freeze-substituted in methanol/0.5% uranyl acetate and low temperature embedded in Lowicryl HM 20. The section was stained only with uranyl acetate. The illustrated area was recorded at 100 kV in a Zeiss 912 Omega electron microscope under low dose conditions with a beam intensity of 83 electrons/nm^2·sec. × 35,000.

Cellular details such as membranes and mitochondria have a low contrast. Ribosomes are barely visible.

FIGURE 10.22B A low magnification image from the same section as in Fig. 10.22A recorded with low beam intensity. The area in Fig. 10.22A (rectangle) is located within the circular area LD, which was recorded with the low dose. Focusing was done on the adjacent circular area F. The circumferences of the circular areas F and LD intersect at the asterisks. × 2500.

In area F, which was exposed to an intense electron beam during focusing, the density of the section is low. Area LD was only exposed to electrons during the recording of Fig. 10.22A but has a lower density than the surrounding and previously completely unexposed area of the section.

FIGURES 10.22C AND 10.22D After additional viewing of area LD with a comparatively high beam intensity for about 1 min a new image was recorded of the boxed area in Fig. 10.22A followed by a survey image (Fig. 10.22D) of areas F and LD. × 35,000 and × 2500 respectively.

After this prolonged exposure to the electron beam there is an increased contrast of the biological structures, e.g., the ribosomes have become visible (Fig. 10.22C). The difference between the electron density of cells in area LD and in the surrounding tissue section is considerably greater than in Fig. 10.22B.

COMMENTS

There is a difference in electron density and contrast of areas exposed during minimum dose conditions (Fig. 10.22A) and the same area exposed to a much larger electron dose (Fig. 10.22C). It is apparent that the first exposure to electrons (Fig. 10.22A) already results in etching of material from the section and a consequently greater transparency to electrons. After additional exposure of area LD, with the purpose of focusing the image to record Fig. 10.22C, additional material evaporates from the section, thus causing area LD to appear even more transparent than the surrounding parts of the section and similar to area F, which was used for focusing for Fig. 10.22A. This experiment demonstrates that the embedding medium starts to sublimate immediately when the section is exposed to the electron beam, even when the dose is low. Thereby the biological structures appear with greater contrast in the section but at the same time begin to undergo radiation damage. The effect can be seen with all embedding media, although to somewhat variable degrees, and with methylcellulose used for preserving cryosections during immunocytochemistry.

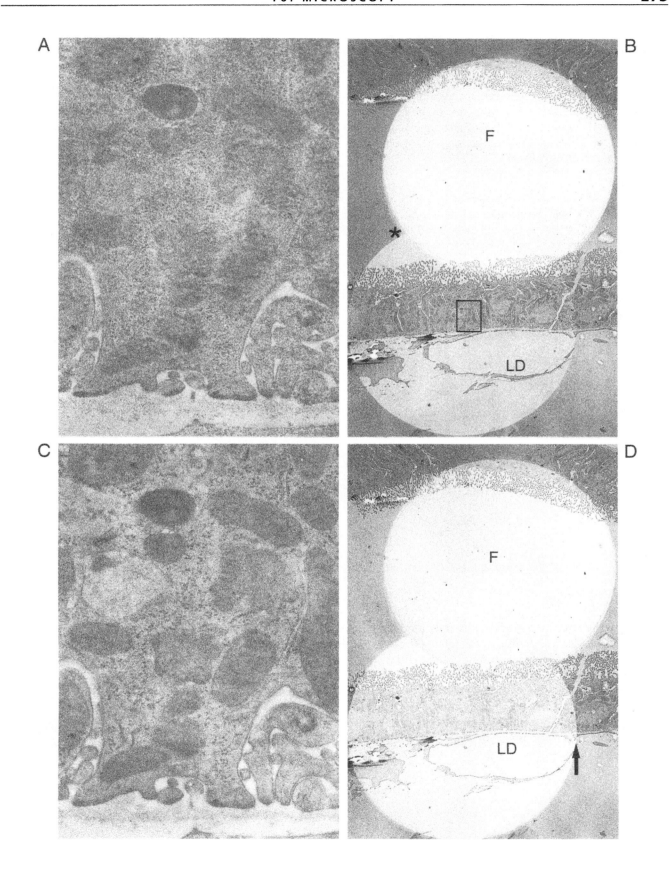

23. Spectroscopic Imaging: Thin Film

FIGURE 10.23A Holey film prepared by evaporation of carbon on mica. The film was imaged at 100 kV in a Zeiss 912 Omega electron microscope with global imaging, i.e., utilizing both elastically and inelastically scattered electrons. × 15,000.

The micrograph shows rounded holes in the support film.

FIGURE 10.23B Same area as in Fig. 10.23A imaged after exclusion of the inelastically scattered electrons by means of a 15 eV filter slit in the microscope. × 15,000.

The image is essentially identical to that in Fig. 10.23A, except that some background structures with circular contours appear slightly more visible.

FIGURE 10.23C Same area imaged with inelastically scattered electrons of the plasmon peak at 25 ± 7.5 eV. × 15,000.

As compared with the two previous pictures the image is reversed—all electrons that reach the photographic emulsion derive from the carbon film. Electrons that pass through the holes are filtered away and do not expose the photographic film.

FIGURE 10.23D Same area imaged with inelastically scattered electrons with the spectrometer set at 120 ± 7.5 eV. × 15,000.

As compared with the image in Fig. 10.23C obtained by plasmon electrons, the image obtained from the carbon film is reduced greatly in intensity. The same structural details are visible as in Fig. 10.23B, but with inverted contrast. Similar to Fig. 10.23C, there are no electrons derived from the holes.

COMMENTS

Because the specimen is very thin, only a minor portion of electrons are inelastically scattered. Therefore, there is essentially no difference in sharpness between Figs. 10.23A and 10.23B and only very minute contrast differences in the support film. When the image is formed by inelastically scattered electrons of the plasmon peak, the only electrons that reach the photographic film are those from the carbon film itself; the holes therefore become completely dark. The contrast differences are diminished in Fig. 10.23D because at 120 eV the carbon film will scatter electrons less than at 25 eV. A longer exposure time (or higher beam intensity) is hence required to make the exposure.

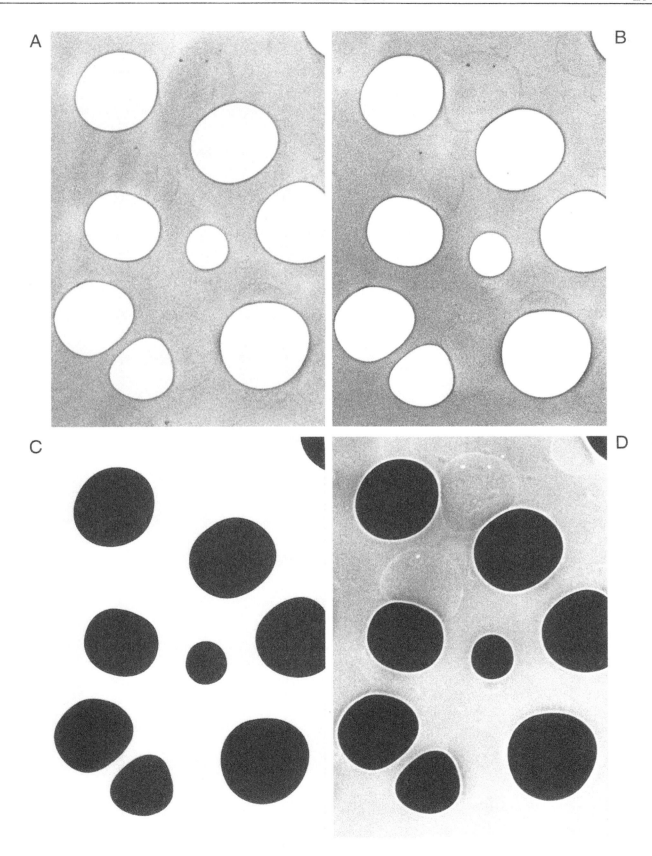

24. Spectroscopic Imaging: Thick Section

FIGURE 10.24A Thick section (1 μm) from a rat kidney cortex showing part of glomerular capillaries. Conventional glutaraldehyde and osmium tetroxide fixation and *en bloc* staining with uranyl acetate were followed by Epon embedding. The section was imaged in a Zeiss 912 Omega electron microscope at 100 kV using global imaging, i.e., utilizing all scattered electrons. × 11,000.

The capillary walls have soft contours and appear somewhat fuzzy, as does the contour of the erythrocyte on the right side of the figure.

FIGURE 10.24B Same field as in Fig. 10.24A but after insertion of the Omega filter slit aperture (15 eV) that removes the inelastically scattered electrons. × 11,000.

Both contrast and resolution are somewhat improved as seen from the contour of the erythrocyte and the capillary wall at the arrow.

FIGURE 10.24C Same area imaged by the inelastically scattered electrons of the plasmon peak around 25 eV and with a slit width of approximately 15 eV. × 11,000.

In this image some structural details appear more crisp than in the other two micrographs and a few new structural details have become visible.

COMMENTS

In a thick specimen, part of the electrons become inelastically scattered and tend to make the image appear somewhat blurred; in other words, chromatic aberration will degrade the image quality. With the aid of spectroscopic imaging the inelastically scattered electrons can be filtered out, in this case by a built-in Omega filter of the microscope. This results in a somewhat improved image as seen in Fig. 10.24B. A further improvement is gained if the spectrometer is set in such a way that only the inelastically scattered electrons of the plasmon peak contribute to the image.

The application of spectroscopic imaging of thick sections can be advantageous in stereology when large 0.5- to 1-μm-thick sections may be useful. In the present case the thickness was 1 μm. However, section thickness is likely to be reduced considerably when exposed to the electron beam (compare Cosslett, 1961). It should be noted that the amount of inelastically scattered electrons represents only a fraction of all electrons hitting the specimen; in order to record images by means of inelastically scattered electrons the exposure time or beam intensity has to be greater than when using all electrons as in global imaging. Another factor limiting the usefulness of the thick section is the superposition effect, i.e., images of several structures falling on top of each other.

A

B

C

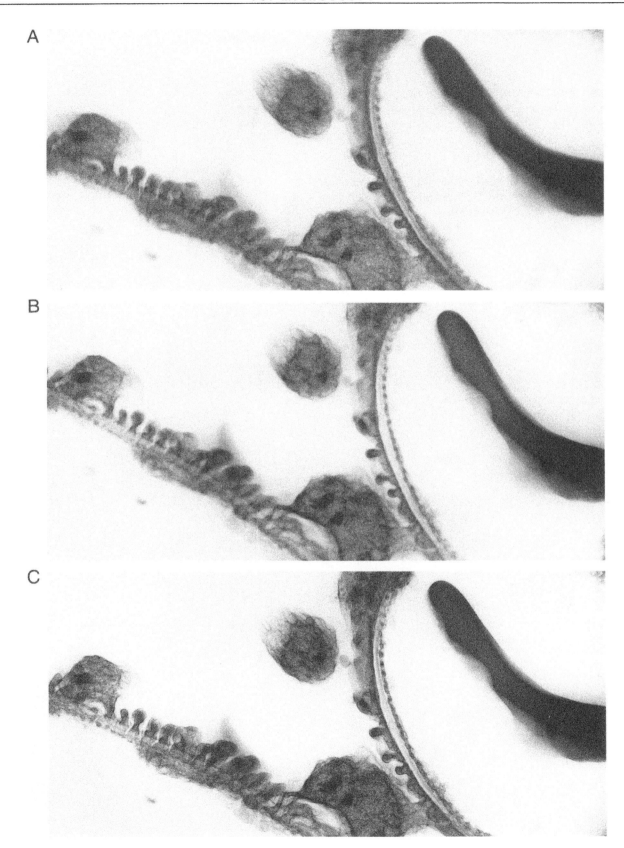

25. Spectroscopic Imaging: Carbon

FIGURE 10.25A Global imaging of a carbon support film recorded in a Zeiss 912 Omega electron microscope. × 20,000.

The image shows open holes in the thin support film.

FIGURE 10.25B Electron energy loss spectrum (EELS) recorded from the carbon film. The spectrum between 270 and 460 eV is shown.

The major peak around 300 eV observed in the spectrum corresponds to the carbon edge.

FIGURES 10.25C–10.25E Electron spectroscopic images (ESI) of the same carbon film shown in Fig. 10.25A but recorded at 220 eV (Fig. 10.25C), 279 eV (Fig. 10.25D), and 303 eV (Fig. 10.25E).

These three images show the same features except that the inelastically scattered electron from the carbon film is low at 220 and 279 eV and peaks at 303 eV.

FIGURE 10.25F Distribution of carbon in the specimen as calculated from the intensity of Fig. 10.25E after subtraction of the background, which was computer calculated from Figs. 10.25C and 10.25D.

Only the carbon film gives a signal.

COMMENTS

Electron spectroscopic imaging of thin sections allows a chemical mapping of the object. In this specimen the only element present is carbon, which results in a distinct carbon K edge in the EELS spectrum (Fig. 10.25B). The signal from carbon (Fig. 10.25F) was computed from Fig. 10.25E by subtracting the background using the software program analySIS (Soft Imaging System). Thus, the intensity observed in the calculated image 10.25F corresponds exclusively to carbon. A large number of elements can be detected in the same way, although with varying sensitivity, depending on the location of the absorption edges.

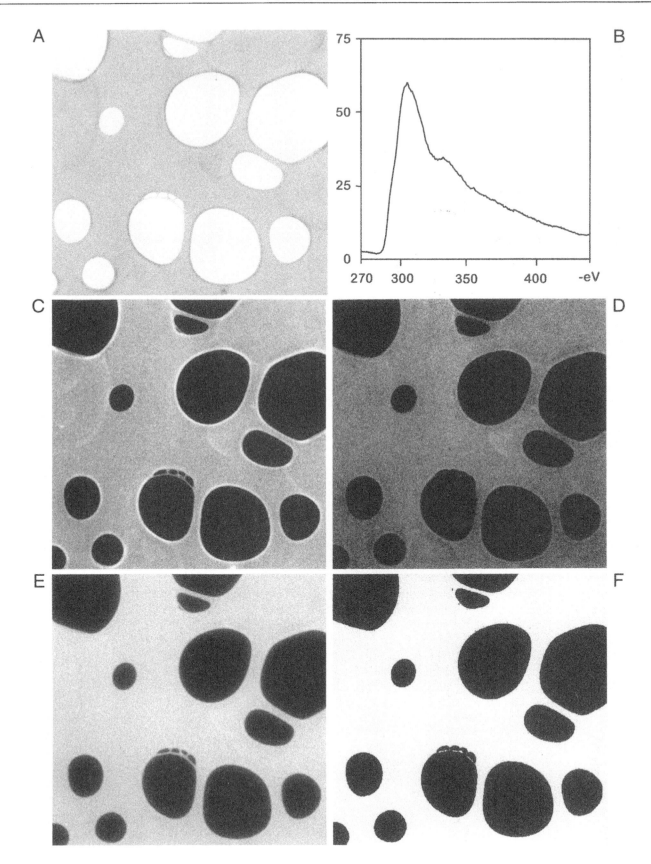

26. Spectroscopic Imaging: Calcium

FIGURE 10.26A An osteoclast-like cell in the femoral condyle of a dog with an inserted apatite-coated joint implant. The tissue was fixed in formaldehyde and glutaraldehyde, postfixed in osmium tetroxide, and embedded in Epon. The section was examined unstained in a Zeiss 912 Omega energy-filtering transmission electron microscope. × 8000.

The micrograph shows the unfiltered image of a cell with a large electron-dense crystalline deposit.

FIGURE 10.26B An electron energy loss spectrum from the intracellular crystal in Fig. 10.26A and from the adjacent cytoplasm that serves as background.

The EELS spectrum from the intracellular crystals gives a distinct peak at 346 eV.

FIGURES 10.26C–10.26E Electron spectroscopic images of the same area as in Fig. 10.26A but recorded at electron losses 338 eV (Fig. 10.26C), 340 eV (Fig. 10.26D), and 358 eV with a slit width of 15 eV (Fig. 10.26E). × 14,000.

The signals from the crystals are weak in Figs. 10.26C and 10.26D but strong in Fig. 10.26E.

FIGURE 10.26F Distribution of calcium calculated by subtraction of background from Fig. 10.26E.

The signal representing calcium is distinctly localized to the intracellular crystals.

COMMENTS

The electron spectroscopic imaging allows the localization of calcium to defined regions of cells and tissues. Background subtraction was carried out using the analySIS software. Two electron energy loss images in front of the element edge in question are determined and then used to extrapolate the background at the calcium edge. In the present case the analysis is qualitative, but quantitative estimations of the elements are also possible. The sensitivity of the method varies for the elements, and for certain heavy atoms only a small number of atoms are required. The spatial resolution in electron spectroscopic imaging is in the range of a few nanometers down to 0.5 nm depending on the element; in this respect, and considering the time required for acquisition, it is superior to X-ray microanalysis, which can also give elemental distribution maps and quantitative data on elements.

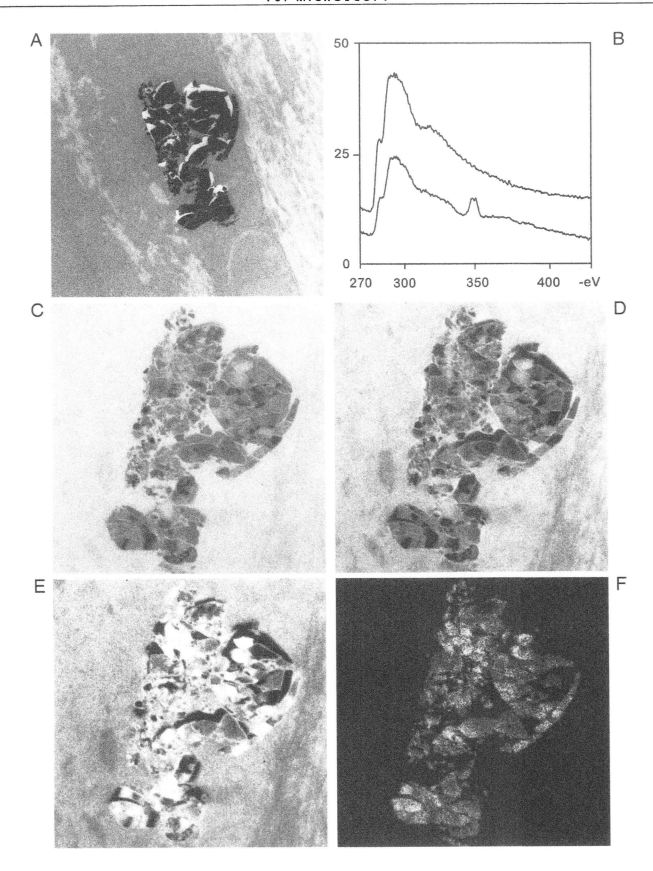

27. Spectroscopic Imaging: Contrast Changes

FIGURE 10.27A Image of a liver cell cytoplasm after conventional glutaraldehyde fixation, osmium tetroxide postfixation, and Epon embedding. The section was mounted without supporting film on the grid and section stained with lead citrate. A conventional transmission electron micrograph was used. × 50,000.

Normal appearance of a liver cell cytoplasm.

FIGURES 10.27B–10.27D Electron spectroscopic imaging of the same area of the section as in Fig. 10.27A using the inelastic electrons of 25 eV (Fig. 10.27B), 120 eV (Fig. 10.27C), and 330 eV (Fig. 10.27D). The images were recorded in a Zeiss (now LEO) 912 Omega energy-filtering transmission electron microscope.

The appearance of the cytoplasm changes dramatically when going from imaging at 25 to 330 eV. At 25 eV mainly the glycogen particles are observed, whereas mitochondria, ribosomes, and nuclear components are only faintly discernible, if at all. At 120 eV the contrast of membranes, glycogen particles, ribosomes, and nuclear components is reversed. There are no cellular components in Fig. 10.27C that are not also visible in Fig. 10.27A. At 330 eV no cellular components are observable, but shadows are observed from three contaminating particles also seen in Fig. 10.27B.

COMMENTS

The inelastic electrons emitted from a section provide some information about the composition of the object. At 25 eV the outstanding feature in the image is the appearance of glycogen granules, which do not scatter inelastic electrons, presumably because of their high content of lead from section staining. In Fig. 10.27C, most cell components emit elastic electrons, probably due to a mixture of their contents of phosphorous and sulfur. In contrast, at 330 eV the main contribution comes from carbon both in the embedding medium and in the biological components, which results in a uniform emission of inelastic electrons and thus no specific structural patterns.

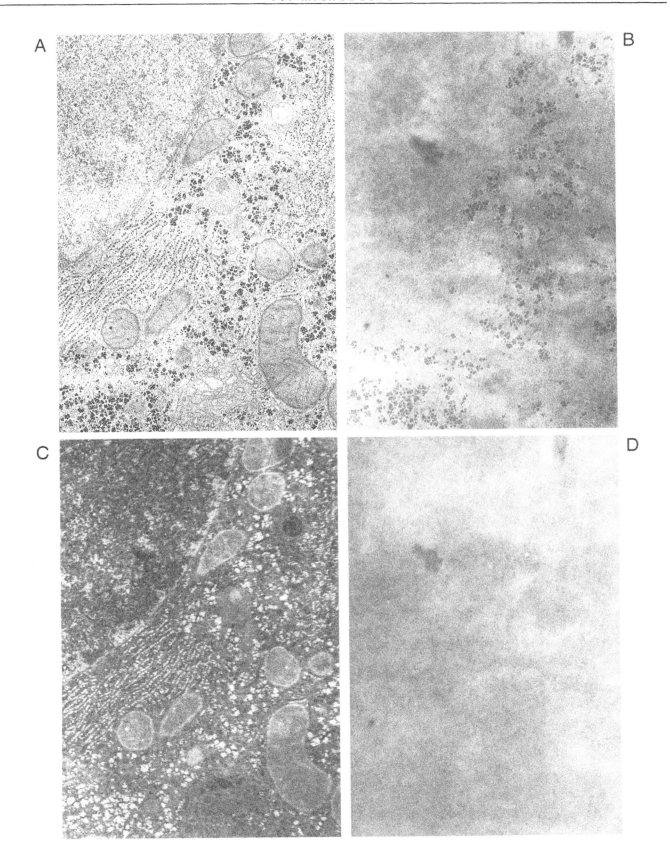

28. Cryoelectron Microscopy: Na,K-ATPase Crystals

FIGURES 10.28A AND 10.28B High magnification electron micrographs of unfixed and unstained Na,K-ATPase membranes embedded in amorphous ice. The enzyme protein was induced to form crystalline arrays before freezing. A suspension of the purified membranes was placed on a carbon-coated triafoil film with open holes 1–2 μm in diameter. Excess fluid was withdrawn from the grid, which was then plunge-frozen in ethane cooled with liquid nitrogen. The grid was transferred at liquid nitrogen temperature to a Zeiss 902A transmission electron microscope with a built-in energy filter, and an elastic bright-field electron micrograph was recorded at 80 kV in the low-dose mode (Maunsbach *et al.*, 1991). Focusing was done on an adjacent area of the grid immediately before exposure. Direct magnification $\times$ 32,500; photographic enlargement to $\times$ 270,000.

The membrane fragment in 10.28A shows rows of electron-dense particles that are often arranged in pairs or forming semicrystalline arrays. The specimen area recorded in Fig. 10.28B shows partly overlapping crystalline Na,K-ATPase membrane fragments recorded as in Fig. 10.28A. The fragment to the right shows crystalline regions oriented parallel to the plane of the grid in its central areas. The edge of the membrane is bent and is, in the upper right, oriented approximately parallel to the electron beam. The bent edge exhibits two types of patterns: one consists of a triple-layered structure (arrows) approximately 5 nm in total thickness, whereas the intermediate part of the membrane is close to 14 nm in thickness (arrowheads). In the latter, thick region there are transverse densities in the membrane as well as two longitudinal strata that are continuous with the dense layers of the thin membranes lateral to it. The ice outside the membrane fragments is essentially structureless as is the area next to the crystalline regions in the center of the membrane.

COMMENTS

This plate illustrates that biological specimen can be embedded in vitreous ice and examined in a completely unfixed and unstained state. The resulting electron micrographs therefore show protein molecules in their native state, which thus approach the ideal artifact-free preservation method. Proteins in this type of frozen-hydrated preparation appear as dark particles in contrast to the situation in negative staining preparations with heavy metal salts (see Chapter 13). The thickness of the vitreous ice in the opening of the triafoil film is variable and may be 100 nm or more. The ice films contain membrane fragments, most of which are observed *en face*, as seen in Fig. 10.28B. Such thin regions are often present toward the center of the holes in the film. In slightly thicker ice films the membrane fragments may also be bent and in part be oriented at a right angle to the ice film. In these places the membranes are imaged in what corresponds to a cross section. In Fig. 10.28B the arrows indicate the lipid bilayer of the membrane without any ATPase proteins; the thick segment in between (arrowheads), however, is interpreted as containing Na,K-ATPase protein spanning the membrane and projecting from the lipid bilayer at both sides.

A

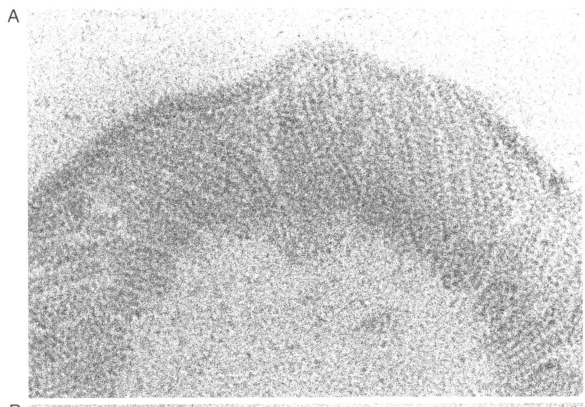

B

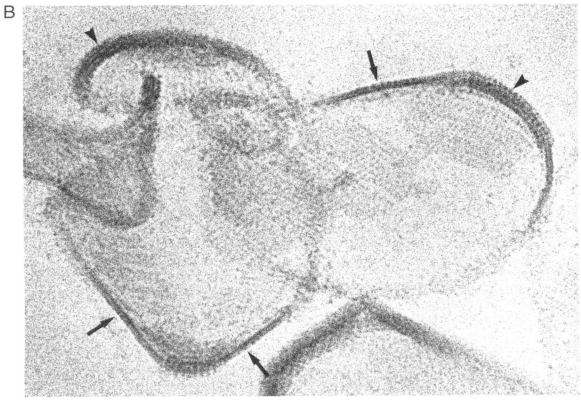

29. Defects in Cryoelectron Micrographs

FIGURE 10.29A A frozen-hydrated suspension of unfixed and unstained Na,K-ATPase membrane crystals recorded at a temperature below −170°C. Freezing and transfer were the same as in Fig. 10.28A. The specimen was recorded in the low-dose mode in a Zeiss 902A energy-filtering transmission microscope. × 72,000.

Most of the micrograph illustrates the vitrified ice in a hole of the triafoil membrane net. A few membrane fragments are seen with low contrast in the frozen water at the edge of the hole.

FIGURE 10.29B A diffraction pattern of an ice film in a hole adjacent to that in Fig. 10.29A.

This pattern is characterized by diffuse, concentric rings but no sharp rings or spots.

FIGURE 10.29C Micrograph taken immediately after Fig. 10.29A. × 72,000.

White spots (holes) have appeared in the membrane fragments.

FIGURES 10.29D AND 10.29E Unfixed and unstained Na,K-ATPase membrane crystals recorded at −170°C in a Zeiss 912 Omega cryoelectron microscope. Figure 10.29D is the first image in a series of micrographs of the same object recorded under low-dose conditions whereas Fig. 10.29E is the third image. × 150,000.

Figure 10.29D shows small irregular spots over most of the image, and the crystalline patterns in the areas marked with asterisks are not visible. In Fig. 10.29E there is a decrease in the amount of dense dots over the specimen, and crystalline patterns are indicated in the membrane fragments.

COMMENTS

There are several crucial steps in the cryoelectron procedure. The first consists of rapid freezing of the specimen in a thin ice film. Provided that freezing is sufficiently rapid, the ice is in a vitreous state, as demonstrated in the diffraction pattern in Fig. 10.29B. During transfer of the specimen to the cold stage of the microscope, a low temperature must be maintained in order to prevent the transformation of vitrous ice to cubic ice. Radiation damage of the specimen is kept at a minimum by scanning the specimen exclusively at very low magnifications and at low intensity levels. The contrast in this type of specimen is inherently low but is enhanced in this case by means of the energy filter, which excludes inelastically scattered electrons from the image. Focusing is carried out at the desired magnification, but on an area adjacent to that which is recorded as done in the low-dose procedure. Except for the small electron dose during the initial scanning of the specimen at low magnification, the photographic image is formed exclusively by the first electrons passing through the specimen. If the specimen is exposed further in the electron beam the ice film will sublimate (Fig. 10.29C).

The particles sprinkled over the object in Fig. 10.29D represent water contamination of the specimen and originate either from the transfer of the frozen membranes into the microscope or from contamination in the column. Some of the contamination has disappeared after the two first low-dose exposures of the specimen.

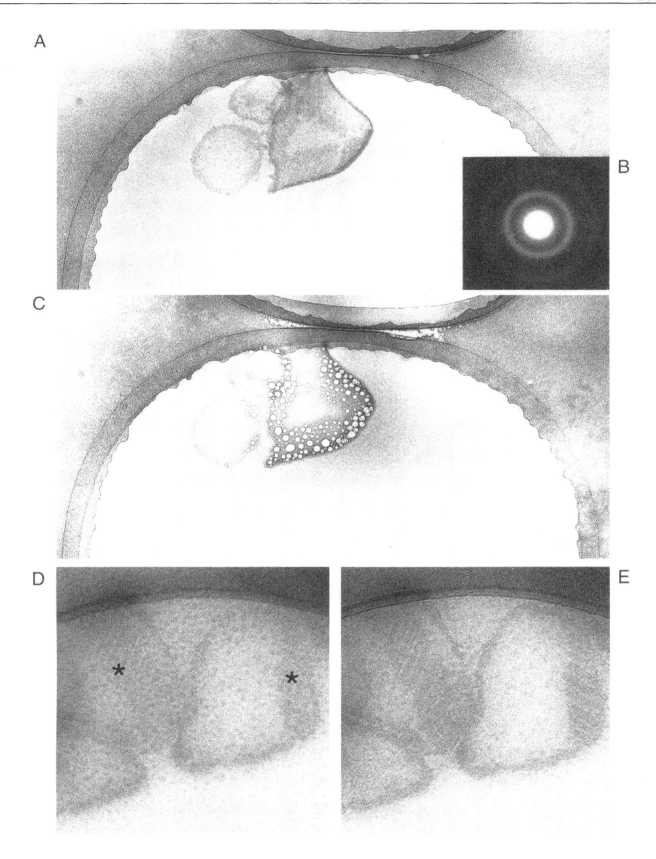

References

Agar, A. W., Alderson, R. H., and Chescoe, D. (1974). Principles and practice of electron microscope operation. *In* "Practical Methods in Electron Microscopy" (A. M. Glauert, ed.), Vol. 2, pp. 1–345. North-Holland, Amsterdam.

Bahr, G. F., and Zeitler, E. (1965). The determination of magnification in the electron microscope. II. Means for the determination of magnification. *Lab. Invest.* **14,** 142–153.

Bauer, R. (1988). Electron spectroscopic imaging: An advanced technique for imaging and analysis in transmission electron microscopy. *In* "Methods in Microbiology" (F. Mayer, ed.), Vol. 20, pp. 113–146. Academic Press, London.

Baumeister, W., and Herrmann, K.-H. (1990). High-resolution electron microscopy in biology: Sample preparation, image recording and processing. *In* "Biophysical Electron Microscopy: Basic Concepts and Modern Techniques" (P. W. Hawkes and U. Valdrè, eds.), pp. 109–130. Academic Press, London.

Bennett, P. M. (1974). Decrease in section thickness on exposure to the electron beam; the use of tilted sections in estimating the amount of shrinkage. *J. Cell Sci.* **15,** 693–701.

Bigelow, W. C. (1994). Vacuum methods in electron microscopy. *In* "Practical Methods in Electron Microscopy" (A. M. Glauert, ed.), Vol. 15. Portland Press, London.

Carlemalm, E., Colliex, C., and Kellenberger, E. (1985). Contrast formation in electron microscopy of biological material. *Adv. Electron. Electr. Phys.* **63,** 269–334.

Cermola, M., and Schreil, W.-H. (1987). Size changes of polystyrene latex particles in the electron microscope under controlled physical conditions. *J. Electr. Microsc. Techn.* **5,** 171–180.

Chescoe, D., and Goodhew, P. J. (1990). "The Operation of Transmission and Scanning Electron Microscopes: Microscopy Handbooks 20." Oxford Univ. Press, Royal Microscopical Society.

Cosslett, A. (1961). The effect of the electron beam on thin sections. *In* "Proceedings of the Second European Regional Conference on Electron Microscopy," Vol. 2, pp. 678–681. Nederlandse Vereniging voor Electronenmicroscopie, Delft.

Cyrklaff, M., Adrian, M., and Dubochet, J. (1990). Evaporation during preparation of unsupported thin vitrified aqueous layers for cryo-electron microscopy. *J. Electr. Microsc. Techn.* **16,** 351–355.

Dubochet, J. (1995). High-pressure freezing for cryoelectron microscopy. *Trends Cell Biol.* **5,** 366–368.

Dubochet, J., Adrian, M., Chang, J.-J., Homo, J.-C., Lepault, J., McDowall, A. W., and Schultz, P. (1988). Cryo-electron microscopy of vitrified specimens. *Q. Rev. Biophys.* **21,** 129–228.

Echlin, P. (1992). "Low-Temperature Microscopy and Analysis." Plenum Press, New York.

Elfvin, L.-G. (1961). Electron microscopic investigation of the plasma membrane and myelin sheath of autonomic nerve fibers in the cat. *J. Ultrastruct. Res.* **5,** 388–407.

Fahrenbach, W. H., and Luft, J. H. (1988). Critical underfocus in low-magnification electron microscopy: Rationale and practice. *J. Electr. Microsc. Techn.* **8,** 433–439.

Frederik, M., Stuart, M. C. A., Bomans, P. H. H., Busing, W. M., Burger, K. N. J., and Verkleij, A. J. (1990). Perspective and limitations of cryo-electron microscopy. *J. Microsc.* **161,** 253–262.

Haine, M. E., and Mulvey, T. (1954). The regular attainment of very high resolving power in the electron microscope. *In* "Proceedings of the Third International Congress of Electron Microscopy, London, pp. 698–705.

Hawkes, P. W., and Kasper, E. (1989). "Principles of Electron Optics," Vol. 1. Academic Press, London.

Haydon, G. B. (1969b). An electron-optical lens effect as a possible source of contrast in biological preparations. *J. Microsc. (Oxford)* **90,** 1–13.

Johansen, B. V. (1977). High resolution bright field electron microscopy of biological specimens. *Ultramicroscopy* **2,** 229–239.

Johansen, B. V. (1975). Optical diffractometry. *In* "Principles and Techniques of Electron Microscopy" (M. A. Hayat, ed.), Vol. 5, pp. 114–173. Van Nostrand-Reinhold, New York.

Hendzel, M. J., and Bazett-Jones, D. P. (1995). Probing nuclear ultrastructure by electron spectroscopic imaging. *J. Microsc.* **182,** 1–14.

Kellenberger, E. (1986). The ups and downs of beam damage, contrast and noise in biological electron microscopy. *Micron Microsc. Acta* **17,** 107–114.

Luther, P. K. (1992). Sample shrinkage and radiation damage. *In* "Electron Tomography" (J. Frank, ed.), pp. 39–60. Plenum Press, New York.

Luther, P. K., Lawrence, M. C., and Crowther, R. A. (1988). A method for monitoring the collapse of plastic sections as a function of electron dose. *Ultramicroscopy* **24,** 7–18.

Massover, W. H. (1993). Ultrastructure of ferritin and apoferritin: A review. *Micron* **24,** 389–437.

Maunsbach, A. B., Hebert, H., and Kavéus, U. (1992). Cryo-electron microscope analysis of frozen-hydrated crystals of Na,K-ATPase. *Acta Histochem. Cytochem.* **25,** 279–285.

Maunsbach, A. B., Zellmann, E., and Skriver, E. (1991). Cryoelectron microscope analysis of frozen-hydrated two-dimensional Na,K-ATPase crystals. *Micron Microsc. Acta* **22,** 57–58.

Ottensmeyer, F. P. (1986). Elemental mapping by energy filtration: Advantages, limitations, and compromises. *Ann. N. Y. Acad. Sci.* **483,** 339–351.

Overgaard, S., Lind, M., Josephsen, K., Maunsbach, A. B., Bünger, C., and Søballe, K. (1998). Resorption of hydroxyapatite and fluorapatite ceramic coatings on weight-bearing implants: A quantitative and morphological study in dogs. *J. Biomed. Mater. Res.* **39,** 141–152.

Probst, W., Zellmann, E., and Bauer, R. (1989). Electron spectroscopic imaging of frozen-hydrated sections. *Ultramicroscopy* **28,** 312–314.

Reimer, L., Fromm, I., Hirsch, U., and Rennekamp, R. (1992). Combination of EELS modes and electron spectroscopic imaging and diffraction in an energy-filtering electron microscope. *Ultramicroscopy* **46,** 335–347.

Reimer, L., and Ross-Messemer, M. (1990). Contrast in the electron spectroscopic imaging mode of a TEM. II. Z-ratio, structure-sensitive and phase contrast. *J. Microsc. (Oxford)* **159,** 143–160.

Schröder, R. R., Hofmann, W., and Ménétret, J.-F. (1990). Zero-loss energy filtering as improved imaging mode in cryoelectronmicroscopy of frozen-hydrated specimens. *J. Struct. Biol.* **105,** 28–34.

Sjöstrand, F. S. (1967). "Electron Microscopy of Cells and Tissues: Instrumentation and Techniques." Academic Press, New York.

Thon, F. (1966). Zur defokussierungsabhängigkeit des phasenkontrastes bei der elektronenmikroskopischen abbildung. *Z. Naturforsch. A* **21,** 476–478.

Williams, D. B., and Carter, C. B. (1996). "Transmission Electron Microscopy: A Textbook for Materials Science." Plenum Press, New York.

Wrigley, N. G. (1968). The lattice spacing of crystalline catalase as an internal standard of length in electron microscopy. *J. Ultrastruct. Res.* **24,** 454–464.

Zeitler, E. (1990). Radiation damage in biological electron microscopy. *In* "Biophysical Electron Microscopy: Basic Concepts and Modern Techniques" (P. W. Hawkes and U. Valdrè, eds.), pp. 289–308. Academic Press, London.

Zierold, K. (1990). Low-temperature techniques. *In* "Biophysical Electron Microscopy: Basic Concepts and Modern Techniques" (P. W. Hawkes and U. Valdrè, eds.), pp. 309–346. Academic Press, London.

IMAGE RECORDING

The image formed by the optical system in the electron microscope is made visible on the fluorescent screen. If a permanent document of the image is desired, it has to be recorded electronically or, more commonly, on photographic material. Another reason for recording the image is that many features of the image cannot be observed directly on the screen for reasons of contrast or screen granularity; many patterns also require extensive analyses. Furthermore, the specimen usually undergoes changes in the electron beam, particularly due to radiation damage and contamination, and it is essential to be able to analyze the object as it appears initially in the electron microscope rather than after prolonged exposure to the electron beam.

In order to record the image, the photographic material is located within the evacuated electron microscope column. This situation provides nearly ideal recording conditions and is limited only by the characteristics of the photographic emulsion. It is also possible to record the image outside the column if it is transmitted by means of a fiber-optic system. This avoids the necessity of introducing the photographic film in the vacuum, but introduces instead a risk of losing resolution and fine detail.

This chapter illustrates some characteristics of recordings on photographic emulsions. Furthermore, it illustrates digital image recordings with charge-coupled device (CCD) cameras or image plates connected to printers. Digital image acquisition is increasingly used because it facilitates subsequent image processing.

The discovery that a regular photographic plate or film is perfectly useful for recording images in the electron microscope must have come as somewhat of a relief to the scientists who built the first instruments. However, there are some important differences. The 60-kV electron penetrates only about 15 μm in the emulsion, hence the photographic emulsion on films or plates used for electron microscopic recording can be made (and indeed are) quite thin. A single electron is sufficient to activate a silver grain or even several grains, whereas the combined action of many light quanta is required to make a grain developable. As a consequence hereof, all photographic material will appear as having nearly the same grade of contrast. Robin Valentine (1965) tested various film materials and found contrast to be more or less the same but film speed to differ. He has also explained film granularity as an effect of the random fusion of several silver grains. Although photographic recordings are still used worldwide, methods for digital image recordings are developing rapidly. Thus, charge-coupled device (CCD) cameras were applied in electron microscopy in the 1980s, and in the 1990s the image plate was refined to be a powerful, albeit expensive, image recording system.

Valentine. R. C. (1965). *Lab. Invest.* **14,** 1334–1340.

1. Exposure Time

FIGURES 11.1A–11.1D A kidney epithelium fixed in glutaraldehyde, followed by osmium tetroxide–potassium ferrocyanide, embedded in Epon, and section stained with uranyl acetate and lead citrate. The four micrographs were recorded in the electron microscope with gradually decreasing beam intensities and with correspondingly increasing exposure times. The total exposure dose was controlled by the exposure meter in a JEOL 100 CX microscope and was constant as judged by the equal density of the developed photographic negatives. The exposure time for Fig. 11.1A was 0.25 sec, whereas those for Figs. 11.1B, 11.1C, and 11.1D were 1, 4, and 16 sec, respectively. × 60,000.

There is no detectable difference in image contrast among the four electron micrographs.

COMMENTS

These electron micrographs show that differences in the length of the exposure do not detectably influence the contrast in the final image, provided that the total electron dose is the same. The minimum time is limited in practical work by the design of the exposure meter and the shutter, by the necessity to avoid excessive electron doses on sensitive biological specimens, and, at high magnifications, by an insufficient illumination of the specimen. The length of exposure may also be limited due to the risk of specimen drift during prolonged exposure. In fact, most conventional specimens, if recorded with long exposure times, e.g., over 5 sec at high magnification, may show evidence of specimen drift (see Fig. 10.16A).

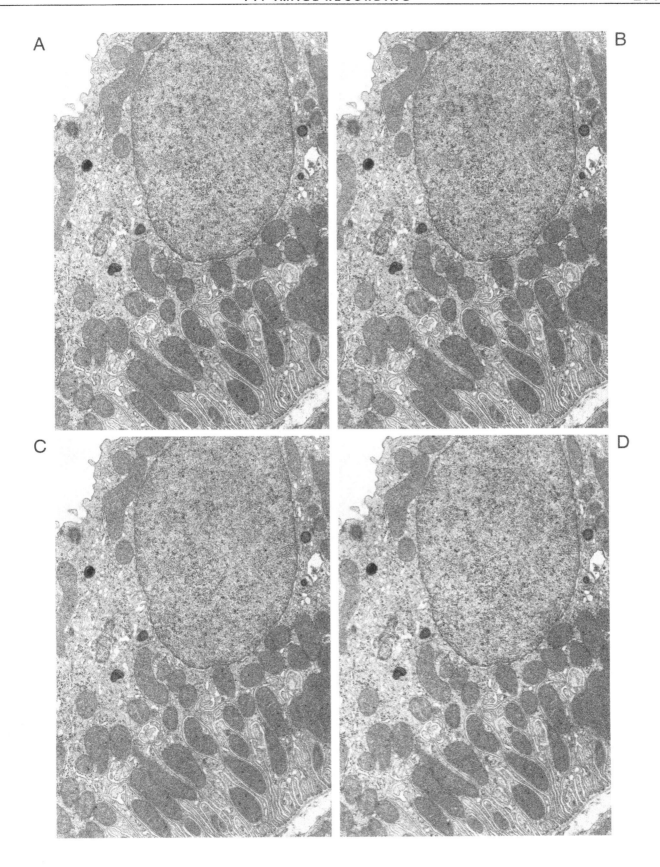

2. Over/Underexposure

FIGURES 11.2A–11.2D A rat liver tissue fixed with glutaraldehyde and postfixed with osmium tetroxide. The tissue was embedded in Epon and the section was stained with uranyl acetate and lead citrate. The microscope was operated at 60 kV with an electron optical magnification of 15,000 times. The images were recorded on Ilford N 40 plates using different exposure times, and the negatives enlarged on the same grade of photographic paper.

Figure 11.2A was given a standard exposure of 2.5 sec in the microscope. During enlargement the photographic paper was exposed for 18 sec. Figure 11.2B was taken with the same beam intensity, but with an exposure time of only 0.5 sec. The photographic paper was exposed for only 1 sec. Figure 11.2C was also taken with the same beam intensity as Fig. 11.2A, but the exposure was 8 sec. The overexposed negative required an exposure of 12 min in the enlarger. Figure 11.2D is identical to Fig. 11.2C except that the overexposed negative was bleached in potassium ferrocyanide ("Farmer's reducer") to give a negative of more suitable density. The bleached negative required 30 sec exposure in the enlarger. × 45,000.

Figure 11.2A can be regarded as a micrograph of normal contrast. Figure 11.2B has a distinctly lower contrast than Fig. 11.2A, whereas Fig. 11.2C has a slightly higher contrast. The overexposed and bleached negative in Fig. 11.2D has a contrast rather similar to that in Fig. 11.2A.

COMMENTS

When the photographic plate or film is given an optimal exposure and has been properly developed, the emulsion density varies within a wide range. No large areas of either black or white are found unless the specimen contains uniformly electron-dense or electron-translucent regions. A loss in contrast is evident in underexposed negatives and may lead to a loss of information; the negatives are too thin and enlargements will lack contrast. By increasing the exposure, the density of the negative increases, which results in an enhanced image contrast as well as a higher signal-to-noise ratio. When, however, the negative is overexposed, the photographer loses much time making the prints. The dense negatives can, however, be modified by different types of reducers. Some of these, such as Farmer's reducer, bleach the negatives with a retained degree of contrast; with other reducers, contrast may be gained or lost.

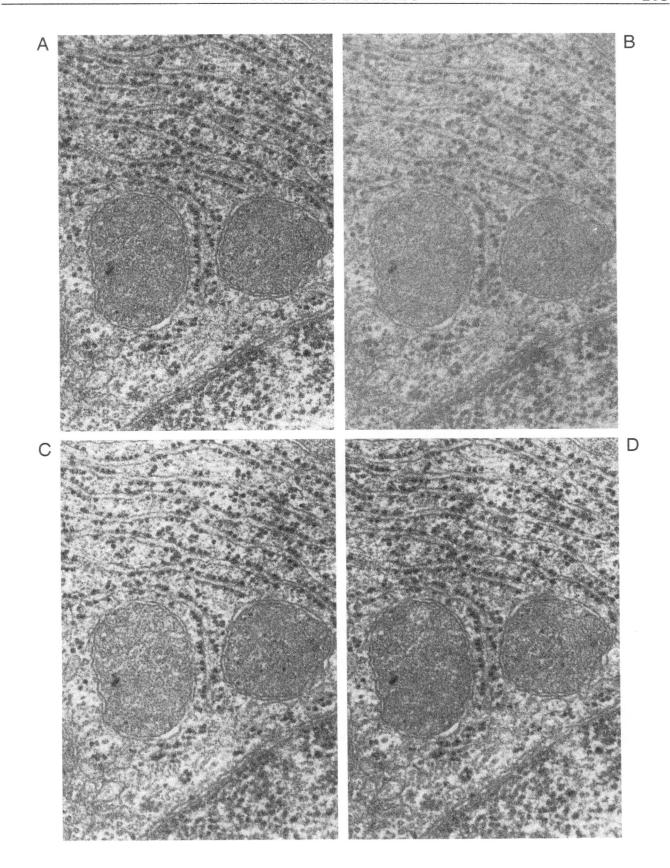

3. Effects of Development

FIGURES 11.3A–11.3F Electron micrographs of the same section of rat liver as in Fig. 11.2 recorded with a fixed exposure time in the microscope but treated differently during development of the plates. The microscope was operated at 60 kV and the negatives were enlarged on the same grade of photographic paper.

Figure 11.3A is made from a negative developed for 3 min and fixed for 5 min. The exposure time in the enlarger was 8 sec.

Figure 11.3B is from a negative processed in a developer that had been left in the open developing tray exposed to air for 24 hr. The negative was developed for 3 min and fixed for 5 min. The exposure time in the enlarger was 0.6 sec.

Figure 11.3C is from a negative that was immersed gradually in the developer and developed for 3 min without agitation and fixed for 5 min. The exposure time in the enlarger was 2 sec.

Figure 11.3D is from a negative that was developed for only 40 sec and then fixed for 5 min. The resulting thin negative was exposed for 0.6 sec in the enlarger.

Figure 11.3E is from a negative that was left in the developer for 30 min and then fixed for 5 min. The resulting negative was so dense that the exposure time in the enlarger had to be 36 min.

Figure 11.3F is from a negative that was developed for 3 min and fixed for 70 hr. This long fixation bleached the plate and reduced the exposure time in the enlarger by a factor of five, thus to 1.6 sec. × 14,000.

Figure 11.3A shows optimal contrast. On the contrary Figs. 11.3B, 11.3D, and 11.3F show decreased contrast. Fig. 11.3C is uneven while Fig. 11.3E resembles Fig. 11.3A, but is even slightly more contrasted.

COMMENTS

The recommendations given by the manufacturers of photographic material should be followed. It is particularly important that adequate time be given for the development and that the developer is fresh. In a negative that is underdeveloped (Fig. 11.3D) or developed in an old developer (Fig. 11.3B), information will be lost. The overdeveloped negative (Fig. 11.3E), however, becomes too dark; to use such negatives is a waste of time when making the prints. They can be bleached by Farmer's reducer, albeit with a risk for loss in image quality. A bleaching effect is also obtained if the negatives are left too long in the fixing solution (Fig. 11.3F). The negatives should be agitated during development or else negative density will become uneven. If the developer has not wetted the entire negative within seconds, it will show regions of lower density (Fig. 11.3C).

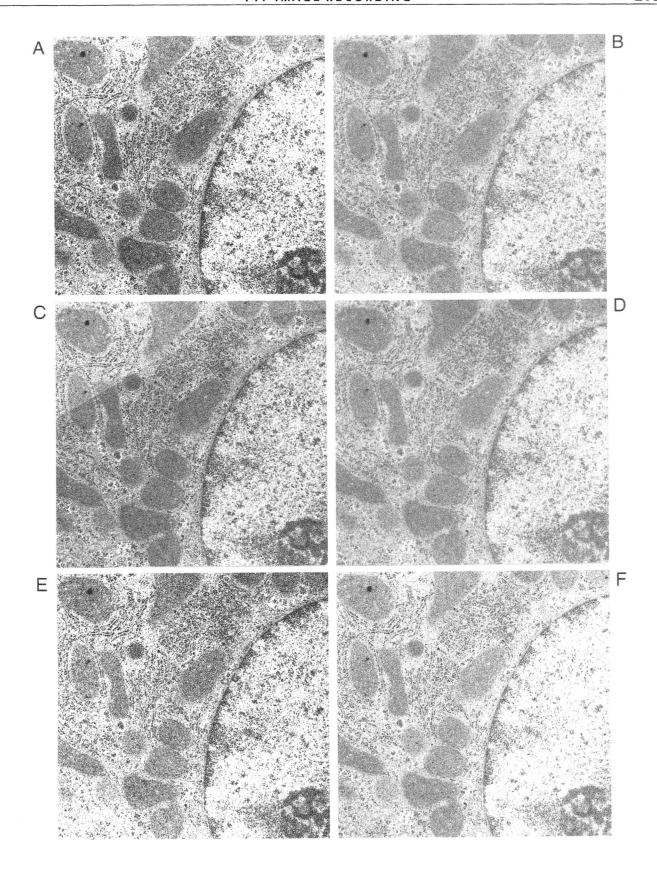

4. Exposure Dose Adjustment

FIGURES 11.4A–11.4C These three electron micrographs of a thin section from kidney (for specifications, see Chapter 10.4) were recorded with the same objective aperture but with different electron doses on Kodak electron microscope film 4489 film by setting the film sensitivity on a JEOL 100 CX electron microscope to sensitivities 10, 8, and 6 for Figs. 11.4A, 11.4B, and 11.4C, respectively. All three negatives were taken with an exposure time in the electron microscope of 1 sec. The resulting negatives were enlarged on normal photographic paper using the same aperture under identical conditions to provide prints of approximately the same overall density. Exposure times in the enlarger to obtain these prints were 2.4 sec (Fig. 11.4A), 12 sec (Fig. 11.4B), and 145 sec (Fig. 11.4C). × 60,000.

The three micrographs show a considerable variation in contrast. The print in Fig. 11.4A, which was made from a very light negative, has a low contrast, whereas Fig. 11.4C, which has been printed from a very dense negative, has a high contrast.

COMMENTS

Modern electron microscopes have an exposure meter with the possibility of adjusting the exposure dose to film sensitivity. In this particular case an adjustment of the sensitivity setting to a more sensitive film than that actually used resulted in underexposure and a very thin negative that gave the low contrast photographic print in Fig. 11.4A. In contrast, for Fig. 11.4C, the sensitivity setting of the microscope was adjusted to a much slower film than the one actually used. The negative consequently became overexposed, very dense, and required a long exposure in the enlarger. However, its contrast is increased relative to that in Fig. 11.4B and is considerably higher than that in Fig. 11.4A.

If radiation-resistant specimens are overexposed in the microscope, the resulting prints may show an improved contrast, although at the expense of long and tedious exposure times in the enlarger. Underexposure of specimens in the microscope is sometimes required, mainly for recording radiation-sensitive objects such as frozen-hydrated material. Underexposure of the negative can, in some cases, be counterbalanced by using more sensitive photographic material and/or developing conditions. Additionally, some underexposed recordings of crystalline materials can be analyzed by computer-based image analysis procedures (Unwin and Henderson, 1975).

A

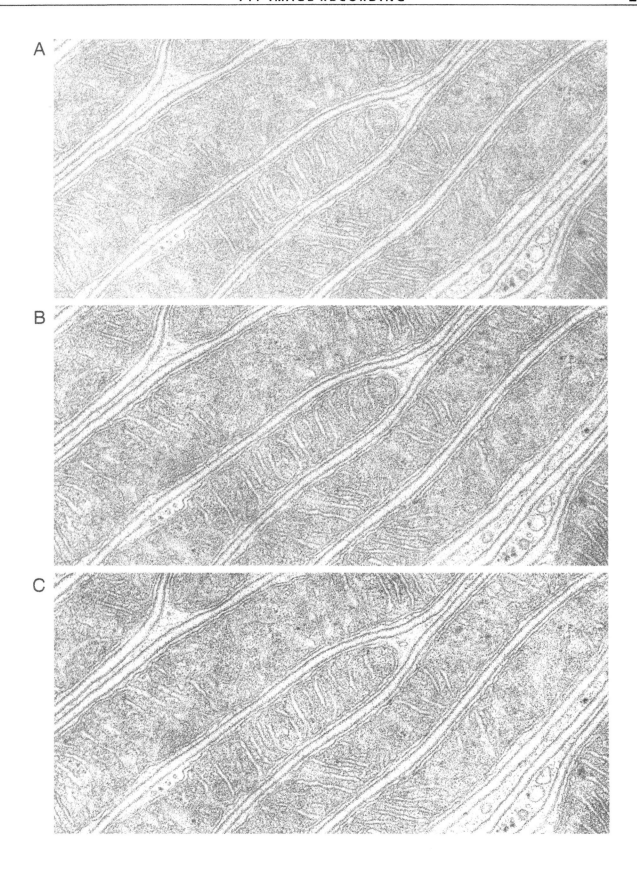

B

C

5. Primary Magnification

FIGURE 11.5A Parts of distal tubule cells from a rat kidney fixed with glutaraldehyde–formaldehyde and postfixed in osmium tetroxide–potassium ferrocyanide. The section was stained with uranyl acetate and lead citrate. The image was recorded at a magnification of $\times$ 8300 and enlarged photographically to $\times$ 25,000.

The micrograph has a sharp and crisp appearance.

FIGURE 11.5B The same negative as Fig. 11.5A, but the illustrated area was enlarged photographically to a final magnification of $\times$ 100,000.

Many details appear grainy rather than crisp. The triple-layered structure of the plasma membranes and mitochondrial membranes are resolved only in places.

FIGURE 11.5C The same area as in Fig. 11.5B, but recorded in the electron microscope with a direct magnification of $\times$ 33,000 and then further enlarged photographically to a final magnification of $\times$ 100,000. Figures 11.5B and 11.5C were printed on the same grade photographic paper.

The micrograph shows well-defined structural details, including triple-layered membranes.

COMMENTS

A point in the specimen is not represented by a single photographic grain but by a group of adjacent photographic grains with a total diameter of up to 20 μm. In order to resolve adjacent points in the micrograph, it is therefore necessary to use a sufficiently high electron optical magnification. With too low a magnification, adjacent image points will excite essentially the same areas of the negative. Because of the electron optical noise, a minimum direct magnification must be used in order to make use of the full resolving power of the microscope. If, for example, a point-to-point resolution of 0.5 nm is required, the direct electron optical magnification should be at least 50,000 times.

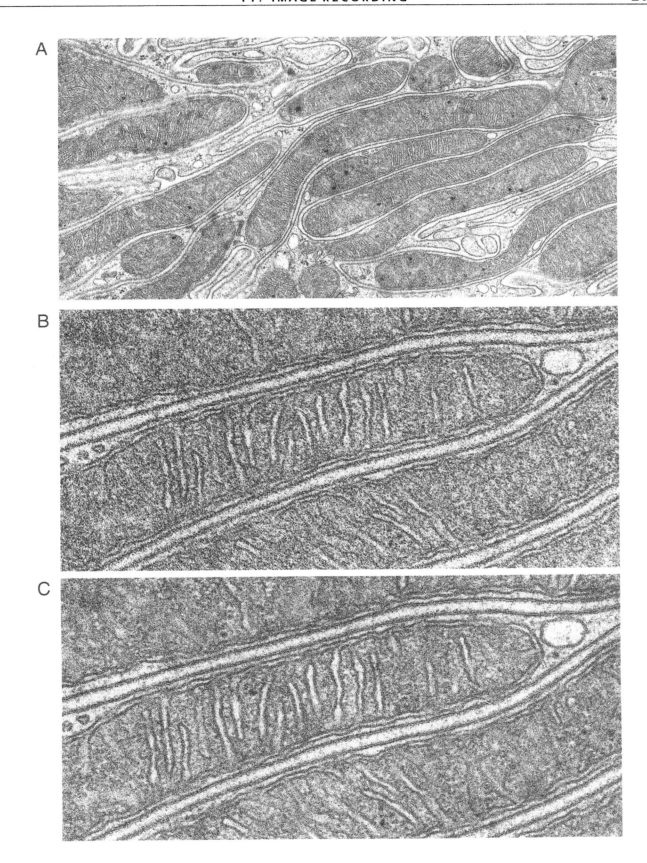

6. Damage to Negatives

FIGURES 11.6A AND 11.6B Electron micrographs of a section of brown adipose tissue from a hamster recorded on two separate photographic plates. Figure 11.6A was taken on a photographic plate that had been stored in a dry place for less than 6 months. Figure 11.6B was taken on the same kind of plate except that it had been stored for some months in a humid place. Photographic magnification × 8 and final magnification × 80,000.

The micrographs in Figs. 11.6A and 11.6B show a mitochondrion to the right and an apparently empty cytoplasm to the left. Figure 11.6A is without apparent flaws, whereas Fig. 11.6B shows numerous small black spots. These are particularly disturbing in the light areas, which correspond to dark areas in the photographic negative.

FIGURE 11.6C Electron micrograph of a section from a rat liver. Photographic magnification × 4 and final magnification × 60,000.

Dark parallel lines cross the picture.

FIGURE 11.6D Electron micrograph of a portion of the nucleus of a sea urchin larva. Photographic magnification × 6 and final magnification × 45,000.

The nucleoplasm shows a pattern consisting of semicircular dotted lines.

COMMENTS

The spots in Fig. 11.6B are due to holes in the emulsion made by microorganisms. Photographic material should be stored in a cold and dry place; the cardboard box with the glass plates used for Fig. 11.6B was moistened with condensed water and all plates were defective. The photographic negative in Fig. 11.6C was scratched on purpose before being inserted in the microscope. Scratches form easily on roll film used as negative material in the electron microscope, on glass plates, or, as in this case, films that are loaded in cassettes with a tight-fitting lid. The pattern in Fig. 11.6D is a fingerprint mark on the photographic negative.

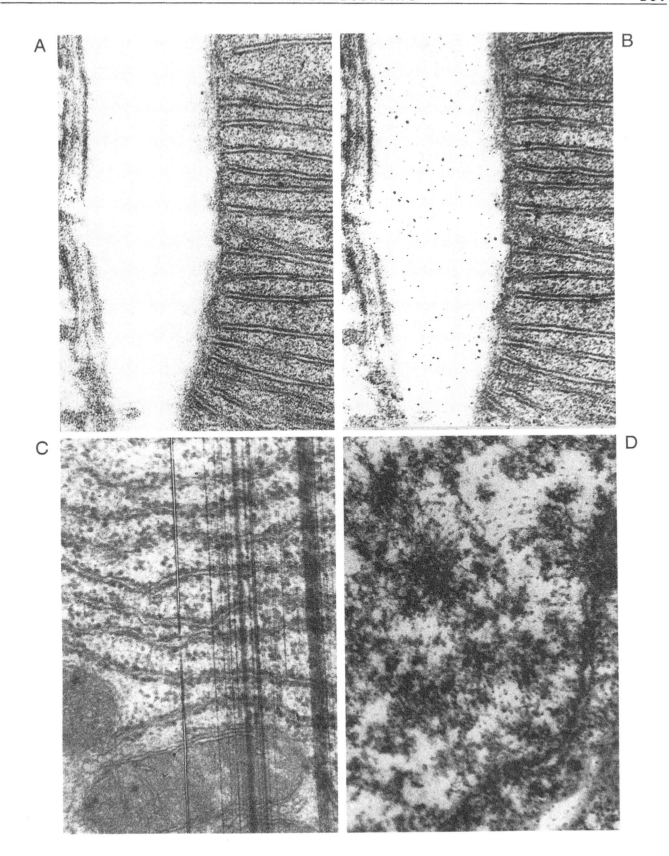

7. Damage to Wet Negatives

FIGURES 11.7A AND 11.7B Electron micrographs of cross-sectioned jellyfish sperm flagella. The photographic negatives (emulsion on glass plate basis) were processed and dried differently. One of the negatives (Fig. 11.7B), but not the other (Fig. 11.7A), was exposed to water splashed on its emulsion side during drying. × 120,000.

Figure 11.7B shows a number of diffuse dark spots distributed irregularly over and between cross-sectioned flagella. Similar spots are not present in Fig. 11.7A.

FIGURES 11.7C AND 11.7D Electron micrographs of a rat liver cytoplasm fixed and processed according to standard procedures. During the rinse in tap water the negative of Fig. 11.7D (on a glass plate) was dipped into water of about 60°C for about 1 min and then again rinsed in cold water of 0°C. The photographic enlargement is 8 times and the final magnification × 120,000.

A reticular pattern of interconnected lines is visible in Fig. 11.7D and hides many structures of the cytoplasm.

COMMENTS

Before the photographic emulsion of the negative has dried and hardened it is sensitive to insults of various kinds. It should hence be put into a dust-free place for drying rather than kept in the darkroom, where there is a risk of water splashes. Exposure of the negative to hot water is more unusual but hot water is present in most darkrooms and a mix-up of the water taps may lead to heating of the negative to a forbidding temperature. The emulsion may then detach completely from the underlying glass or become unduly soft and floppy. If the plate is then transferred to cold water, the emulsion will form numerous small wrinkles of the type shown in Fig. 11.7D. However, there is no risk of damage to the emulsion from a prolonged rinse in cold water, e.g., an overnight rinse.

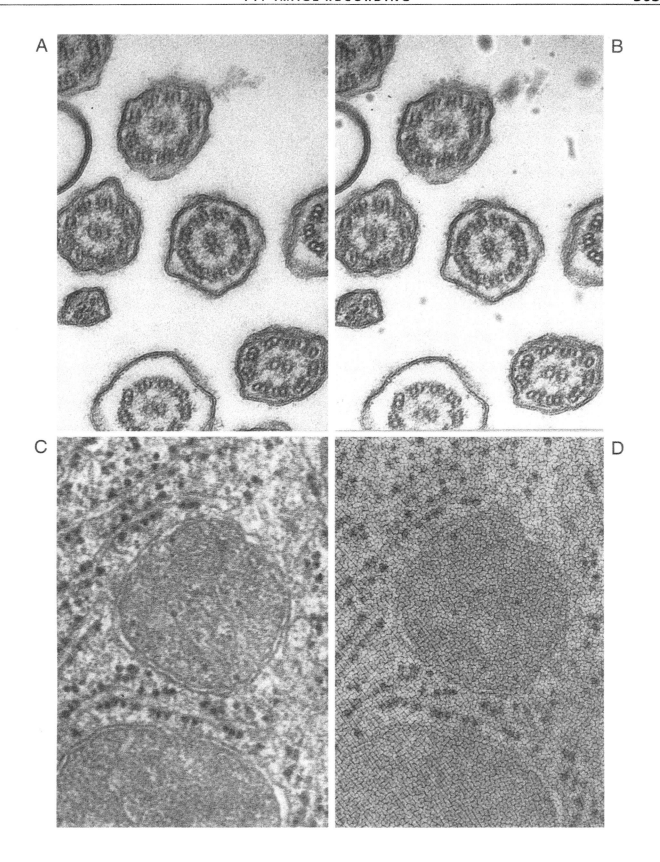

8. Film/Imaging Plate/Charge-Coupled Device (CCD) Camera

FIGURE 11.8A Part of a liver cell following standard double fixation with glutaraldehyde and osmium tetroxide, embedding in Epon, and section staining with uranyl acetate and lead citrate. The image was recorded in a Zeiss 912 Omega electron microscope on Kodak electron microscope film at a primary magnification of 8000 times and photographically enlarged to 42,000 times on grade 2 photographic paper.

This field shows a typical part of a liver cytoplasm.

FIGURE 11.8B This image was recorded on a Fujifilm imaging plate (IP) placed in the electron microscope immediately after the photographic negative used for Fig. 11.8A. The same microscopic settings were used, including focus and magnification. The digitized image was read in a Fujifilm FDL 5000 reader and the image was electronically enlarged 5.2 times to a final magnification of 42,000 times and printed by a 400 dpi (dots per inch) Fuji Pictrography 3000 printer.

With respect to contrast, sharpness, and general appearance, the image is very similar to the image recorded on photographic film.

FIGURE 11.8C The same section recorded on a Proscan HSS 1024 slow-scan CCD camera in a Zeiss 912 Omega electron microscope. The magnification of the microscope was set at 8000, which gives a magnification at the level of the CCD chip (19×19 mm^2) of $\times 4000$. The image was electronically enlarged about 10.5 times and printed on a Fuji Pictrography 3000 printer at a magnification of $\times 42,000$.

The general appearance of this micrograph is the same as in Figs. 11.8A and 11.8B.

COMMENTS

CCD cameras and imaging plates represent two ways of recording micrographs in digital form. These techniques therefore open possibilities for immediate image processing and analysis (see Chapter 19). The exposure time required for an image is about the same in these systems as with photographic negatives. The quality of the images at the final magnifications illustrated here is quite similar. Minor differences in contrast among the three recordings can be contributed to the characteristics of the paper and processing of the photographic paper (Fig. 11.8A) and to the characteristics of the printer (Figs. 11.8B and 11.8C). With respect to image resolution, a photographic emulsion has a slight advantage. Recording by means of an imaging plate or a CCD camera evidently eliminates the need for a darkroom facility and work. Another advantage of digitized images is that they can be stored electronically for future processing. In the case of a CCD camera, an additional advantage is that repeated exposures can be made without having to open the microscope, thus avoiding contamination and securing a high vacuum.

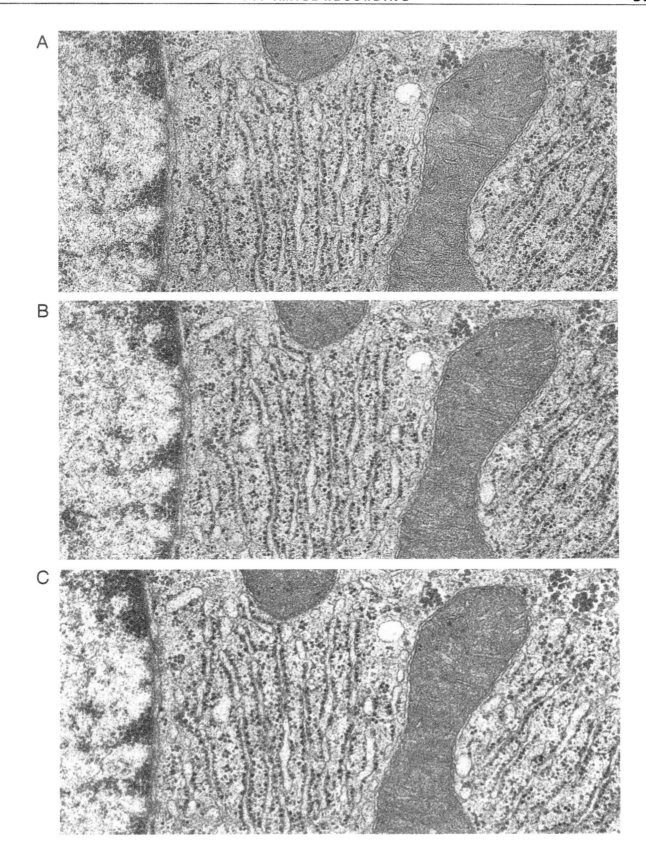

9. Enlarged Digital Recordings

FIGURE 11.9A A 19 times photographic enlargement of the negative used for Fig. 11.8A. × 152,000.

The ground substance has a homogeneous appearance, although photographic grains and clusters are evident at this high photographic enlargement.

FIGURE 11.9B A 19 times electronic enlargement of the Fujifilm imaging plate used for Fig. 11.8B.

The overall impression is that of an ordinary micrograph of a cell cytoplasm with well-defined membranes and ribosomes, but the enlargement reveals that the entire image is made up of small square pixels with different shades from black to white.

FIGURE 11.9C An electronic enlargement of the CCD recording in Fig. 11.8C. The total enlargement relative to the chip is about 38 times.

Although the general appearance of the cellular structures can be recognized, the pixels making up the image are clearly visible.

COMMENTS

Micrographs recorded in the electron microscope on photographic films, imaging plates, or CCD chips and enlarged 10 times or less show only minor differences in image quality (compare Figs. 11.8A–11.8C). However, if the original image is enlarged further (as shown in Figs. 11.9A–11.9C), the digital recording modes show characteristic differences, notably with respect to the appearance of pixels in the images.

The imaging plate has a pixel size of $25 \times 25\ \mu m^2$ whereas the pixel size of the CCD camera is $19 \times 19\ \mu m^2$ for a 1024×1024 pixel chip. Obviously a smaller pixel size means a higher resolution, but the smaller the pixel size the higher the cost of the chip. To contain the same area of the specimen, the CCD image on the chip must be enlarged higher than the corresponding image on the imaging plate (38 versus 19 times), and pixels in the enlarged CCD image are therefore larger than in the IP micrograph. In photographic material the size of photographic grain clusters has a diameter in the order of $20\ \mu m$. Thus with the recordings made here, the photographic film is still very competitive, and the area of the photographic film, e.g., $6.5 \times 9\ cm^2$, is about 16 times greater than that of a CCD chip.

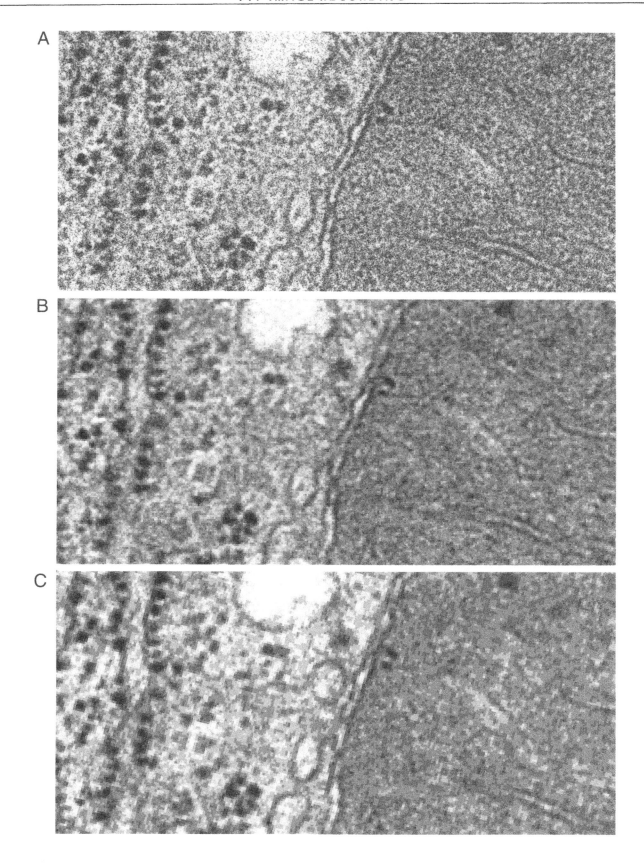

10. Variation in Electron Dose

FIGURES 11.10A–11.10H Transmission electron micrographs of a liver cell cytoplasm following fixation in glutaraldehyde and osmium tetroxide, resin embedding, and section staining with uranyl acetate and lead citrate. The microscope camera was loaded alternatively with casettes containing Agfa Scientia film and imaging plates from Fujifilm. All eight micrographs were recorded with the same focus setting and electron beam intensity, which was adjusted to give optimal results on film with an exposure time of 1 sec.

Figures 11.10A–11.10D were recorded on film with exposure times of 0.2, 1.0, 5.0, and 20 sec, respectively. Figures 11.10E–11.10H were recorded in parallel on imaging plates with exposure times of 0.2, 1.0, 5.0, and 20 sec, respectively. × 35,000.

Figure 11.10A shows very low contrast despite being printed on the highest grade contrast paper available. Figure 11.10B shows normal contrast and density whereas Figs. 11.10C and 11.10E are blanks due to greatly overexposed and completely dense negatives. In contrast, Figs. 11.10E–11.10H appear very similar in density, although Fig. 11.10E is somewhat less contrasted than the other recordings.

COMMENTS

These two series of micrographs recorded in parallel clearly demonstrate the differences in electron dose linearity between commonly used film material and imaging plates. Although a micrograph of suitable contrast and density is obtained on film with an exposure of 1 sec under the present conditions, the image loses contrast greatly if underexposed by a factor of 5 and is completely useless if overexposed by a factor of 5 or 20. The imaging plate, however, tolerates underexposure to some extent and easily accommodates overexposure by a factor of 20. Thus the linearity of the imaging plate in relation to the electron dose is much greater than in film material.

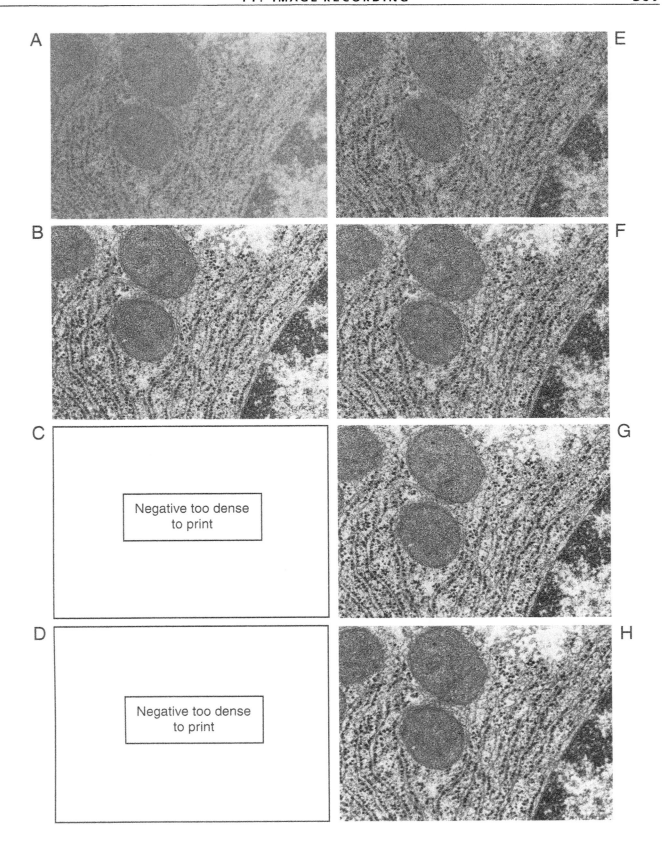

A

E

B

F

C

Negative too dense to print

G

D

Negative too dense to print

H

11. Corrections of CCD Camera

FIGURE 11.11A An electron beam in a Zeiss 912 Omega electron microscope recorded on a Proscan HSS 1024 × 1024 CCD chip. The illustrated image covers approximately 12 × 12 μm^2 of the 19 × 19-mm^2 chip. No background correction has been applied.

The image is uniform except for a few weak, diagonally oriented streaks indicating a faint honeycomb pattern.

FIGURE 11.11B Same image after the subtraction of background ("flat-field correction").

The image is now completely uniform in density.

FIGURE 11.11C Similar image as in Fig. 11.11A except that the CCD camera has been opened, whereupon the chip apparently acquired some undefined surface contamination.

The image shows many irregularities in the form of diffuse densities and distinct spots.

FIGURE 11.11D The image in Fig. 11.11C after background subtraction ("flat-field correction").

The image is now essentially free of background noise.

FIGURES 11.11E AND 11.11F A recording of mitochondria and cytoplasm of a rat liver cell with a CCD camera in a Zeiss 912 Omega microscope at high magnification before (Fig. 11.11E) and after (Fig. 11.11F) background correction ("flat-field correction").

In the uncorrected micrograph there are fine parallel stripes all over the image. The stripes appear in patches and vary somewhat in orientation. In the corrected image such stripes are absent.

COMMENTS

The possibility of correcting background irregularities is essential in CCD camera recordings as illustrated here. Uncorrected images of biological objects may show patterns that interfere with the biological interpretations.

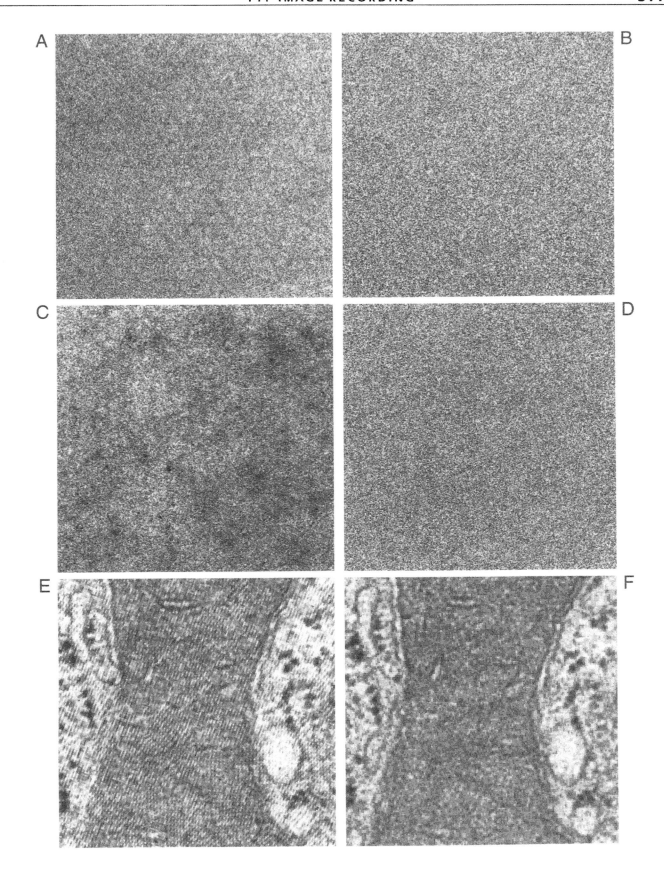

References

Aikens, R. S., Agard, D. A., and Sedat, J. W. (1989). Solid-state imagers for microscopy. *In* "Methods in Cell Biology" (Y.-L. Wang and D. L. Taylor, eds.), Vol. 29, pp. 291–313. Academic Press, San Diego.

Carlemalm, E., and Weibull, C. (1975). The response to electrons and developing conditions of two photographic films. *J. Ultrastruct. Res.* **53,** 298–305.

Chiu, W., and Glaeser, R. M. (1980). Evaluation of photographic emulsions for low-exposure imaging. *In* "Electron Microscopy at Molecular Dimensions" (W. Baumeister and W. Vogell, eds.), pp. 194–199. Springer, Berlin.

De Ruijter, W. J. (1995). Imaging properties and applications of slow-scan charge-coupled-device cameras suitable for electron microscopy. *Micron Microsc. Acta* **26,** 247–275.

Farnell, G. C., and Flint, R. B. (1975a). Exposure level and image quality in electron micrographs. *J. Microsc. (Oxford)* **103,** 319–332.

Farnell, G. C., and Flint, R. B. (1975b). Photographic aspects of electron microscopy. *In* "Principles and Techniques of Electron Microscopy: Biological Applications" (M. A. Hayat, ed.), Vol. 5, pp. 19–61. Van Nostrand-Reinhold, New York.

Hamilton, J. F., and Marchant, J. C. (1967). Image recording in electron microscopy. *J. Opt. Soc. Am.* **57,** 232–239.

Katoh, M., Fukushima, K., and Adachi, M. (1982). Application of Fuji Miniposi fiche film to electron microscope image recording. *J. Electr. Microsc.* **31,** 98–100.

Kodak (1993). "Electron Micrography. Using Electrons Effectively." Eastman Kodak Company, New York.

Kujawa, S., and Krahl, D. (1992). Performance of a low-noise CCD camera adapted to a transmission electron microscope. *Ultramicroscopy* **46,** 395–403.

Matricardi, V., Wray, G., and Parsons, D. F. (1972). Evaluation of emulsions and other recording media for 100 and 1000 kV electron microscopes. *Micron* **3,** 526–538.

Sherman, M. B., Brink, J., and Chiu, W. (1996). Performance of a slow-scan CCD camera for macromolecular imaging in a 400 kV electron cryomicroscope. *Micron* **27,** 129–139.

Spring, K. R., and Lowy, R. J. (1989). Characteristics of low light level television cameras. *In* "Methods in Cell Biology" (Y.-L. Wang and D. L. Taylor, eds.), Vol. 29, pp. 269–289. Academic Press, San Diego.

Unwin, P. N. T., and Henderson, R. (1975). Molecular structure determination by electron microscopy of unstained crystalline specimens. *J. Mol. Biol.* **94,** 425–440.

Valentine, R. C. (1965). Characteristics of emulsions for electron microscopy. *In* "Quantitative Electron Microscopy" (G. F. Bahr and E. Zeitler, eds.), pp. 596–602. Williams and Wilkins, Baltimore.

Valentine, R. C. (1966). The response of photographic emulsions to electrons. *In* "Advances in Optical and Electron Microscopy" (R. Barer and V. E. Cosslett, eds.), pp. 180–203. Academic Press, New York.

Zeitler, E. (1992). The photographic emulsion as analog recorder for electrons. *Ultramicroscopy* **46,** 405–416.

Zeitler, E., and Hayes, J. R. (1965). Electrography. *In* "Quantitative Electron Microscopy" (G. F. Bahr and E. Zeitler, eds.), pp. 586–595. Williams and Wilkins, Baltimore.

PHOTOGRAPHIC AND DIGITAL PRINTING

Electron micrographs are processed either photographically or electronically. The photographic treatment resembles ordinary darkroom work in many ways and involves enlargement of the negatives on photographic paper. Because of the special characteristics of the electron micrograph, certain considerations are necessary. Thus, the processing may differ for electron micrographs that are used for visual analysis, for measurements, or for publication. It will also be different for negatives with a high and a low inherent contrast and for negatives with a wealth of fine details or for those with only a few grades of density. Even excellent electron micrographs will lose much of their information by inadequate photographic processing, whereas electron micrographs of suboptimal quality may turn out to be useful if skillfully treated.

A large number of different automatic development systems exist for darkroom work. The type of photographic paper used in these machines is in several aspects different from classical photographic paper. A new system has been developed whereby the contrast level of the image is regulated not by the choice of different grades of paper, but by the color of the light illuminating the negative.

In addition to photographic printing of negatives, the classic way in electron microscopy, electron micrographs are now often printed electronically. The micrographs either are recorded digitally in the microscope or are first processed in the usual way photographically and the photographic image recorded with a digital scanner. The image is then printed digitally with the aid of any of the several types of electronic printers now available.

However, the digital printers vary considerably with respect to image characteristics, including resolution, contrast, and speed of printing. The use of digital printing of micrographs was expensive initially, but prices of equipment and paper are decreasing rapidly and their use is increasing correspondingly.

This chapter illustrates some of the modifications that electron micrographs may undergo as a result of different treatments during the photographic printing procedure and illustrates some aspects of digital printing of micrographs.

The classical histologists Camillo Golgi, Santiago Ramon y Cajal, and Gustaf Retzius were masters in making drawings and had to be: Their light microscope observations could rarely, if ever, be recorded on photographic material. Electron microscopists do not need this skill and are actually regarded with suspicion if they present their findings with drawings only rather than with micrographs. However, they have to be good photographers and be familiar with all procedures in the darkroom. The increasing use of digital image recordings in the microscope and digitalization of micrographs in recent years are gradually changing the situation; many electron microscope images now bypass the photographic darkroom entirely as the enlarger is replaced by a computer and digital printer.

1. Photographic Paper of Different Grades

FIGURES 12.1A–12.1C Parts of cross-sectioned epithelial cells from a rat colon, fixed in osmium tetroxide, and embedded in Epon. The section was stained with lead hydroxide, and the electron microscope negative was enlarged on soft (Fig. 12.1A), normal (Fig. 12.1B), and hard (Fig. 12.1C) paper. × 10,000.

In comparison with the micrograph printed on normal paper (Fig. 12.1B), the soft print (Fig. 12.1A) has a generally gray tone. Thus, different types of dense bodies appear less black and the cytoplasmic vacuoles less white; as a result the picture appears less crisp than in Fig. 12.1B. In contrast, the micrograph printed on hard paper (Fig. 12.1C) shows a wider range of photographic densities. The dense bodies are almost black, the vacuoles almost white, and the ground cytoplasm is very light. As a result the micrograph appears more crisp. The matrix granules of the mitochondria stand out less distinctly than in Fig. 12.1B or even less than in Fig. 12.1A. There is a difference in general density between the left and the right part of the micrograph, and this difference appears more pronounced in the contrasty Fig. 12.1C than in the soft Fig. 12.1A.

COMMENTS

Considerable variations in the appearance of an electron micrograph are imposed by the choice of photographic paper, particularly by its grade. A print with adequate contrast should span essentially the entire black and white scale. The present examples do not represent the extremes in the range of photographic grades. Although micrographs on hard paper usually appear sharp and contrasty and seem to have much information, fine details are often hidden in the very light or very dark regions, e.g., the mitochondrial granules to the right in Fig. 12.1C.

For routine purposes, micrographs on the soft side are generally preferred as they allow the detection of all objects in the image. For publication purposes, micrographs of slightly higher contrast are preferable, particularly in journals having a nonglossy paper. Too contrasty prints could, however, give published micrographs a washed-out appearance of the cytoplasm.

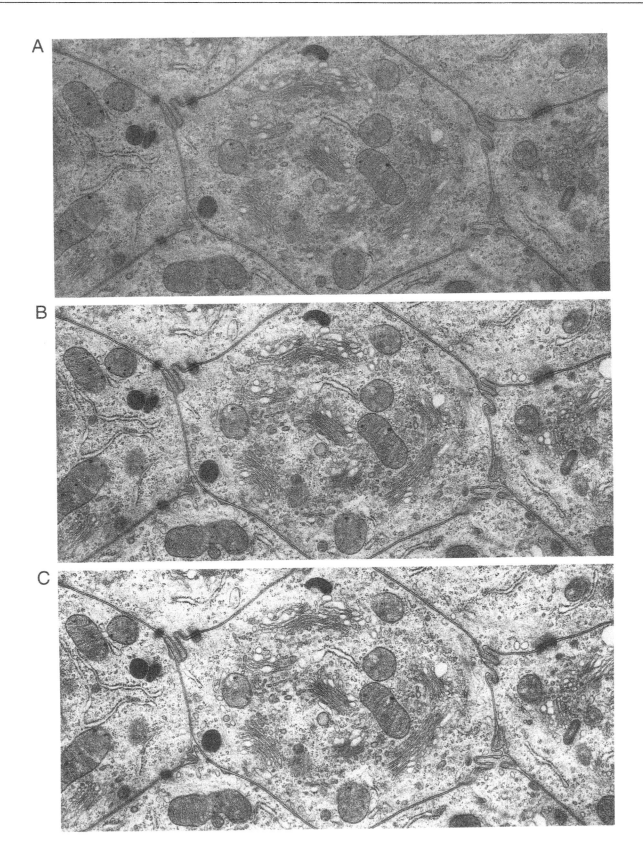

2. Multigrade Paper

FIGURES 12.2A–12.2C The same negative used for Figs. 12.1A–12.1C was enlarged here on so-called multigrade paper using the Ilford multigrade 500H system. In Fig. 12.2A the exposure was made with an overweight of green light, Fig. 12.2B with a mixture of blue and green light, and Fig. 12.2C with predominantly blue light. × 10,000.

The contrast in these three images ranges from low contrast in Fig. 12.2A to high contrast in Fig. 12.2C with Fig. 12.2B in between.

COMMENTS

The contrast differences between micrographs printed on multigrade paper depend on the fact that the emulsion of the paper consists of a mixture of an emulsion sensitive to blue light and an emulsion sensitive to blue light and also sensibilized for green light. Thus one part of the emulsion is only sensitive to blue light, but the other part is sensitive to both blue and green light. Both emulsions have the same sensitivity for blue light, but the nonsensibilized has a very low sensitivity for green light. The combined property of the photographic emulsion is that blue light will give pictures with high contrast whereas exposure with green light will result in low-contrast prints. Thus by changing the proportion between blue and green light, the contrast of the print will change between high and low contrast. The variation between the light properties of the light can be done either with filters or by using different types of lamps. The advantage in the routine photographic laboratory is that a single type of photographic paper suffices for the darkroom work. The softest images on multigrade paper obtained with the green light are similar to prints on what is usually labeled grade 1 paper. However, in our experience it is still necessary to use high-contrast paper corresponding to grades 4 1/2 or 5 in order to obtain prints of very high contrast.

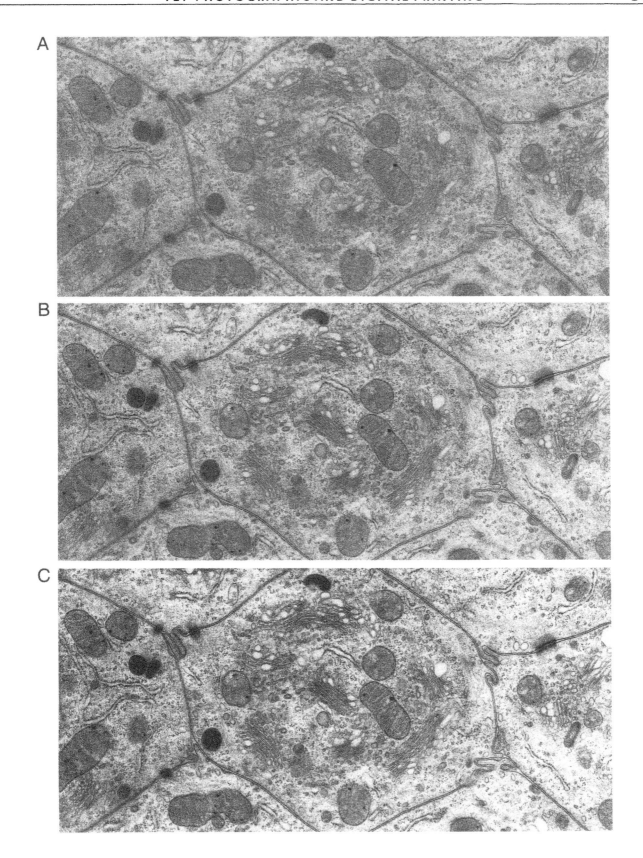

3. Exposure and Development

FIGURES 12.3A–12.3F Sperm flagella fixed in glutaraldehyde, postfixed in osmium tetroxide, and embedded in Epon. The section was stained with uranyl acetate and lead citrate. These figures are enlargements from the same negative using the same photographic paper but with different combinations of exposure time and development. All enlargements were made with aperture 4.5 in the enlarger. × 170,000.

Figure 12.3A shows a print given an exposure of 2.5 sec and hence underexposed. The paper was developed under standard conditions (3 min at 20°C). Many structural details such as the connections between the microtubules are almost absent.

Figure 12.3B was also made with an exposure of 2.5 sec, but the paper was developed for 10 min, which is more than three times the normal time. This micrograph shows much more detail, contrast, and density than Fig. 12.3A.

Figure 12.3C was also underexposed, but was developed for 3 min at high temperature (45°C) to accelerate the developing process. The micrograph has a much higher contrast and density than Fig. 12.3A and contains more details.

Figure 12.3D shows a normally exposed (in this case 5 sec) print processed under standard conditions (3 min). This micrograph was judged to give an optimal structural information in a single micrograph.

Figure 12.3E is overexposed (10-sec exposure) and developed under standard conditions. The entire micrograph is very dark and much detail of the object has been lost.

Figure 12.3F was overexposed (10 sec), but developed for only 30 sec to prevent the structures from becoming too dark. The micrograph is almost similar in density to Fig. 12.3D. The structural information obtained from this micrograph is essentially the same as in Fig. 12.3D.

COMMENTS

Much structural information is lost in suboptimally processed prints. Modifications of the developing procedures can, at least in part, compensate for too short or too long exposure times (Figs. 12.3B, 12.3C, and 12.3F). These modifications can also be used to improve parts of the enlargement when manual development is used instead of a developing machine. It is, for example, possible to overdevelop part of a photograph that appears too light by prolonging the development of that part of the paper while the photographic process in the rest of the paper is terminated in a stop bath. In the opposite situation the developmental process can be accelerated by increasing the temperature, e.g., by warming part of the paper by rubbing it with a finger during development.

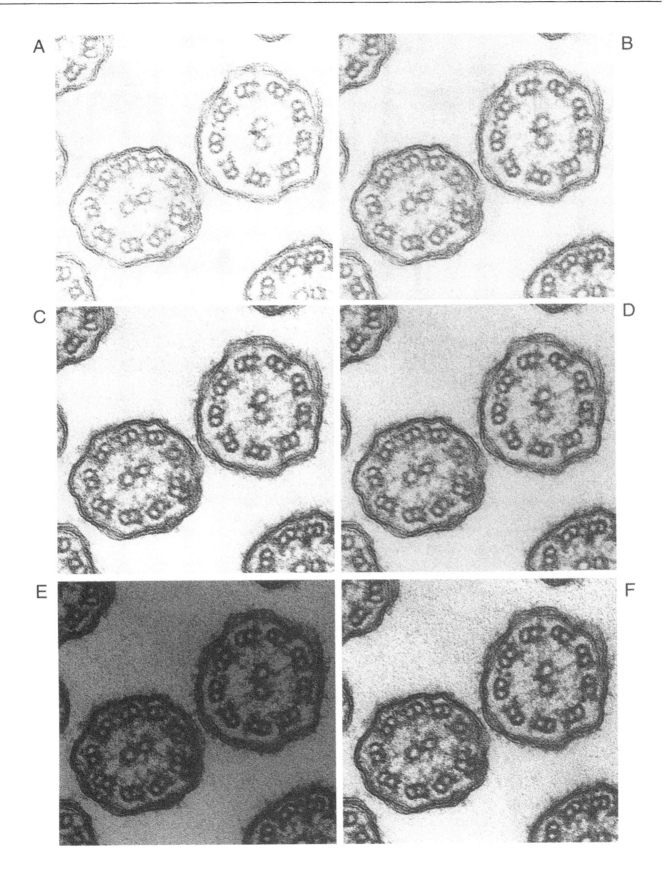

4. Enlargement of Micrograph Details

FIGURES 12.4A–12.4F Myelin-like material in urinary sediment fixed with osmium tetroxide, embedded in Epon, and section stained with lead citrate. Figures 12.4A–12.4E are all enlargements from a negative with an electron optical magnification of 25,000 times, whereas Fig. 12.4F is enlarged from a negative with an electron optical magnification of 15,000 times. All prints were made on the same grade of paper, developed, and fixed in the same way and all have a final magnification of about × 375,000.

In Fig. 12.4A the negative was enlarged 15 times in an enlarger with a regular tungsten lamp bulb and using aperture f16. In Fig. 12.4-B the negative was enlarged as in Fig. 12.4A except that the aperture was opened to f5.6 and the exposure time was shortened correspondingly. In Figs. 12.4C and 12.4D the enlarged part of the negative was placed off center in the enlarger and the exposure was carried out with apertures f16 (Fig. 12.4C) and f5.6 (Fig. 12.4D). In Fig. 12.4E the negative was enlarged using a point-source lamp. In Fig. 12.4F the negative was enlarged 25 times using the same condition of enlargement as in Fig. 12.4A.

A close inspection of these six micrographs reveals a number of differences with respect to definition of structural details. In comparison with Fig. 12.4A, which is the preferred print, the structural definition in Fig. 12.4B is slightly less clear and Fig. 12.4C is even more so. Figure 12.4D appears somewhat blurred compared to Fig. 12.4A, and the points indicated by arrows in Fig. 12.4A are not resolved clearly. Figure 12.4E is similar to Fig. 12.4A and actually appears as more crisp, but a white line is present at the upper right corner. Much less structural detail is present in Fig. 12.4F than in Fig. 12.4A.

COMMENTS

With a good objective lens in the enlarger, there are hardly any differences in the image quality of prints made with different apertures between f5.6 and f16 (Figs. 12.4B and 12.4A). However, there is an appreciable deterioration in image quality if the negative is placed off center in the enlarger, particularly when a large aperture is used (Fig. 12.4D). When a point-source lamp is used the image is very sharp, but dust on the negative shows up easily (Fig. 12.4E).

Comparison between the two last micrographs in this series shows that fewer details are demonstrable in the micrograph with low electron optical magnification and high photographic enlargement (Fig. 12.4F) than with the reverse combination. Normally, electron optical noise and film granularity start to become apparent above a photographic enlargement of 10–15 times.

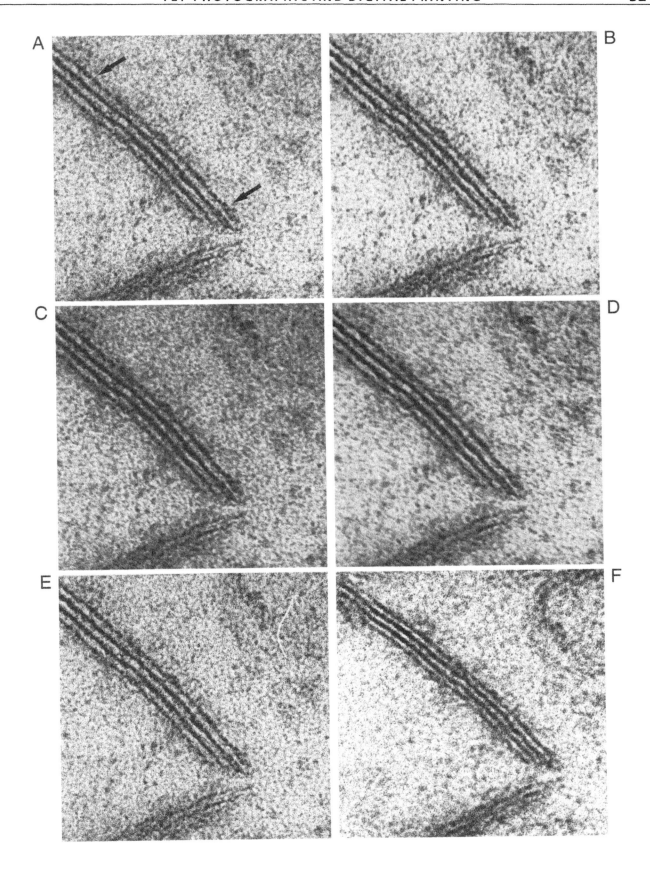

5. Objective Lens in Enlarger

FIGURES 12.5A–12.5C These three enlargements were made from the same negative, which shows a kidney tubule cell. Figures 12.5A and 12.5B were made with an objective lens of 50 mm focal length, whereas Fig. 12.5C was made with a lens of 105 mm focal length. All prints were exposed to the same density, but the opening of the lens was f2.8 in Fig. 12.5A, f16 in Fig. 12.5B, and f8 in Fig. 12.5C. × 85,000.

The exposure is uneven in Fig. 12.5A, with the periphery of the enlargement much lighter than its center. This is to some degree also noticeable in Fig. 12.5B, although to a lesser extent, whereas illumination in Fig. 12.5C is even over the entire field.

COMMENTS

The focal length of the objective lens in the enlarger must be chosen to accommodate the size of the negative to be enlarged. If the focal length of the objective lens is too short, the periphery of the field will be underexposed, especially with a relatively large aperture. With a long focal length of the objective an even illumination of the same field will already be obtained with an intermediate aperture but not with a maximum opening of the aperture. As a rule of thumb, the focal length of the lens should be approximately equal to a diagonal in the negative.

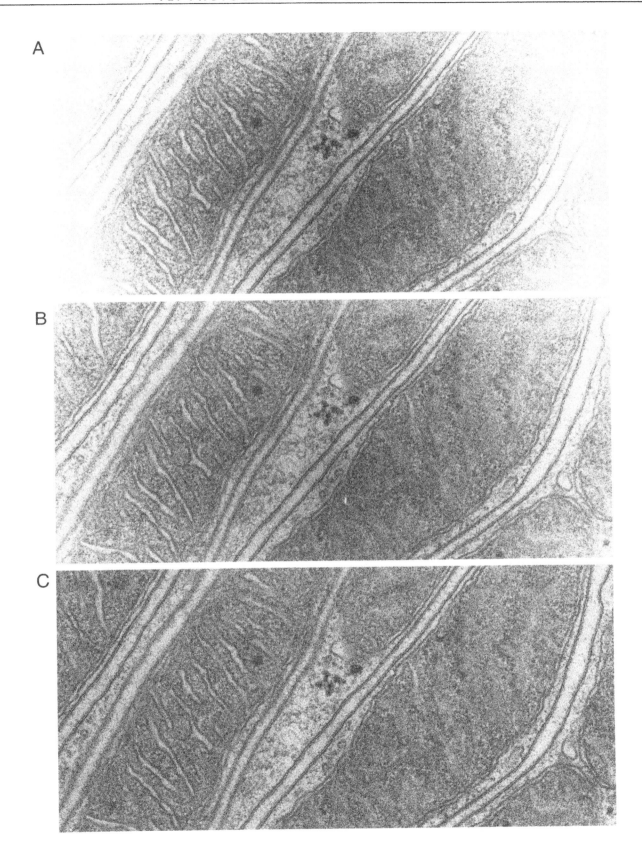

6. Focusing of Enlarger

FIGURES 12.6A–12.6D These four enlargements were made from the same negative and illustrate a thin section of kidney tubule cells. The objective lens of the enlarger had a focal length of 105 mm and the prints were exposed to the same density. Figure 12.6A was made with an aperture setting of f2.8 and was focused carefully, whereas the aperture setting in Fig. 12.6B was f16 without any readjustment of the focus. Figure 12.6C was made with the aperture setting of f2.8 and was slightly overfocused, whereas Fig. 12.6D was made at the same focus, although with the aperture shut down to f16. × 130,000.

Figures 12.6A, 12.6B, and 12.6D are indistinguishable even on close examination of the original prints, whereas Fig. 12.6C has a blurred and soft appearance.

COMMENTS

It is evident that the aperture setting of the enlarger has only little influence on the quality of the final print if the negative is well focused. Thus, if the negative is excessively dense, i.e., overexposed, it can be enlarged with an open aperture in the enlarger to cut down the exposure time, provided that the enlarger is focused properly. However, the open aperture will result in loss of even illumination (compare Fig. 12.5). A large aperture combined with incorrect focusing of the negative invariably gives a blurred image, but with a small aperture a certain amount of defocusing is tolerated.

Various aids in focusing the negative in the enlarger exist; they provide an extra magnification and make the photographic granules visible.

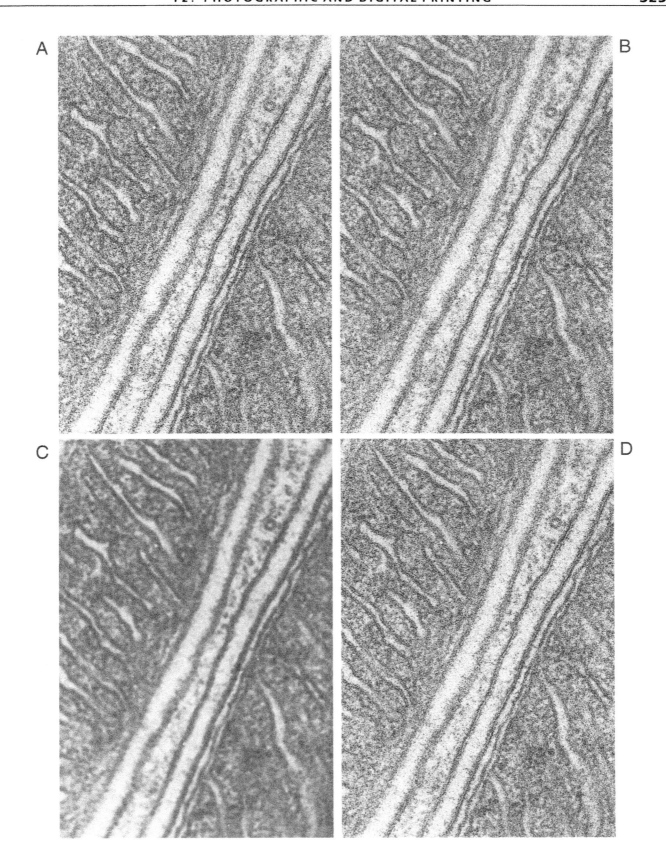

7. Intermediate Diapositive

FIGURE 12.7A A muscle cytoplasm fixed with glutaraldehyde and osmium tetroxide, embedded in Epon, and examined unstained in the electron microscope. The original photographic negative was enlarged on high-contrast photographic paper. × 25,000.

The micrograph has low contrast.

FIGURE 12.7B A negative print of the same cytoplasmic field as in Fig. 12.7A. The original electron micrograph negative was copied onto an Agfa Gevaert Scientia film (i.e., the same film that was used to record the original negative in the microscope), and a print was made from this film. × 25,000.

The print resembles the original photographic negative recorded in the microscope.

FIGURE 12.7C A new negative was made from the intermediate diapositive used for Fig. 12.7B and finally the photographic print in Fig. 12.7C was made from it. × 25,000.

By photographing the negatives onto one or more intermediate films a considerable gain in contrast has been achieved along with some loss in resolution and gradation of image density. The three figures have all been printed on the same grade of photographic paper and processed at the same time in the same printing machine.

FIGURE 12.7D Polystyrene latex spheres shadow cast with platinum at an angle of about 35°. The polystyrene particles have a diameter of 0.088 μm. × 120,000.

The latex spheres have white shadows.

FIGURE 12.7E The same polystyrene particles as in Fig. 12.7D but in a negative print. The print was made from an intermediate diapositive of Agfa Gevaert Scientia film. × 120,000.

Structures that appear black in Fig. 12.7D appear white in Fig. 12.7E and vice versa.

COMMENTS

Intermediate photographic films have two principal applications in the electron microscopical technique. One is the possibility of producing a photographic print of high contrast from an original negative with a very low contrast, as exemplified in Figs. 12.7A and 12.7C. Two intermediate photographic films have to be made (unless a positive-direct type of film material is used) and the method is hence somewhat tedious. In the other application, only one intermediate diapositive is used and a reversal of the black and white scale is obtained. This is sometimes desired in presentations of freeze-fracture images. In case of metal shadowing, a negative print is usually easier to interpret as it will show dark shadows.

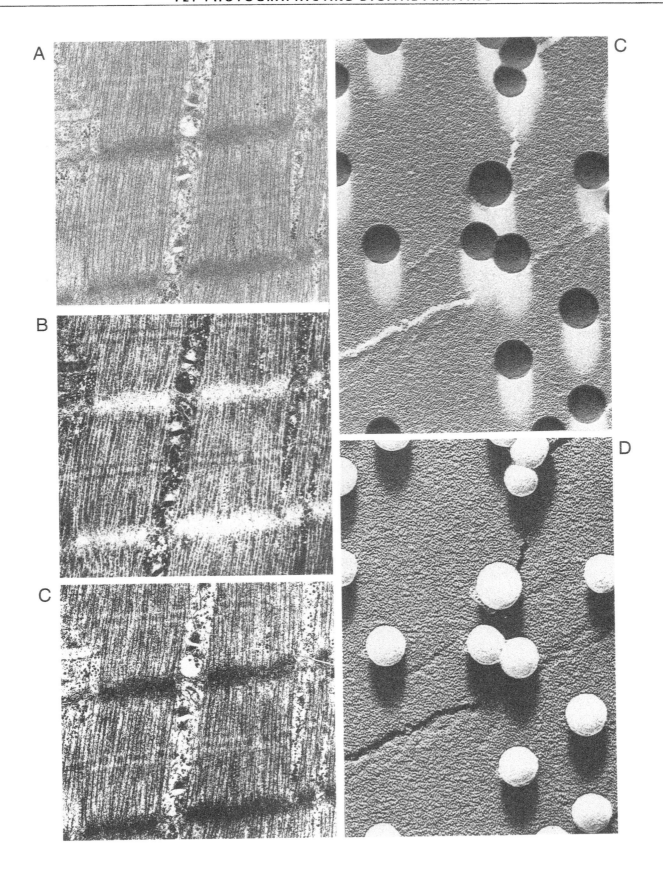

8. Errors in Photographic Printing

FIGURES 12.8A–12.8C Part of a renal glomerulus from a rat kidney fixed in osmium tetroxide, embedded in Epon, and section stained with lead citrate. Before exposure in the enlarger, the photographic paper was splashed with drops of water (Fig. 12.8A), developer (Fig. 12.8B), or fixing solution (Fig. 12.8C). × 8000.

A comparison among these three prints shows that splashed water caused light spots (Fig. 12.8A), developer caused dark spots (Fig. 12.8B), and fixing solution caused white spots (Fig. 12.8C).

FIGURES 12.8D–12.8F Prints from the same negative used for Figs. 12.8A–12.8D, although a somewhat different area was enlarged. In Fig. 12.8D the paper was left with the emulsion side up in the developer without agitation and a central band on the paper surfaced. When printing Fig. 12.8E, several exposed papers were developed at the same time without proper agitation. In Fig. 12.8F the developed paper was placed in the fixing solution with its emulsion side up. The paper was submerged insufficiently in the solution and part of the paper touched the surface. × 8000.

In Fig. 12.8D the white central band is independent of the glomerular structures and, similarly, the upper right portion of Fig. 12.8E is less blackened than in the control picture. Figure 12.8F has an irregular shadow in the center, which actually has a violet tone in the original print.

COMMENTS

The different types of photographic defects illustrated here are caused largely by sloppiness in the darkroom and are easily recognized as artifacts: insufficient development of the surfaced part of the paper, underdevelopment due to partial adherence of the papers in the developer, or too little fixing solution reaching the emulsion. Incomplete development of a micrograph can, however, also be used as an aid to suppress or even remove dense parts of the micrographs, e.g., by intentionally shortening or eliminating the development of these parts of the photographic paper.

Defects similar to those shown here may be difficult to interpret when superimposed on complex biological patterns, but will only rarely be seen in laboratories where the processing of the paper is performed by an automated machine. Such a processing, however, sometimes gives rise to another family of defects in the final print, consisting primarily of parallel streaks and lines.

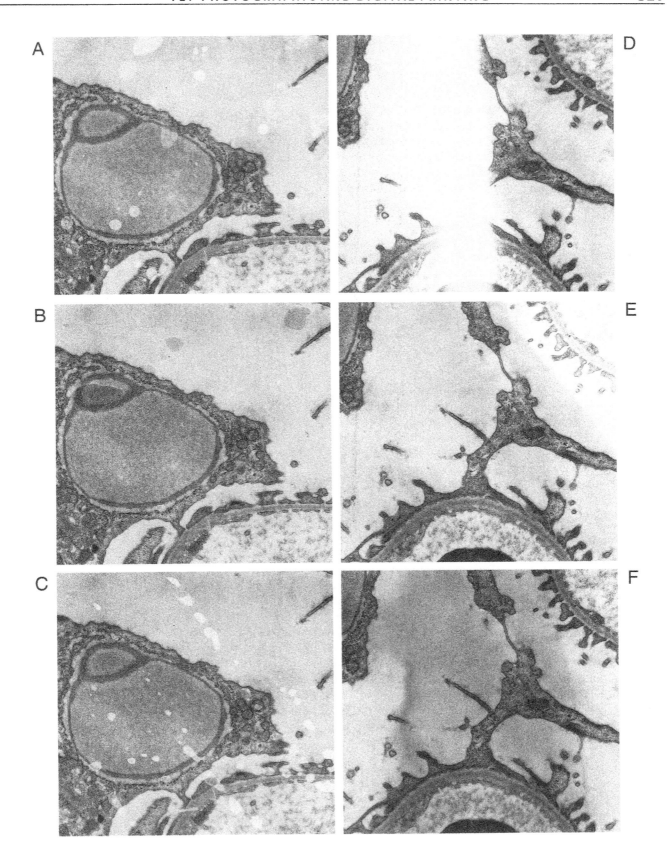

9. Retouch

FIGURES 12.9A–12.9C Parts of kidney cells from negative printed before (Fig. 12.9A) and after (Fig. 12.9B) some contaminating particles had been put onto the negative. A second print (Fig. 12.9C) was used for retouch of the white spots. First the thin surface layer of the paper over the white spots was carefully scraped off with a knife, then the resulting rough surface was dotted with a sharp, soft pencil. × 75,000.

The micrograph shows normal cellular structures without obvious technical faults in Fig. 12.9A, and with the contaminating particles appearing as white spots in Fig. 12.9B. In Fig. 12.9C the white spots are absent.

FIGURES 12.9D AND 12.9E A conventional transmission electron micrograph of a liver cell cytoplasm before (Fig. 12.9D) and after (Fig. 12.9E) the two irregular, dense spots have been removed with a sharp knife and the resulting white areas darkened with a sharp, soft pencil. × 25,000.

In addition to ribosomes and mitochondria Fig. 12.9D shows two irregular, very electron-dense spots of unknown origin and composition. The areas where the dense spots were located are barely, if at all, distinguishable from the adjacent cytoplasm.

FIGURES 12.9F AND 12.9G Part of kidney cells showing normal cell organelles including ribosomes and mitochondria with matrix granules (Fig. 12.9F) and the same field after all electron-dense areas have been removed in the same way as in Fig. 12.9E. × 75,000.

Figure 12.9F shows normal cell structures but in Fig. 12.9G the sites where the ribosomes and granules have been removed can hardly be distinguished from the unretouched parts.

COMMENTS

Retouching requires no fancy equipment or great skill and can be done either with a sharp knife and pencil or with professional retouch color. The first consideration though is whether the dense or light objects are indeed technical artifacts or whether they are part of the biological structure. It is evident that the white spots in Fig. 12.9B are contaminants on the negative in the enlarger. The dark spots in Fig. 12.9D, however, were already visible in the microscope on the section, but there is little doubt that they are contaminants because they are dissimilar to known organelles and because they overlie the cytoplasm in a random fashion. Removal of such white or black spots has no consequence for the interpretation of the biological information, yet the resulting images may seem more pleasing to the eye, and retouching in these particular cases is therefore mainly an aesthetic question. The situation is quite different when biological information is modified or removed; the retouching in Fig. 12.9G is therefore completely unacceptable. Such a retouch should be considered scientific fraud.

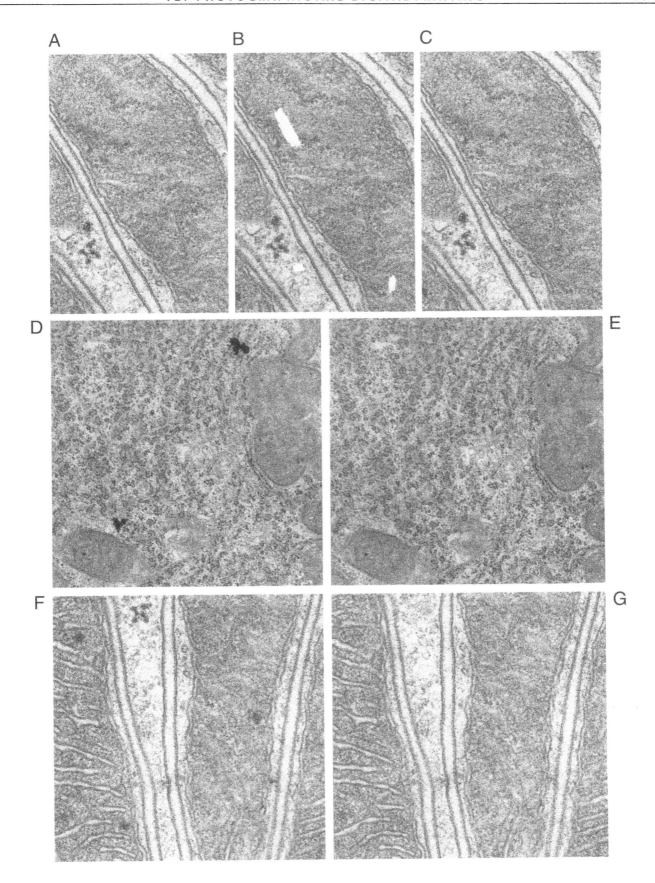

10. Comparison of Printers: Low Magnification

FIGURES 12.10A–12.10D Micrographs of a liver cytoplasm. The photographic print (Fig. 12.10A) was first scanned at 400 dpi (dots per inch) with a UMAX Vista-S8 scanner and then printed with the aid of three different printers: a 300-dpi Kodak 8650 PS color printer (Fig. 12.10B), a 300-dpi Tektronix Phaser 440 (Fig. 12.10C), and a 600-dpi HP4 laser scanner (Fig. 12.10C). The same file was used for the three different prints. × 40,000.

Certain differences can be observed among Figs. 12.10B–12.10D. The contrast of Fig. 12.10C appears slightly greater than in Fig. 12.10B. Figure 12.10D, however, is similar in contrast to Fig. 12.10A, but is not as sharp as Figs. 12.10B and 12.10C.

COMMENTS

At a low magnification the printed images of Figs. 12.10B–12.10D in this particular file are reasonably similar, although the laser printer gives a less sharp image. With the naked eye it is difficult to distinguish the photographic print (Fig. 12.10A) from the scanned and printed figures (Figs. 12.10B and 12.10C). Only upon close examination does Fig. 12.10A seem somewhat sharper than Figs. 12.10B and 12.10C, although no pixels can be resolved. Even if the laser printer does not give quite as crisp an image as the photographic print (Fig. 12.10A) or the prints in Figs. 12.10B and 12.10C, the very low price of such a print makes it attractive for different types of routine work.

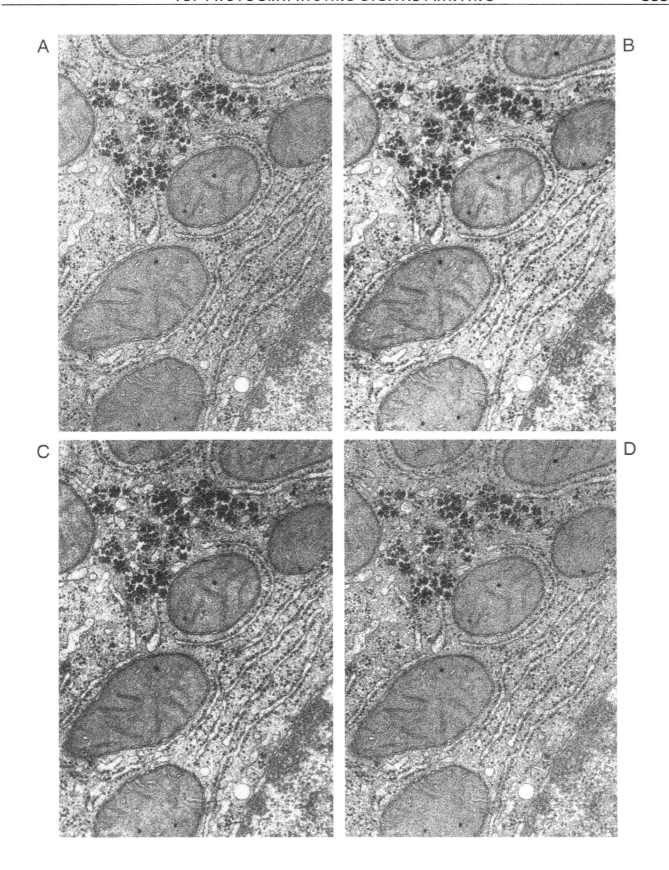

11. Comparison of Printers: Enlarged Prints

FIGURES 12.11A–12.11D The central regions of Figs. 12.9A–12.9D were microphotographed and enlarged 14 times. Total magnification: × 560,000.

Considerable differences are apparent among the enlarged prints. A distinct pixel pattern is evident in Fig. 12.11C and consists of light lines separating the pixels. A horizontal faint line pattern is also observable in Fig. 12.11B, but the pixels are not as sharply outlined as in Fig. 12.11C and structural contours appear smoother. The 14 times enlargement from the laser print (Fig. 12.11D) shows a very coarse pattern consisting of dots of different diameters and lacks sharp contours of the biological structures.

COMMENTS

At the great enlargements illustrated in this plate, distinct differences exist among the printed images shown at low magnification in Figs. 12.10A–12.10D. Although there is a faint line pattern in the Kodak 8650 printer, the biological objects appear well defined, almost as in a photographic print. The Tektronics printer visualizes the pixels more clearly and also provides a reasonable definition of biological structures such as the ribosomes. The enlargement of the laser print shows poorly defined objects such as membranes and ribosomes. The photographic print (Fig. 12.11A) shows more detail, but due to the 42-fold enlargement (× 3 from negative, ×14 from photographic print), the image has a grainy appearance.

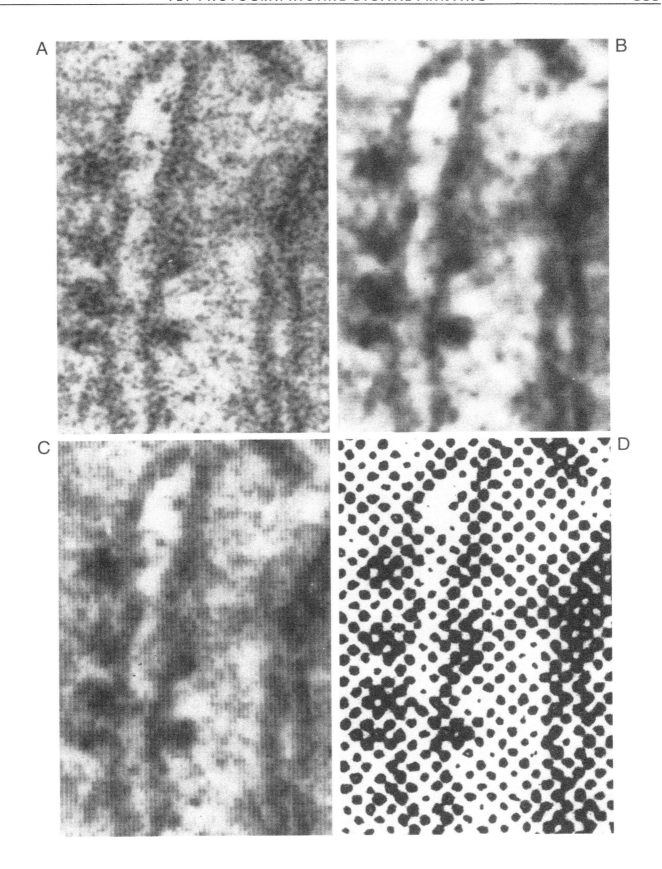

12. Pixel Size at Printing

FIGURES 12.12A–12.12F An electron micrograph of a liver cell cytoplasm after conventional specimen preparation with glutaraldehyde, osmium tetroxide, and Epon embedding. This image was scanned from the original photographic enlargement using 400 dpi with a UMAX Vista-S 8 scanner and then printed with a Kodak 8650 PS color printer using 300 dpi (Fig. 12.12A), 150 dpi (Fig. 12.12B), 75 dpi (Fig. 12.12C), 30 dpi (Fig. 12.12D), 15 dpi (Fig. 12.12E), and 7 dpi (Fig. 12.12F). × 32,000.

Micrographs A–F show a gradually lower resolution. The image deterioration between the micrographs printed at 300 and 150 dpi is hardly noticeable, but the image already appears blurred at 75 dpi (Fig. 12.12C). At 30 dpi the structure is hardly discernable, and at 15 dpi, and even more so at 7 dpi, identification of the object has become impossible.

COMMENTS

To the naked eye, Fig. 12.12A, which is printed with 300 dpi, has almost the same quality as a photographic print, such as that seen in Fig. 12.10A. With decreasing dots per inch, the image quality deteriorates as shown in this plate, but a print at 150 dpi is just acceptable even to a person with keen eyesight. When looked at from a distance of some 2–3 m, however, the four first figures appear fairly similar, and in the fifth (Fig. 12.12E) and even in the sixth (Fig. 12.12F), some structural patterns that are not appreciated at normal reading distance become vaguely indicated.

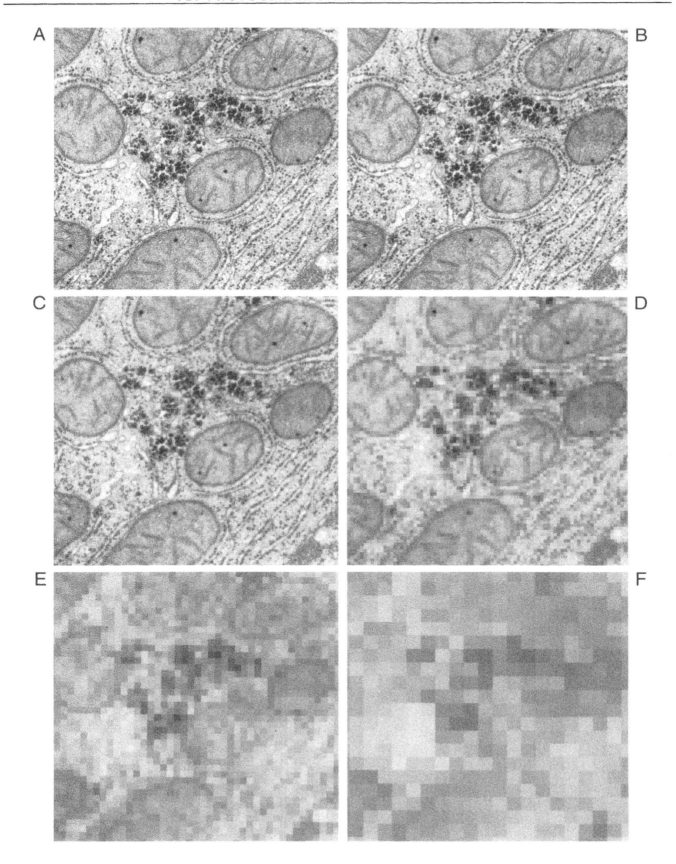

References

Farnell, G. C., and Flint, R. B. (1975). Photographic aspects of electron microscopy. *In* "Principles and Techniques of Electron Microscopy" (M. A. Hayat, ed.), pp. 19–61. Van Nostrand-Reinhold, New York.

Gonzales, F. (1962). A masking technique for contrast control in electron micrographs. *J. Cell Biol.* **15,** 146–150.

Hamilton, J. F. (1968). Use of unsharp masking in printing electron micrographs. *J. Appl. Phys.* **39,** 5333–5334.

James, T. H. (1977). "The Theory of the Photographic Process," 4th Ed. Macmillan, New York and Collier Macmillan, London.

Johannessen, J. V. (1978). Photographic techniques. *In* "Electron Microscopy in Human Medicine: Instrumentation and Techniques" (J. V. Johannessen, ed.), Vol. 1, pp. 311–325. McGraw-Hill, New York.

Mees, C. E. K., and James, H. (1966). "The Theory of the Photographic Process," 3rd Ed. Macmillan, New York.

Premsela, H. F., Nieuwenhuizen, J. M., and Schotanus, B. (1982). "Photographic Techniques for Electron Microscopy," 2nd revised Ed. Philips Industries, Eindhoven, The Netherlands.

Simonsberger, P., Lametschwandtner, A., Salzer, G., Albrecht, U., and Adam, H. (1977). Eine einfache Reisstechnik zur Herstellung grossflächiger Fotomontage von licht-und elektronenmikroskopischen Aufnahmen. *Mikroskopie (Wien)* **33,** 277–282.

Wergin, W. P., and Pooley, C. D. (1988). Photographic and interpretive artifacts. *In* "Artifacts in Biological Electron Microscopy" (R. F. E. Crang and K. L. Klomparens, eds.), pp. 175–204. Plenum Press, New York/London.

NEGATIVE STAINING

Negative staining was introduced in the 1950s and rapidly became the method of choice for the analysis of macromolecules, membranes, organelles, and microorganisms. It is a simple and rapid technique that makes it possible to characterize small structural features of biological objects in a reproducible manner. Its usefulness is based on the fact that it requires little instrumentation and yields high contrast micrographs with a high resolution. The interactions of large particles or particle complexes with the negative stain are quite variable and while many viruses seem to retain their viability, other structures such as mitochondria are deformed considerably. The resolution obtainable with negative staining is limited by the graininess of the stain but may reach about 15–20 Å in optimal preparation, thus almost invariably better than the object resolution of 20–40 Å achievable in sectioned material.

The method of negative staining has several variables that affect the outcome, including choice of stain and its pH and interaction or inertness toward the biological object, properties of the support film, procedure for applying the stain to the object, and concentration of the object. The functions of the negative stain solutions are twofold: To provide contrast against which objects can be seen and to encapsulate small objects so that they will stand the surface tension forces during drying of the specimen; this last function alone is used when small objects are encapsulated with glucose, trihalose, or tannic acid. Additionally, microscopy of the specimen requires precautions in order to reduce the influence of beam damage of the object or of the stain.

This chapter exemplifies variables associated with the procedures in negative staining, including a comparison of methods to make the support film hydrophilic.

The first paper describing a negative staining method in some detail was published in 1959 by Sidney Brenner and Robert (Bob) Horne. However, the usefulness of the method was first realized by Hugh Huxley in his paper from the Stockholm Conference on Electron Microscopy in 1956: "A very dilute solution of tobacco mosaic viruses was placed on carbon-coated electron microscope grids and stained with 40% phosphotungstic acid in the manner described by Cecil Hall (1955). The results at first seemed disappointing, for even when the resolution was better than 10 Å, no regular internal structure was visible in the particles. However, a curious effect came to light that revealed one feature of the internal structure." This internal structure was a hollow core in the virus rods, which themselves became outlined by the electron-dense phosphotungstic acid. Huxley concluded: "The outlining technique would appear to be quite a useful one for this type of specimen, particularly as it is so simple and gives excellent contrast and resolution." So Cecil Hall in 1955, before Hugh Huxley, had used what was later called the negative staining method, without pointing out its value, and Friedrich Krause had actually seen a similar outlining phenomenon already in 1937.

Brenner, S., and Horne, R. W. (1959). *Biochim. Biophys. Acta* **34**, 103–110.
Hall, C. E. (1955). *J. Biophys. Biochem. Cytol.* **1**, 1–12.
Huxley, H. E. (1956). "Proceedings of the First European Regional Conference on Electron Microscopy, Stockholm" (F. S. Sjöstrand and J. Rhodin, eds.), pp. 260–261.
Krause, F. (1937). *Naturwissenschaften* **25**, 817–825.

1. Negative Staining Methods

FIGURES 13.1A AND 13.1B Na,K-ATPase membranes negatively stained with uranyl acetate by a "stain-on-grid" method. A 5-μl droplet (0.1 mg protein/ml) was first applied onto the grid for 1 min before excess specimen was withdrawn. The grid was then placed on a droplet of 1% uranyl acetate in distilled water. After 1 min excess stain was withdrawn with filter paper and the grid dried. Figure 13.1A × 30,000, Fig. 13.1B × 120,000.

Membrane fragments are distributed evenly on the support film and embedded in a thin layer of electron-dense stain.

FIGURES 13.1C AND 1.31D The same membrane sample as in Fig. 13.1A stained by a "grid-on-stain" method. The membranes were first adsorbed on the grid for 1 min, the excess was removed, the grid rinsed on two drops of distilled water, and after removal of excess water the grid was placed onto a droplet of 1% uranyl acetate stain solution for 1 min. Figure 13.1C × 30,000, Fig. 13.1D × 120,000.

The distribution of membranes and the characteristics of the staining are similar to that seen with the procedure in Figs. 13.1A and 13.1B.

FIGURES 13.1E AND 13.1F The same membrane sample as in Fig. 13.1A stained by a one-step method. Five microliters of 1% uranyl acetate was mixed with 5 μl of specimen solution (0.1 mg protein/ml). After 1 min excess fluid was withdrawn and the grid dried. Figure 13.1E ×30,000, Fig. 13.1F × 120,000.

In this one-step method there is a tendency for membranes in some samples to aggregate as seen in Fig. 13.1E, but the quality of staining of single membrane fragments is essentially similar to that in Figs. 13.1D and 13.1E.

FIGURES 13.1G AND 13.1H. Latex particles (diameter 0.088 μm) sprayed twice with a commercial vaporizer: once in distilled water and once in 2% sodium phosphotungstate. × 57,000.

The right portion of Fig. 13.16 shows an aggregate of unstained particles that were sprayed in distilled water. Latex particles sprayed in sodium phosphotungstate can be seen to the left. Particles in Fig. 13.1H appear smaller in the center than at the periphery of the negative stain.

COMMENTS

A great repertoire of different modifications of the negative staining technique has been developed over the years; four are illustrated here. In the beginning, spraying techniques predominated but they may present health hazards as they distribute viruses and toxic agents.

A comparison of Figs. 13.1A and 13.1C shows similar distributions of the membrane fragments; few differences, if any, are observed even at high magnifications. The "grid-on-stain" procedure can be extended with several intermediate steps, such as washing the grid in distilled water to remove, for example, salts of the specimen solution (Bremer *et al.*, 1998) or to include exposure to antibodies in order to label specific antigens, so-called immunonegative staining (see Fig. 16.17).

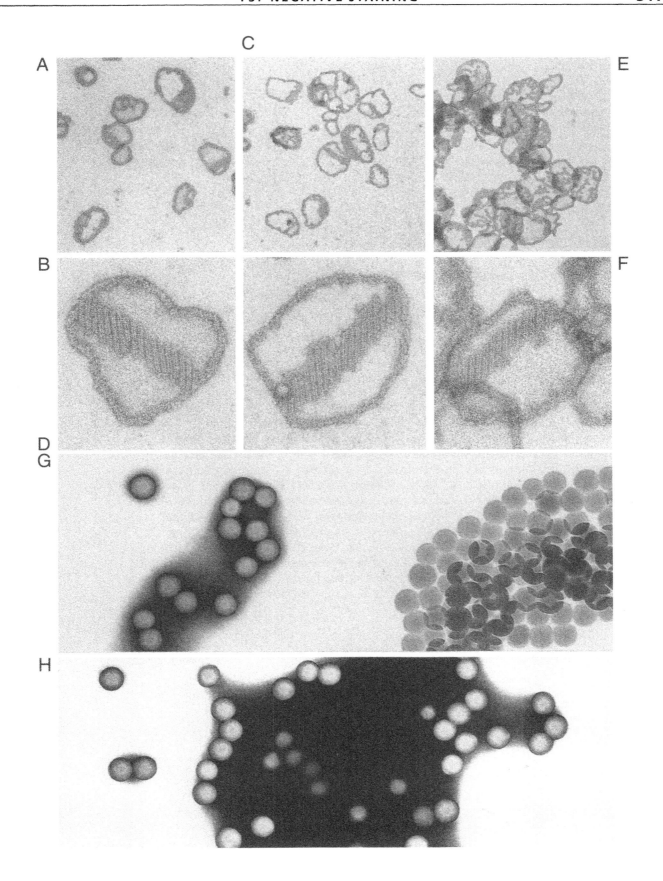

2. Properties of Support Film

FIGURES 13.2A–13.2D Ferritin molecules negatively stained with sodium phosphotungstate, pH 7.2, on carbon films that are untreated (Fig. 13.2A), glow discharged (Fig. 13.2B), following addition of 30 μg/ml bacitracin (Fig. 13.2C), or 2 mM octylglucoside (Fig. 13.2D) to the sample. $\times$ 150,000.

The distribution of ferritin on the untreated carbon film is more irregular than following the other three treatments. The most uniform distribution is observed after glow discharge alone or glow discharge followed by octylglycoside treatment (Fig. 13.2D).

FIGURES 13.2E–13.2H Ferritin molecules negatively stained with 1% uranyl acetate on carbon films treated in different ways: untreated (Fig. 13.2E), glow discharged (Fig. 13.2F), inclusion of 30 μg/ml bacitracin in the stain solution (Fig. 13.2G), and addition of 2 mM octylglucoside (M_r 292) to the stain solution (Fig. 13.2H). $\times$ 160,000.

The distribution of ferritin molecules varies only slightly among these four micrographs. In Fig. 13.2E the molecules are more aggregated than they are in Figs. 13.2G and 13.2H. The stain is uniform in Fig. 13.2F but somewhat uneven in Fig. 13.2G.

COMMENTS

During negative staining it is essential to obtain an even distribution of test object and stain on the support film. In most cases this requires that the surface of the support film be hydrophilic. Old carbon films are usually hydrophilic and can be used without further treatment. Fresh carbon films are more hydrophobic with the result that objects and stain do not distribute evenly on the surface. The remedy is then either to glow discharge the surface (Namork and Johansen, 1982; Aebi and Pollard, 1987) or to include a wetting agent in either the stain or the specimen solution. Glow discharge is a simple procedure that allows the simultaneous preparation of multiple grids with the desired surface properties. Bacitracin has long been recognized as such an agent (Gregory and Pirie, 1973), but due to its rather high molecular mass (M_r 1411) it may disturb the imaging of various small objects. In fact, a faint background granularity is observed in Figs. 13.2C and 13.2G. The most reproducible wetting agent in our experience is octylglucoside, which almost invariably results in an even distribution of objects and stain.

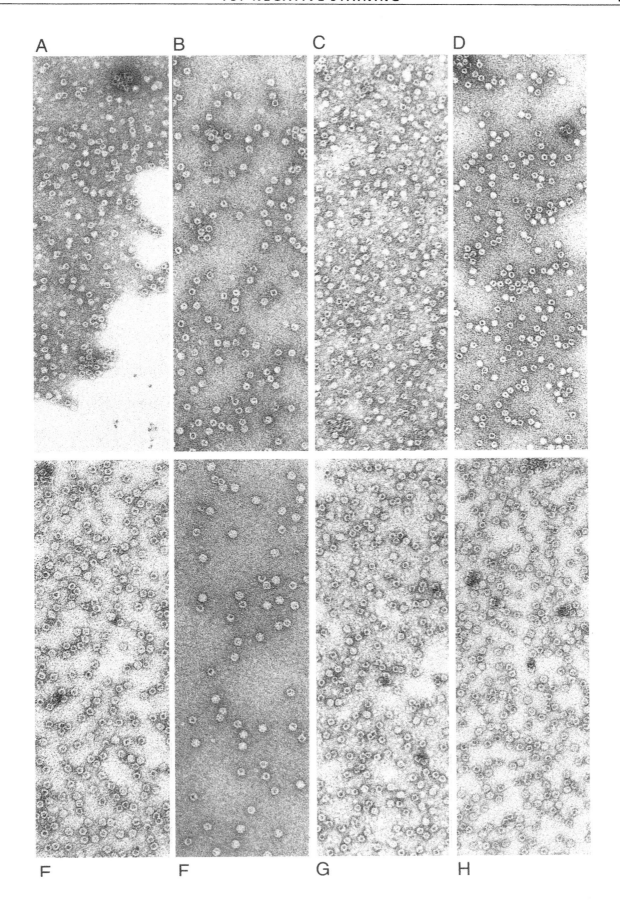

3. Comparison of Stains

FIGURE 13.3A Ferritin molecules negatively stained with the 1% uranyl formate in distilled water. × 200,000.

Ferritin molecules are distributed evenly and have an apparent diameter of about 12 nm. The negative stain forms a thin layer on the support film and surrounds each molecule as a ring. All ferritin molecules have a central electron-dense core with a density that varies from hardly noticeable to very dense. No ferritin molecules are completely devoid of a central density.

FIGURE 13.3B Ferritin molecules stained with 1% aurothioglucose. × 200,000.

Ferritin molecules are distributed fairly evenly and appear either as white areas (arrowheads) or as white areas with a central electron-dense core (arrows). Note that the aurothioglucose stain forms a uniform background density that is distinctly less electron dense than the cores in some of the ferritin molecules.

FIGURES 13.3C–13.3E Membrane crystals of Na,K-ATPase negatively stained with different solutions: 2% potassium phosphotungstate, pH 7.2 (Fig. 13.3C), 2% sodium phosphotungstate, pH 7.2 (Fig. 13.3D), and 2% sodium phosphotungstate, pH 5.2 (Fig. 13.3E). × 160,000.

There is no noticeable difference between the membrane crystals when stained with potassium (Fig. 13.3C) or sodium (Fig. 13.3D) phosphotungstate at pH 7.0. However, when membranes are stained with sodium phosphotungstate at low pH the linear arrays of Na,K-ATPase protein units in many membranes appear disordered (Fig. 13.3E).

COMMENTS

Several of the negative stains give essentially the same images of molecules, membranes, or viruses. However, there are some differences as explored on various objects (Bremer *et al.*, 1992; Massover, 1993; Harris, 1997).

Ferritin solutions normally contain a mixture of ferritin with an iron core and apoferritin, which is a protein shell without an iron core. In Fig. 13.3B this difference is evident: ferritin molecules with and without an electron-dense core. The negative stain in this case has not penetrated the protein shell of the apoferritin, whereas in Fig. 13.3A essentially all molecules contain a central density that consists of either iron or a negative stain that has penetrated the empty cavity of the apoferritin molecule. Thus these two negative stains give distinctly different patterns of the analyzed molecules.

The greatest contrast is usually attributed to uranyl formate, but the differences are small for most objects. Ammonium molybdate has the unique property in that it is approximately isotonic in a 2% solution, which is why cell organelles may remain in their native formation after negative staining (Muscatello and Horne, 1968; Munn, 1968).

The stain may also influence specimen structure in different ways, such as seen when staining Na,K-ATPase membrane crystals. These crystals have been introduced with the enzyme system in the potassium (E_2) conformation, and the presence of a negative stain based on a sodium salt tends to disorganize the aligned protein arrays (Fig. 13.3E).

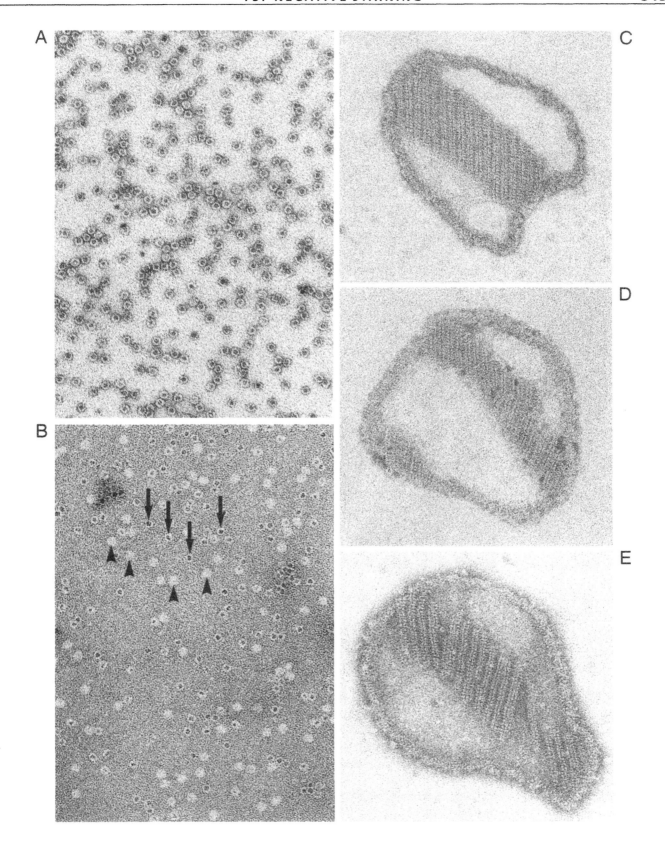

4. Thickness of Stain

FIGURE 13.4A A membrane crystal of Na,K-ATPase on thin carbon film and not exposed to negative stain. × 180,000.

The micrograph shows a faint density with the shape of a membrane fragment.

FIGURES 13.4B–13.4D Membrane crystals of Na,K-ATPase on thin carbon films stained with 1% uranyl acetate in distilled water (procedure as in Fig. 13.3A). In Fig. 13.4B as much stain as possible was withdrawn before drying the grid. In Fig. 13.4C the "standard" procedure for withdrawing stain from the grid was applied, and in Fig. 13.4D relatively more stain solution was left on the grid before drying. × 180,000.

The membrane fragment in Fig. 13.4B is only stained faintly. There is no rim of stain at the periphery of the membrane and only few details are seen on the surface of the membrane. In Fig. 13.4C, however, the membrane fragment is well outlined by the stain and surface details are distinct. The membrane fragment in Fig. 13.4D is well defined, but compared to Fig. 13.4C, surface details are somewhat less crisp and there is a thick layer of stain outside the membrane fragment.

FIGURE 13.4E Bacteriophage T4 on carbon film following negative staining with 1% uranyl acetate. × 150,000.

The head of the phage is symmetrical and stained intensely. The tail is faintly stained and its periodicity is only vaguely indicated. There is no negative stain surrounding the phage.

FIGURES 13.4F–13.4H Bacteriophage T4 stained negatively with uranyl acetate. Figure 13.4G was stained with the "standard" procedure whereas in Fig. 13.4 F more and in Fig. 13.4H less stain was withdrawn before drying. × 150,000.

Structural details in Fig. 13.4G, including the periodicity of the tail and the tail fibers, are well recognized. In contrast, the phage in Fig. 13.4H is surrounded by a thick layer of electron stain, rendering tail periodicity and tail fibers invisible.

COMMENTS

The thickness of the negative stain surrounding an object influences greatly the structural information obtainable. Visual inspection of unstained objects reveals no structural details (Fig. 13.4A). If as much stain as possible is withdrawn from the grid, only a faint image is observed, yet with a crisp outline (Fig. 13.4B). Such images may represent not only negative staining but a combination of positive and negative staining. Figure 13.4C is close to optimal, whereas structural details in Fig. 13.4D have become partly obscured by the stain.

The bacteriophage T4 in Fig. 13.4G shows considerable structural detail in the tail, including its tail fibers. An increased amount of stain around the specimen immediately results in loss of structural detail (Fig. 13.4H). In Fig. 13.4F there is not sufficient stain around the bacteriophage to outline its details, whereas in Fig. 13.4E the bacteriophage was probably positively stained and stabilized in the stain before becoming attached to the support film. The smaller the specimen, the more critical is the adjustment of stain thickness.

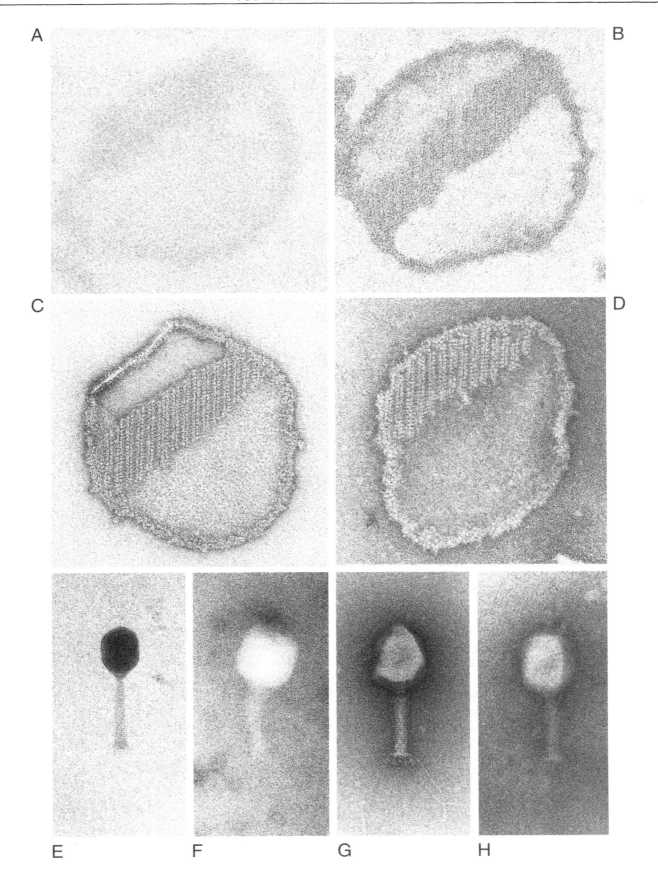

5. Concentration of Specimen

FIGURES 13.5A AND 13.5B Na,K-ATPase membranes at a concentration of 0.1 mg protein/ml following staining on glow-discharged carbon film with 1% uranyl acetate in distilled water. In Fig. 13.5A the sample solution was efficiently removed before staining, whereas more sample was left in Fig. 13.5B. × 18,000.

The membrane fragments are distributed evenly and not overlapping in Fig. 13.5A, whereas in Fig. 13.5B, many membranes are superimposed and in places form clusters.

FIGURE 13.5C A higher magnification of the same preparation as in Fig. 13.5B. × 150,000.

The membrane fragments form layers and their crystalline arrays are superimposed.

FIGURE 13.5D Bacteriophage T4 negatively stained with uranyl acetate on a carbon film that was not made hydrophilic. × 37,000.

All phages at the border of the droplet are oriented with their heads toward the center of the droplet whereas those within the droplet are arranged almost randomly.

FIGURE 13.5E A sample of diluted calf feces negatively stained with uranyl acetate. × 150,000.

In two places (arrows) there is a structure resembling a virus.

FIGURES 13.5G–13.5I Negative staining with uranyl acetate of rotaviruses isolated from calf feces. × 150,000.

The micrographs show round viruses with and without a central core.

COMMENTS

The concentration of the object in the solution used for negative staining is an important parameter. For many proteins and membrane preparations a concentration of 0.1–0.2 mg protein/ml usually results in a suitable frequency of molecules or membranes. If the concentration is low, the objects are of course more rarely encounted and the investigation more time-consuming. However, if the specimen concentration is too high or the sample is not efficiently withdrawn with filter paper, the structures are usually superimposed and difficult to analyze in detail (Fig. 13.5C).

Even in suitably diluted samples irregular concentrations of objects may occur when the fluid of the sample evaporates and the droplet shrinks (Fig. 13.5D). In this case there may be a specific orientation of objects, as illustrated in Fig. 13.5D, where probably the support film was too hydrophobic. The object of interest may sometimes be partly obscured or represent only a small fraction of the sample and therefore be difficult to identify by negative staining, as illustrated in Fig. 13.5E. It may still be possible to identify if compared with a purified sample (Figs. 13.5F–13.5H). When analyzing an unknown sample, it is often helpful to stain a dilution series of the specimen.

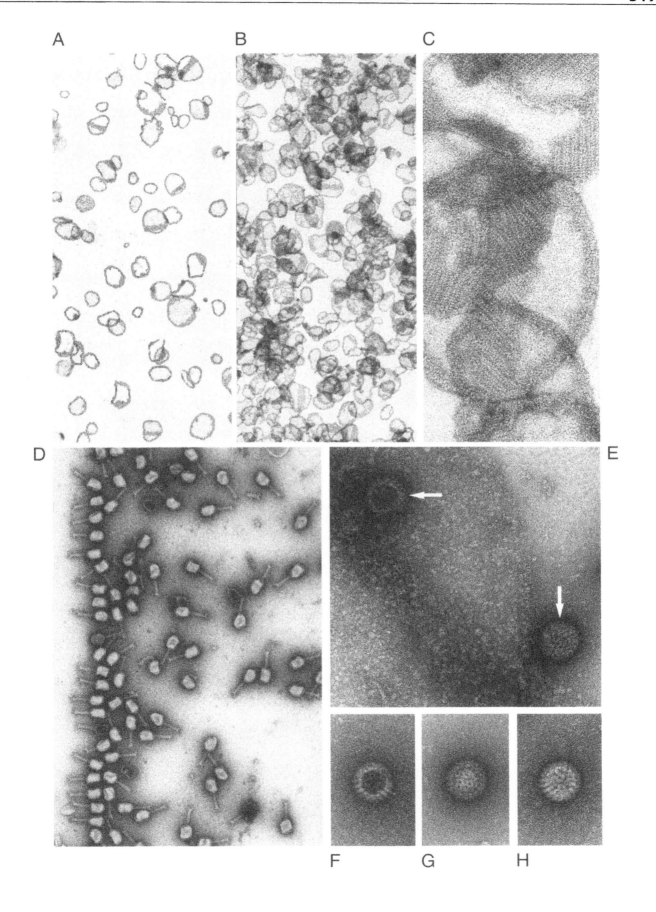

6. Deformation of Specimen

FIGURES 13.6A–13.6D Membrane crystals of Na,K-ATPase negatively stained with 1% uranyl acetate on glow-discharged carbon films. All micrographs originate from the same sample solution. × 160,000.

The micrographs illustrate different appearances of the membranes. In Fig. 13.6A the membrane is intact, with negative stain surrounding the entire membrane fragment and outlining an intact and smooth central lipid region (L).

In Fig. 13.6B the stain likewise surrounds the membrane fragment. It also vaguely delimits the area of the lipid bilayer (L). The upper lipid area (asterisk) has ruptured and is now limited by snake-like extensions (arrows) and small pieces of membranes (arrowheads). Note that the density of the upper area (asterisks) is the same as outside the membrane fragment whereas the lipid area (L) has a slightly greater density than the naked support film surrounding the membrane fragment.

Figure 13.6C shows a membrane that has disintegrated into three regions joined by narrow connections.

Figure 13.6D illustrates a pronounced fragmentation of membranes, which appear here as small membrane fragments with surface particles, some of which may be aligned in linear arrays.

COMMENTS

Figure 13.6A shows what can be considered an optimal preparation of this type of membrane. A purified Na,K-ATPase membrane exhibits surface particles in random order (see Fig. 1.11C). Following the induction of linear arrays of Na,K-ATPase molecules with the aid of vanadate, the proteins in the membrane form one or more confluent crystalline areas and leave large lipid regions (L in Fig. 13.6A) devoid of protein. Such intact membranes are observed when the membranes are distributed evenly on the support film and not overlapping. Figure 13.6B shows a membrane where part of the lipid region of the membrane has ruptured and the limiting protein-rich band is transformed into what appears as snake-like tubular extensions. The transformation may be considered a mechanical damage of the membrane, which is further exaggerated in Figs. 13.6C and 13.6D. All preparations were made at the same time and the main difference among these micrographs is that the best preserved membranes were found in those parts of the grids where the membranes were solitarily and evenly distributed.

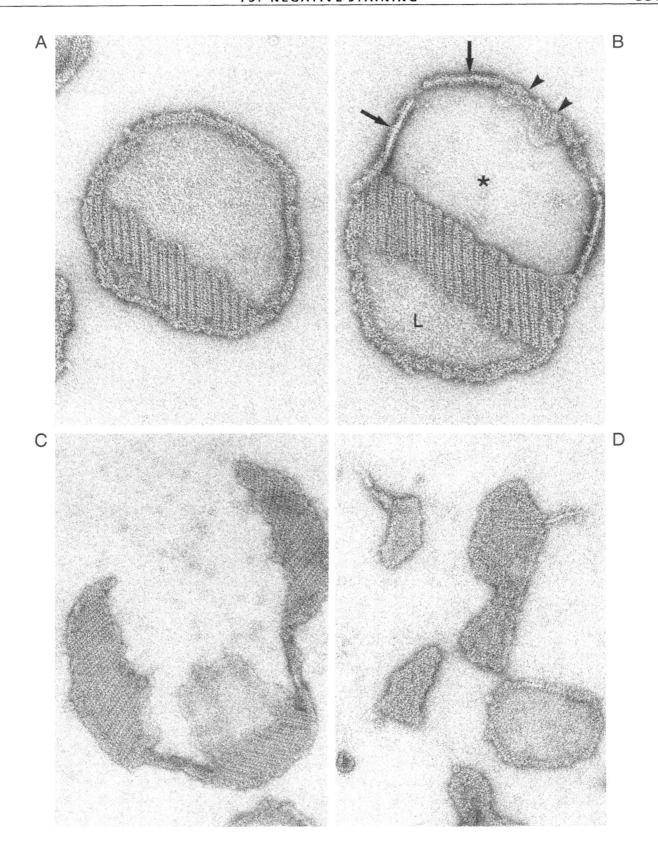

7. Radiation Damage

FIGURE 13.7A An electron micrograph of purified, membrane-bound Na,K-ATPase following negative staining. × 150,000.

The negative stain of the background around the membrane fragments is uniform except for a few less dense spots and is also uniform over the membrane fragments, where it outlines surface particles of $\alpha\beta$ protomers of the enzyme.

FIGURE 13.7B Same specimen as in Fig. 13.7A after an additional 2-min exposure to the electron beam. × 150,000.

The negative stain over and around the membrane fragments is uneven; it displays many small and irregular (some elongated) regions that evidently contain no electron-dense material. The individual surface particles on the membrane are hardly, if at all, discernable.

FIGURE 13.7C Bacteriophage T4 stained negatively with 1% uranyl acetate and recorded in the electron microscope immediately after being introduced into the electron beam. × 175,000.

The phage shows structural details such as periodicity of the tail, a base plate, and tail fibers (arrowheads).

FIGURE 13.7D Same bacteriophage after further exposure to the electron beam. × 175,000.

Compared to Fig. 13.7C the bacteriophage has lost structural details, notably parts of the tail fibers (arrowheads). In addition, the electron-dense stain surrounding the bacteriophage has become granular and acquired numerous small electron-transparent holes.

COMMENTS

Radiation may severely effect some biological specimens, in this case negatively stained, unfixed membranes and bacteriophages. These specimens are only protected to a limited degree by the uranyl acetate stain, part of which has been lost, and the delicate structural details of the membranes have become distorted. Electron microscopy of very radiation-sensitive objects should therefore ideally be performed under "low-dose conditions" (see Fig. 10.22).

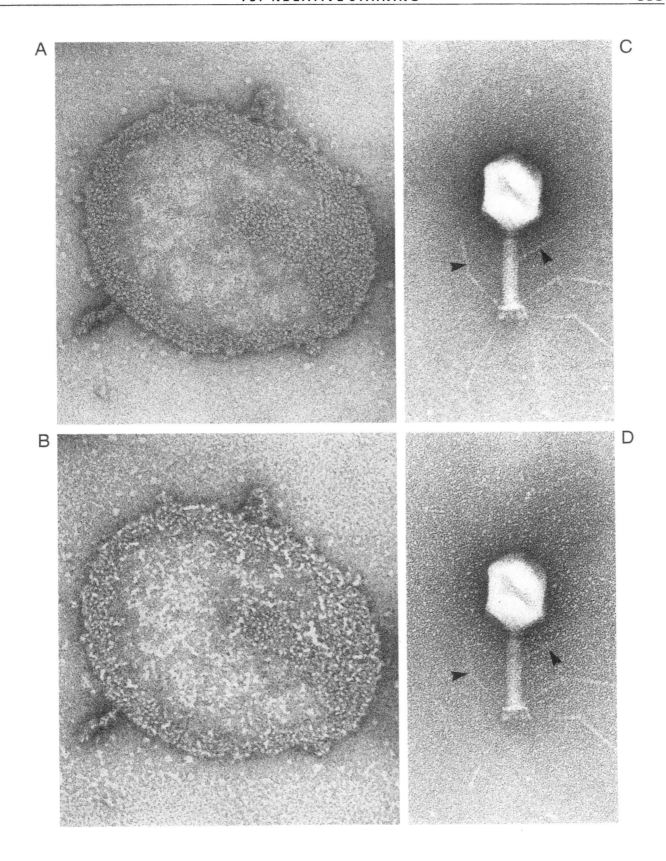

References

Aebi, U., and Pollard, T. D. (1987). A glow discharge unit to render electron microscope grids and other surfaces hydrophilic. *J. Electr. Microsc. Techn.* **7**, 29–33.

Bremer, A., Henn, C., Engel, A., Baumeister, W., and Aebi, U. (1992). Has negative staining still a place in biomacromolecular electron microscopy? *Ultramicroscopy* **4**, 85–111.

Bremer, A., Häner, M., and Aebi, U. (1998). Negative staining *In* "Cell Biology: A Laboratory Handbook" (J. E. Celis, ed.), Vol. 3, pp. 277–284. Academic Press, San Diego.

Brenner, S., and Horne, R. W. (1959). A negative staining method for high resolution electron microscopy of viruses. *Biochim. Biophys. Acta* **34**, 103–110.

Carrascosa, J. L. (1988). Immunoelectronmicroscopical studies on viruses. *Electr. Microsc. Rev.* **1**, 1–16.

Deguchi, N., Jørgensen, P. L., and Maunsbach, A. B. (1977). Ultrastructure of the sodium pump: Comparison of thin sectioning, negative staining, and freeze-fracture of purified, membrane-bound (Na$^+$,K$^+$)-ATPase. *J. Cell Biol.* **75**, 619–634.

Frank, J. (1989). Image analysis of single molecules. *Electr. Microsc. Rev.* **2**, 53–74.

Frank, J. (1996). "Three-Dimensional Electron Microscopy of Macromolecular Assemblies." Academic Press, San Diego.

Frank, J., and Radermacher (1992). Three-dimensional reconstruction of single particles negatively stained or in vitreous ice. *Ultramicroscopy* **4**, 241–262.

Gordon, C. N. (1972). The use of octadecanol monolayers as wetting agents in the negative staining technique. *J. Ultrastruct. Res.* **39**, 173–185.

Gregory, D. W., and Pirie, B. J. S. (1973). Wetting agents for biological electron microscopy. *J. Microsc. (Oxford)* **99**, 261–278.

Harris, J. R. (1991). The negative staining-carbon film procedure: Technical considerations and a survey of macromolecular applications. *Micron Microsc. Acta* **22**, 341–359.

Harris, J. R. (1997). "Negative Staining and Cryoelectron Microscopy: The Thin Film Techniques." Bios Scientific Publishers, Oxford.

Harris, J. R., and Horne, R. W. (1994). Negative staining: A brief assessment of current technical benefits, limitations and future possibilities. *Micron* **25**, 5–13.

Harris, J. D., Gegenbauer, W., and Markl, J. (1995). Keyhole limpet haemocyanin: Negative staining in the presence of trehalose. *Micron* **26**, 25–33.

Horne, R. W. (1967). Electron microscopy of isolated virus particles and their components. *Methods Virol.* **3**, 521–574.

Horne, R. W. (1973). Contrast and resolution from biological objects examined in the electron microscope with particular reference to negative stained specimens. *J. Microsc. (Oxford)* **98**, 286–298.

Horne, R. W. (1991). Early developments in the negative staining technique for electron microscopy. *Micron Microsc. Acta* **4**, 321–326.

Horne, R. W., and Pasquali Ronchetti, I. (1974). A negative staining-carbon film technique for studying viruses in the electron microscope. I. Preparative procedures for examining icosahedral and filamentous viruses. *J. Ultrastruct. Res.* **47**, 361–383.

Hyatt, A. (1991). Immunonegative staining. *In* "Electron Microscopy in Biology: A Practical Approach" (J. R. Harris, ed.), pp. 59–81. IRL Press, Oxford.

Johansen, B. V. (1997). High resolution bright field electron microscopy of biological specimens. *Ultramicroscopy* **2**, 229–239.

Johansen, B. V. (1978). Negative staining. *In* "Electron Microscopy in Human Medicine" (J. V. Johannessen, ed.), Vol. 1, pp. 84–98. McGraw-Hill, New York.

Kiselev, N. A., Sherman, M. B., and Tsuprun, V. L. (1990). Negative staining of proteins. *Electr. Microsc. Rev.* **3**, 43–72.

Kjeldsberg, E. (1986). Immunonegative stain techniques for electron microscopic detection of viruses in human faeces. *Ultrastruct. Pathol.* **10**, 553–570.

Massover, W. H. (1978). The ultrastructure of ferritin macromolecules. III. Mineralized iron in ferritin is attached to the protein shell. *J. Mol. Biol.* **123**, 721–726.

Massover, W. H. (1993). Ultrastructure of ferritin and apoferritin: A review. *Micron* **24**, 389–437.

Massover, W. H., and Marsh, P. (1997). Unconventional negative stains: Heavy metals are not required for negative staining. *Ultramicroscopy* **69**, 139–150.

Munn, E. A. (1974). The application of the negative staining technique to the study of membranes. *In* "Methods in Enzymology" (S. Fleischer and L. Packer, eds.), Vol. XXXII, pp. 20–35. Academic Press, New York.

Muscatello, U., and Horne, R. W. (1968). Effect of the tonicity of some negative-staining solutions on the elementary structure of membrane-bounded systems. *J. Ultrastruct. Res.* **25**, 73–83.

Namork, E., and Johansen, B. V. (1982). Surface activation of carbon film supports for biological electron microscopy. *Ultramicroscopy* **7**, 321–330.

Nermut, M. V. (1972). Negative staining of viruses. *J. Microsc. (Oxford)* **96**, 351–362.

Small, J. V. (1988). The actin cytoskeleton. *Electr. Microsc. Rev.* **1**, 155–174.

Small, J. V., and Sechi, A. (1998). Whole-mount electron microscopy of the cytoskeleton: Negative staining methods. *In* "Cell Biology: A Laboratory Handbook" (J. E. Celis, ed.), 2nd Ed., Vol. 3, pp. 285–291. Academic Press, San Diego.

Small, J. V., Herzog, M., Häner, M., and Aebi, U. (1994). Visualization of actin filaments in keratocyte lamellipodia: Negative staining compared with freeze-drying. *J. Struct. Biol.* **113**, 135–141.

Steven, A. C., and Navia, M. A. (1980). Fidelity of structure representation in electron micrographs of negatively stained protein molecules. *Proc. Natl. Acad. Sci. USA* **77**, 4721–4725.

Steven, A. C., and Navia, M. A. (1982). Specificity of stain distribution in electron micrographs of protein molecules contrasted with uranyl acetate. *J. Microsc.* **128**, 145–155.

Tranum-Jensen, J. (1988). Electron microscopy: Assays involving negative staining. *In* "Methods in Enzymology" (S. Harshman, ed.), Vol. 165, pp. 357–374. Academic Press, New York.

Unwin, P. N. T. (1972). Negative staining of biological specimens using mixture salts. *In* "Fifth European Congress on Electron Microscopy, Manchester," pp. 232–233. The Institute of Physics, London and Bristol.

CHAPTER 14

AUTORADIOGRAPHY

Electron microscope autoradiography (EMAR) is an efficient method used to study certain dynamic processes in cells and tissues. It enables the investigators to localize, for example, the site of synthesis and intracellular transport of proteins and the absorption pathways for labeled substances. The technique is based on the fact that many large biological molecules can be labeled with radioactive elements, particularly tritium, iodine 125, carbon 14, phosphorus 32, and sulfur 35. Molecules biosynthesized with any of these radioactive elements behave in the cell exactly as their unlabeled counterparts.

The technique requires the consideration of several technical factors, including not only the choice of isotope, but also the choice of tissue preparation procedure, choice of photographic emulsion and its application onto the ultrathin sections, type of development, and evaluation of qualitative and/or quantitative results. Additionally, an evaluation must be made beforehand of the possibility of achieving high enough concentrations of the isotopes in the tissue. In other words, the experimental system has to be analyzed with respect to the possibility of obtaining results within a reasonable time, e.g., using small experimental animals or *in vitro* systems. The technique requires several types of controls and a determination of whether some of the isotopes have been extracted during the preparatory steps or have diffused within the tissue. Water-soluble labeled substances, such as ions and many small molecules cannot be localized in the tissues by standard procedures.

Among all the techniques used in biological electron microscopy, autoradiography is unique in providing information on the localization of synthesis, turnover, and transport of metabolites. It is based on the fact that radioisotopes behave chemically the same way as the nonradioactive counterpart but can be revealed by a photographic emulsion. It got an early start when London (1904) put a frog that had been immersed in water containing radium on a photographic plate. Later Lacassagne and Lattes developed the technique of making autoradiographs of tissues embedded in paraffin blocks; they used polonium as the tracer. The usefulness of autoradiography was expanded greatly when George de Hevesy produced radioisotopes of biologically important atoms such as P^{32} and used them among others for autoradiography. He received the 1943 Nobel prize in chemistry "for his work on the use of isotopes as tracers in the study of chemical processes." Liquier-Milward (1956) was able to scale down the technique so that it became useful in biological electron microscopy.

Liquier-Milward, J. (1956). *Nature* **177**, 619.
Hevesy, G. (1948). *Cold Spring Harbor Symp. Quant. Biol.* **13**, 129–140.
Lacassagne, A., and Lattes, J. (1924). *C. R. Acad. Sci. Paris* **90**, 352.
London, E. S. (1904). *Arch. Eléctr. Méd.* **12**, 363–372.

1. Undeveloped Emulsion

FIGURE 14.1A A survey electron micrograph of Ilford L4 autoradiographic emulsion deposited onto Formvar film with the loop method. × 3000.

The silver halide crystals are distributed in a uniform monolayer. × 3000.

FIGURE 14.1B A high magnification of a monolayer of silver halide crystals of Ilford L4 emulsion. × 15,000.

The crystals are evenly distributed and densely packed, although essentially no crystals touch each other.

FIGURE 14.1C Silver halide crystals in L4 emulsion made with a slightly diluted emulsion. × 15,000.

The silver halide crystals are distributed irregularly and leave large empty areas in places.

FIGURE 14.1D Similar preparation as in Fig. 14.1B but the emulsion used to form the autoradiographic layer was more concentrated. × 15,000.

The grains are densely but unevenly packed and form double or multilayers in places.

COMMENTS

The first step in EMAR when forming an emulsion on a grid is to make sure that the emulsion consists of an evenly distributed layer of silver halide crystals (Figs. 14.1A and 14.1B). For quantitative studies, where grain densities are determined over cell organelles, this is a particularly important requirement. If the emulsion shows an irregular distribution of the halide crystals with empty areas, it is evident that radioactivity emitted in such areas will not be detected. If, however, the emulsion is thick and consists of multilayers in places, radioactivity may be detected but the resolution of the emulsion will be decreased for geometric reasons. A large number of procedures have been described for the application of the autoradiographic emulsion onto the ultrathin tissue sections on the grids. Several of these procedures result in equally useful preparations. Whatever method is used, it is important that the method be strictly standardized so that essentially uniform silver crystal layers are applied when large series of grids are used. In the Ilford L4 emulsion the undeveloped silver halide crystals average about 0.14 μm in diameter. Other emulsions have smaller diameter grains, which potentially increases the resolution.

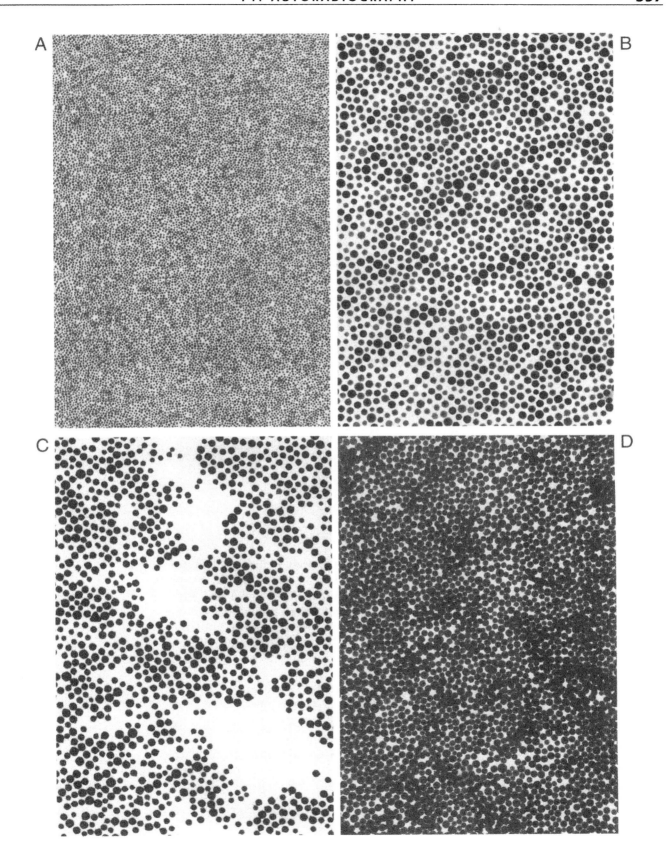

2. Developed Emulsion

FIGURE 14.2A An emulsion layer similar to that in Fig. 14.1B following development with D19 for 90 sec without prior exposure to any type of radiation, including visible light. × 15,000.

This field shows one single developed silver halide crystal. The frequency of developed grains was so low that only about 1 out of 25 similarly sized areas showed a single developed grain.

FIGURE 14.2B An autoradiographic preparation on a grid that had been exposed to a low dose of gamma radiation and developed for 90 sec in D19. × 15,000.

This field shows a rather even distribution of developed silver halide crystals. Note that the developed silver grains vary greatly in size and shape and on average are larger than the undeveloped silver halide crystals (compare with Fig. 14.1B).

FIGURE 14.2C An autoradiographic preparation similar to that in Fig. 14.1B but fully exposed to light and developed for 90 sec in D19. × 15,000.

This field contains a dense, irregular mesh of developed silver grains.

FIGURE 14.2D Ilford L4 emulsion similar to that in Fig. 14.1B exposed to a low dose of gamma radiation and developed with a "physical" developer (Lumière *et al.*, 1911; Caro and Tubergen, 1962) containing sodium sulfite and *p*-phenylenediamine for 90 sec. × 60,000.

The silver grains are very small and are either round or slightly elongated.

COMMENTS

A basic requirement in electron microscope autoradiography is that the emulsion applied to the ultrathin sections is free from background, i.e., silver halide crystals inadvertently exposed to light or irrelevant radiation or originating from an aged emulsion. As seen in Fig. 14.2A, the frequency of developed grains in an optimal preparation is very low. Preparations exposed to a weak radiation for a short time show a random distribution of mostly nontouching silver grains (Fig. 14.2B). However, after full exposure to radiation or light (Fig. 14.2C), individual silver grains cannot be recognized any longer.

The size of individual silver grains depends on the method of development. Conventional development with D19 results in silver grains that are invariably larger than the original undeveloped silver halide crystals, which average 0.14 μm. However, development with a so-called "physical developer" containing *p*-phenylenediamine results in silver grains that are much smaller, as illustrated in Fig. 14.2D (note that the magnification in Fig. 14.2D is four times higher than in Figs. 14.2A–14.2C). In some situations, the small size of these developed grains contributes to an increased resolution in the autoradiographic preparation.

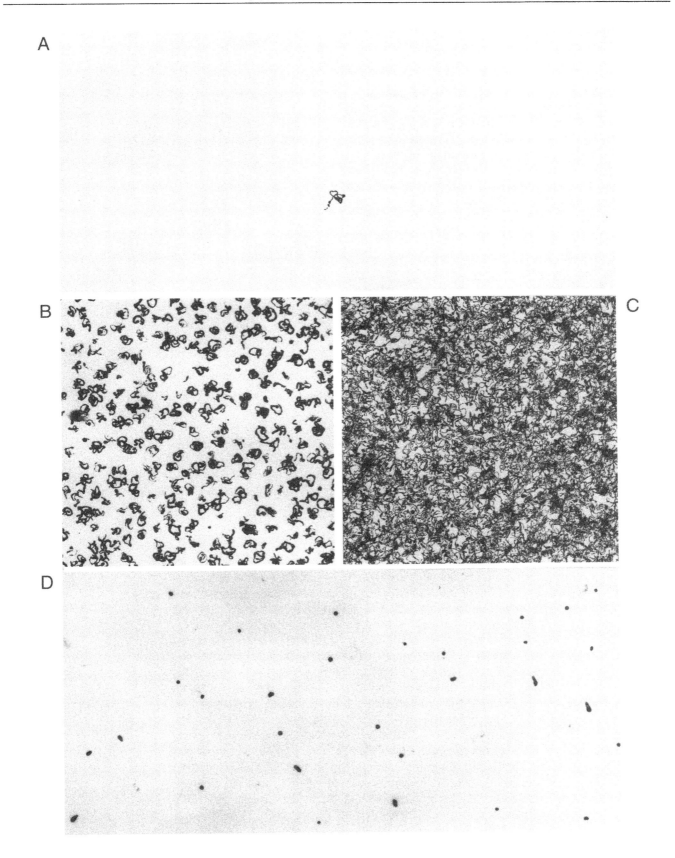

3. Resolution

FIGURE 14.3A An electron microscope autoradiograph of a renal proximal tubule cell that has absorbed albumin labeled with ^{125}I. The tubule was fixed with glutaraldehyde that was microperfused through the tubule lumen shortly after the start of protein absorption. The section was stained, carbon coated, and covered by Ilford L4 emulsion. The preparation was stored at 4°C and then developed in Kodak D19. × 36,000.

Autoradiographic silver grains are observed in the endocytic vacuole over the inner aspects of its limiting membrane. One grain is associated with an apical endocytic invagination.

FIGURE 14.3B Similar preparation as in Fig. 14.3A except that the preparation was developed with the "physical" developer Caro and van Tubergen, 1962) containing sodium sulfite and p-phenylenediamine. × 36,000.

The silver grains (arrows) are considerably smaller than in Fig. 14.3A and more closely associated with the limiting membrane. Some grains may originate from the same silver bromide crystal.

FIGURE 14.3C Proximal tubule cell fixed 1 hr after the start of absorption of ^{125}I-labeled albumin. × 36,000.

Most grains are located over the lysosome in the center of the field; a few lie over adjacent lysosomes.

FIGURE 14.3D Apical part of a renal proximal tubule cell in the process of absorbing ^{125}I-labeled albumin. Preparation as in Fig. 14.3A. × 36,000.

Some grains are located over the vacuoles whereas others are on their limiting membranes or the surrounding cytoplasm.

COMMENTS

The resolution in electron microscope autoradiography is limited by several factors, as discussed extensively in the literature: thickness of the tissue section, energy of the radioactive isotope, size of the undeveloped silver halide crystals, distribution of the crystals in the applied emulsion, and method of development (Caro and Van Tubergen, 1962; Maunsbach, 1966; Salpeter et al., 1969, 1978). Figure 14.3B illustrates the improved resolution obtainable with fine grain development. However, such development may result in more than one developed small silver particle for each crystal, which complicates quantitative studies. In Fig. 14.3C the distinct concentration of labels over one lysosome clearly resolves this cell organelle as a site of absorbed albumin. The few grains immediately outside the lysosomal membrane are predictable on the basis of the geometric error caused by the thickness of the section. Note that there is no label over other parts of the cytoplasm. The apical endocytic vacuoles in Fig. 14.3D are small in comparison with the size of the developed autoradiographic grains. For this reason the association between silver grains and small vacuoles becomes more difficult to analyze. In such cases one must resort to statistical analyses based on grain counts and morphometrically determined areas of cellular compartments (Williams, 1977).

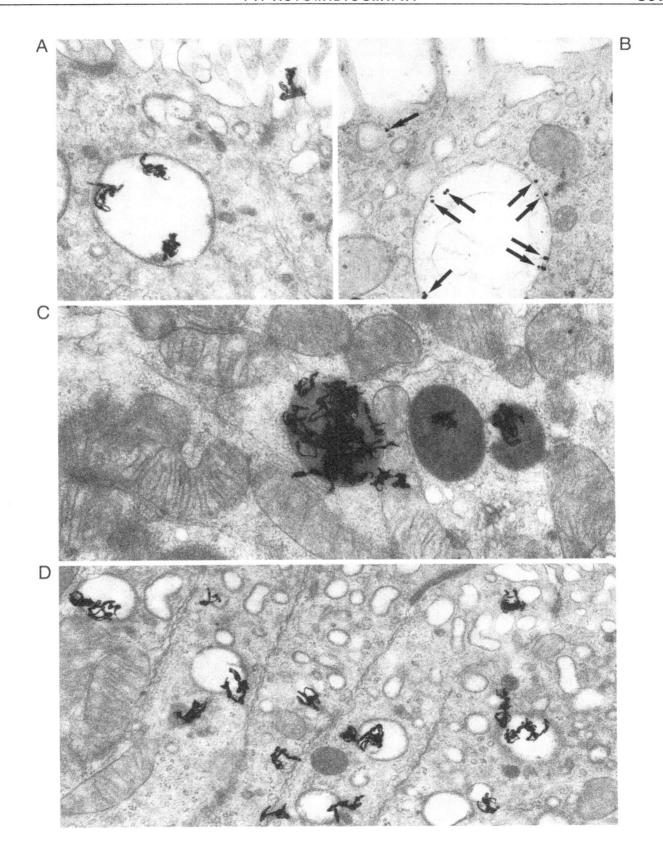

4. Quantitation

FIGURE 14.4A An electron microscope autoradiograph showing parts of two thyroid follicles from a mouse that had received an injection with ^{125}I. The tissue was fixed in osmium tetroxide and covered with Ilford L4 emulsion. After exposure the preparation was developed in D19, fixed, and observed unstained in the electron microscope. × 6000.

At the upper left and lower right of the figure are parts of thyroid follicle colloids. These are labeled extensively with developed silver grains. Between the colloids are thyroid epithelial layers with only a few developed grains. A capillary (C) is present between the follicles.

FIGURE 14.4B An electron microscope autoradiograph of a proximal tubule cell 1 hr after absorbtion of ^{125}I-labeled albumin. Preparation as in Fig. 14.3A. × 30,000.

Only a single autoradiographic grain is observed and is associated with a lysosome in the cytoplasm. The rest of the cytoplasm is unlabeled.

FIGURE 14.4C An electron microscope autoradiograph of part of a renal proximal tubule cell 1 hr after absorption of radioactive albumin. Prior to embedding the glutaraldehyde-fixed tissue was incubated for acid phosphatase. No postfixation in osmium tetroxide was applied. The autoradiographic preparation was performed as in Fig. 14.3A. × 40,000.

Two lysosomes are identified by the reaction product for acid phosphatase (arrows) and show autoradiographic labeling at the same time. The lysosome to the left is labeled so extensively that no grain counts can be made. The lysosome to the right is labeled less intensely and shows eight grains. The light spots in the background represent sites of unexposed and fixative-removed silver bromide crystals.

COMMENTS

Electron microscope autoradiography is a powerful method used to study dynamic processes in cells; usually it is desirable to obtain not only qualitative but also quantitative estimates as to where the isotope is located. In some situations quantitation is unnecessary as the distribution of silver grains is easy to define and the number of grains overwhelming, as in Fig. 14.4A. In other systems the amount of radioactive isotope may be very small indeed and the number of developed grains per unit area minute, as in Fig. 14.4B. Nevertheless, such preparations can be analyzed quantitatively provided sufficiently large tissue areas are analyzed. In practice, this means that a large number of micrographs must be taken randomly from many cells. If, however, the frequency of grains is very high, accurate grain counts become impossible, such as in Fig. 14.4C, but qualitative data are then usually quite convincing.

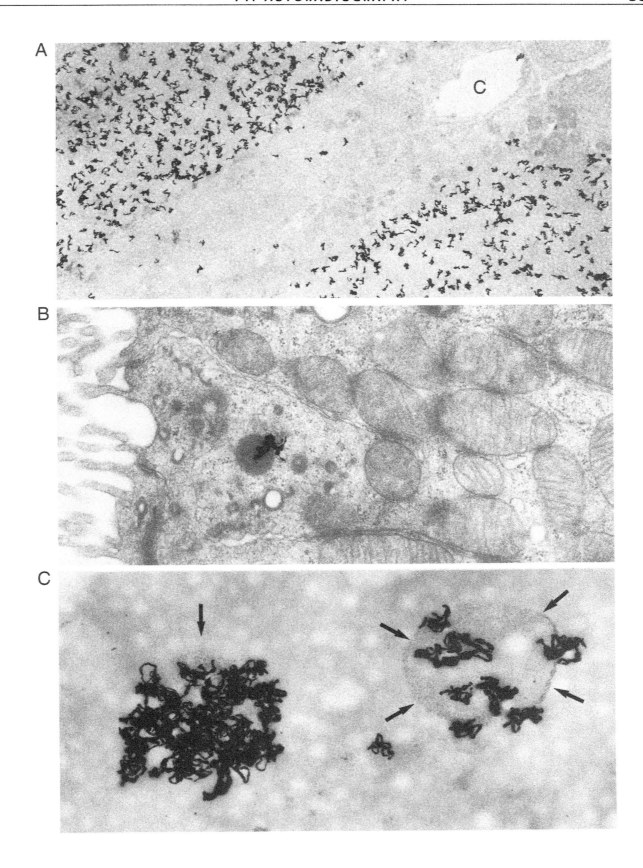

5. Preparatory Defects

FIGURE 14.5A Part of an intestinal epithelial cell from a rat that had been given a fat emulsion containing ^{3}H-labeled oleic acid by stomach tubing. The tissue was fixed with glutaraldehyde, postfixed in osmium tetroxide, and embedded in Epon, and the sections were coated with Ilford L4 emulsion. Following exposure the grid was developed in D19. × 60,000.

Autoradiographic silver grains are present all over the cytoplasm, including mitochondria, endoplasmic reticulum, cell membranes, and cytoplasmic matrix but appear more frequently over the Golgi region.

FIGURE 14.5B Part of a proximal tubule cell following absorption of ^{125}I-labeled albumin. The tissue was prepared for electron microscope autoradiography as in Fig. 14.5A. × 20,000.

A labeled lysosome is located in the upper left corner. A long track of developed autoradiographic grains is seen to the right in the micrograph.

FIGURE 14.5C Similar preparation as shown in Fig. 14.5B. The sections were stained with uranyl acetate and lead citrate and carbon coated before application of emulsion. × 20,000.

In the center of the micrograph there is a region with irregular masses of dense material but no section staining. The surrounding cytoplasm has a normal appearance and includes lysosomes, some of which are labeled.

COMMENTS

The autoradiographic technique at the electron microscopical level has several inherent pitfalls. Perhaps the most deleterious artifact is a movement or an extraction of the radioactively labeled substance. In Fig. 14.5A the radioactively labeled fatty acid has been absorbed by the intestinal cells and remains in part in the tissue. However, it is not known whether the radioactivity located by the autoradiographic emulsion is representative for the absorbed fatty acid or whether some, even a major part, has been extracted during the preparatory procedure and displaced within the tissue. The fact that the label seems concentrated over the Golgi apparatus indicates that some is firmly bound there, whereas the scattered grains over the rest of the cytoplasm may not be representative of the original location of the label.

Very rarely long tracks of exposed silver grains are seen in the autoradiographic preparations (Fig. 14.5B). The origin may be radiation from uranium in the stain or possibly external radiation or chemical interaction from the section. The weak radiation from iodine 125 is unlikely to cause long tracks. Defects like those in Fig. 14.5C are caused by imperfections in the carbon film evaporated on the stained section and lead to the penetration of developer or fixative into the stain and interference with the stain.

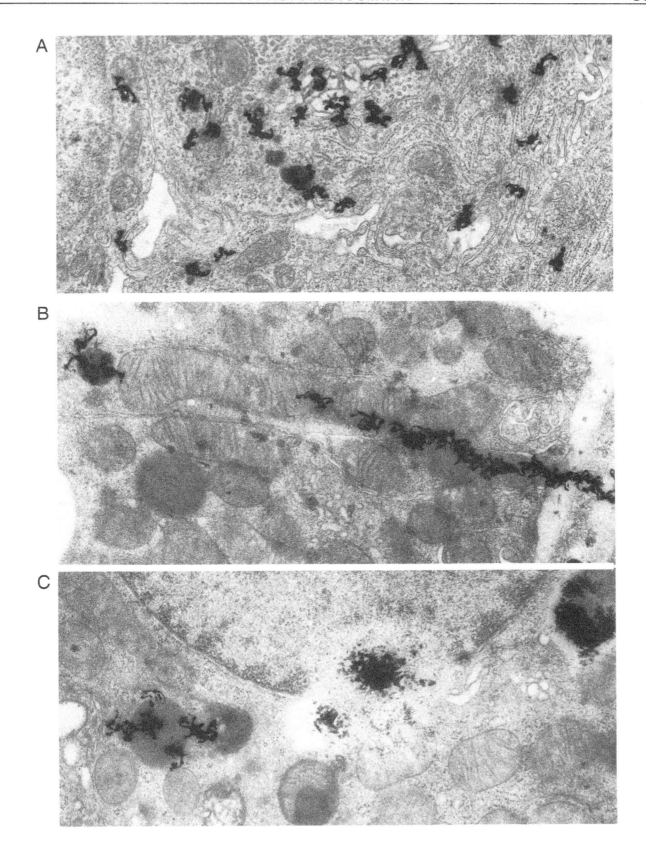

References

Bachmann, L., and Salpeter M. M. (1967). Absolute sensitivity of electron microscope radioautography. *J. Cell Biol.* **33**, 299–305.

Caro, L. G. (1962). High-resolution autoradiography. II. The problem of resolution. *J. Cell Biol.* **15**, 189–199.

Caro, L. G., and van Tubergen, R. P. (1962). High resolution radioautography. I. Methods. *J. Cell Biol.* **15**, 173–188.

Christensen, A. K., and Paavola, L. G. (1972). The use of frozen thin sections of fresh tissue for autoradiography of diffusible substances. *Proc. 4th Intl. Congr. Histochem. Cytochem.* **41–42.**

Christensen, E. I. (1976). Rapid protein uptake and digestion in proximal tubule lysosomes. *Kidney Int.* **10**, 301–310.

Christensen, E. I., and Maunsbach, A. B. (1974). Intralysosomal digestion of lysozyme in renal proximal tubule cells. *Kidney Int.* **6**, 396–407.

Comer, J. J., and Skipper, S. J. (1954). Nuclear emulsions for electron microscopy. *Science* **119**, 441–442.

Hay, E. D., and Revel, J. P. (1963). The fine structure of the DNP component of the nucleus: An electron microscopic study utilizing autoradiography to localize DNA synthesis. *J. Cell Biol.* **16**, 29–51.

Jamieson, J. D., and Palade, G. E. (1967). Intracellular transport of secretory proteins in the pancreatic exocrine cell. II. Transport to condensing vacuoles and zymogen granules. *J. Cell Biol.* **34**, 597–615.

Lumière, A., Lumière, A., and Seyewetz, A. (1911). Sur le développement des images photographiques après fixage. *Compt. Rend. Acad. Sci.* **153**, 102–110.

Madsen, K. M., and Christensen, E. I. (1978). Effects of mercury on lysosomal protein digestion in the kidney proximal tubule. *Lab. Invest.* **38**, 165–174.

Maunsbach, A. B. (1966). Absorption of I¹²⁵-labeled homologous albumin by rat kidney proximal tubule cells: A study of microperfused single proximal tubules by electron microscopic autoradiography and histochemistry. *J. Ultrastruct. Res.* **15**, 197–241.

Maunsbach, A. B. (1966b). Albumin absorption by renal proximal tubule cells. *Nature* **212**, 546–547.

Miller, A., and Maunsbach, A. B. (1966). Electron microscopic autoradiography of rabbit reticulocytes active and inactive in protein synthesis. *Science* **151**, 1000–1001.

Mizuhira, V., Shiihashi, M., and Futaesaku, Y. (1981). High-speed electron microscope autoradiographic studies of diffusible compounds. *J. Histochem. Cytochem.* **29**, 143–160.

Nagata, T., and Murata, F. (1977). Electron microscopic dry mounting radioautography for diffusible compounds by means of ultracryotomy. *Histochemistry* **54**, 75–82.

Nagata, T. (1994). Electron microscopic radioautography with cryofixation and dry-mounting procedure. *Acta Histochem. Cytochem.* **27**, 471–489.

Nielsen, S., Nielsen, J. T., and Christensen, E. I. (1987). Luminal and basolateral uptake of insulin in isolated, perfused, proximal tubules. *Am. J. Physiol.* **253** (*Renal Fluid Electrolyte Physiol* 22), F857–F867,

Ottosen, P. D. (1978). Reversible peritubular binding of a cationic protein (lysozyme) to flounder kidney tubules. *Cell Tissue Res.* **194**, 207–218.

Peters, T., Jr., and Ashley, C. A. (1967). An artefact in radioautography due to binding of free amino acids to tissues by fixatives. *J. Cell Biol.* **33**, 53–60.

Salpeter, M. M., and Bachmann, L. (1964). Autoradiography with the electron microscope: A procedure for improving resolution, sensitivity, and contrast. *J. Cell Biol.* **22**, 469–477.

Salpeter, M. M., Bachmann, L., and Salpeter, E. E. (1969). Resolution in electron microscope radioautography. *J. Cell Biol.* **41**, 1–20.

Salpeter, M. M., McHenry, F. A., and Salpeter, E. E. (1978). Resolution in electron microscope autoradiography. IV. Application to analysis of autoradiographs. *J. Cell Biol.* **76**, 127–145.

Stein, O., and Stein, Y. (1967). Lipid synthesis, intracellular transport, storage, and secretion. I. Electron microscopic radioautographic study of liver after injection of tritiated palmitate or glycerol in fasted and ethanol-treated rats. *J. Cell Biol.* **33**, 319–339.

Williams, M. A. (1977). Quantitative methods in biology. *In* "Practical Methods in Electron Microscopy" (A. M. Glauert, ed.), Vol. 6. North-Holland, Amsterdam.

Wisse, E., and Tates, A. D. (1968). A gold latensification-elon ascorbic acid developer for Ilford L4 emulsion. *In* "Fourth European Regional Conference on Electron Microscopy" (S. D. Bocciarelli, ed.), Vol. 2, pp. 465–466. Tipografia Poliglotta Vaticana, Rome.

CYTOCHEMISTRY

Electron microscope cytochemistry aims at identifying the chemical or enzymatic nature of various molecules in cells and tissues. There is a vast repertoire of techniques that have been developed continuously from an early start in the 1950s, which range from simple methods for a more-or-less specific staining to complex procedures for specific identification of enzymes. The field of electron microscope cytochemistry also includes procedures based on biotin–avidin interactions and on the specific affinity of various lectins for certain carbohydrates. *In situ* hybridization, which is based on very specific interactions between nucleic acids, can be included as a cytochemical method in a broad sense, as can immunocytochemistry, which is treated separately in Chapter 16.

Successful electron microscope cytochemistry invariably requires a consideration of the initial methods for tissue processing, particularly the fixation procedure. Preservation of enzyme activities usually requires a mild fixation, for instance, with formaldehyde or low concentrations of glutaraldehyde, as strong fixatives, i.e., osmium tetroxide or high concentrations of glutaraldehyde, usually inhibit enzyme activity. Weak fixatives, however, may lead to less optimal structural preservation, and a suitable balance has to be reached between adequate fixation and ultrastructural preservation. The preparatory procedures may therefore differ for different enzymes or for other substances.

This chapter exemplifies the ultrastructural application of a few cytochemical procedures and illustrates some of the general factors that influence the results.

The early applications of electron microscope cytochemistry were based on the same principles as the pioneering light microscope work on phosphatase localization by George Gomori (1939) and J. Takamatsu (1939). Indeed, the first electron microscope application of enzyme cytochemistry, pursued in 1955 by Huntington Sheldon, Hans Zetterqvist, and David Brandes, was the demonstration of phosphatases in intestinal cells. These investigations were carried out on tissue fixed briefly (8 min) with osmium tetroxide, but in subsequent studies Alex Novikoff (1959) used formaldehyde and David Sabatini, Klaus Bensch, and Russell Barrnett (1963) used glutaraldehyde. The introduction of the mild aldehyde fixatives from thereon enhanced the development of electron microscope cytochemistry greatly.

Gomori, G. (1939). *Proc. Soc. Exp. Biol.* **42,** 23–26.
Novikoff, A. B. (1959). *J. Biophys. Biochem. Cytol.* **6,** 136–138.
Sabatini, D. S., Bensch, K., and Barrnett, R. J. (1963). *J. Cell Biol.* **17,** 19–58.
Sheldon, H., Zetterqvist, H., and Brandes, D. (1955). *Exp. Cell Res.* **9,** 592–596.
Takamatsu, J. (1939). *Japan. Pathol.* **29,** 492–496.

1. Influence of Fixation

FIGURE 15.1A Demonstration of acid phosphatase activity in lysosomes of a rat proximal tubule cell. The kidney was perfusion fixed for 3 min with 3% glutaraldehyde in 0.1 M sodium cacodylate buffer, pH 7.2, and small tissue blocks were fixed for another 2 hr by immersion in the same fixative. Nonfrozen sections (20 μm) were prepared with the tissue sectioner (Smith and Farquhar, 1963) and were incubated for 10 min at 37°C in a freshly prepared Gomori medium for acid phosphatase, rinsed for 2 min in 0.05 M acetate buffer, pH 5.0, postfixed in 1% in osmium tetroxide, and embedded in Epon. Sections were stained with uranyl acetate. × 48,000.

The electron-dense reaction product is limited to cytoplasmic lysosomes, which are sharply delimited from the cytoplasm where there is no reaction product.

FIGURE 15.1B Cells from the medullary thick ascending limb (mTAL) of the loop of Henle in a rat kidney, which were perfusion fixed with 2% formaldehyde in 0.1 M sodium cacodylate buffer, pH 7.2, containing 0.2 M sucrose. Tissue blocks were sectioned as in Fig. 15.1A at 40 μm and incubated in the cytochemical medium (Mayahara *et al.*, 1980) for the demonstration of the potassium-dependent p-nitrophenylphosphatase activity of the Na,K-ATPase. The thick sections were postfixed in osmium tetroxide and embedded in Epon. Ultrathin sections were stained with lead citrate. × 60,000.

Basolateral cell membranes are lined with an electron-dense reaction product. Mitochondria are completely free of precipitates.

FIGURE 15.1C Tissue from the same kidney prepared as in Fig. 15.1B except that the tissue was also fixed for 30 min in 1% glutaraldehyde before cytochemical incubation and posttreatment. × 38,000.

There is no cytochemical reaction products in any parts of the cells.

COMMENTS

The choice of fixative and fixation procedure is crucial to the outcome of enzyme cytochemistry as enzymes vary considerably with respect to their sensitivity to different fixatives. Potassium-dependent p-nitrophenylphosphatase, which is a partial enzyme activity of Na,K-ATPase, can only be preserved with formaldehyde fixatives of low strength as shown in Fig. 15.1B, but is inhibited by glutaraldehyde fixation (Fig. 15.1C). In contrast, acid phosphatase is much more resistent and is active even after fixation in 3% glutaraldehyde for 2 hr as seen in Fig. 15.1A.

The absence of cytochemical reaction products does not prove the absence of enzyme; the enzyme might have been inactivated by the fixative or other steps in the preparatory procedure preceding the incubation. Biochemical determinations are useful for establishing to what extent an enzyme is sensitive to fixatives, as illustrated in model experiments with various enzymes (Fahimi and Drochmans, 1968; Sabatini *et al.*, 1963).

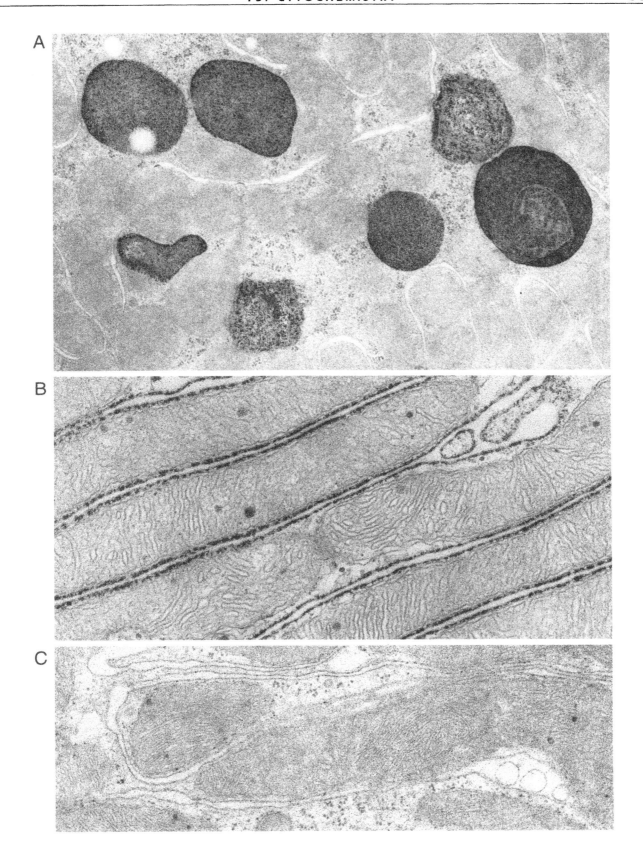

2. Preincubation Treatment

FIGURE 15.2A A cytochemical demonstration of catalase in peroxisomes in a rat liver with the diaminobenzidine method. The rat liver was perfusion fixed for 3 min with 1% glutaraldehyde in 0.1 *M* sodium cacodylate buffer, postfixed for 2 hr in the same solution, and washed overnight in cadodylate buffer. Sections 25 μm in thickness were prepared on a Vibratome and incubated for catalase according to Fahimi (1969). After postfixation in osmium tetroxide and Epon embedding, ultrathin sections were stained with lead citrate. $\times$ 5800.

The micrograph shows a complete cross section of the tissue section (23 μm between the asterisks) and illustrates numerous stained peroxisomes throughout the section.

FIGURE 15.2B A cytochemical demonstration of peroxisomes from the same liver as in Fig. 15.2A. In this case, about 200-μm-thick sections were incubated as above. One-micron sections were observed unstained in the light microscope. The distance between the asterisks is 215 μm. $\times$ 650.

Peroxisomes are stained in the outer thirds of the thick section, but unstained in the middle third. The one-micron section was cut at right angle to the thick section.

FIGURES 15.2C AND 15.2D Similar preparation as in Fig. 15.2B. Figure 15.2C is from the surface of the thick section, Fig. 15.2D is from its center $\times$ 12,000.

Figure 15.2C shows several stained peroxisomes whereas in Fig. 15.2D the peroxisomes remain unstained (arrowheads).

FIGURE 15.2E A cytochemical demonstration of lysosomes in kidney proximal tubule cells (segment S3). The kidney was perfusion fixed with 1% glutaraldehyde. Thick sections (40 μm) were prepared without freezing using the Smith and Farquhar (1963) tissue sectioner, incubated in the Gomori medium at pH 5.0, rinsed in 0.05 *M* acetate buffer, postosmicated, and embedded in Epon. Ultrathin sections were double-stained with uranyl acetate and lead citrate. $\times$ 15,000.

The cells are well preserved and contain several acid phosphatase-positive lysosomes.

COMMENTS

Tissue blocks used for the incubation in cytochemical media must not exceed a certain thickness in order to allow penetration of the incubation medium. Preparation of nonfrozen thick sections, either with the Smith and Farquhar tissue sectioner or the Vibratome, results in well-preserved cell fine structure. For phosphatase or diaminobenzidine reactions the thickness of the sections used for incubation should not exceed 30–40 μm. If thicker sections are used the distribution of enzyme reactivity tends to be uneven, with the center of the section being incompletely reactive or unreactive (Fig. 15.2D). Penetration of the incubation medium into thick sections may be improved if the tissue is frozen before incubation or pretreated with weak detergents or dimethyl sulfoxide.

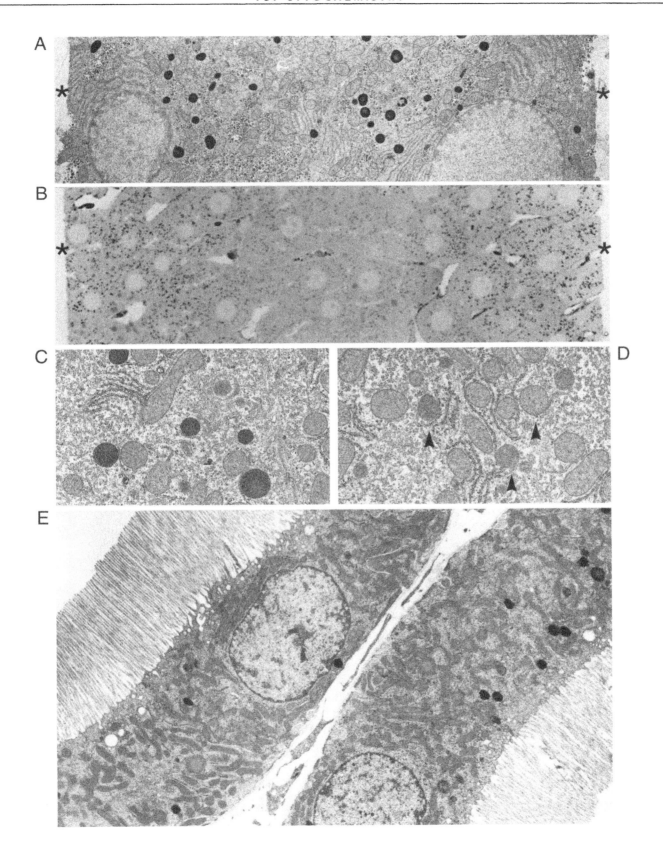

3. Appearance of Reaction Product

FIGURES 15.3A AND 15.3B Lysosomes in proximal tubule cells identified by the Gomori medium for acid phosphatase. The tissue was fixed by perfusion with 3% purified glutaraldehyde in 0.1 M cacodylate buffer. Nonfrozen sections prepared with a thickness of 20 μm were incubated for 7 min in fresh Gomori medium made up of DL-β-glycerophosphate. After rinsing in 0.05 M acetate buffer the tissue was embedded without osmication in Epon and the sections were observed unstained in the electron microscope. Figure 15.3A × 46,000, Fig. 15.3B × 140,000.

The lysosome is outlined by its electron-dense reaction product for acid phosphatase. There is no reaction product whatsoever in the cytoplasm outside the lysosome. At high magnification the reaction product appears as finely granular, electron-dense material (Fig. 15.2B).

FIGURES 15.3C AND 15.3D Same preparation as in Figs. 15.3A and 15.3B except that the tissue was osmicated after incubation. Figure 15.3C × 46,000, Fig. 15.3D × 140,000.

The reaction product is equally well contained in the lysosome as in Fig. 15.3A and has the same appearance at high magnification, although osmication renders the matrix of the lysosome electron dense.

FIGURES 15.3E AND 15.3F Similar preparation as in Fig. 15.3A except that the incubation medium was made up of DL-β-glycerophosphate containing 75% β isomer. Figure 15.3E × 46,000, Fig. 15.3F × 140,000.

The reaction product in this lysosome has a more uneven distribution than in Figs. 15.3A and 15.3C and there is a fine sprinkling of reaction products in the cytoplasm. At high magnification the reaction product has in part the appearance of a fine needle-like material.

COMMENTS

The precise localization of the reaction product in Figs. 15.3A–15.3D is related to the fact that the incubation medium contained almost 100% of the β isomer of glycerophosphate whereas Figs. 15.3E and 15.3F were obtained with a medium containing a mixture of α and β isomers. The precise outcome of the cytochemical reaction is also based on adequate choices of tissue fixation, thickness of tissue slice for incubation (only 20 μm), and incubation time. Each of these factors may influence the cytochemical reaction, as further illustrated in Fig. 15.6. Figures 15.3C and 15.3D illustrate that the reaction product is not modified by short postfixation in osmium tetroxide, although the additonal staining of the cell renders the reaction product less well defined. However, postosmication and staining permit the identification surrounding cytoplasmic features.

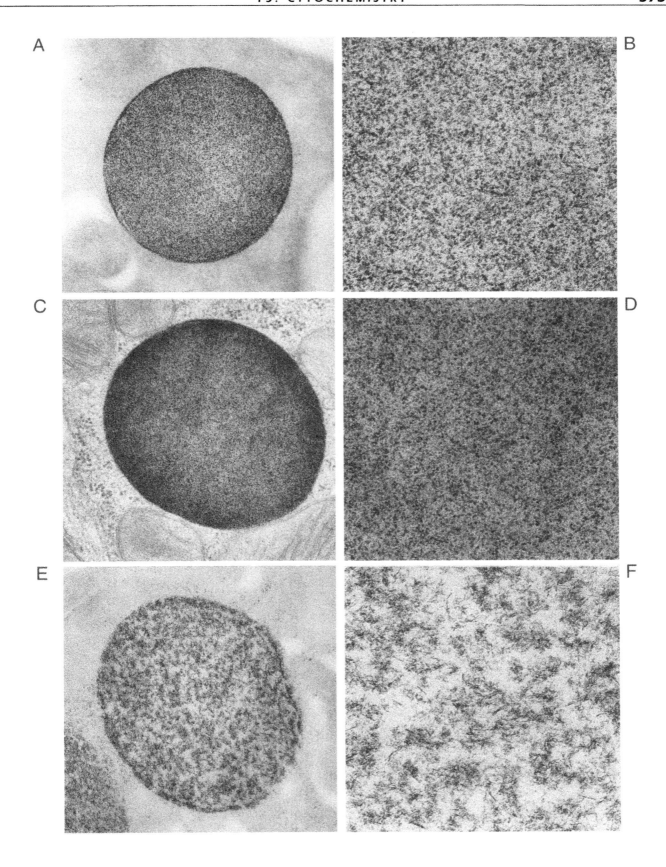

4. Composition of Incubation Medium

FIGURE 15.4A Basal part of cells in isolated flounder tubules incubated in a solution of horseradish peroxidase. Tubules were fixed in 1% glutaraldehyde in 0.1 M sodium cacodylate buffer, rinsed in buffer, and incubated in the diaminobenzidine medium for peroxidase, pH 7.6 (Graham and Karnovsky, 1966). The tissue was postfixed in osmium tetroxide and embedded in Epon, and the sections were analyzed unstained in the electron microscope. × 30,000.

An electron-dense reaction product from the peroxidase is present in the basement membrane (left) and has penetrated into the lateral intercellular spaces but not into the cytoplasm.

FIGURE 15.4B A liver cell incubated in the diaminobenzidine medium adjusted to 8.5 (Fahimi, 1969). The tissue was postfixed in osmium tetroxide and Epon embedded, and ultrathin sections were stained with lead citrate. × 30,000.

The electron-dense reaction product is associated exclusively with peroxisomes. There is no reaction product in other parts of the cytoplasm.

FIGURES 15.4C–15.4E Basal parts of cells from the mTAL in the rat kidney after perfusion fixation with 2% paraformaldehyde in 0.1 M cacodylate buffer containing 2% sucrose. After buffer rinse, 30- to 40-μm sections were prepared unfrozen on a Vibratome and incubated in three modifications of the Mayahara *et al.* (1980) medium for p-nitrophenylphosphatase. In Fig. 15.4C the incubation was carried out in complete medium, in Fig. 15.4D the substrate was sodium p-nitrophenylphosphate, and in Fig. 15.4E 10 mM ouabain was added to the regular complete medium. × 30,000.

In Figure 15.4C there is an intense reaction product associated with the basolateral membranes. In Figs. 15.4D and 15.4E there is essentially no reaction product associated with these membranes.

COMMENTS

In enzyme cytochemistry the composition of the medium is of crucial importance for the outcome. By changing the pH or the ionic composition or by adding inhibitors, greatly different results are obtained. Thus in the classical diaminobenzidine procedure for light microscopy, which was invented by Strauss (1959) and applied for electron microscopy by Graham and Karnovsky (1966), a change in pH means that peroxidase activity is detected close to neutral pH, whereas at high pH the catalase activity in peroxisomes is reactive. Additionally, if the same medium is somewhat modified and used at a low pH, mitochondrial membranes are labeled due to their cytochrome oxidase content. The medium developed by Mayahara *et al.* (1980) detects the potassium-dependent p-nitrophenylphosphatase activity of the Na,K-ATPase (Fig. 15.4C). If the substrate is used in its sodium form instead of its potassium form there is no reaction (Fig. 15.4D). Likewise if the enzyme is inhibited by ouabain the enzyme activity disappears (Fig. 15.4E).

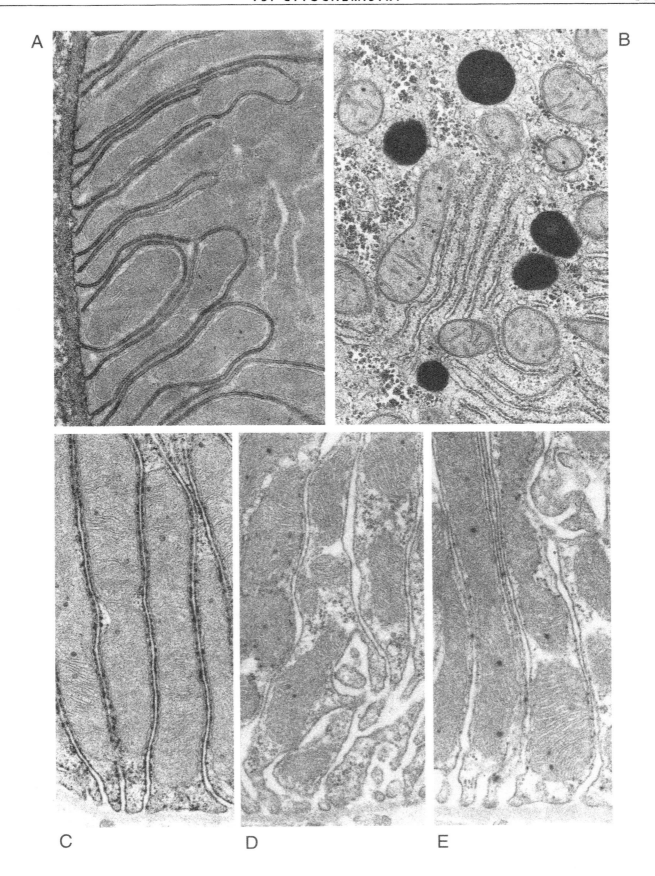

5. Cytochemical Resolution

FIGURES 15.5A AND 15.5B Lysosomes in proximal tubule cells prepared as in Fig. 15.3A except that the tissue was postfixed in osmium tetroxide and the thin sections were stained with lead citrate (Fig. 15.5A) or double-stained with uranyl acetate and lead citrate (Fig. 15.5B). × 225,000 and 180,000, respectively.

The reaction products in these lysosomes are present up to the inner leaflet of the lysosomal membrane. Neither the outer leaflet of the lysosomal membrane (arrows) nor the surrounding cytoplasm shows any enzyme reaction product. Double-staining of the section tends to mask the texture of the reaction product in the matrix of the lysosome.

FIGURE 15.5C Interdigitating cells from the distal nephron (mTAL cells) incubated for potassium-dependent *p*-nitrophenylphosphatase activity as in Fig. 15.1B. × 150,000.

The enzyme reaction product is associated with the cytoplasmic leaflet of the membrane (arrows). The thickness of the deposited reaction product varies between 5 and 15 nm.

FIGURE 15.5D Similar preparation as in Fig. 15.5C except that the incubation time was 50% longer. × 40,000.

The reaction product almost fills the entire cytoplasm and obscures the membranes. Additionally, there is some precipitate in the mitochondria.

COMMENTS

The precision in the localization of reaction products in cytochemical preparations, and thus the cytochemical resolution, is influenced by several steps in the preparation procedure, notably fixation, composition of medium, incubation time, and postincubation treatment. Prolonged incubation times may lead to the formation of large amounts of enzyme reaction products falsely associated with various parts of the cytoplasm, such as seen in Fig. 15.5D. The decision of whether an observed enzyme localization is correct must be based on other known properties of the system. The potassium-dependent *p*-nitrophenylphosphatase activity is a partial reaction of Na,K-ATPase, and is known biochemically to be present in the basolateral membranes of mTAL cells and absent in mitochondria, compatible with the observed localization of enzyme reaction products on the inner side of the cytoplasmic leaflet of the cell membrane in Fig. 15.5C.

With respect to the localization of the reaction product in Figs. 15.5A and 15.5B, it is well established that the lysosomal enzymes are present inside the lysosomes and not in the cytoplasm. However, it is uncertain whether there is a gradient in enzyme activity from the periphery to the center of the lysosome (Fig. 15.5A); this pattern may be related to a greater availability of substrate at the periphery of the organelle.

Some of the factors just discussed can also be applied to numerous other enzyme systems that have been studied by electron microscope cytochemistry over the last decades and which have been summarized, e.g., in the treatises by Lewis and Knight (1992), Ogawa and Barka (1993), and Pearse (1996).

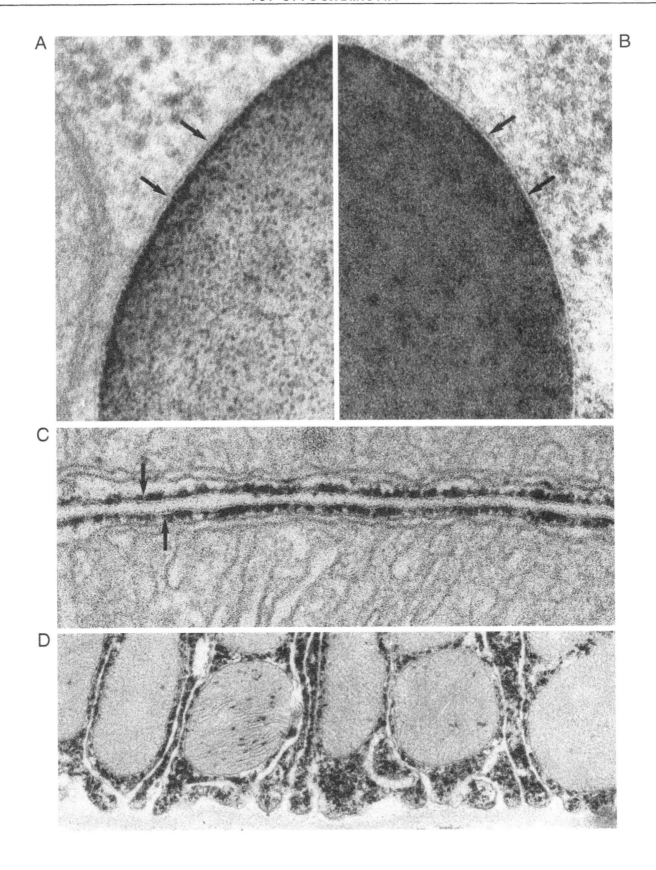

6. Unspecific Staining

FIGURE 15.6A Lysosomes in proximal tubule cells incubated for acid phosphatase according to the Gomori method. Details of the procedure are as in Fig. 15.1A except that the incubation time was longer and the thin section was unstained. × 45,000.

The lysosomes contain a finely granular electron-dense reaction product. In addition, some reaction product is located outside the lysosomes (upper right corner) and in the cleft between the two left lysosomes (arrows).

FIGURE 15.6B Part of a cytoplasm of proximal tubule cells incubated in acid phosphatase. Conditions are as in Fig. 15.1C except that in this experiment the Gomori medium was not freshly made but had already developed a foggy precipitate before incubation. × 20,000.

This field shows two lysosomes with an irregular distribution of the reaction product. In addition there is an extensive sprinkling of electron-dense precipitate all over the cytoplasm.

FIGURE 15.6C Similar preparation as in Fig. 15.6A except that the buffer rinse after incubation but before postosmication was very brief. × 35,000.

The reaction product is present in a lysosome and in the Golgi apparatus. A fine precipitate is also seen in the cytoplasm.

FIGURE 15.6D Part of a proximal tubule cell with an apical cytoplasm (lower left) and brush border in tissue incubated in acid phosphatase with the Gomori medium at pH 5.0. × 35,000.

A fine sprinkling of electron-dense precipitate is associated with the membrane of the brush border whereas little precipitate is present is the apical cytoplasm.

COMMENTS

Cytochemical reactions sometimes cause deposits of precipitates in unexpected locations. In many cases these precipitates can be identified as artifacts by considering the conditions of preparation and their relations to known biological characteristics. Thus in Fig. 15.6A the cytoplasmic precipitate is incompatible with the biochemical characteristics of lysosomes but may be caused by too long an incubation. This may lead to a leakage of released phosphate ions out of the lysosomes before they are reached by the capture reagent, i.e., lead ions.

Incubation media should generally be prepared immediately before use. Usually, a freshly prepared Gomori medium will gradually acquire a foggy precipitate and, during prolonged incubation, unspecific precipitates will deposit in the tissue (Fig. 15.6B). A more stable incubation medium for acid phosphatase is that of Barka and Anderson (1962).

A principally different type of precipitate is illustrated in Fig. 15.5D where some reaction product is associated with the plasma membrane of the brush border. This precipitate may originate from a splitting of the substrate by the alkaline phosphatase present in the brush border membranes, although this enzyme has a pH optimum well above that of the Gomori medium. An alternative explanation is that it is caused by acid phosphatase expelled from the tubule cells.

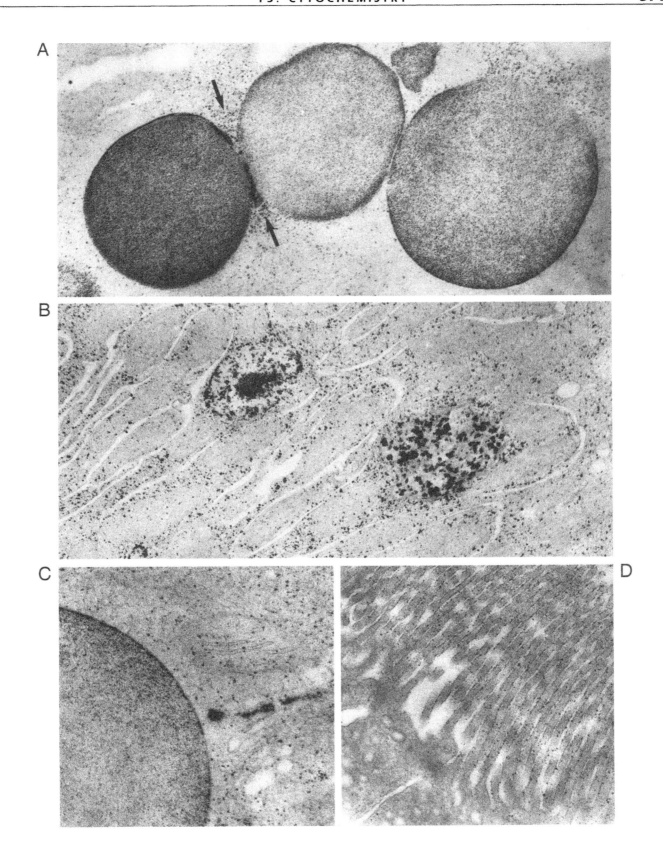

7. Extraction of Reaction Product

FIGURE 15.7A Brush border of a renal proximal tubule following incubation for alkaline phosphatase according to Mayahara *et al.* (1967). After ultramicrotomy the section was immediately collected from the fluid and stained with lead citrate. × 20,000.

The reaction product is associated with the surface of the microvilli.

FIGURE 15.7B Section from the same tissue block as in Fig. 15.7A. During ultramicrotomy this section was left on the surface of the water in the trough of the knife for 2 hr before it was collected on a Formvar-coated grid. The section was stained with lead citrate. × 20,000.

Much of the electron-dense precipitate seen in Fig. 15.7A is absent; instead there are holes in the section at the level of the brush border membranes.

FIGURE 15.7C Kidney cells incubated for acid phosphatase as in Fig. 15.1C. Following incubation the thick sections were rinsed for 30 sec in 2% acetic acid. The ultrathin section was stained with lead citrate. × 30,000.

The lysosome (L) shows an electron-dense reaction product partly associated with the periphery (arrows). Other parts of the lysosome are devoid of reaction products (asterisk). There is a faint sprinkling of reaction product in the Golgi apparatus but not in other parts of the cytoplasm.

FIGURE 15.7D Periphery of a lysosome in a similar preparation as in Fig. 15.7C. × 150,000.

The reaction product is primarily associated with the periphery of the lysosome and only in small patches in the lysosomal matrix.

COMMENTS

Because many lead-capture reactions in electron microscope cytochemistry may cause unspecific staining it has sometimes been customary to remove such precipitates by a brief wash in acetic acid. However, this also leads to the absence of reaction product in parts of the lysosomes as illustrated in Figs. 15.7C and 15.7D. Thus such a step in the cytochemical procedures is undesirable as it leads to uncertainty about the real site of enzyme activity. Reaction products can sometimes also be lost during ultramicrotomy (Fig. 15.7B). In some sections the reaction product already disappears after a few minutes in the water. Because this disappearance is only observed in some blocks, the phenomenon appears related to the degree of polymerization of the epoxy resin. Similar observations have also been made with other enzyme reaction products and is similar to the extraction during sectioning of calcium apatite crystallites (see Fig. 8.18A).

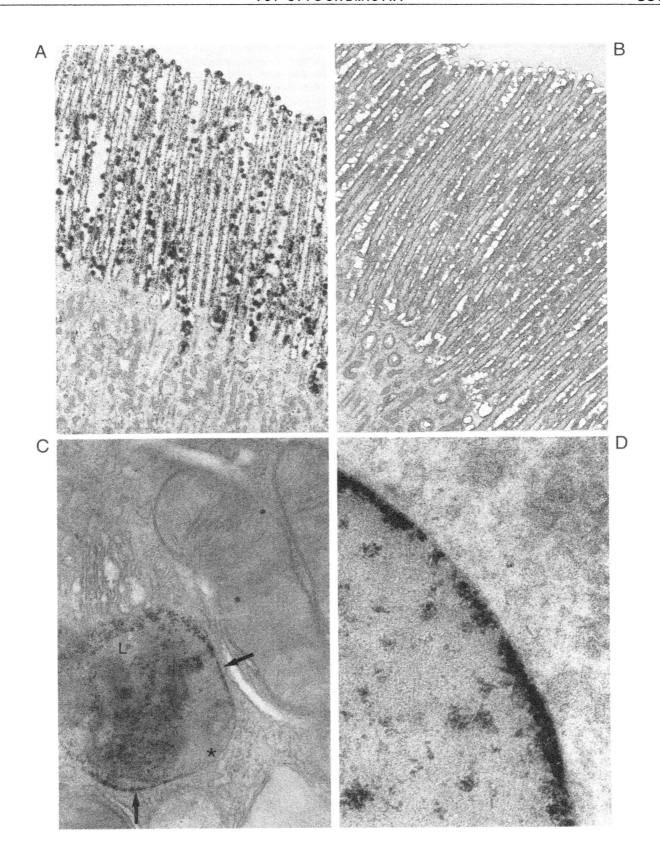

References

Angermüller, S., and Fahimi, H. D. (1981). Selective cytochemical localization of peroxidase, cytochrome oxidase and catalase in rat liver with 3,3'-diaminobenzidine. *Histochemistry* **71**, 33–44.

Barka, T., and Anderson, P. F. J. (1962). Histochemical methods for acid phosphatase using hexazonium pararosanilin as coupler. *J. Histochem. Cytochem.* **10**, 741–752.

Bernhard, W., and Avrameas, S. (1971). Ultrastructural visualization of cellular carbohydrate components by means of concanavalin A. *Exp. Cell Res.* **64**, 232–236.

Brooks, S. A., Leathem, A. J. C., and Schumacher, U. (1997). "Lectin Histochemistry: A Concise Practical Handbook." Bios Scientific, Oxford.

Danscher, G. (1981). Localization of gold in biological tissue: A photochemical method for light and electron microscopy. *Histochemistry* **71**, 81–88.

Deimann, W., Angermüller, S., Stoward, P. J., and Fahimi, H. D. (1991). Peroxidases. *In* "Histochemistry" (P. J. Stoward and A. G. Everson Pearse, eds.), Vol. 3, pp. 135–159, Churchill Livingstone, London.

Desmet, V. J. (1962). The hazard of acid differentiation in Gomori's method for acid phosphatase. *Stain Technol.* **37**, 373–376.

Ernst, S. A., and Hootman, S. R. (1981). Microscopical methods for the localization of Na⁺, K⁺-ATPase. *Histochem. J.* **13**, 397–418.

Fahimi, H. D. (1969). Cytochemical localization of peroxidatic activity of catalase in rat hepatic microbodies (peroxisomes). *J. Cell Biol.* **43**, 275–288.

Fahimi, H. D., and Drochmans, P. (1968). Purification of glutaraldehyde, its significance for preservation of acid phosphatase activity. *J. Histochem. Cytochem.* **16**, 199–204.

Friedenberg, R. M., and Seligman, A. M. (1972). Acetylcholinesterase at the myoneural junction: Cytochemical ultrastructure and some biochemical considerations. *J. Histochem. Cytochem.* **20**, 771–792.

Graham, R. C., and Karnovsky, M. J. (1966). The early stages of absorption of injected horseradish peroxidase in the proximal tubules of mouse kidney: Ultrastructural cytochemistry by a new technique. *J. Histochem. Cytochem.* **14**, 291–302.

Holt, S. J., and Hicks, R. M. (1961). The localization of acid phosphatase in rat liver cells as revealed by combined cytochemical staining and electron microscopy. *J. Biophys. Biochem. Cytol.* **11**, 47–66.

Hopsu, V. K., Arstila, A. U., and Glenner, G. G. (1965). The electron microscopic localization of aryl-sulfatase activity. *J. Histochem. Cytochem.* **13**, 711–712.

Kalicharan, D., Hulstaert, C. E., and Hardonk, M. J. (1985). Prevention of penetration hindrance in cerium-based glucose-6-phosphatase cytochemistry by freezing tissue in melting nitrogen. *Histochemistry* **82**, 287–292.

Kalimo, H. O., Helminen, H. J., Arstila, A. U., and Hopsu-Havu, V. K. (1968). The loss of enzyme reaction products from ultrathin sections during the staining for electron microscopy. *Histochemie* **14**, 123–130.

Kawano, J. I., and Akiwa, E. (1987). Ultrastructural localization of arylsulfatase C activity in rat kidney. *J. Histochem. Cytochem.* **35**, 523–530.

Lewis, P. R., and Knight, D. P. (1992). Cytochemical staining methods for electron microscopy. *In* "Practical Methods in Electron Microscopy" (A. M. Glauert, ed.), Vol. 14. Elsevier, Amsterdam.

Maunsbach, A. B. (1966). Observations on the ultrastructure and acid phosphatase activity of the cytoplasmic bodies in rat kidney proximal tubule cells. *J. Ultrastruct. Res.* **16**, 197–238.

Maunsbach, A. B., Skriver, E., Söderholm, M., and Hebert, H. (1986). Electron microscopy of the Na,K-ion pump. *In*: "Proceedings of the Eleventh Congress on Electron Microscopy," pp. 1801–1806. Kyoto.

Mayahara, H., Fujimoto, K., Ando, T., and Ogawa, K. (1980). A new one-step method for the cytochemical localization of ouabain-sensitive, potassium-dependent p-nitrophenylphosphatase activity. *Histochemistry* **67**, 125–138.

Mayahara, H., Hirano, H., Saito, T., and Ogawa, K. (1967). The new lead citrate method for the ultracytochemical demonstration of activity of non-specific alkaline phosphatase (orthophosphoric monoester phosphohydrolase) *Histochemie* **11**, 88–96.

Murata, F., Tsuyama, S., Kanae, T., Ihida, K. and Yang, D.-H. (1994). Sites of glycosylation observed by postembedding and preembedding lectin staining. *Acta Histochem. Cytochem.* **27**, 607–612.

Neiss, W. F. (1988). Enhancement of the periodic acid-Schiff (PAS) and periodic acid-thiocarbohydrazide-silver proteinate (PA-TCH-SP) reaction in LR White sections. *Histochemistry* **88**, 603–612.

Novikoff, A. B., and Goldfischer, S. (1961). Nucleosidediphosphatase activity in the Golgi apparatus and its usefulness for cytological studies. *Proc. Natl. Acad. Sci. USA* **47**, 802–810.

Novikoff, A. B., and Goldfischer, S. (1969). Visualization of peroxisomes (microbodies) and mitochondria with diaminobenzidine. *J. Histochem. Cytochem.* **17**, 675–680.

Novikoff, A. B., Novikoff, P. M., Quintana, N., and Davis, C. (1972). Diffusion artifacts in 3,3'-diaminobenzidine cytochemistry. *J. Histochem. Cytochem.* **20**, 745–749.

Ogawa, K., and Barka, T. (eds.) (1993). "Electron Microscopic Cytochemistry and Immunocytochemistry in Biomedicine." CRC Press, Boca Raton, FL.

Ogawa, K., Saito, T., and Mayahara, H. (1968). The site of ferricyanide reduction by reductases within mitochondria as studied by electron microscopy. *J. Histochem. Cytochem.* **16**, 49–57.

Ottosen, P. D., and Maunsbach, A. B. (1973). Transport of peroxidase in flounder kidney tubules studied by electron microscope histochemistry. *Kidney Int.* **3**, 315–326.

Pihl, E. (1968). Recent improvements of the sulfide-silver procedure for ultrastructural localization of heavy metals. *J. Microsc. (Paris)* **7**, 509–520.

Pinto da Silva, P., Parkison, C. and Dwyer, N. (1981). Fracture-label: Cytochemistry of freeze-fracture faces in the erythrocyte membrane. *Proc. Natl. Acad. Sci. USA* **78**, 343–347.

Rambourg, A., and Leblond, C. P. (1967). Electron microscope observations on the carbohydrate-rich cell coat present at the surface of cells in the rat. *J. Cell Biol.* **32**, 27–53.

Revel, J.-P. (1964). A stain for the ultrastructural localization of acid mucopolysaccharides. *J. Microsc. (Paris)* **3**, 535–544.

Roth, J. (1983). Application of lectin-gold complexes of electron microscopic localization of glycoconjugates on thin sections. *J. Histochem. Cytochem.* **31**, 987–999.

Sabatini, D. D., Bensch, K., and Barrnett, R. J. (1963). Cytochemistry and electron microscopy: The preservation of cellular ultrastructure and enzymatic activity by aldehyde fixation. *J. Cell Biol.* **17**, 19–58.

Saxton, R., and Hall, J. L. (1991). Enzyme cytochemistry. *In* "Electron Microscopy of Plant Cells" (J. L. Hall and C. Hawes, eds.), pp. 105–180. Academic Press, London.

Smith, R. E., and Farquhar, M. G. (1963). Preparation of thick sections for cytochemistry and electron microscopy by a non-freezing technique. *Nature* **200**, 691.

Stoward, P. J., and Pearse, A. G. E. (1991). "Histochemistry: Theoretical and Applied," Vol. 3. Churchill Livingstone, Edinburgh.

Strauss, W. (1959). Rapid cytochemical identication of phagosomes in various tissues of the rat and their differentiation from mitochondria by the peroxidase method. *J. Biophys. Biochem. Cytol.* **5**, 193–204.

Thiéry, J. P. (1967). Mise en évidence des polysaccharides sur coupes fines en microscopie électronique. *J. Microsc. (Paris)* **6**, 987–1018.

IMMUNOCYTOCHEMISTRY

Immunoelectron microscopy is a very powerful tool in cell biology. Its goals are identifying and localizing specified biological substances at the ultrastructural level. The great potential and applicability of these methods are due to the fact that practically all antigens can be localized to their cellular sites if suitable antibodies are available. In direct methods, the antigen is detected by an antibody labeled with an electron-dense marker, such as ferritin or colloidal gold. In indirect methods, which usually show a greater sensitivity and specificity, the primary antibody is unlabeled but is detected with a secondary or tertiary, labeled antibody.

A further distinction can be made between preembedding methods, where the primary antibody reacts with surface antigens or antigens in permeabilized tissue before embedding and sectioning, and postembedding procedures, where thin sections are labeled. The outcome of immunoelectron microscope methods not only depends on the type of antibody used, but is influenced by the methods of tissue preparation. The fixative has to be carefully chosen in order not to reduce or eliminate tissue antigenicity. The embedding procedure must not induce denaturation of the antigen. Immunoelectron microscope procedures, although very simple in principle, are therefore associated with many practical and methodological problems and occur in a great variety of sophisticated forms.

This chapter illustrates some of the main techniques in immunoelectron microscopy and the influence of pre-paratory variations. Some of the involved procedures are addressed in more detail in other chapters, particularly fixation (Chapters 2–4), freezing and low-temperture embedding (Chapter 6), and cryosectioning (Chapter 8).

The principles of electron microscope immunocytochemistry can be traced to the pioneering immunofluorescence work in the laboratory of Albert Coons in the 1940s (Coons et al., 1941). The step to the level of ultrastructure was first taken by S. Jonatan Singer (1959) who introduced ferritin as an antibody marker in electron microscope immunocytochemistry. Over the next several years additional refinements in the preparation techniques were gradually introduced, such as the now widespread colloidal gold labeling technique by W. Page Faulk and G. Malcolm Taylor in 1971 and the cryoultramicrotomy procedures by Kiyoteru T. Tokuyasu in 1973. Immunoelectron microscopy has now developed into one of the most expanding and useful methods in biological electron microscopy.

Coons, A. H., Creech, H. J., and Jones, R. W. (1941). *Proc. Soc. Exp. Biol. Med.* **47**, 200–202.
Faulk, W. P., and Taylor, G. M. (1971). *Immunochemistry* **8**, 1081–1088.
Singer, S. J. (1959). *Nature* **183**, 1523–1527.
Tokuyasu, K. T. (1973). *J. Cell Biol.* **57**, 551–565.

1. Fixation of Sensitive Antigens

FIGURE 16.1A Immunolabeling of Na,K-ATPase in a medullary thick ascending limb (mTAL) in rat kidney. The kidney was perfusion fixed with 4% paraformaldehyde in 0.1 M sodium cacodylate buffer containing 0.2 M sucrose. After 3 min of perfusion and 2 hr of postfixation in the same solution, small pieces of tissue were dissected out, rinsed in the same buffer, cryoprotected in sucrose, frozen in liquid nitrogen, and cryosectioned. Following preincubations the sections were incubated for 1 hr at room temperature on a droplet of antibody solution consisting of PBS with 1% BSA and mouse monoclonal antibody against the α-subunit of Na,K-ATPase (Cat. No. 05-369, Upstate Biotechnology, Lake Placid, NY). The primary antibody was detected by goat anti-mouse IgG conjugated to 10 nm colloidal gold. After rinses in PBS and finally distilled water the grid was stained for 10 min on 1.8% methylcellulose containing 0.3% uranyl acetate. × 50,000.

Colloidal gold particles are associated with basolateral cell membranes. Most colloidal gold particles are located on the cytoplasmic side of the membrane, and only very few gold particles are seen over the remaining cytoplasm or over mitochondria. Several factors have contributed to prevent unspecific labeling: (i) rinses of the cryosection on phosphate buffered saline (PBS), (ii) preincubation on a blocking solution consisting of PBS containing 0.05% glycine and 1% bovine serum albumin (BSA), (iii) three rinses on droplets containing PBS with 1% BSA, after the incubation on primary antibodies, and (iv) appropriate dilution of primary antibodies (1:100) and secondary probe (1:50).

FIGURE 16.1B Similar preparation as in Fig. 16.1A except that the fixative was 4% paraformaldehyde plus 0.1% glutaraldehyde. × 50,000.

Labeling is associated with basolateral membranes as in Fig. 16.1A, but the labeling is less intense than in Fig. 16.1A.

FIGURE 16.1C Similar preparation as in Fig. 16.1A but the fixative was 1% glutaraldehyde. × 50,000.

No specific labeling is observed.

FIGURE 16.1D Similar section as in Fig. 16.1A incubated with non-immune mouse IgG. × 50,000.

There is no labeling.

COMMENTS

Antigens vary considerably with respect to their sensitivity to the method of fixation and tissue processing. Many exhibit a greatly reduced antigenicity following glutaraldehyde fixation, although they may tolerate formaldehyde fixatives. Figures 16.1A–16.1C illustrate that Na,K-ATPase can be immunolabeled following formaldehyde fixation alone, but that even a small concentration of glutaraldehyde in the fixative will reduce the antigenicity and that 1% glutaraldehyde will completely eliminate the labeling. The first step is therefore to determine the tolerance of the antigen to fixation. The N-terminus of the α-subunit is known to be located on the cytoplasmic side of the membrane. The monoclonal antibody used here is directed against an epitope among the first 20 amino acids of the N-terminus (M. Caplan, personal communication), which is consistent with the observed labeling of the cytoplasmic side of the membrane (Fig. 16.1A).

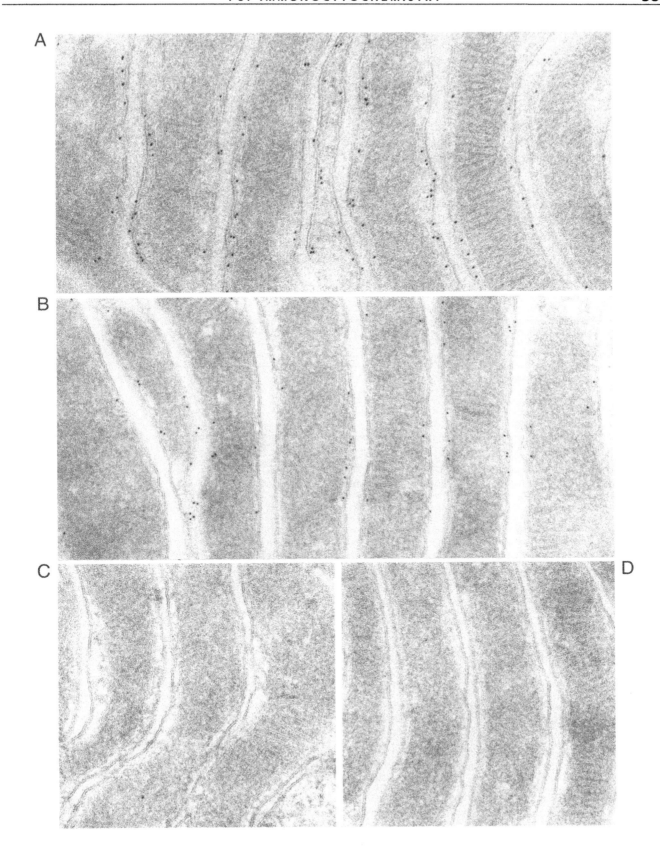

2. Fixation of Insensitive Antigens

FIGURE 16.2A A somatotrope cell in rat hypophysis perfusion fixed with 4% paraformaldehyde in 0.1 M sodium cacodylate buffer. After fixation for 2 hr in the same solution the tissue was dehydrated in ethanol without postfixation in osmium tetroxide and embedded in Epon. Ultrathin sections were preincubated with PBS containing 0.05% glycine and 1% BSA and immunolabeled using polyclonal rabbit anti-human growth hormone antiserum (A 0570, DAKO, Glostrup, Denmark). The primary antibody was detected with goat anti-rabbit IgG conjugated to 10 nm colloidal gold. The section was stained weakly with uranyl acetate. × 60,000.

Secretory granules are labeled extensively with colloidal gold particles.

FIGURE 16.2B Same preparation as in Fig. 16.2A except that the tissue was postfixed in 1% osmium tetroxide for 1 hr before embedding in Epon. × 60,000.

Secretory granules are also labeled here, although less so than without postfixation in osmium tetroxide. There is no background labeling over the cytoplasm.

FIGURE 16.2C Similar preparation as in Fig. 16.2A except that the fixative was 1% glutaraldehyde in sodium cacodylate buffer. The tissue was dehydrated and Epon embedded without osmium tetroxide postfixation. × 60,000.

Secretory granules are labeled extensively without obvious background in the cytoplasm.

FIGURE 16.2D Same preparation as in Fig. 16.2C except that the tissue was postfixed in osmium tetroxide before Epon embedding. × 60,000.

Secretory granules are labeled, but less so than in the nonosmicated tissue. The labeling is also more sparse than in tissue fixed with formaldehyde and osmium tetroxide in Fig. 16.2B.

COMMENTS

The antigenicity of growth hormone, similar to that of other peptide hormones, is quite resistant to different types of fixation and/or embedding media. In the present example the antigenicity is preserved after both formaldehyde and glutaraldehyde fixation even when followed by postfixation in osmium tetroxide and subsequent Epon embedding. It should be noted that glutaraldehyde fixation at the same concentration as used in Fig. 16.2C completely eliminates the antigenicity of Na,K-ATPase (Fig. 16.1C) and that the same result is observed if osmium tetroxide fixation is attempted or if the tissue is embedded in Epon. Comparisons between immunolabelings of Na,K-ATPase and growth hormone clearly illustrate that pronounced differences exist between different antigens with respect to their sensitivities to procedures of fixation, postfixation, and embedding.

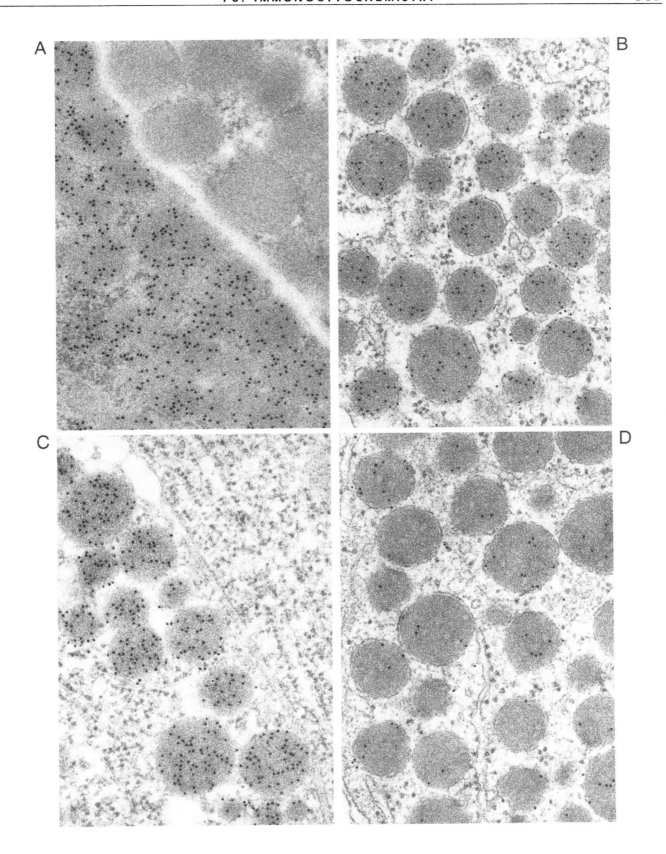

3. Comparison of Embedding Media

FIGURE 16.3A Cells in medullary thick ascending limb (mTAL) from a rat kidney perfusion fixed with 4% paraformaldehyde and cryosectioned. The cryosection was incubated with a mouse monoclonal antibody against the α-subunit of Na,K-ATPase as in Fig. 16.1A. The primary antibody was detected with goat anti-mouse IgG conjugated to 10 nm colloidal gold. $\times$ 50,000.

Basolateral membranes are labeled extensively, predominantly on their cytoplasmic side.

FIGURES 16.3B–16.3D Tissue from the same preparation as in Fig. 16.3A, although freeze-substituted and low-temperature embedded in Lowicryl HM20 (Fig. 16.3B) or K4M (Fig. 16.3C) or low-temperature embedded in LR White (Fig. 16.3D). The sections were immunolabeled with the same antibodies as in Fig. 16.3A used at the same dilution. The sections were stained with uranyl acetate. $\times$ 50,000.

Figures 16.3B–16.3D all show scattered labeling of the basolateral cell membrane. The labeling intensity is comparable on these three micrographs, although a strict quantitative determination has not been performed. The labeling intensity of the plastic sections is clearly lower than that on the cryosection.

COMMENTS

The fixative and the embedding medium influence the labeling intensity. The three embedding media give a lower labeling intensity than cryosections. In this example there is no obvious difference in labeling intensity among these three embedding media, but for some other antigens differences exist, e.g., among the different types of Lowicryls.

Notice that the label in Fig. 16.3A is predominantly located on the cytoplasmic side of the plasma membranes indicating that the epitope is located on that side. It seems likely that cross-sectioned membranes in a cryosection are elevated over adjacent cytoplasm and that this will cause this probe to remain on the cytoplasmic side. In Figs. 16.3B–16.3D on the other hand, the surfaces of the plastic sections may be smoother and will allow the probe attached to the primary antibody to dry down in random orientations relative to the cell membrane. An alternative explanation, namely that the probes in Figs. 16.3B–16.3D are displaced during, e.g., the drying of the grids seems less likely since many probes would then be expected at larger distances from the cell membrane.

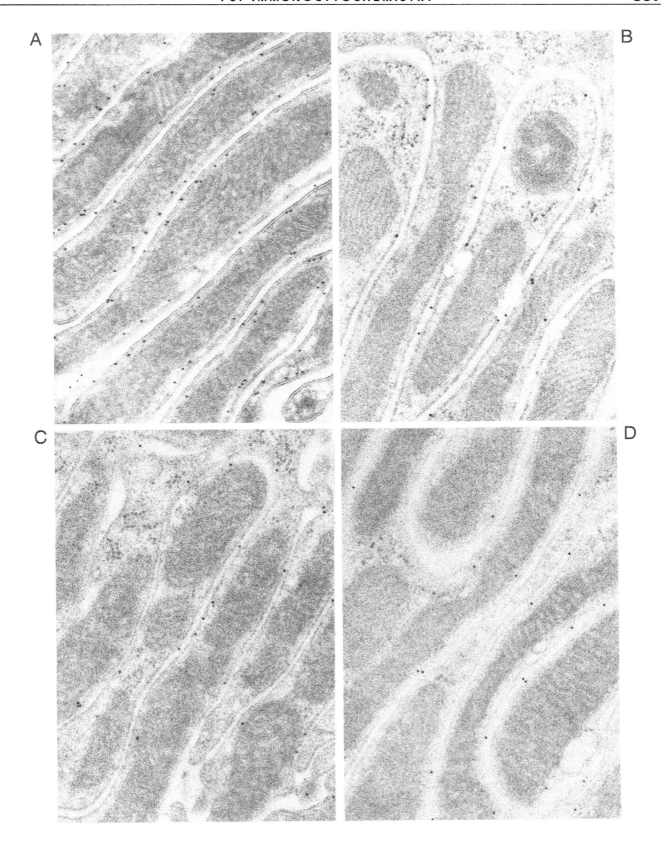

4. Influence of Preincubation Solutions

FIGURE 16.4A Cryosection of medullary thick ascending limb in a rat kidney prepared as in Fig. 16.1A. The section was preincubated in PBS containing 0.05% glycine and 5% BSA (as compared to 1% BSA in Fig. 16.1A). The section was labeled with rabbit anti-Na,K-ATPase antiserum. Before the grid was transferred to the droplet with the primary antibody, most of the preincubation solution was removed carefully with a filter paper without drying the grid. The secondary probe was goat anti-rabbit IgG on 10 nm colloidal gold. × 50,000.

Basolateral membranes are labeled extensively.

FIGURE 16.4B Same preparation as in Fig. 16.4A except that the last preincubation solution was not removed before the grid was placed on a small droplet with primary antibody. × 50,000.

Basolateral membranes are labeled, although much less so than in Fig. 16.4A.

FIGURE 16.4C Same preparation as in Fig. 16.4A except that the preincubation solutions consisted of only PBS without glycine or albumin. × 50,000.

Basolateral membranes are labeled. There are also scattered probes over mitochondria, but only slightly more than in Fig. 16.4A.

COMMENTS

Before sections are labeled with the primary antibody, they are usually preincubated on a solution containing aldehyde-blocking agents, such as glycine, and albumin or other substances that prevent unspecific binding of the antibodies to the section. The tendency for primary antibodies to bind unspecifically to the section varies greatly depending on the tissue, its preparation, and the specificity and affinity of the antibodies to the antigen. For these reasons it is difficult to predict how intense the unspecfic labeling will be.

The decreased labeling in Fig. 16.4B as compared to that in Fig. 16.4A may be caused by an incomplete removal from the grid of the concentrated preincubation rinsing solution. This may result in an unstirred layer on the grid surface which reduces the access of the primary antibody to the section during incubation. Even small modifications of the technique may thus result in distinct differences in labeling efficiency.

In the extensive literature on immunolabeling, one can find many recommendations of blocking and rinsing solutions used for preincubation. However, Figure 16.4C shows that at least in some cases unspecific labeling may be insignificant even if the rinsing solution contains no blocking agent at all.

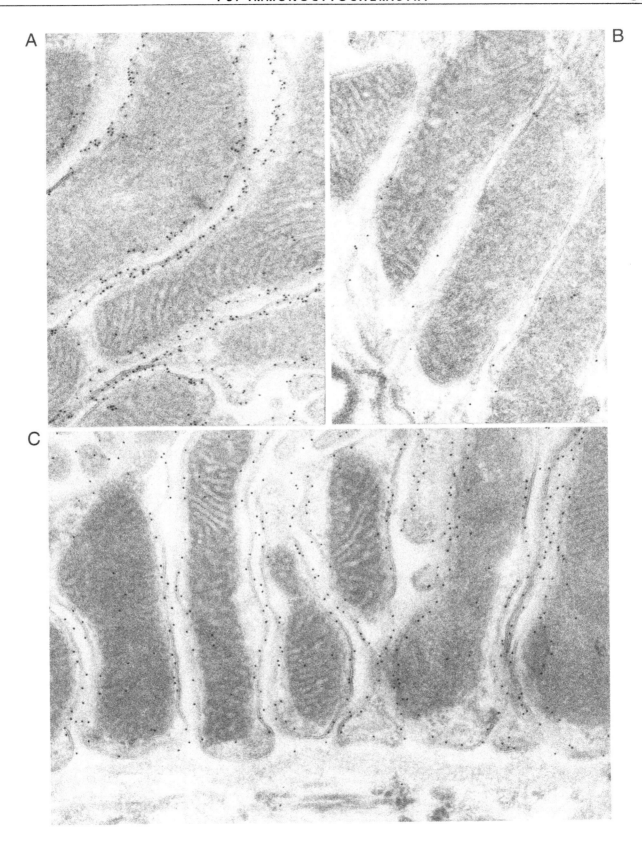

5. Comparison of Primary Antibodies

FIGURE 16.5A Immunolabeling of Na,K-ATPase in medullary thick ascending limb in a rat kidney. The cryosection was labeled with a mouse monoclonal antibody against the α-subunit of Na,K-ATPase as in Fig. 16.1A. The primary antibody was detected with goat anti-mouse IgG on 10 nm colloidal gold. $\times$ 50,000.

Colloidal gold particles are associated with basolateral membranes but are not observed over mitochondria.

FIGURE 16.5B Similar preparation as in Fig. 16.5A except that the primary antibody was detected with protein A on 10 nm gold. The dilution of the primary antibody was the same as in Fig. 16.5A.

Compared to Fig. 16.5A there are much fewer colloidal gold particles observed over the section.

FIGURE 16.5C Cryosection from the same block as in Figs. 16.5A and 16.5B but incubated with rabbit polyclonal anti-Na,K-ATPase IgG. $\times$ 50,000.

Basolateral membranes are labeled, but there is also a fair amount of colloidal gold particles over mitochondria.

FIGURE 16.5D Same preparation as in Fig. 16.5C except that the primary antibody was detected with protein A gold. $\times$ 50,000.

Basolateral membranes show a similar labeling as in Fig. 16.5C.

COMMENTS

The primary rabbit antibodies in Figs. 16.5C and 16.5D were detected equally well with gold-conjugated goat anti-rabbit antibodies and with protein A. However, when cryosections were labeled with mouse monoclonal antibodies at the same concentration in Figs. 16.5A and 16.5B, the labeling intensity was distinctly greater with gold-conjugated rabbit anti-mouse IgG than with protein A. Thus the choice of gold probes requires consideration of the species of the primary antibodies. Notably, protein A gold is less suitable for the detection of monoclonal mouse antibodies than for polyclonal rabbit antibodies.

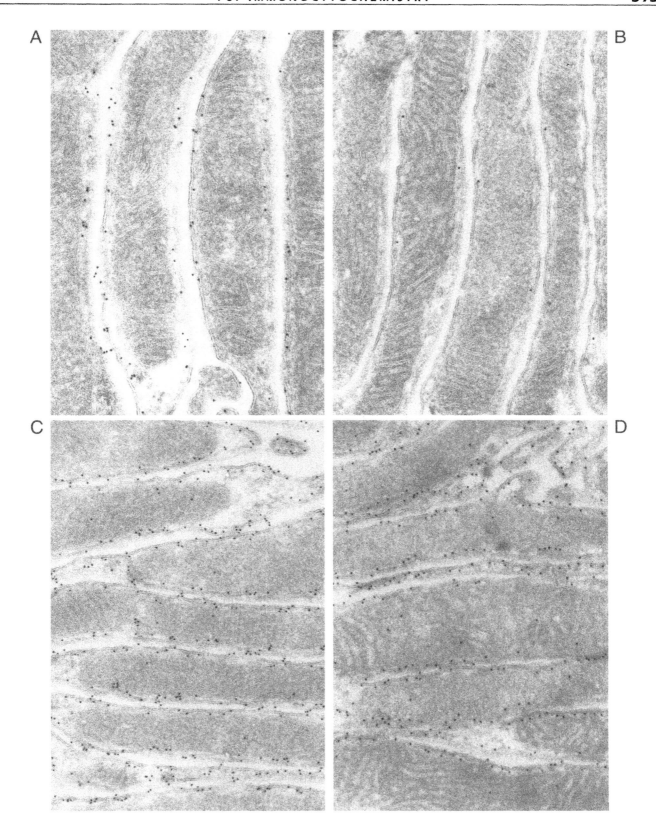

6. Dilution of Primary Antibody

FIGURES 16.6A–16.6E Immunolabeling of secretory granules in somatotrope cells in rat hypophysis. The rat was perfusion fixed through the heart with 4% paraformaldehyde in cacodylate buffer, and the hypophysis postfixed for 2 hr in the same fixative. After rinsing in buffer, small tissue blocks were cryprotected in 2.3 M sucrose, frozen in liquid nitrogen, freeze-substituted in methanol containing 0.5% uranyl acetate, and low temperature embedded in Lowicryl HM20. Thin sections were preincubated for 10 min with PBS containing 0.05% glycine and 1% BSA, rinsed three times on PBS containing 1% BSA, and then incubated with rabbit anti-human growth hormone antiserum (hGH, DAKO, Glostrup, Denmark) diluted with PBS containing 0.1% BSA. The dilutions were 1:25 (Fig. 16.6A), 1:100 (Fig. 16.6B), 1:400 (Fig. 16.6C), 1:1600 (Fig. 16.6D), and 1:6400 (Fig. 16.6E). The primary antibodies were detected with goat anti-rabbit IgG conjugated to 10 nm colloidal gold. × 60,000.

The six figures show gold labeling associated with the secretory granules in somatotrope cells. The density of gold particles in Figs. 16.6A and 16.6B seems (without actual quantitation) approximately similar. However, from Fig. 16.6B to Fig. 16.6E there is a gradual decrease in the labeling density, corresponding to the increasing dilution of the antibody.

FIGURE 16.6F Similar preparation as in Fig. 16.6A except that an irrelevant rabbit IgG was used instead of rabbit anti-human growth hormone. × 60,000.

There is no immunolabeling of this cell or of any other cells in the section.

COMMENTS

The intensity of immunolabeling decreases with increasing dilution of the antibody. However, the labeling with the two most concentrated dilutions, 1:25 and 1:100, is similar and a difference is not obvious without detailed counts of gold particles. Thus it appears that the antibody at the dilution 1:100 detects most of the available antigenic sites and that dilution 1:25 does not clearly increase the labeling intensity. Between dilutions 1:100 and 1:6400 there is a falling labeling intensity approximately corresponding to the dilution steps. There is very little background labeling over the cytoplasm or nuclei in this or other cell types (not shown).

The degree of dilution of the primary antibodies must be related to the affinity of antibodies to the antigen. If the affinity is weak the dilution of the antiserum may be low with a risk of unspecific labeling. However, if the affinity is very strong, as in the case of polyclonal sheep anti-rat megalin antiserum the dilution may be as high as 1:50,000 and yet result in optimal labeling with a minimum of unspecific labeling (Christensen *et al.*, 1995). In the case of purified monoclonal or polyclonal antibodies the degree of dilution will depend upon the concentration of IgG obtained by the producer (yourself, your colleagues, or a commercial company). A typical concentration of an affinity purified IgG is 0.1–1 mg IgG/ml solution. The degree of dilution will then depend upon the concentration of the antigen in the section as well as the affinity between the antigen and antibody.

A

B

C

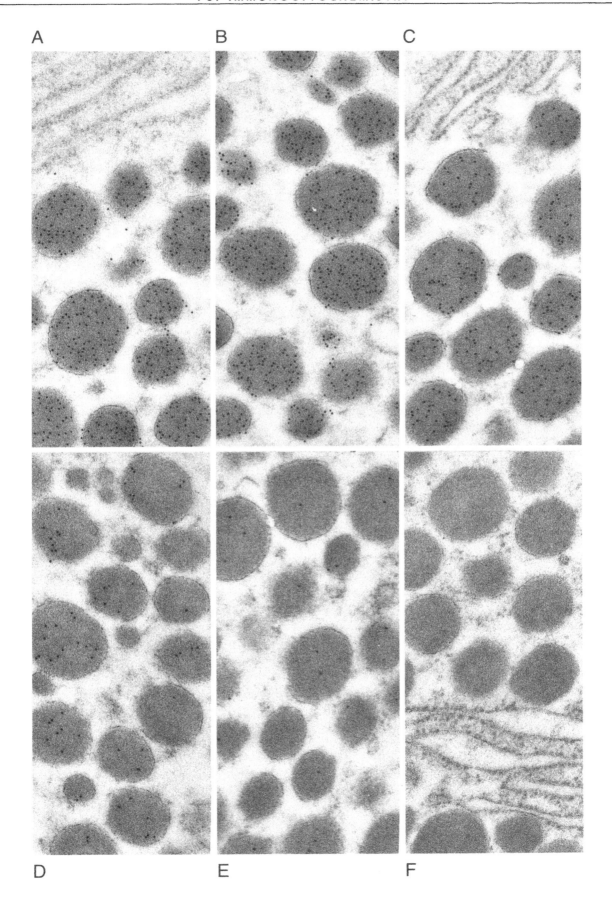

D

E

F

7. Quantitation of Gold Particles

FIGURES 16.7A–16.7F Same micrographs as in Fig. 16.6 following digital scanning (UMAX Vista-S8 scanner) and printing (Kodak 8650PS printer). Scanned images were processed using the AnalySIS program (Soft Imaging System, Münster, Germany). The gray values were selected to eliminate light areas and to retain the stained core of the secretory granules as gray areas. The analysis program then calculated the area of each individual granule, determined its number of gold particles, and finally averaged the number of particles per μm^2. Gold particles that were not surrounded by gray areas, i.e., background grains, were not counted. $\times$ 60,000.

In these six images the average number of gold particles per μm^2 was 0.85, 0.86, 0.39, 0.17, 0.04, and 0.

COMMENTS

Gold particles observed against a less contrasted background can be counted automatically in digitized images. To be able to determine their frequency over specific organelles, these must be well defined with respect to electron density, as in this case. It should be emphasized that the numbers obtained here from these individual images are not statistically representative as this would require a random sampling. However, they agree with the subjective evaluation of the labeling intensity of many somatotrope cells and suggest that the maximum labeling intensity in this experiment is obtained with the dilution of 1:100 and that the frequency of gold particles falls with a factor of 2–4 for each fourfold dilution of the antibody.

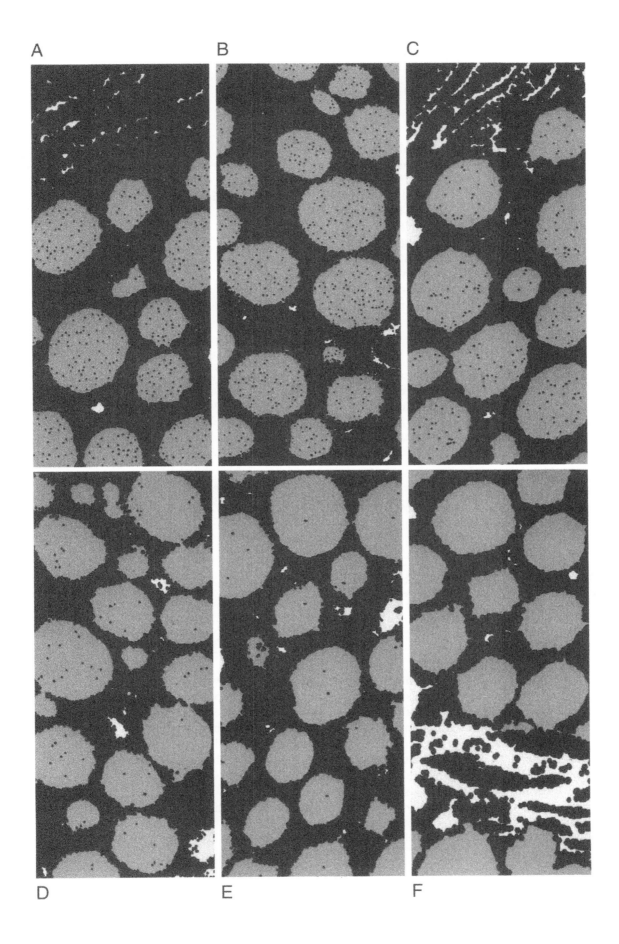

8. Controls

FIGURE 16.8A Immunocytochemical localization of growth hormone in rat hypophysis perfusion fixed with 4% formaldehyde in 0.1 *M* cacodylate buffer. No postfixation. Small tissue blocks were embedded in Epon. Thin sections were incubated with rabbit anti-human growth hormone antibodies. Western blot analysis of this antibody showed that it cross-reacted strongly with an approximate 22-kDa protein in a homogenate of untreated rat hypophysis (not shown). The primary antibody was detected with goat anti-rabbit IgG conjugated to 10 nm colloidal gold and the section stained with uranyl acetate. × 60,000.

Granules in the left cell are labeled uniformly and intensely, whereas those in the right cell are unlabeled. The border between the two cells is indistinct as it has been sectioned obliquely.

FIGURE 16.8B Section from the same series as in Fig. 16.8A. The labeling procedure was exactly as in Fig. 16.8A except that the antibody, before incubation with the section, was preabsorbed overnight at 4°C with an excess of human growth hormone. × 60,000.

The micrograph shows the same two cells as in Fig. 16.8A and several of the same granules can be identified. The main difference is that the left cell, which was labeled intensely in Fig. 16.8A, is essentially unlabeled.

FIGURE 16.8C Somatotrope cell from the same preparation as in Fig. 16.8A. × 60,000.

Secretory granules are labeled extensively.

FIGURE 16.8D Same cell as in Fig. 16.8C identified in a serial section and incubated with non-immune rabbit IgG. × 60,000.

No labeling is present over the secretory granules.

COMMENTS

Control experiments are crucial in immunocytochemistry. Without relevant controls the biological conclusions of labeling results remain inconclusive. A simple control is to leave out the primary antibody. Another simple control is to incubate the section with unrelated control serum or non-immune IgG (Fig. 16.8D). None of these controls is necessarily conclusive and does not exclude that the observed labeling is unspecific. A more specific control consists of preincubating the antibody solution with the antigen (compare Figs. 16.8A and 16.8B). If preimmunization serum is available, it is also a suitable control for antibody-containing serum.

A basic information in all immunocytochemical studies is whether the antigen is actually present in the tissue. Western blot analyses are important in order to determine if the antibody has the ability to react with one (or more) protein in the tissue and if antibodies raised against an antigen from one species cross react with the corresponding antigen from other species.

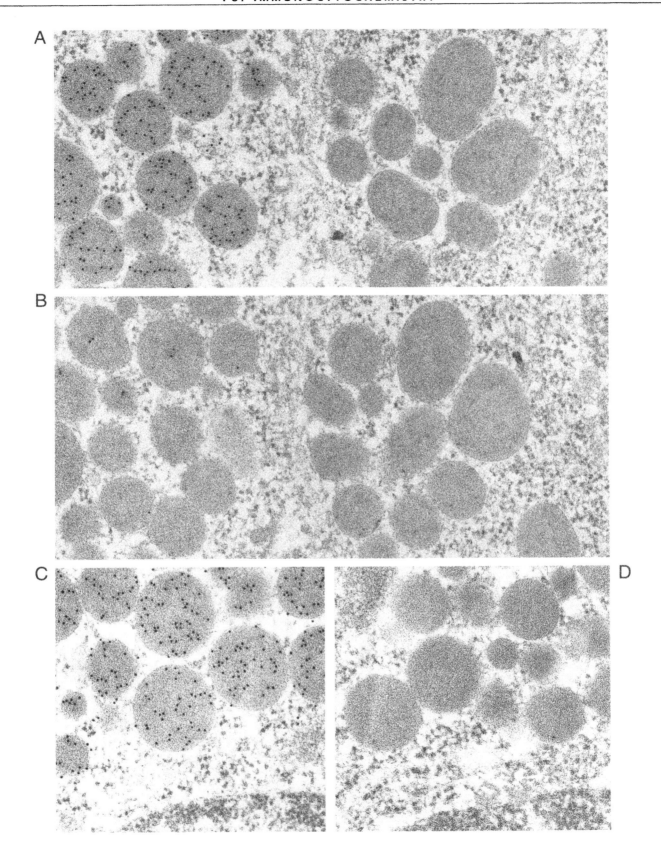

9. Comparison of Gold Probes

FIGURES 16.9A–16.9G Somatotrope cells in rat hypophysis following perfusion fixation with 4% paraformaldehyde in cacodylate buffer. After immersion fixation for 2 hr, small tissue blocks were cryoprotected in 2.3 M sucrose, frozen in liquid nitrogen, freeze-substituted and low-temperature embedded in HM20. Ultrathin sections were immunolabeled with rabbit anti-human growth hormone antibodies. After rinses in PBS containing 1% BSA, the primary antibodies were detected with goat anti-rabbit IgG conjugated to 5 nm colloidal gold (Fig. 16.9A), 10 nm colloidal gold (Fig. 16.9B), 15 nm colloidal gold (Fig. 16.9C), and in Fig. 16.9D to 10 nm colloidal gold from another manufacturer than used for Fig. 16.9B. In Fig. 16.9E goat anti-rabbit Fab fragments were conjugated to 10 nm colloidal gold. Figure 16.9F shows a fresh commercial preparation of protein A conjugated to 10 nm colloidal gold and Fig. 16.9G protein A conjugated to 10 nm colloidal gold after storage of the probe for 5 years in the refrigerator. Uranyl acetate staining. × 60,000.

All secondary probes have labeled granules in somatotrope cells, although differently with respect to frequency, size, and clumping. Less intense labeling is seen with 15-nm gold probes (Fig. 16.9C) than with smaller gold particles (Figs. 16.9A and 16.9B). Comparison between goat anti-rabbit IgG and protein A on 10 nm gold (Figs. 16.9B and 16.9F) shows that with this particular antibody more labeling occurred with goat anti-rabbit probes. Old protein A probes (Fig. 16.9G) show less labeling than fresh probes (Fig. 16.9F). Although most probes show little clumping, similar probes from other manufacturers may exhibit considerable clumping (Fig. 16.9D).

COMMENTS

The best probe for routine preparation in this comparison was goat anti-rabbit IgG on 10 nm gold (Fig. 16.9B). It was superior with respect to visibility (better than 5 nm, Fig. 16.9A), frequency of probes (greater than goat anti-rabbit IgG on 15-nm probes), and absence of clumping (much less than with another brand, Fig. 16.9D). This comparison emphasizes the need to choose a relevant secondary probe carefully. Small probes, 5 nm or less, increase the sensitivity but are obviously difficult to observe in low magnification micrographs. Clumped probe solutions give less label and patches of gold that decrease resolution and are useless in attempts to quantify the labeling. In the present system, less label is obtained with protein A than with goat anti-rabbit IgG. However, with other antigens this may be less obvious or reversed (compare Figs. 16.5C and 16.5D).

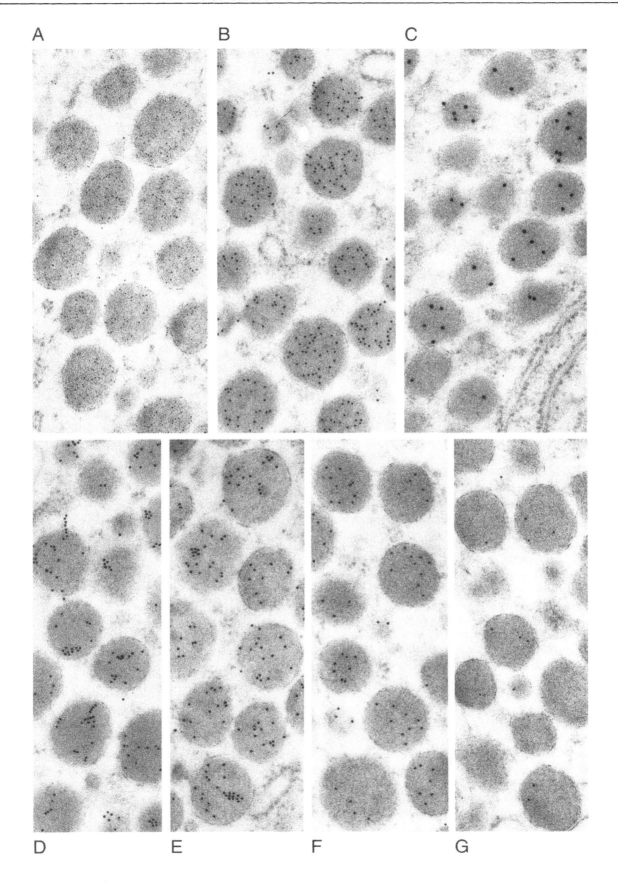

10. Amplification of Gold Particles

FIGURE 16.10A Localization of α-subunit of Na,K-ATPase in mTAL of a rat kidney prepared as in Fig. 16.1A, except that the secondary probe was goat anti-mouse IgG conjugated to 1 nm colloidal gold. $\times$ 50,000.

Mitochondria and cell membranes are visualized clearly but no colloidal gold particles are observable.

FIGURE 16.10B Same preparation as in Fig. 16.10A except that the dried section was placed for 30 min on a droplet of a silver amplification solution (R-gent, Aurion, Wageningen, The Netherlands) to amplify the 1-nm colloidal gold particles. $\times$ 50,000.

Numerous large electron-dense particles are associated with the basolateral membranes.

FIGURE 16.10C AND 16.10D Same preparations as in Fig. 16.10A except that the primary antibody was a rabbit anti-Na,K-ATPase antibody detected with goat anti-rabbit IgG on 5 nm colloidal gold (Fig. 16.10C) or 5-nm gold particles amplified with silver as in Fig. 16.10D. $\times$ 50,000.

Numerous small colloidal gold particles are associated with the basolateral membranes in Fig. 16.10C. Also in Fig. 16.10D the amplified (20–30 nm) particles are associated with basolateral membranes but they are less numerous than the small particles in Fig. 16.10C. The fine structure of the cells is well preserved.

FIGURE 16.10E Localization of growth hormone in somatotrope cells. The hypophysis was fixed as in Fig. 16.5A and embedded in Lowicryl HM20. The thin section was incubated with rabbit anti-human growth hormone antibodies, which were detected with goat anti-rabbit IgG conjugated to 1 nm colloidal gold. The section was stained with uranyl acetate, dried, placed on a droplet of silver amplification solution, rinsed, and dried. $\times$ 20,000.

The micrograph shows three different cell types in the hypophysis. One cell type contains labeled secretory granules with a diameter of about 400 nm. One cell with larger (about 600 nm) granules and one cell type with much smaller (about 150 nm) granules are also observed and are both unlabeled.

COMMENTS

Colloidal gold particles with a diameter of 10 nm represent a practical compromise between detectability and frequency. Because the frequency of secondary gold probes, i.e., the sensitivity, increases with decreasing diameter of the gold particle it is often an advantage to use small gold particles. Whereas 5-nm gold particles require at least an original magnification of 20,000 for easy recording, 1-nm colloidal gold particles cannot be visualized normally at all, at least not in stained sections. The silver enhancement technique (Danscher et al., 1993) can then be used to increase the size of the gold particles by adding electron-dense silver to the gold probes. Such enhanced grains are then readily visible even at low magnifications (Figs. 16.10B and 16.10D). The final size of the amplified particles is determined by the diameter of the original particles, the time and temperature of the incubation, and the composition of the silver solution.

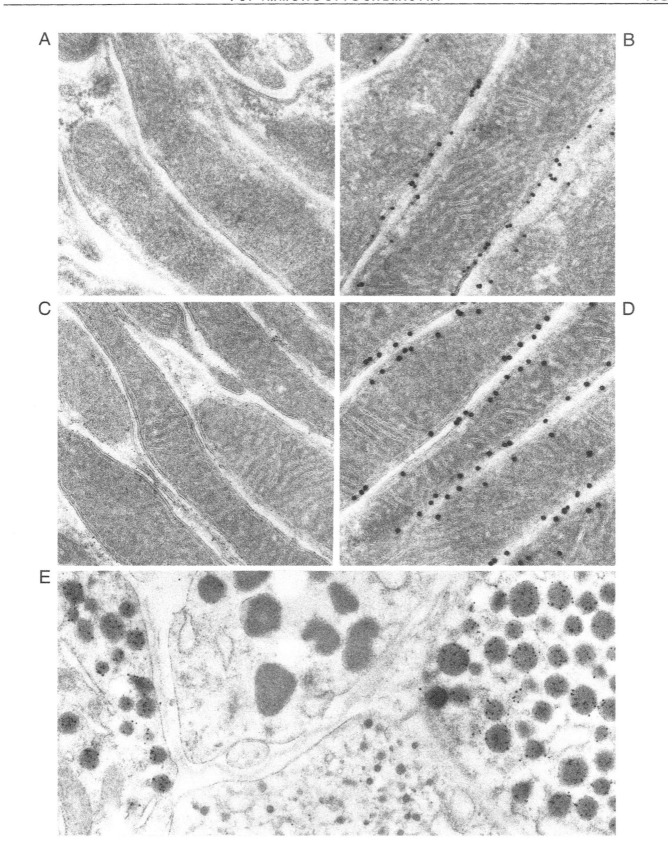

11. Section Staining

FIGURE 16.11A Cells in a medullary thick ascending limb in a rat kidney medulla perfusion fixed with paraformaldehyde and embedded in Lowicryl K4M by progressively lowering the temperature (PLT) (see Chapter 5). The section was immunolabeled with rabbit anti-Na,K-ATPase antiserum which was detected with goat anti-rabbit IgG conjugated to 5 nm colloidal gold. The section was stained with 0.3% uranyl acetate in 1.8% methylcellulose (Roth *et al.*, 1990). × 50,000.

Labeling is associated with basolateral membranes, which are stained and often appear triple layered. Mitochondrial staining is less distinct.

FIGURES 16.11B–16.11E Cells in mTAL in a rat kidney medulla perfusion fixed with paraformaldehyde, freeze-substituted in Lowicryl HM20, and immunolabeled with rabbit anti-Na,K-ATPase antiserum. Primary antibodies were detected with goat anti-rabbit IgG conjugated to 10 nm gold. Sections were unstained in Fig. 16.11B, stained with uranyl acetate in Fig. 16.11C, stained with lead citrate in Fig. 16.11D, and double stained with uranyl acetate and lead citrate in Fig. 16.11E. × 50,000.

In Fig. 16.11B the contrast of the tissue section is low and the gold probes stand out. In Fig. 16.11C the gold probes are also easily recognized but there is more staining of the cells. In Fig. 16.11D the gold probes are also clearly recognized against the cytoplasmic stain, but in Fig. 16.11E the contrast of the tissue is increased and the gold probes are identified less readily against cytoplasmic membranes or other structures.

COMMENTS

In immunoelectron microscopy it is desirable to be able to clearly recognize both gold probes and cellular features. Unstained plastic sections provide little information about the cells; however, a double-stained section tends to have a low visibility of the gold probes. For many purposes uranyl acetate staining alone gives a balance between visibility of the probes and identification of cellular features.

A special situation exists for Lowicryl K4M, which can be stained with uranyl acetate in methylcellulose, providing a clear membrane staining used for defining the location of gold probes (Fig. 16.11A).

Although biological electron micrographs in general are reproduced photographically with high contrast, this is not always desirable for immunoelectron micrographs. In fact, the best visibility of gold probes is often obtained if the micrographs are reproduced with fairly low density, which nevertheless permits the colloidal gold to stand out distinctly. In particular, when publishing immunoelectron micrographs, it is important that the density, contrast, and magnification of the image are adjusted to allow the probes to be clearly observed.

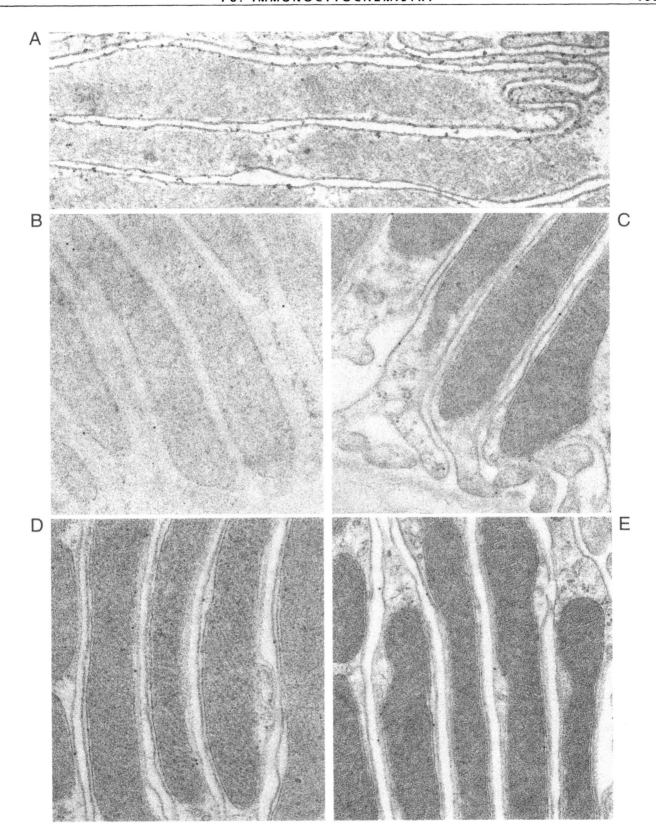

12. Resolution

FIGURE 16.12A Localization of Na,K-ATPase in a thick ascending limb of a rat kidney perfusion fixed with paraformaldehyde, freeze-substituted, and embedded in HM20. The ultrathin section was incubated with monoclonal antibodies against the α-subunit of Na,K-ATPase, and the primary antibodies were detected with goat anti-mouse IgG on 10-nm gold particles. $\times$ 100,000.

Colloidal gold particles are associated closely with, or directly overlying, the basolateral membranes in this thin section.

FIGURE 16.12B Same tissue and fixation as in Fig. 16.12A except that the tissue was cryosectioned. The section was incubated with a mouse monoclonal antibody against the Na,K-ATPase α-subunit. Primary antibodies were detected with goat anti-mouse IgG on 5 nm colloidal gold. $\times$ 75,000.

Colloidal gold particles are associated with basolateral membranes and, in most cases, located close to the inner surface of the plasma membranes. Only a few particles are located on the extracellular side.

FIGURE 16.12C Same preparation as in Fig. 16.12B and incubation with the same primary antibody. The secondary antibody was goat anti-mouse IgG on 10 nm colloidal gold. $\times$ 120,000.

Colloidal gold particles are associated closely with the plasma membrane, which transverses the micrograph and are in places associated with its cytoplasmic side. Below the center of the micrograph is a series of gold particles that follow an obliquely sectioned plasma membrane.

COMMENTS

The resolution in immunoelectron micrographs depends on several factors. One is the dimensions of the IgG and protein A molecules. The maximum distance from the antigenic site to the center of a colloidal gold particle, if antibodies are detected with a protein A–gold conjugate, can be expected to be in the order of 15–18 nm and possibly slightly more with an IgG gold probe (Griffiths, 1993; Matsubara *et al.*, 1996). However, in practice the distance is usually much less, as illustrated in Fig. 16.12A, where many gold particles are directly overlying or closely associated with the membrane containing the antigen. The resolution can be somewhat improved by using Fab fragments instead of IgG molecules and by using very small gold particles.

The localization of the gold label on the inner surface of the membrane in Fig. 16.12B suggests that this antibody binds to the cytoplasmic side of the membrane. In this cryosection the membrane probably protrudes above the surrounding section surface, and the probe therefore has a great tendency to "fall down" on the cytoplasmic side during drying, thus improving the resolution.

The section thickness also influences the resolution. In Fig. 16.12C the relationship between gold probes and the plasma membrane is distinct, where the membrane is sectioned approximately perpendicularly, but if the membrane is cut obliquely the precison in the localization of the antigenic sites becomes impaired greatly as seen in the lower third of Fig. 16.12C, where the plasma membrane appears only as a faint band observed above the gold particles.

A

B

C

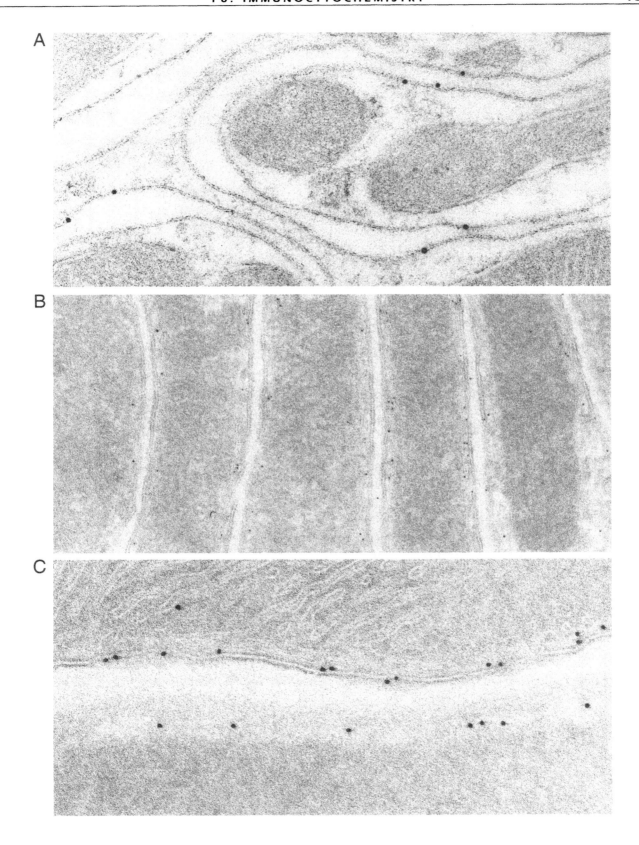

13. Background Labeling

FIGURE 16.13A Rye leaf fixed in 4% formaldehyde and 0.25% glutaraldehyde, freeze-substituted in methanol/uranyl acetate, and embedded in Lowicryl HM20. The thin section was immunolabeled with rabbit antiserum against a rye glucanase-like protein (32 kDa) that was detected with goat anti-rabbit IgG on 10 nm gold (Pihakaski-Maunsbach *et al.*, 1996). × 40,000.

Cell walls show intense and even distribution of gold probes whereas cytoplasms of the mesophyll cells are practically unlabeled, as are the cell vacuoles (top and lower left) and the intercellular space (right).

FIGURE 16.13B Thick ascending limb in a rat outer medulla fixed with 4% formaldehyde, cryosectioned, and immunolabeled with rabbit anti-Na,K-ATPase antiserum diluted 1:100. Primary antibodies were detected with goat anti-rabbit IgG on 10 nm gold. × 50,000.

Gold probes are associated with the basolateral membranes but are to a large extent also present over mitochondria and cytoplasmic matrix.

FIGURE 16.13C Section of a renal capillary in tissue fixed in 4% formaldehyde and freeze-substituted into Lowicryl HM20, which during sectioning showed evidence of incomplete polymerization. The section was incubated with nonimmune rabbit IgG and probed with goat anti-rabbit IgG on 10 nm gold. × 50,000.

Clusters of gold particles are present over the capillary lumen and are associated mainly with amorphous material.

COMMENTS

Unspecific immunolabeling can be caused by a number of different factors, including incomplete blocking, too high a concentration of antibodies, insufficient rinsing, and nonoptimal pH or salt concentrations in antibody or rinsing solutions, as well as "sticky spots" on the sections. In Fig. 16.13A the cell walls are labeled intensely and there are very few gold particles located over other structures. This specific labeling was absent in controls incubated with irrelevant antibodies or preimmune serum.

In Fig. 16.13B, however, the background labeling is extensive. Contributing factors may be that the section was incubated with antiserum rather than with purified antibodies, that the concentration of antibodies was too high, and that the rinsing after incubation was incomplete. Although many gold probes are associated with the basolateral membrane, the general background labeling renders this preparation less useful.

In Fig. 16.13C the gold probes are associated with a stainable material located in a capillary. The origin of such "sticky spots" is not known, but it sometimes seems related to incomplete polymerization of the resin.

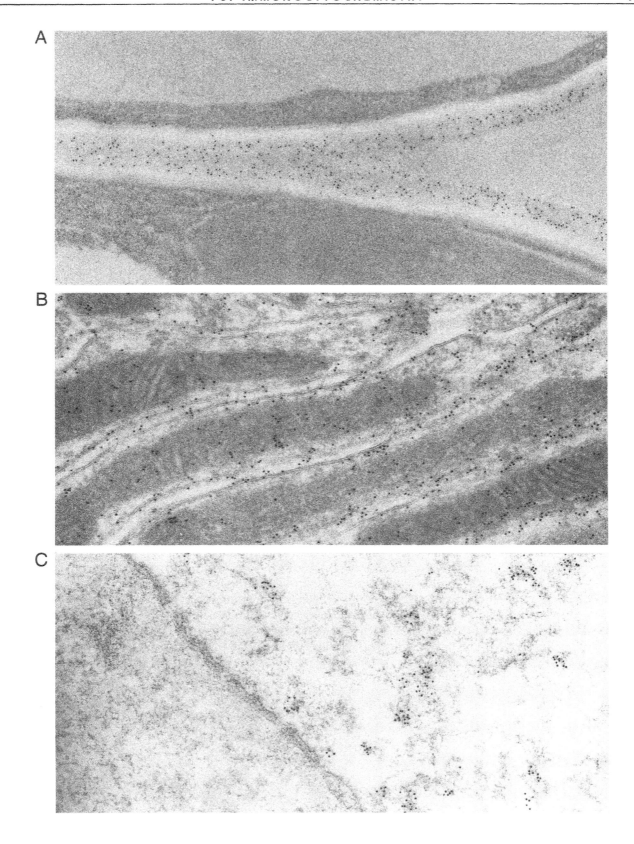

14. Antigen Retrieval by Etching

FIGURE 16.14A Cells in mTAL in a rat kidney perfusion fixed with 4% paraformaldehyde and freeze-substituted in Lowicryl HM20. The section was labeled with rabbit anti-Na,K-ATPase antiserum, and the primary antibody was detected with goat anti-rabbit IgG conjugated to 10 nm colloidal gold. Section staining with uranyl acetate. × 50,000.

Cells are well preserved and a sparse gold labeling is associated with the basolateral membranes.

FIGURE 16.14B Section from the same tissue block as in Fig. 16.14A and treated in the same way except that it was etched (Matsubara *et al.*, 1996) before labeling. The etching solution consisted of saturated sodium hydroxide in ethanol. The grid was placed briefly on a droplet of the etching solution then rinsed rapidly in PBS. × 50,000.

Compared with Fig. 16.14A, the immunolabeling is much increased. An increased frequency of gold probes is associated with basolateral membranes. The fine structure of the cells is modified in the sense that structures show increased granularity and that there is uneven density of the entire section.

FIGURE 16.14C Survey micrograph of a kidney tubule from the same section as in Fig. 16.14B. × 7,500.

At this low magnification folds in the section are seen; these have appeared during the etching. In between the folds the section appears smooth, but in places there is contaminating material.

COMMENT

Mild etching of the section with basic ethanolic solutions can remove part of the embedding medium and expose epitopes in the tissue, thus increasing the sensitivity of the labeling. At the same time there is a gradual modification and deterioration of tissue fine structure, and the duration of the etching is critical. In many cases only a few seconds contact between the section and the etching solution is sufficient and must be followed by a thorough rinse. The effects on immunolabeling of the etching procedure vary between embedding media and between cellular antigens. If the method is applied it is thus necessary to determine carefully in each case the optimal conditions for the etching. If the etching is too extensive, immunolabeling will decrease rather than increase as compared to the untreated section.

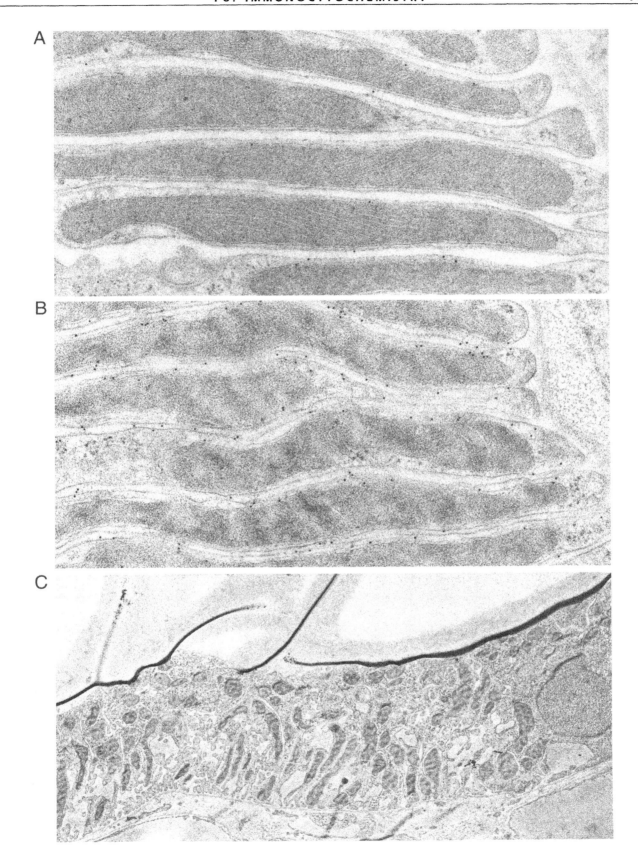

15. Antigen Retrieval with Sodium Dodecyl Sulfate

FIGURE 16.15A Immunolabeling of Na,K-ATPase in mTAL in a rat kidney perfusion fixed with 4% paraformaldehyde, cryosectioned and preincubated as in Fig. 16.1A. The grid was then placed for 1 min on a solution consisting of 1% sodium dodecyl sulfate (SDS) in PBS containing 1% BSA, thereafter rinsed carefully on three droplets in PBS/BSA, and incubated for 1 hr with rabbit anti-Na,K-ATPase serum. After rinses and labeling with goat anti-rabbit IgG conjugated to 10 nm colloid gold the section was stained with methylcellulose/uranyl acetate. × 24,000.

Extensive gold labeling is observed along basolateral membranes (arrowheads) whereas no labeling is present on the apical cell membrane (upper right) or in the nucleus (upper left).

FIGURE 16.15B Similar area to that in Fig. 16.15A. × 50,000.

The gold particles follow two parallel paths between the mitochondria at the locations of basolateral membranes. There is some indication of mitochondrial substructures, but the cell membranes are not well defined.

FIGURE 16.15C Similar preparation as in Figs. 16.15A and 16.15B except that the incubation with SDS lasted 5 min. × 60,000.

As in Fig. 16.15B the gold particles are arranged in paths between what corresponds in dimensions and location to mitochondria. Cytoplasmic structures appear more extracted than in Fig. 16.15B.

FIGURE 16.15D Survey micrograph of similar tubules as in Fig. 16.15C. × 5,000.

The epithelium is identifiable but there are several dense streaks representing folds or elevations in the section.

COMMENTS

Treatment of the ultrathin cryosections with the detergent SDS increases the extent of labeling of the cell dramatically, although at the same time the cellular fine structure is compromised. A similar antigen retrieval has also been observed in 5 μm cryostat sections for membrane proteins (Brown *et al.*, 1996). From the appearance of the tissue structure, it seems that the detergent dissolves or modifies the lipid composition of the membranes, thereby exposing more enzyme epitopes. There is undoubtly much room to refine the procedure for different antigens with respect to detergent concentration, time of exposure, as well as choice of detergent.

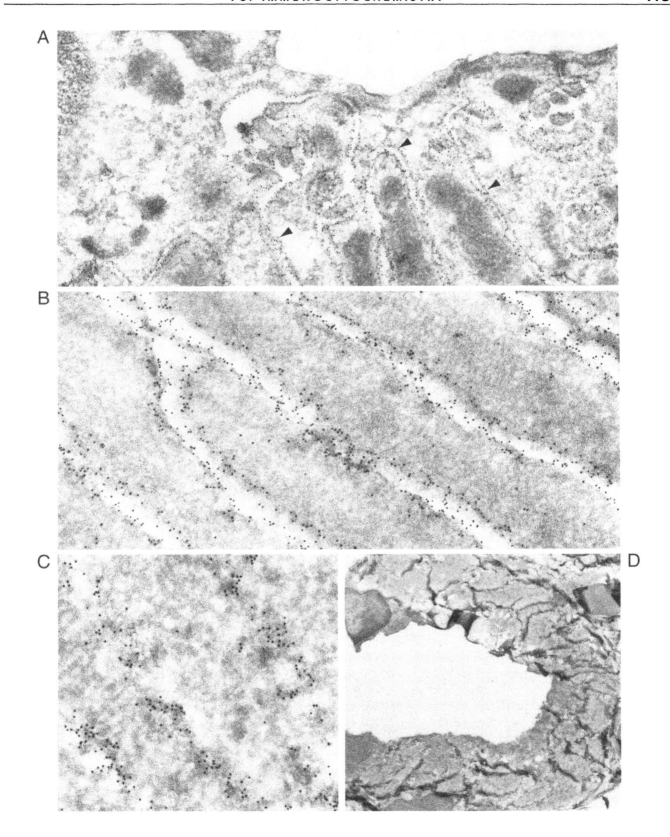

16. Double Labeling

FIGURE 16.16A Third segment (S3) of a proximal tubule in an inner rat cortex double labeled for aquaporin-1 and megalin. Cryosections were incubated with a mixture of sheep anti-rat megalin antiserum diluted 1:50,000 and affinity-purified rabbit anti-aquaporin-1 antibodies diluted 1:200. Incubation was performed for 60 min at room temperature. The primary antibodies were detected with a mixture of goat anti-rabbit IgG on 10 nm gold and donkey anti-sheep antibodies on 15 nm gold (both probes diluted 1:50). × 10,000.

Part of a tubule wall showing microvilli on top and basement membrane below. At this magnification the gold probes are hardly visible.

FIGURE 16.16B Higher magnification of an area corresponding to part of the upper left corner of Fig. 16.16A. × 67,000.

The cell membrane of the microvilli is labeled both with small gold particles representing aquaporin-1 and large ones representing megalin.

FIGURE 16.16C Lateral cell membranes of a similar proximal tubule as in Fig. 16.16A. × 67,000.

Only small colloidal gold particles are associated with basolateral membranes.

FIGURE 16.16D Basal cell surface of a tubule cell from a similar tubule as in Fig. 16.16A. × 67,000.

Only small colloidal gold particles are associated with the basal cell membrane.

COMMENTS

Simultaneous labeling of two antigens can be performed if the antibodies have been raised in different species. In this case, antibodies against the aquaporin-1 water channel were raised in rabbit whereas antibodies against megalin were raised in sheep. Thus secondary gold-labeled antibodies had to be directed against rabbit and sheep, respectively, and conjugated to colloidal gold particles of different diameters.

In Fig. 16.16B, both large (15 nm) and small (10 nm) gold particles are present and are associated mainly with the cell membrane, whereas in Figs. 16.16C and 16.16D, only small gold probes detecting the aquaporin are observed. Thus this polarized epithelium exhibits megalin at the apical cell membrane only whereas aquaporin-1 is exhibited at the apical, lateral, and basal cell membrane. Despite the high dilution of the sheep anti-rat megalin antiserum, a distinct labeling of the membrane was observed. There is some variation in the diameter of the gold particles, a few probes may therefore be difficult to classify. For this reason a better design of this experiment would have been to use 15 and 5 nm gold particles for the two proteins, respectively.

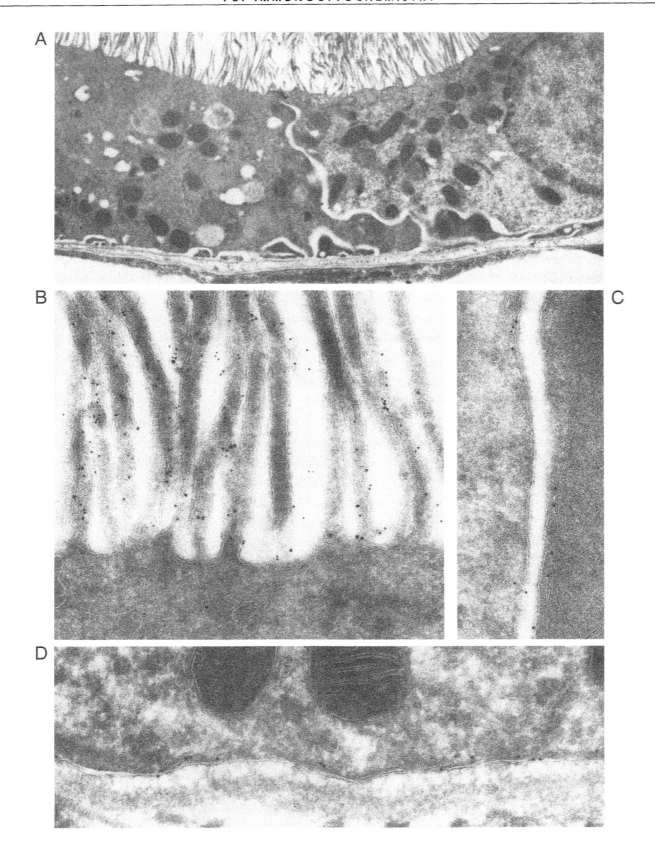

17. Immunonegative Staining

FIGURE 16.17A Purified basolateral membranes from pig kidney medulla incubated with vanadate before negative staining with 1% uranyl acetate. × 130,000.

The membrane shows crystalline Na,K-ATPase arrays separated by smooth lipid regions.

FIGURES 16.17B–16.17E Similar membrane fragments as in Fig. 16.17A labeled in different ways with oligopeptide-specific polyclonal antibodies against the Na,K-ATPase α-subunit, which were detected with protein A conjugated either to 5 or 10 nm colloidal gold (Ning *et al.*, 1993). The membranes were negatively stained with uranyl acetate. In Fig. 16.17B the membrane was incubated with preimmune serum and then with protein A on 5 nm gold particles and in Fig. 16.17C with antibodies against the N terminus and then with protein A on 10 nm gold. The membrane in Fig. 16.17D was first incubated with C terminus antibodies, which were detected with protein A on 5 nm gold, and then with N terminus antibodies, which were probed with protein A on 10 nm gold. In Fig. 16.17E the membrane preparation was likewise double labeled, first with antibodies raised against a peptide corresponding to amino acids 815–828 of the Na,K-ATPase α-subunit, which were detected with protein A on 5 nm gold, and then with antibodies against amino acids 889–903, which were detected with protein A on 10 nm gold. × 130,000.

No labeling is seen when the membrane is incubated with preimmune serum (Fig. 16.17B), but a distinct labeling over the crystalline arrays of Na,K-ATPase molecules occurs with antibodies against the N terminus (Fig. 16.17C). The membrane fragment in Fig. 16.17D shows labeling with both large and small gold particles, demonstrating the presence of both epitopes on the same side of the membrane. However, in Fig. 16.17E one membrane fragment is labeled only with the small gold probes and the other exclusively with the large gold probe, which demonstrates that these two membranes expose opposite sides to the incubation medium.

FIGURES 16.17F AND 16.17G The membrane was fixed for 2 min with 4% formaldehyde in Fig. 16.17F and for 60 sec with 1% glutaraldehyde in Fig. 16.17G before negative staining and labeling with identical antibodies. × 130,000.

The intensity of labeling on the formaldehyde-fixed membrane is similar to that on unfixed membranes, whereas even brief glutaraldehyde fixation has reduced the labeling.

COMMENTS

Immunogold labeling combined with negative staining allows the identification of epitopes on isolated membranes. When two antibodies are applied in sequence, double immunonegative staining, topological information can be obtained with regard to the sidedness of the epitopes. If the two antibodies have affinities to epitopes on opposite sides of the membrane and are detected with different sized gold probes, some membranes will be labeled with small gold probes and others with large gold probes, depending on which side of the membrane that faces the incubation medium. Membranes fixed with formaldehyde before labeling show essentially no reduction in labeling as compared to unfixed membranes, but if membranes are fixed even briefly with glutaraldehyde, a reduction in labeling is observed, similar to the observation on cryosections (compare with Fig. 16.1D).

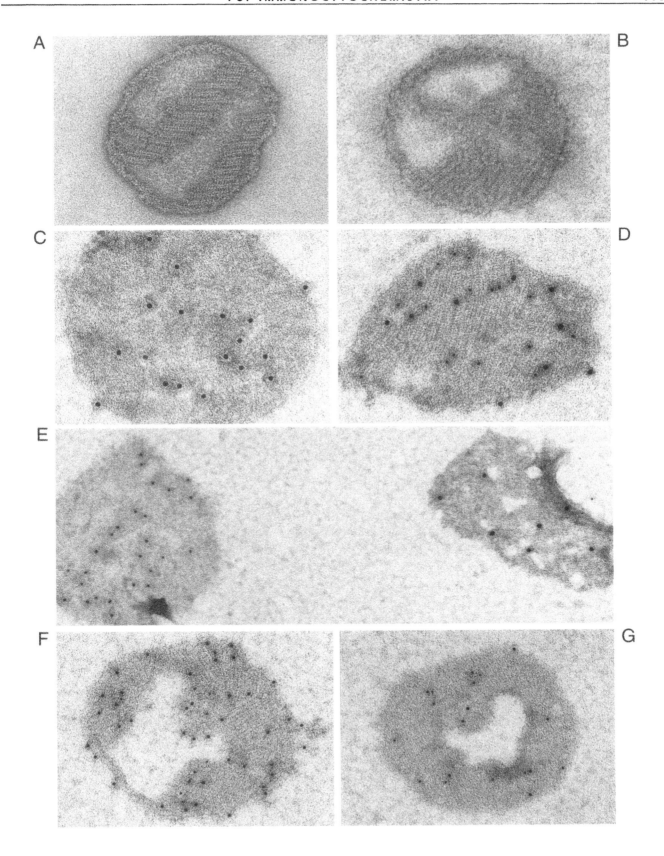

18. Freeze-Fracture Replica Labeling

FIGURE 16.18A Electron micrograph of a basolateral plasma membrane of an epithelial cell from a medullary thick ascending limb of rat kidney processed with the sodium dodecyl sulfate-digested freeze-fracture replica labeling (SDS-FRL) technique (Fujimoto, 1995) to determine the sidedness of Na,K-ATPase α-subunit epitopes. Unfixed tissue slices of rat kidney medulla were quick-frozen by contact with a copper block cooled with liquid helium, freeze-fractured, and replicated. After digestion with SDS to solubilize membrane-associated cytoplasm, the platinum/carbon replica, along with attached cytoplasmic and exoplasmic membrane halves, was processed for immunocytochemistry using an oligopeptide-specific rabbit antibody against the N terminus of the Na,K-ATPase α-subunit (Fujimoto et al., 1996). The primary antibody was detected with goat anti-rabbit IgG conjugated to 10 nm colloidal gold. × 80,000.

Primary antibodies detected with 10 nm gold particles conjugated to goat anti-rabbit IgG antibodies show dense labeling of the P face of the basolateral plasma membrane (left), whereas the E face (right) is essentially unlabeled.

FIGURE 16.18B Distal tubule cells processed with the SDS-FRL method with antibodies raised against the oligopeptide 889–903 of the Na,K-ATPase α-subunit. × 120,000.

Immunogold particles are associated exclusively with cross-fractured adjacent plasma membranes. Both P and E faces are unlabeled.

COMMENTS

Integral membrane proteins that partition with the inner (cytoplasmic) half of the plasma membrane can be labeled with antibodies against their cytoplasmic sites (Fig. 16.18A) while their extracellular sites remain unlabeled. However, extracellular aspects of such proteins are exposed partially in cross-fractured membranes (Fig. 16.18B) and can be labeled with antibodies against their extracellular epitopes. The SDS-FRL method allows an analysis of the sidedness of epitopes on an integral membrane protein. Thus, Fig. 16.18A demonstrates that the N terminus is located on the cytoplasmic surface of the plasma membrane whereas Fig. 16.18B shows that the oligopeptide-specific antibodies raised against the oligopeptide 889–903 are located on the extracellular side of the membrane.

A

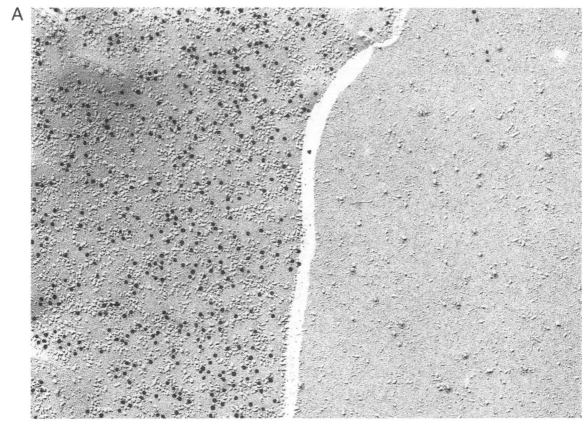

B

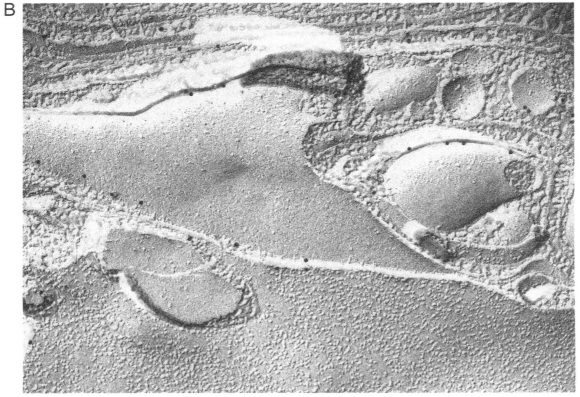

19. Preembedding Labeling

FIGURE 16.19A Immunofluorescence localization of Na,K-ATPase in the proximal tubule of the salamander *Ambystoma tigrinum*. The tissue was fixed with 4% periodate-lysine-paraformaldehyde (PLP) for 2 hr. Cryosections (about 1 μm) were incubated with mouse monoclonal antibodies against the α-subunit of Na,K-ATPase. The primary antibody was detected with fluorescein-conjugated rabbit anti-mouse IgG. $\times$ 650. [*See also color insert.*]

Immunolabeling of Na,K-ATPase is observed along the basolateral cell membrane, including its complex basal folds and projections, but not at the luminal cell membrane.

FIGURE 16.19B Low magnification electron micrograph of the proximal tubule epithelium in *A. tigrinum* fixed with PLP fixative as in Fig. 16.19A and postfixed in 1% osmium tetroxide. $\times$ 2000.

Cells show a uniform brush border and a basolateral membrane that is extensively folded toward the peritubular space.

FIGURE 16.19C Basal part of an *Ambystoma* proximal tubule fixed with PLP fixative as in Fig. 16.19A, treated with 10% dimethyl sulfoxide (DMSO), and cryosectioned at 30 μm. Thick sections were incubated overnight at room temperature with a mouse monoclonal antibody against the Na,K-ATPase α-subunit. After rinses the primary antibody was detected with peroxidase-labeled goat anti-mouse IgG, which was visualized with the diaminobenzidine reaction. After postfixation in 1% osmium tetroxide the tissue was embedded in Epon and the section stained with lead citrate. $\times$ 25,000.

The electron-dense diaminobenzidine reaction product is associated with folds and projections of the basal plasma membrane. Intracellular structures are unlabeled as is the capillary endothelium to the right.

FIGURE 16.19D Higher magnification of part of Fig. 16.19C. $\times$ 60,000.

The reaction product is present on the basal plasma membrane, but some reaction product is also observed in the cytoplasm inside the folds but not in the cell body.

COMMENTS

Preembedding labeling is an alternative to postembedding labeling with colloidal gold in immunoelectron microscopy. It has been applied extensively to surface antigens in various tissues and is a sensitive procedure. A drawback is that the diaminobenzidine reaction product has a tendency to diffuse from the original site of the peroxidase-conjugated antibody as illustrated in Figs. 16.19C and 16.19D. It also requires that the incubated tissue be sufficiently thin to allow the antibodies to penetrate the cells. Thus negative results may be due to poor penetration of the primary antibodies or the peroxidase-conjugated secondary antibody. To improve penetration, the tissue may be either frozen and/or treated with substances that permeabilize the membranes, such as DMSO.

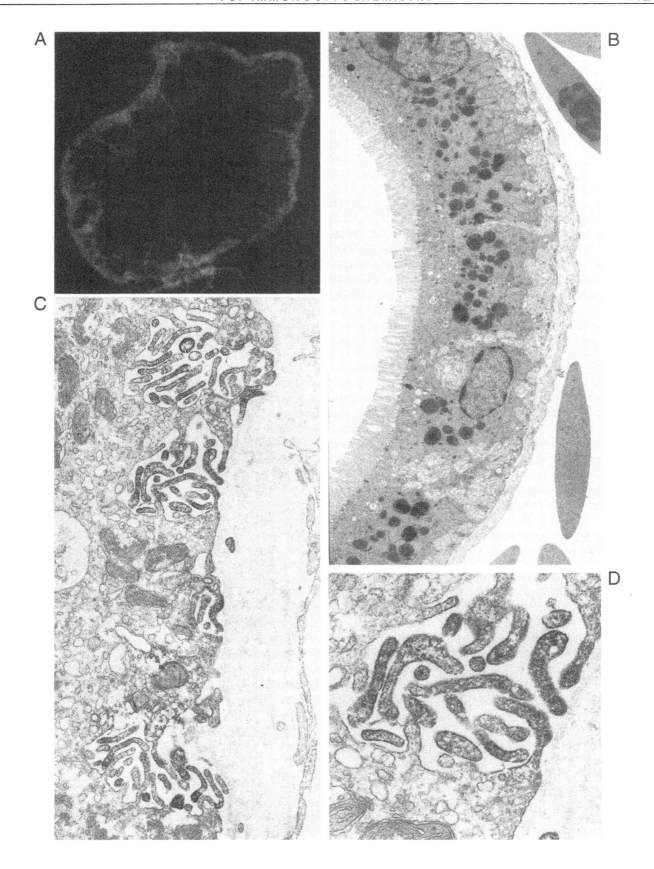

20. Semithin Light Microscopic Sections

FIGURES 16.20A–16.20C Light microscopic immunocytochemical localization of the aquaporin-1 (AQP1) water channel in human kidney cortex and medulla. Surgical specimens from nonpathological tissue (kidney with a tumor at the opposite pole) were immersion fixed with 8% paraformaldehyde in 0.1 *M* cacodylate buffer, infiltrated with 2.3 *M* sucrose, and frozen rapidly in liquid nitrogen. Cryosections (0.8 μm in thickness) were preincubated with PBS containing 1% BSA and 0.05 *M* glycine and incubated with affinity-purified rabbit polyclonal anti-AQP1 antibodies (Agre *et al.*, 1993), which were visualized with horseradish perioxidase-conjugated goat anti-rabbit antibodies and incubation with diaminobenzidine. Figure 16.20A, × 800; Figs. 16.20B and 16.20C, × 1600. [*See also color insert.*]

Figure 16.20A illustrates a cross-sectioned proximal tubule with distinct AQP1 localization at the brush border and basolateral membrane as well as in the peritubular capillary endothelium (lower right corner). Figure 16.20B shows complex interdigitations of the basolateral membranes at higher magnification. In Fig. 16.20C the water channel protein is associated with the descending thin limb in the outer medulla, whereas surrounding structures, including a thick ascending limb (to the right in the middle), are unlabeled. The diaminobenzidine reaction product is visualized against a counterstain, which was Mayer stain.

FIGURES 16.20D AND 16.20E Immunoelectron microscope labeling of aquaporin-1 in a human renal proximal tubule (Fig. 16.20D) and a thin descending limb in inner medulla (Fig. 16.20E). The tissue was fixed and frozen as in Figs. 16.20A–16.20C but was then freeze-substituted and embedded in Lowicryl HM20. Ultrathin sections were incubated overnight at 4°C with the same antibody as in Fig. 16.20A, labeling was visualized with goat anti-rabbit IgG conjugated to 10 nm colloidal gold, and the section was stained with uranyl acetate. × 60,000.

Figure 16.20D illustrates that microvilli of the brush border in the proximal tubule are labeled extensively for AQP1 whereas little label is associated with cytoplasmic structures. In Fig. 16.20E, antibodies against AQP1 distinctly label the plasma membrane of cells in the descending thin limb in the renal medulla.

COMMENTS

Light microscopic immunocytochemistry of semithin sections (less than 1 μm in thickness) provides an excellent overview of the general distribution of antigen in tissues. It is recommended that such preparations be analyzed before immunoelectron microscopy is performed on ultrathin frozen sections or sections prepared by freeze-substitution and plastic embedding. The light microscopic procedure is sensitive and allows very precise cellular localization and may, in some cases, even eliminate the need for electron microscope immunocytochemistry.

Figures 16.20A and 16.20B demonstrate the localization of aquaporin-1 in the brush border and in basolateral membranes along with less intense labeling of some peritubular capillaries. The brush border localization is extended in the immunoelectron micrograph in Fig. 16.20D. The light microscope overview in Fig. 16.20C is helpful for the identification of AQP1-containing tubule segments in the renal medulla whereas fine structures of the thin walled tubular segments that express AQP1 can only be appreciated in ultrathin cryosections or ultrathin sections of freeze-substituted and plastic embedded tissue (Fig. 16.20E).

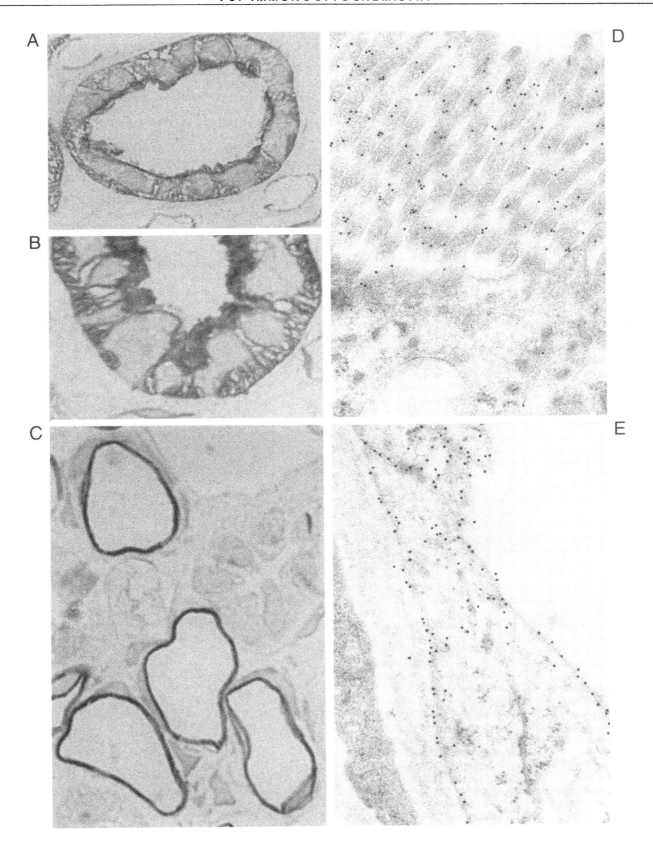

References

Agre, P., Preston, G. M., Smith, B. L., Jung, J. S., Raina, S., Moon, C., Guggino, W. B., and Nielsen, S. (1993). Aquaporin CHIP: The archetypal molecular water channel. *Am. J. Physiol.* **265**, F463–F476.

Armbruster, B. L., Garavito, R. M., and Kellenberger, E. (1983). Dehydration and embedding temperatures affect the antigenic specificity of tubulin and immunolabeling by the protein A-colloidal gold techniques. *J. Histochem. Cytochem.* **31**, 1380–1384.

Beesley, J. E. (1989). Colloidal gold: Immunonegative staining method. *In* "Colloidal Gold: Principles, Methods, and Applications" (M. A. Hayat, ed.), Vol. 2, pp. 243–254. Academic Press, San Diego.

Beesley, J. E. (ed.) (1993). "Immunocytochemistry: A Practical Approach." IRL Press, Oxford University Press, Oxford.

Behnke, O., Ammitzbøll, T., Jessen, H., Klokker, M., Nilausen, K., Tranum-Jensen, J., and Olsson, L. (1986). Non-specific binding of protein-stabilized gold sols as a source of error in immunocytochemistry. *Eur. J. Cell Biol.* **41**, 326–338.

Bendayan, M. (1982). Double immunocytochemical labeling applying the protein A-gold technique. *J. Histochem. Cytochem.* **30**, 81–85.

Bendayan, M. (1984). Protein A-gold electron microscopic immunocytochemistry: Methods, applications, and limitations. *J. Electr. Microsc. Tech.* **1**, 243–270.

Bendayan, M., Nanci, A., and Kan, F. W. K. (1987). Effect of tissue processing on colloidal gold cytochemistry. *J. Histochem. Cytochem.* **35**, 983–996.

Birrell, G. B., Hedeberg, K. K., and Griffith, O. H. (1987). Pitfalls of immunogold labeling: Analysis by light microscopy, transmission electron microscopy, and photoelectron microscopy. *J. Histochem. Cytochem.* **35**, 843–853.

Bonnard, C., Papermaster, D. S., and Kraehenbuhl, J.-P. (1984). The streptavidin-biotin bridge technique: Application in light and electron microscope immunocytochemistry. *In* "Immunolabelling for Electron Microscopy" (J. M. Polak and I. M. Varndell, eds.), pp. 95–111, Elsevier, Amsterdam.

Brorson, S.-H. (1996). Improved immunogold labeling of epoxy sections by use of propylene oxide as additional agent in dehydration, infiltration and embedding. *Micron* **27**, 345–353.

Brorson, S.-H., and Skjørten, F. (1996). Improved technique for immunoelectron microscopy: How to prepare epoxy resin to obtain approximately the same immunogold labeling for epoxy sections as for acrylic sections without any etching. *Micron* **27**, 211–217.

Brown, D., Lydon, J., McLaughlin, M., Stuart-Tilley, A., Tyszkowski, R., and Alper, S. L. (1996). Antigen retrieval in cryostat tissue sections and cultured cells by treatment with sodium dodecyl sulfate. *Histochem. Cell Biol.* **105**, 261–267.

Carlemalm, E. (1990). Lowicryl resins in microbiology. *J. Struct. Biol.* **104**, 189–191.

Christensen, E. I., Nielsen, S., Moestrup, S. K., Borre, C., Maunsbach, A. B., Heer, E. de, Ronco, P., Hammond, T. G., and Verroust, P. (1995). Segmental distribution of the endocytosis receptor gp330 in renal proximal tubules. *Eur. J. Cell Biol.* **66**, 349–364.

Christensen, E. I., Birn, H., Verroust, P., and Moestrup, S. K. (1998). Membrane receptors for endocytosis in the renal proximal tubule. *Intl. Rev. Cytol.* **180**, 237–284.

Courtoy, P. J., Picton, D. H., and Farquhar, M. G. (1983). Resolution and limitations of the immunoperoxidase procedure in the localization of extracellular matrix antigens. *J. Histochem. Cytochem.* **31**, 945–951.

Danscher, G. (1981). Localization of gold in biological tissue: A photochemical method for light and electronmicroscopy. *Histochemistry* **71**, 81–88.

Danscher, G., Hacker, G. W., Grimelius, L., and Nørgaard, J. O. R. (1993). Autometallographic silver amplification of colloidal gold. *J. Histotechnol.* **16**, 201–207.

Danscher, G., Hacker, G. W., Hauser-Kronberger, C., and Grimelius, L. (1995). Trends in autometallographic silver amplification of colloidal gold particles. *In* "Immunogold-Silver Staining: Principles, Methods, and Applications" (M. A. Hayat, ed.), pp. 11–45. CRC Press, Boca Raton, FL.

de Graaf, A., van Bergen en Henegouwen, P. M. P., Meijne, A. M. L., Van Driel, R., and Verkleij, A. J. (1991). Ultrastructural localization of nuclear matrix proteins in HeLa cells using silver-enhanced ultrasmall gold probes. *J. Histochem. Cytochem.* **39**, 1035–1045.

de May, J. (1986). The preparation and use of gold probes. *In* "Immunocytochemistry, Modern Methods and Applications" (J. M. Polak and S. van Noorden, eds.), pp. 115–145, Butterworth-Heinemann, Oxford.

Faulk, W. P., and Taylor, G. M. (1971). An immunocolloid method for the electron microscope. *Immunochemistry* **8**, 1081–1083.

Fahimi, H. D., Reich, D., Völkl, A., and Baumgart, E. (1996). Contributions of the immunogold technique to investigation of the biology of peroxisomes. *Histochem. Cell Biol.* **106**, 105–114.

Fujimoto, K. (1995). Freeze-fracture replica electron microscopy combined with SDS digestion for cytochemical labeling of integral membrane proteins. *J. Cell Sci.* **108**, 3443–3449.

Fujimoto, K. (1997). SDS-digested freeze-fracture replica labeling electron microscopy to study the two-dimensional distribution of integral membrane proteins and phospholipids in biomembranes: Practical procedure, interpretation and application. *Histochem. Cell Biol.* **107**, 87–96.

Fujimoto, K., Møller, J. V., and Maunsbach, A. B. (1996). Epitope topology of Na,K-ATPase α subunit analyzed in basolateral cell membranes of rat kidney tubules. *FEBS Lett.* **395**, 29–32.

Geuze, H. J., Slot, J. W., van der Ley, P. A., and Scheffer, R. C. T. (1981). Use of colloidal gold particles in double-labeling immunoelectron microscopy of ultrathin frozen tissue sections. *J. Cell Biol.* **89**, 653–665.

Giberson, R. T., and Demaree, R. S., Jr. (1994). The influence of immunogold particle-size on labeling density. *Microsc. Res. Techn.* **27**, 355–357.

Graham, R. C., and Karnovsky, M. J. (1966). The early stages of absorption of injected horseradish peroxidase in the proximal tubules of mouse kidney: Ultrastructural cytochemistry by a new technique. *J. Histochem. Cytochem.* **14**, 291–302.

Griffiths, G. (1993). "Fine Structure Immunocytochemistry." Springer-Verlag, Berlin.

Griffiths, G., McDowall, A., Back, R., and Dubochet, J. (1984). On the preparation of cryosections for immunocytochemistry. *J. Ultrastr. Res.* **89**, 65–78.

Grziwa, A., Baumeister, W., Dahlmann, B., and Kopp, F. (1991). Localization of subunits in proteasomes from *Thermoplasma acidophilum* by immunoelectron microscopy. *FEBS Lett.* **290**, 186–190.

Hacker, G. W., Grimelius, L., Danscher, G., Bernatzky, G., Muss, W., Adam, H., and Thurner, J. (1988). Silver acetate autometallography: An alternative enhancement technique for immunogold-silver staining (IGSS) and silver amplification of gold, silver, mercury and zinc in tissue. *J. Histotechnol.* **11**, 213–221.

Hainfeld, J. F. (1987). A small gold-conjugated antibody label: Improved resolution for electron microscopy. *Science* **236**, 450–453.

Hainfeld, J. F., and Furuya, F. R. (1992). A 1.4-nm gold cluster covalently attached to antibodies improves immunolabeling. *J. Histochem. Cytochem.* **40**, 177–184.

Hainfeld, J. F., and Furuya, F. R. (1995). Silver-enhancement of nanogold and undecagold. *In* "Immunogold-Silver Staining: Principles, Methods, and Applications" (M. A. Hayat, ed.), pp. 71–96. CRC Press, Boca Raton, FL.

Harris, J. R., Gebauer, W., and Markl, J. (1993). Immunoelectron microscopy of hemocyanin from the keyhole limpet (*Megathura crenulata*): A parallel subunit model. *J. Struct. Biol.* **111**, 96–104.

Hobot, J. A. (1989). Lowicryls and low-temperature embedding for colloidal gold methods. *In* "Colloidal Gold: Principles, Methods, and Applications" (M. A. Hayat, ed.), Vol. 2, pp. 75–115. Academic Press, San Diego.

Horisberger, M., and Rosset, J. (1977). Colloidal gold, a useful marker for transmission and scanning electron microscopy. *J. Histochem. Cytochem.* **25,** 295–305.

Humbel, B. M., and Schwarz, H. (1989). Freeze-substitution for immunochemistry. *In* "Immuno-Gold Labeling in Cell Biology" (A. J. Verkleij and J. L. M. Leunissen, eds.), pp. 115–134. CRC Press, Boca Raton, FL.

Humbel, B. M., Sibon, O. C. M., Stierhof, Y.-D., and Schwarz, H. (1995). Ultra-small gold particles and silver enhancement as a detection system in immunolabeling and in situ hybridization experiments. *J. Histochem. Cytochem.* **7,** 735–737.

Ichikawa, M., Sasaki, K., and Ichikawa, A. (1989). Optimal preparatory procedures of cryofixation for immunocytochemistry. *J. Electr. Microsc. Techn.* **12,** 88–94.

Kashgarian, M., Biemesderfer, D., Caplan, M., and Forbush, B., III (1985). Monoclonal antibody to Na,K-ATPase: Immunocytochemical localization along nephron segments. *Kidney Int.* **28,** 899–913.

Kellenberger, E., Dürrenberger, M., Villiger, W., Carlemalm, E., and Wurtz, M. (1987). The efficiency of immunolabel on Lowicryl sections compared to theoretical predictions. *J. Histochem. Cytochem.* **35,** 959–969.

Keller, G.-A., Tokuyasu, K. T., Dutton, A. H., and Singer, S. J. (1984). An improved procedure for immunoelectron microscopy: Ultrathin plastic embedding of immunolabeled ultrathin frozen sections. *Proc. Natl. Acad. Sci. USA* **81,** 5744–5747.

Kerjaschki, D., Sawada, H., and Farquhar, M. G. (1986). Immunoelectron microscopy in kidney research: Some contributions and limitations. *Kidney Int.* **30,** 229–245.

Kjeldsberg, E. (1986). Immunonegative stain techniques for electron microscopic detection of viruses in human faeces. *Ultrastr. Pathol.* **10,** 553–570.

Larsson, L.-I. (1979). Simultaneous ultrastructural demonstration of multiple peptides in endocrine cells by a novel immunocytochemical method. *Nature* **282,** 743–745.

Larsson, L.-I. (1988). "Immunocytochemistry: Theory and Practice." CRC Press, Boca Raton, FL.

Liou, W., Geuze, H. J., and Slot, J. W. (1996). Improving structural integrity of cryosections for immunogold labeling. *Histochem. Cell Biol.* **106,** 41–58.

Matsubara, A., Laake, J. H., Davanger, S., Usami, S.-i., and Ottersen, O. P. Organization of AMPA receptor subunits at a glutamate synapse: A quantitative immunoglod analysis of hair cell synapses in the rat organ of Corti. *J. Neuroscience* **16,** 4457–4467.

Maunsbach, A. B. (1998). Embedding of cells and tissues for ultrastructural and immunocytochemical analysis. *In* "Cell Biology: A Laboratory Handbook" (J. E. Celis, ed.), 2nd Ed., Vol. 3, pp. 260–267. Academic Press, San Diego.

Maunsbach, A. B. (1998). Immunolabeling and staining of ultrathin sections in biological electron microscopy. *In* "Cell Biology: A Laboratory Handbook" (J. E. Celis, ed.), 2nd Ed., Vol. 3, pp. 268–276. Academic Press, San Diego.

Maunsbach, A. B., and Boulpaep, E. L. (1991). Immunoelectron microscope localization of Na,K-ATPase in transport pathways in proximal tubule epithelium. *Micron Microsc. Acta* **22,** 55–56.

Maunsbach, A. B., Marples, D., Chin, E., Ning, G., Bondy, C., Agre, P., and Nielsen, S. (1997). Aquaporin-1 water channel expression in human kidney. *J. Am. Soc. Nephrol.* **8,** 1–14.

Maunsbach, A. B., and Reinholt, F. P. (1992). Ultrastructural immunolocalization of Na,K-ATPase in high-pressure frozen kidney tubules. *Micron Microsc. Acta* **23,** 109–110.

McCann, J. A., Maddox, D. A., Mount, S. L., Hong, R., and Taatjes, D. J. (1996). Cryofixation, cryosubstitution, and immunoelectron microscopy: Potential role in diagnostic pathology. *Ultrastruct. Pathol.* **20,** 223–230.

McLean, I. W., and Nakane, P. K. (1974). Periodate-lysine-paraformaldehyde fixative: A new fixative for immunoelectron microscopy. *J. Histochem. Cytochem.* **22,** 1077–1083.

Merighi, A. (1992). Post-embedding electron microscopic immunocytochemistry. *In* "Electron Microscopic Immunocytochemistry," Vol. 4 (J. M. Polak and J. V. Priestley, eds.). Oxford University Press, Oxford.

Monoghan, P., and Robertson, D. (1990). Freeze-substitution without aldehyde or osmium fixatives: Ultrastructure and implications for immunocytochemistry. *J. Microsc. (Oxford)* **158,** 355–363.

Nakane, P. K. (1968). Simultaneous localization of multiple tissue antigens using the peroxidase-labeled antibody method: A study on pituitary gland of the rat. *J. Histochem. Cytochem.* **16,** 557–560.

Newman, G. R., and Hobot, J. A. (1987). Modern acrylics for post-embedding immunostaining techniques. *J. Histochem. Cytochem.* **35,** 971–981.

Newman, G. R., and Hobot, J. A. (1993). "Resin Microscopy and On-Section Immunocytochemistry." Springer-Verlag, Berlin.

Nielsen, S., and Christensen, E. I. (1989). Insulin absorption in renal proximal tubules: A quantitative immunocytochemical study. *J. Ultrastr. Mol. Struct. Res.* **102,** 205–220.

Nielsen, S., Chou, C.-L., Marples, D., Christensen, E. I., Kishore, B. K., and Knepper, M. A. (1995). Vasopressin increases water permeability of kidney collecting duct by inducing translocation of aquaporin-CD water channels to plasma membrane. *Proc. Natl. Acad. Sci. USA* **92,** 1013–1017.

Nielsen, S., Nagelhus, E. A., Amiry-Moghaddam, M., Bourque, C., Agre, P., and Ottersen, O. P. (1997). Specialized membrane domains for water transport in glial cells: High-resolution immunogold cytochemistry of aquaporin-4 in rat brain. *J. Neurosci.* **17,** 171–180.

Nielsen, S., Pallone, T. L., Smith, B. L., Christensen, E. I., Agre, P., and Maunsbach, A. B. (1995). Aquaporin-1 water channels in short and long loop descending thin limbs and in descending vasa recta in rat kidney. *Am. J. Physiol.* **268,** F1023–F1037.

Nielsen, S., Smith, B. L., Christensen, E. I., Knepper, M. A., and Agre, P. (1993). CHIP28 water channels are localized in constitutively water-permeable segments of the nephron. *J. Cell Biol.* **120,** 371–383.

Ning, G., Maunsbach, A. B., Lee, Y.-J., and Møller, J. V. (1993). Topology of Na,K-ATPase α subunit epitopes analyzed with oligopeptide-specific antibodies and double-labeling immunoelectron microscopy. *FEBS Lett.* **336,** 521–524.

Ning, G., and Maunsbach, A. B. (1994). Evaluation of double immunolabeling methods for epitope mapping on isolated Na,K-ATPase membranes. *Acta Histochem. Cytochem.* **27,** 347–356.

Ogawa, K., and Barka, T. (1993). "Electron Microscopic Cytochemistry and Immunocytochemistry in Biomedicine." CRC Press, Boca Raton, FL.

Oprins, A., Geuze, H. J., and Slot, J. W. (1993). Cryosubstitution dehydration of aldehyde-fixed tissue: A favorable approach to quantitative immunocytochemistry. *J. Histochem. Cytochem.* **42d,** 497–503.

Ottersen, O. P. (1987). Postembedding light and electron microscopic immunocytochemistry of amino acids: Description of a new model system allowing identical conditions for specificity testing and tissue processing. *Exp. Brain Res.* **69,** 167–174.

Ottersen, O. P. (1989). Quantitative electron microscopic immunocytochemistry of neuroactive amino acids. *Anat. Embryol.* **180,** 1–15.

Pihakaski-Maunsbach, K., Griffith, M., Antikainen, M., and Maunsbach, A. B. (1996). Immunogold localization of glucanase-like antifreeze protein in cold acclimated winter rye. *Protoplasma* **191,** 115–125.

Polak, J. M., and Varndell, J. M. (eds.) (1984). "Immunolabelling for Electron Microscopy." Elsevier, Amsterdam.

Polak, J. M., and Van Noorden, S. (1997). "Introduction to Immunocytochemistry," 2nd Ed. Bios Scientific Publishers, Oxford.

Posthuma, G., Slot, J. W., and Geuze, H. J. (1987). Usefulness of the immunogold technique in quantitation of a soluble protein in ultrathin sections. *J. Histochem. Cytochem.* **35**, 405–410.

Priestley, J. V., Alvarez, F. J., and Averill, S. (1992). Preembedding electron microscopic immunocytochemistry. *In* "Electron Microscopic Immunocytochemistry," Vol. 4 (J. M. Polak and J. V. Priestley, eds.). Oxford University Press, Oxford.

Robert, F., Pelletier, G., Serri, O., and Hardy, J. (1988). Mixed growth hormone and prolactin-secreting human pituitary adenomas: A pathologic, immunocytochemical, ultrastructural and immunoelectron microscopic study. *Human Pathol.* **19**, 1327–1334.

Robertson, D., Monaghan, P., Clarke, C., and Atherton, A. J. (1992). An appraisal of low-temperature embedding by progressive lowering of temperature into Lowicryl HM20 for immunocytochemical studies. *J. Microsc. (Oxford)* **168**, 85–100.

Rodman, J. S., Kerjaschki, D., Merisko, E., and Farquhar, M. G. (1984). Presence of an extensive clathrin coat on the apical plasmalemma of the rat kidney proximal tubule cell. *J. Cell Biol.* **98**, 1630–1636.

Roth, J. (1984). The protein A-gold technique for antigen localisation in tissue sections by light and electron microscopy. *In* "Immunolabelling for Electron Microscopy" (J. M. Polak and I. M. Varndell, eds.). Elsevier, Amsterdam.

Roth, J. (1986). Post-embedding cytochemistry with gold-labelled reagents: A review. *J. Microsc. (Oxford)* **143**, 125–137.

Roth, J., Taatjes, D. J., and Tokuyasu, K. T. (1990). Contrasting of Lowicryl K4M thin sections. *Histochemistry* **95**, 123–136.

Roth, J., Taatjes, D. J., and Warhol, M. J. (1989). Prevention of nonspecific interactions of gold-labeled reagents on tissue sections. *Histochemistry* **92**, 47–56.

Roth, J., Zuber, C., Komminoth, P., Sata, T., Li, W.-P., and Heitz, P. U. (1996). Applications of immunogold and lectin-gold labeling in tumor research and diagnosis. *Histochem. Cell Biol.* **106**, 131–148.

Sabolic, I., Valenti, G., Verbavatz, J. M., Van Hoek, A. N., Verkman, A. S., Ausiello, D. A., and Brown, D. (1992). Localization of the CHIP28 water channel in rat kidney. *Am. J. Physiol.* **263**, C1225–C1233.

Schwarz, H., and Humbel, B. M. (1989). Influence of fixatives and embedding media on immunolabelling of freeze-substituted cells. *Scan. Microsc. Suppl.* **3**, 57–64.

Segal, A. S., Boulpaep, E. L., and Maunsbach, A. B. (1996). A novel preparation of dissociated renal proximal tubule cells that maintain epithelial polarity in suspension. *Am. J. Physiol.* **270**, C1843–C1863.

Simon, G. T., Thomas, J. A., Chorneyko, K. A., and Carlemalm, E. (1987). Rapid embedding in Lowicryl K4M for immunoelectron microscopic studies. *J. Elect. Microsc. Techn.* **6**, 317–324.

Slot, J. W., and Geuze, H. J. (1983). The use of protein A-colloidal gold (PAG) complexes as immunolabels in ultrathin frozen sections. *In* "Immunohistochemistry" (A. C. Cuello, ed.), pp. 323–346. Wiley, Chicester.

Slot, J. W., and Geuze, H. J. (1984). Gold markers for single and double immunolabelling of ultrathin cryosections. *In* "Immunolabelling for Electron Microscopy" (J. M. Polak and I. M. Varndell, eds.), pp. 129–142. Elsevier, Amsterdam.

Slot, J. W., and Geuze, H. J. (1985). A new method of preparing gold probes for multiple-labeling cytochemistry. *Eur. J. Cell Biol.* **38**, 87–93.

Slot, J. W., Posthuma, G., Chang, L.-Y., Crapo, J. D., and Geuze, H. J. (1989). Quantitative aspects of immunogold labelling in embedded and nonembedded sections. *Am. J. Anat.* **185**, 271–281.

Steinbrecht, R. A. (1993). Freeze-substitution for morphological and immunocytochemical studies in insects. *Microsc. Res. Techn.* **24**, 488–504.

Stierhof, Y.-D., Humbel, B. M., Hermann, R., Otten, M. T., and Schwarz, H. (1992). Direct visualization and silver enhancement of ultra-small antibody-bound gold particles on immunolabeled ultrathin resin sections. *Scan. Microsc.* **6**, 1009–1022.

Stierhof, Y.-D., Hermann, R., Humbel, B. M., and Schwarz, H. (1995). Use of TEM, SEM and STEM in imaging 1-nm colloidal gold particles. *In* "Immunogold-Silver Staining: Principles, Methods, and Applications" (M. A. Hayat, ed.), pp. 97–118. CRC Press, Boca Raton, FL.

Stirling, J. W. (1990). Immuno- and affinity probes for electron microscopy: A review of labeling and preparation techniques. *J. Histochem. Cytochem.* **38**, 145–157.

Takizawa, T., and Robinson, J. M. (1994). Use of 1.4 nm-immunogold particles for immunocytochemistry on ultra-thin cryosections. *J. Histochem. Cytochem.* **42**, 1615–1623.

Taylor, C. R., Shi, S.-R., and Cote, R. J. (1996). Antigen retrieval for immunohistochemistry: Status and need for greater standardization. *Appl. Immunohistochem.* **4**, 144–166.

Tokuyasu, K. T., and Singer, S. J. (1976). Improved procedures for immuno-ferritin labeling of ultrathin frozen sections. *J. Cell Biol.* **71**, 894–906.

Tokuyasu, K. T. (1986). Application of cryoultramicrotomy to immunocytochemistry. *J. Microsc. (Oxford)* **143**, 139–149.

Torrisi, M. R., and Mancini, P. (1996). Freeze-fracture immunogold labeling. *Histochem. Cell Biol.* **106**, 19–30.

Tsuji, S., Anglade, P., Daudet-Monsac, M., and Motelica-Heino, I. (1992). Cryoultramicrotomy: Electrostatic transfer of dry ultrathin frozen sections on grids applied to the central nervous system. *Arch. Histol. Cytol.* **55**, 423–428.

Usuda, N., Ma, H., Hanai, T., Yokota, S., Hashimoto, T., and Nagata, T. (1990). Immunoelectron microscopy of tissues processed by rapid freezing and freeze-substitution fixation without chemical fixatives: Application to catalase in rat liver hepatocytes. *J. Histochem. Cytochem.* **38**, 617–623.

van Bergen en Henegouwen, P. M. P. (1989). Immunogold labeling of ultrathin cryosections. *In* "Colloidal Gold: Principles, Methods, and Applications" (M. A. Hayat, ed.), Vol. 1, pp. 191–216. Academic Press, San Diego.

van Lookeren Campagne, M., Oestreicher, A. B., van der Krift, T. P., Gispen, W. H., and Verkleij, A. J. (1991). Freeze-substitution and Lowicryl HM20 embedding of fixed rat brain: Suitability for immunogold ultrastructural localization of neural antigens. *J. Histochem. Cytochem.* **39**, 1267–1279.

Van Noorden, S., and Polak, J. M. (1983). Immunocytochemistry today, techniques and practice. *In* "Immunocytochemistry: Practical Applications in Pathology and Biology" (J. M. Polak and S. van Noorden, eds.), pp. 11–42. John Wright & Sons, Bristol.

Verkleij, A. J., and Leunissen, J. L. M. (eds.) (1989). "Immuno-Gold Labeling in Cell Biology." CRC Press, Boca Raton, FL.

Voorhout, W. (1988). "Possibilities and Limitations of Immuno-Electronmicroscopy." Thesis, University of Utrecht, The Netherlands.

Voorhout, W., van Genderen, I., van Meer, G., and Geuze, H. (1991). Preservation and immunogold localization of lipids by freeze-substitution and low temperature embedding. *Scan. Microsc. Suppl.* **5**, S17–S25.

Zaar, K., and Fahimi, H. D. (1991). Immunoelectron microscopic localization of the isozymes of L-α-hydroxyacid oxidase in renal peroxisomes of beef and sheep: Evidence of distinct intraorganellar subcompartmentation. *J. Histochem. Cytochem.* **39**, 801–808.

FREEZE FRACTURING AND SHADOWING

Structural information of objects can be obtained by preparing replicas of their surfaces; this is performed by evaporating metals and carbon onto the specimen in a high vacuum chamber. The specimen is placed in the metal evaporation chamber, which is evacuated to high vacuum, whereupon metal is evaporated at an oblique angle onto the specimen. The specimen may be a macromolecule, a virus particle, a cell organelle, or a large specimen lying on a smooth substrate, such as a mica flake, and shadowed at ambient temperature. It may also be a specimen that has first been frozen and then fractured and shadowed at low temperature. The shadowing procedure creates an effect similar to that of a sunlit object with a shadow. In this way the three-dimensional topography of the specimen becomes visible. Carbon is subsequently deposited onto the specimen and will serve the same function as a regular carbon film. Together with the metal layer, the carbon is stripped off the specimen.

Shadowing techniques were initiated in the early days of electron microscopy and have evolved to include a large variety of specialized procedures for freezing, shadowing, and replica handling. Many different instrumental designs have also been developed, usually of high technical complexity. A wealth of significant biological information has been obtained by these sets of techniques. However, the various procedures also present a number of difficult interpretation problems, in particular because artifactual patterns are commonly encountered.

This chapter includes examples of shadowing procedures of specimens both at close to ambient temperature and after freezing and fracturing.

The first commercial apparatuses that reproducibly permitted fracturing, etching, and replication of biological specimens used to be introduced with strong arguments such as: "This is the first artifact-free specimen preparation technique—no distortion of morphology—dehydration or embedding will not dissolve any substances—the rapid event in the life of the cell will be instantaneously stopped—the microscope will reveal a frozen moment in the life of the cell." Indeed, freeze-fracturing seemed to be the long searched for ideal method. Also, pioneering freeze-fracture replicas by Cecil E. Hall (1950), Russell L. Steere (1957), and Hans Moor et al. (1961) showed new and intriguing aspects of cells and other structures that were not always easy to interpret, but were always rich in details. However, time has shown that it is certainly not the easy, artifact-free preparation method.

Hall, C. E. (1950). *J. Appl. Phys.* **21**, 61–62.
Steere, R. L. (1957). *J. Biophys. Biochem. Cytol.* **3**, 45–60.
Moore, H., Mühlethaler, K., Waldner, H., and Frey-Wyssling A. (1961). *J. Biophys. Biochem. Cytol.* **10**, 1–13.

1. Shadowing of DNA Molecules

FIGURE 17.1A DNA from pBluescript plasmid prepared according to Kleinschmidt (1968). The DNA was spread on water in the presence of cytochrome c, absorbed on a carbon-strengthened Formvar film, and rotary shadowed at an angle of 8° with platinum/carbon. × 75,000.

DNA molecule forming a closed loop. The thickness of the shadowed molecule is about 12 nm. The replica background around this molecule is smooth.

FIGURE 17.1B Several DNA molecules from a different area of the same grid as in Fig. 17.1A. × 75,000.

The DNA molecules have the same appearance as in Fig. 17.1A and form a complex network. The background is covered with particles representing aggregates of cytochrome c.

FIGURE 17.1C Preparation as in Fig. 17.1A except that the concentration of plasmid DNA is 10 times greater. × 75,000.

This field shows DNA strands that form a very dense network. Individual strands cannot be followed for long distances.

FIGURE 17.1D Similar preparation as in Fig. 17.1A except that these molecules were located close to a grid bar. × 75,000.

The appearance of shadowed DNA molecules is greatly dependent on the parameters of the rotary shadowing. Because these molecules were located close to a grid bar, they are mainly shadowed from one direction, causing (white) shadows in the direction upward in this micrograph.

FIGURE 17.1E Similar preparation as in Fig. 17.1B except that these molecules were reached only by very small amounts of shadowing material.

Incomplete shadowing results in almost a complete absence of structural information.

Since the late-1950s, the Kleinschmidt–Zahn technique has provided information about isolated DNA molecules under a large variety of conditions. It has gradually developed into a whole family of preparatory methods with the aim of extracting different types of information about DNA molecules and nucleic acid–protein interactions. Special procedures have been worked out for double-stranded and for single-stranded nucleic acids (see Sommerville and Scheer, 1987).

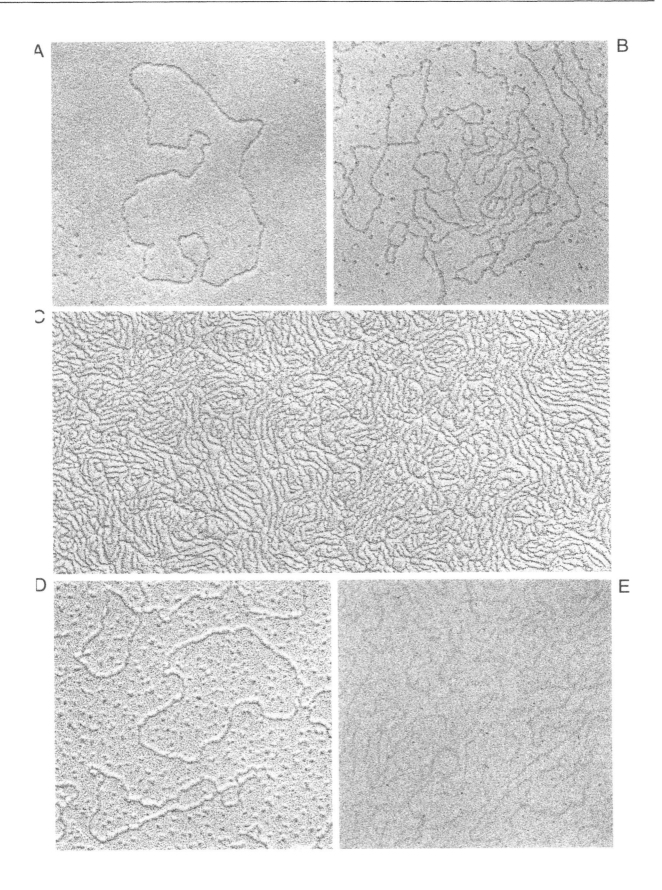

2. Shadowing of Protein Molecules

FIGURES 17.2A AND 17.2B Ferritin molecules rotary shadowed with platinum/carbon. A solution of ferritin in water (Fig. 17.2A) or in 30% ethanol (Fig. 17.2B) was placed between two thin mica flakes. The sandwiched mica preparation was manually plunge-frozen in liquid nitrogen, the mica flakes separated in the liquid nitrogen, transferred to the freeze-fracture apparatus, and freeze-dried. They were rotary shadowed at an angle of 6° with platinum/carbon at −80°C, carbon reinforced, the replica floated out on water, and recovered on a Formvar-coated copper grid. × 200,000.

The ferritin molecules appear as round units with a peripheral metal coating. There is an even sprinkling of platinum grains in the background except for a faint halo around each molecule. The diameter of each shadowed molecule is approximately 14 nm.

FIGURE 17.2C Ferritin molecules prepared as in Fig. 17.2A except that they were unidirectionally shadowed at an angle of 6°. × 200,000.

The ferritin molecules have the same approximate diameter as when rotary shadowed but have an uneven shape related to the direction of shadowing.

FIGURE 17.2D Ferritin molecules dried on mica without freezing and rotary shadowed in the freeze-fracture apparatus at ambient temperature. × 125,000.

Solitary ferritin molecules show uniform pheripheral shadowing but only the upper surfaces of packed molecules are decorated.

FIGURE 17.2E Ferritin molecules prepared as in Fig. 17.2A except that the concentration was more than 10 times greater. × 200,000.

Due to the low angle of shadowing and the dense packing of the molecules, the platinum only decorates the upper surface of the molecules.

FIGURE 17.2F Collagen type II prepared in the same way as the ferritin molecules in Fig. 17.2A and rotary shadowed in the same way. × 200,000.

The collagen molecules appear as thin strings of evaporated platinum. This field shows intact as well as some fragmented molecules.

COMMENTS

Shadowing of protein molecules at low or ambient temperatures provides some information about the structure of the proteins. The degree of details obtained is related to a number of factors, including the method of applying the molecules on the substrate, the type of evaporation unit, and the direction and size of the evaporated grains. Comparisons of Figs. 17.2B, 17.2C, and 17.2E illustrate the clear-cut differences between rotary shadowing, unidirectional shadowing, and a too dense preparation of molecules. Figure 17.2D shows the influence of the method of applying the protein to the substrate and Fig. 17.2F illustrates the difficulty of outlining molecules when the sizes of the evaporated grains are similar or even larger than the diameter of the molecule itself.

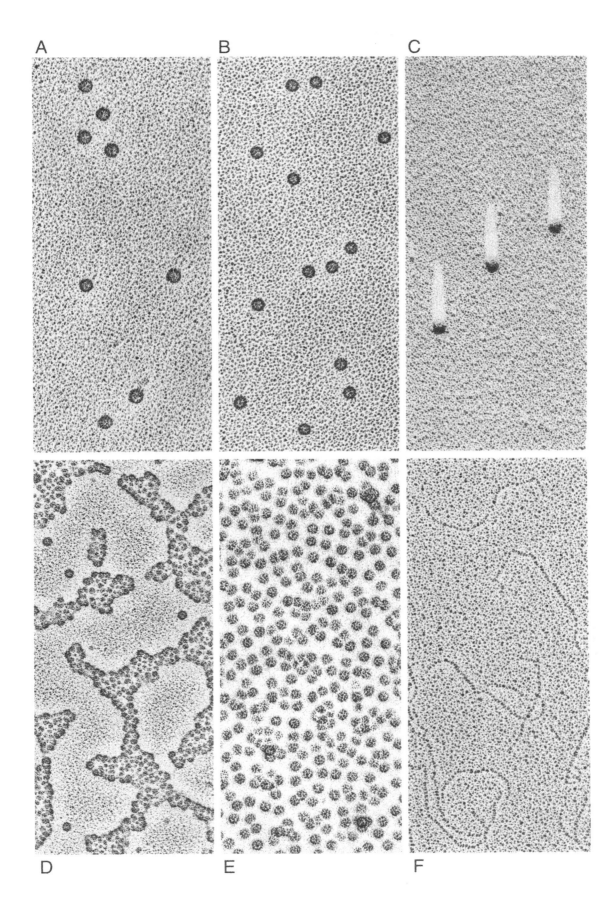

3. Freeze-Fractured Membrane Faces

FIGURE 17.3A Freeze-fractured basolateral membranes of proximal tubule cells in a rat kidney that was perfusion fixed in 1% glutaraldehyde. Small tissue blocks were rinsed in buffer and cryoprotected in 30% glycerol, manually plunge-frozen in Freon-22 cooled with liquid nitrogen, fractured at −110°C, and shadowed with platinum/carbon at an angle of 45° without etching. The replica was cleaned in sodium hypochlorite and picked up on a Formvar-coated copper grid. × 140,000.

Two different types of fractured faces of the basolateral membrane are shown: the particle-rich P face (upper right) and the particle-poor E face (lower left). The light, diagonal zone between the two fractured faces represents the fracture through the intercellular space. The direction of shadowing is from below.

FIGURE 17.3B The P face of a membrane from a lysosome in a kidney cell prepared as in Fig. 17.3A. × 240,000.

The fractured face exhibits numerous distinct particles, which vary in diameter from 5 to 12 nm. The shadowing direction is from below and each particle has a light shadow upward. The fractured surface is slightly concave, which is why the length of the shadow is longer in the lower part than in the upper part of the micrograph. The fractured face between the particles is smooth.

FIGURE 17.3C Same micrograph as in Fig. 17.3B except that the black-and-white scale has been reversed photographically.

All particles now appear light with a dark shadow in much the same way as objects illuminated by light from behind.

COMMENTS

It is widely accepted that freeze-fractured plasma membranes split in the middle into two halves, which are referred to as the protoplasmic face (P) and the exoplasmic face (E). When observing a replica of a P face, the observer is looking in the direction of the cell nucleus. Almost invariably membrane proteins partition with this face, which therefore exhibits numerous particles. When observing the E face, however, the observer is actually looking away from the nucleus. In Fig. 17.3A the E face represents the outer half of the basolateral membrane in one cell and the fracture then passes through the intercellular space and exposes the P face of the adjacent cell. The P and E faces are not strictly complementary in that particles on the P face normally do not correspond to depressions in the E face (see, however, Figs. 17.7A and 17.7B).

Most freeze-fracture micrographs are shown with the shadowing metal dark and the shadows light. For some purposes it may be easier to interpret the micrograph if the contrast is reversed, thus showing shadows dark as we are used to in daily life (see also Figs. 12.7D and 12.7E).

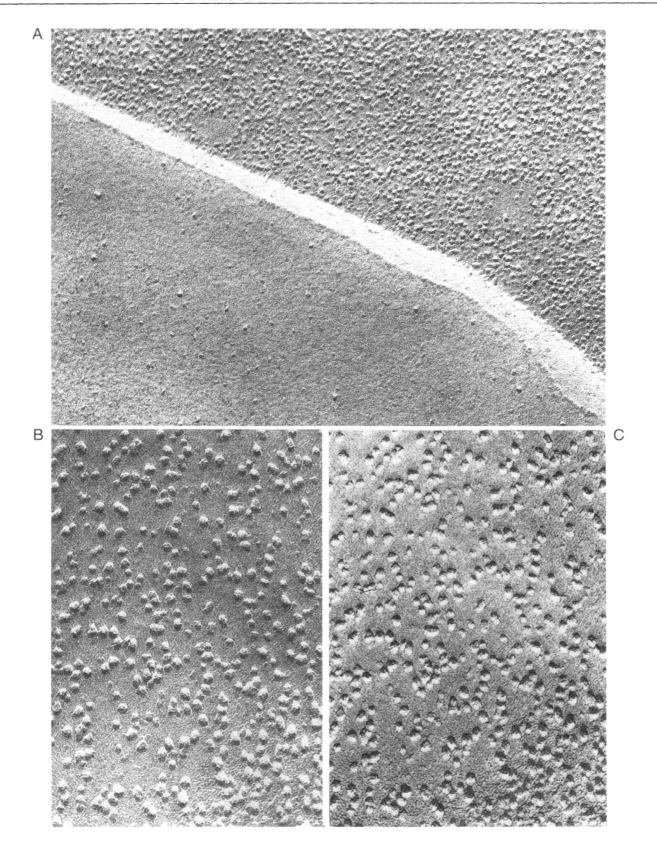

4. Thickness of Replica: Low Magnification

FIGURE 17.4A A freeze-fracture replica of microvilli in proximal tubule from a rat kidney perfusion fixed with 1% glutaraldehyde. Small pieces of tissue were rinsed in buffer and cryoprotected in 30% (v/v) glycerol. The specimen was then frozen rapidly in Freon-22 cooled with liquid nitrogen. Freeze-fracture was carried out in a Balzers BAF 300 freeze-fracture apparatus (Balzers AG, Liechtenstein). The specimens were fractured at −110°C and immediately shadowed at an angle of 45° with platinum/carbon. The thickness of the evaporated metal (<2 nm) was controlled with a quartz crystal thin film monitor. The temperature was then raised to −80°C, and carbon was deposited at an angle of 90° to strengthen the replica, which was cleaned in sodium hypochlorite overnight and collected on a Formvar-coated copper grid. × 60,000.

The replica shows longitudinally fractured microvilli. The replica shows either the protoplasmic or the exoplasmic fracture faces of the plasma membrane. The P face, in contrast to the E face, shows abundant intramembrane particles. This replica is thin and has wide areas devoid of platinum on both the P face and the E face.

FIGURE 17.4B Same preparation as in Fig. 17.4A except that the fractured surface was evaporated with a slightly thicker layer of platinum/carbon. × 60,000.

In this replica the intramembrane particles are well-defined and the fracture face between the microvilli is smooth. On the convex P faces, platinum is only absent on the upper, unshadowed side of the microvilli and the concave E faces are only devoid of platinum at their lower side.

FIGURE 17.4C Same preparation as in Fig. 17.4A except that a thick (>2 nm) layer of platinum/carbon was evaporated on the fractured surface. × 60,000.

The replica surface has a granular appearance both between the microvilli and on the P and E faces. Intramembrane particles are outlined poorly and are partly confluent.

COMMENTS

The appearance of a freeze-fracture replica is influenced greatly by the thickness of evaporated metal. A thin replica reveals an incomplete image of the fracture surface as in Fig. 17.4A whereas a thick replica will obscure surface details as in Fig. 17.4C. Using a quartz film thickness monitor, however, it is a routine procedure to adjust the metal evaporation to a useful thickness, as illustrated in Fig. 17.4B. Nevertheless, any replica may show variations in thickness, depending on the patterns and angle of the fracture face. The strength of the replica is not primarily dependent on the thickness of the evaporated platinum, but rather on the carbon layer evaporated after replication.

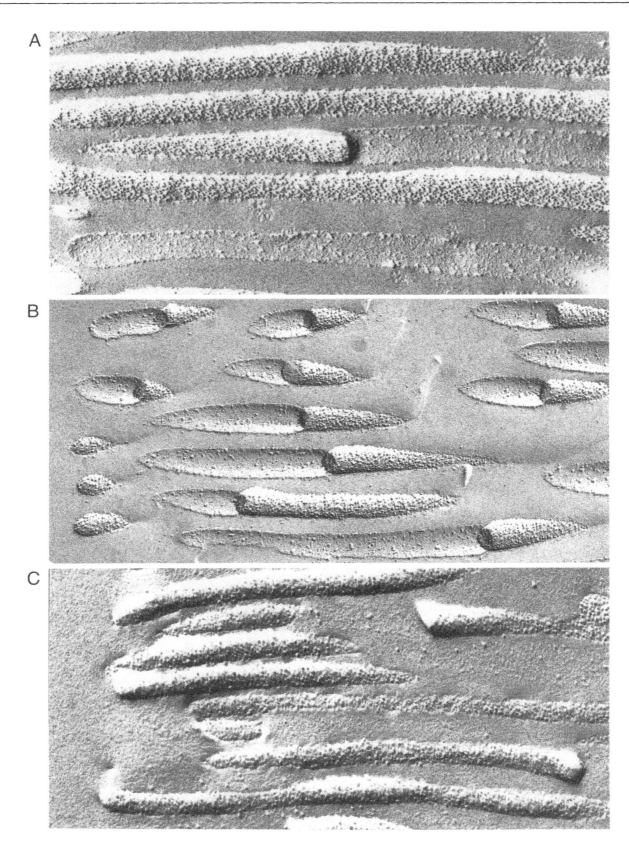

5. Thickness of Replica: High Magnification

FIGURE 17.5A A higher magnification of the replica shown in Fig. 17.4A. × 160,000.

Close inspection of this replica reveals that it does not contain a continuous platinum layer but instead consists of a layer of small electron-dense platinum grains. In the thinnest part of the P face the metal grains have diameters of 10–20 Å (1–2 nm), and many intramembrane particles are only revealed by small clusters of grains on the top of the particle.

FIGURE 17.5B A higher magnification of the same replica as in Fig. 17.4B. × 160,000.

The replica between the microvilli forms a continuous smooth surface. The membrane fracture faces also appear largely continuous, except on the upper side of the convex P face. Intramembrane particles are well defined in most membrane fracture faces.

FIGURE 17.5C A higher magnification of the same replica as in Fig. 17.4C. × 160,000

The surface of the replica between the microvilli is coarse and granular. The membrane fracture faces are uneven and their particles are defined poorly.

COMMENTS

The resolution in a freeze-fracture replica is in part determined by the size of the evaporated metal grains. Another factor limiting the resolution is the movement of platinum atoms on the frozen fracture surface before they coalesce to small grains. In practice, these factors will limit the resolution to 2–3 nm even in the best replicas. Deviation from optimal replica thickness will further lower the resolution such as in Fig. 17.5A, where particles are not fully outlined, or in Fig. 17.5C where they are covered extensively with metal. Any attempts to determine particle dimensions or frequencies on such thin or thick replicas are meaningless. An additonal factor decreasing resolution is a possible rearrangement of platinum grains if the replica is exposed to high beam intensities. A slight improvement in resolution can be obtained using evaporation with tantalum/tungsten but at some expense of replica contrast and reproducibility.

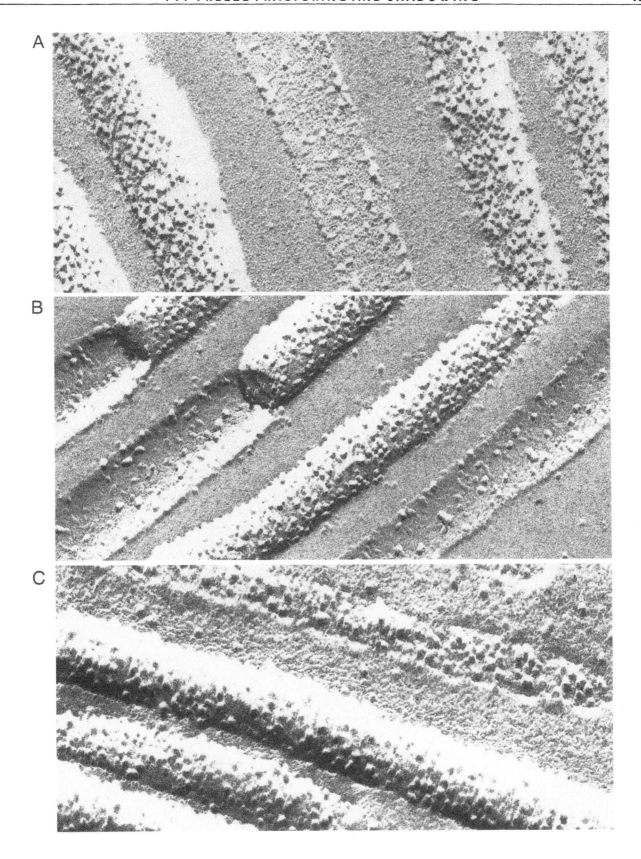

6. Rotary Shadowing

FIGURE 17.6A Suspension of basolateral cell membrane fragments purified with respect to Na,K-ATPase. A microsomal fraction was isolated from the rabbit outer renal medulla, treated with sodium dodecyl sulfate (SDS), and purified on a continuous sucrose gradient in the presence of ATP (Jørgensen, 1974). Samples of the Na,K-ATPase membranes were frozen in Freon-22 cooled with liquid nitrogen and were rotary shadowed in a Balzers BAF 300 freeze-fracture apparatus as described previously (Maunsbach *et al.*, 1979) using platinum/carbon at an angle of 10°. × 38,000.

The membrane fragments are cup-shaped or flat and seen in cross section or tangentially fractured. The tangentially fractured membranes appear either concave (lower left corner) or convex. In the concave fracture face the center is devoid of evaporated metal, whereas on the convex face the periphery has the thickest platinum layer.

FIGURE 17.6B A higher magnification of a rotary-shadowed convex (right) and a concave (left) membrane from the same preparation. × 190,000. The inset shows a gallery of intramembrane particles at a higher magnification. × 585,000.

In the center of the convex faces, where the shadowing angle is at a minimum and the particles are shadowed symmetrically, the replica shows considerable structural details as illustrated in the inset; the gallery of particles is thus derived from the center of a convex fracture face. The particles are about 80 Å in diameter and appear to consist of two subunits.

COMMENTS

Rotary shadowing provides a symmetrical metal evaporation onto the fracture face. On elevated spherical objects the periphery receives a thick platinum coating whereas the coating in the center becomes attenuated, especially if the shadowing angle is small (Fig. 17.6B). The shadow is symmetrical provided the object is round; if it is not round the rotary shadowing will accentuate any asymmetries in objects, such as intramembrane particles. Thus, the asymmetrical, bilobated appearance of the particles in the inset panel of Fig. 17.6B shows that the particles are composed of two subunits. In the illustrated case each intramembrane particle is supposed to correspond to one dimer of Na,K-ATPase and each subunit in the bilobate particles represents one protomer of the enzyme. The middle of the concave fracture face in Fig. 17.6B (lower left) has not been reached by the evaporated platinum/carbon due to the low shadowing angle and hence appears electron lucid.

A

B

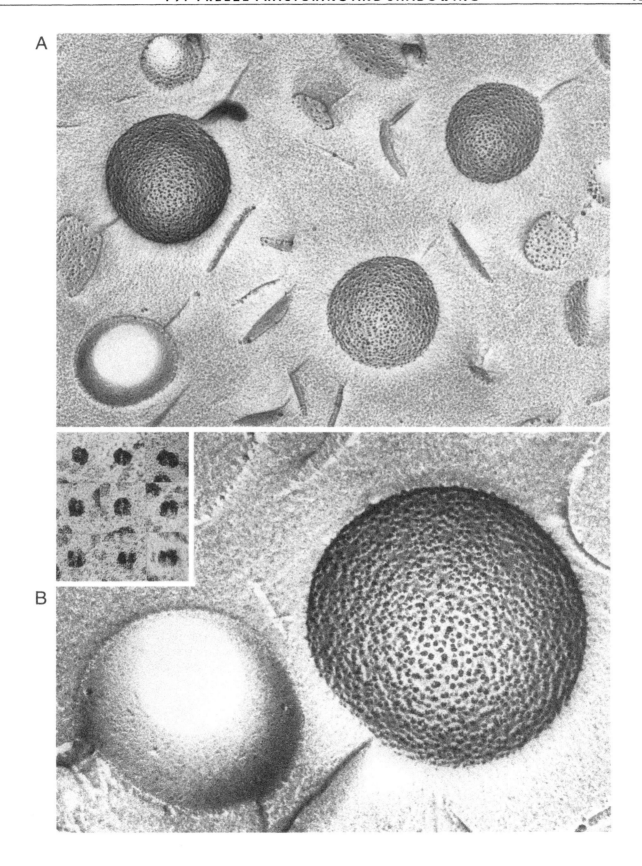

7. Complementary Replicas and Stereo Images

FIGURES 17.7A AND 17.7B These two electron micrographs show complementary freeze-fracture replicas of an apyrene snail spermatozoon (*Melanopsis dufouri*). For preparation (Afzelius *et al.*, 1989), fragments of the snail testes were fixed in 5% glutaraldehyde and 4% paraformaldehyde, washed in buffer, infiltrated in glycerol (10–30%), and frozen in Freon-22 cooled with liquid nitrogen. The tissue was fractured with a Balzers BAF 400 freeze-fracture apparatus equipped with a double replica stage. The two complementary replicas were cleaned and picked up on two separate grids and examined in the electron microscope. × 45.000.

In Fig. 17.7A there are regular plaques or domains of intramembrane particles arranged in short rows. The complementary face (Fig. 17.7B) shows regular parallel furrows, also forming similar plaques.

FIGURES 17.7C AND 17.7D These two micrographs represent a stereo pair recorded from a freeze-fracture replica of microvilli in a renal proximal tubule with the aid of a CompuStage goniometer in a Philips CM100 electron microscope. The angle between the two micrographs is 15° and the tilt axis is vertical. The micrographs are mounted with an interdistance of 63 mm. × 20,000.

When observed in a stereo viewer (or with crossed eyes, as some people are able to do) the freeze-fracture replica stands out in relief with some short crosscuts of the microvilli projecting above the background. In addition, small elevated particles are seen within the crosscut microvilli. The fracture plane has proceeded in steps with the upper part of the micrograph showing the "highest" level.

COMMENTS

The fracture faces in Figs. 17.7A and 17.7B are complementary, as particle rows on the P face in Fig. 17.7A correspond to furrows in the E face on the corresponding membrane in Fig. 17.7B. Globular elevations in Fig. 17.7A correspond to pits or holes in Fig. 17.7B and vice versa. In other replicas, similar plaques may show deviations in the complementarity, as membranes and particles may be plastically deformed during fracturing (see Fig. 17.12).

In some situations, stereo recordings will help in determining the three-dimensional shape of the observed object as in Fig. 17.7C and 17.7D. Stereo micrographs are particularly helpful in situations where the fracture face is devoid of intramembrane particles or other structures that signify the direction of the curvature. When mounting stereo micrographs it is essential that the tilt axis is oriented vertically and that the distance between the same features in the two micrographs coincides with the intereye distance (usually 60–63 mm).

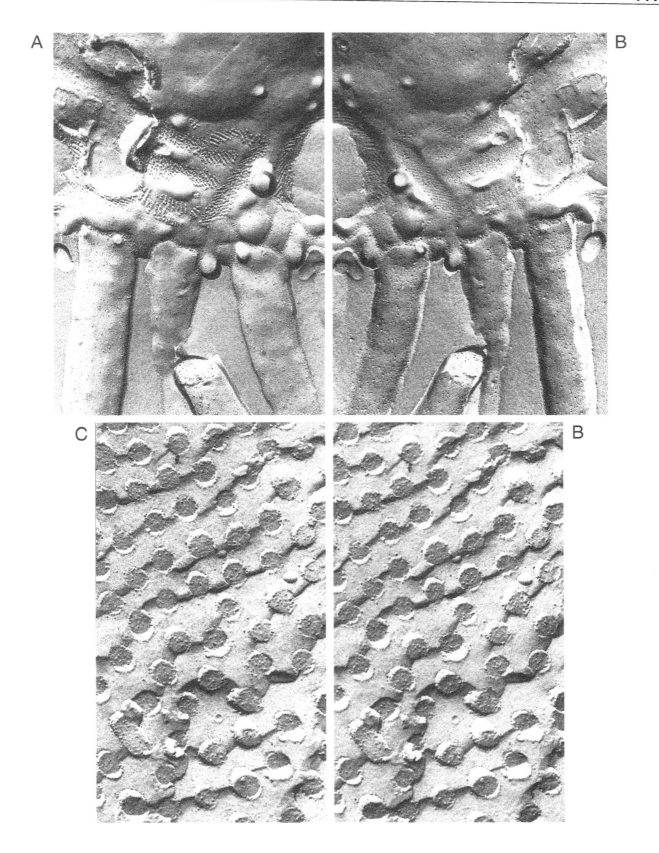

8. Ice Crystals and Etching

FIGURES 17.8A–17.8C Freeze-fracture replicas of a proximal tubule in a rat kidney perfusion fixed with 1% glutaraldehyde. Small pieces of tissue were immersed in 5% (Fig. 17.8A), 15% (Fig. 17.8B), and 30% (Fig. 17.8C) glycerol for 2 hr and quickly frozen in Freon-22 cooled with liquid nitrogen. The tissues were fractured at −100°C and etched for 1 min before shadowing with platinum/carbon at an angle of 45°. × 10,000.

The replica in Fig. 17.8A shows extensive formation of large ice crystals in the tubule lumen and between the microvilli. Ice crystals in the cytoplasm are smaller than outside the cell. In Fig. 17.8B the cytoplasm is largely devoid of ice crystals, but the tubule lumen (top) and the peritubular capillary (bottom) contain small ice crystals. There is no evidence of ice crystals in Fig. 17.8C.

FIGURES 17.8D AND 17.8E Membrane fragments of purified Na,K-ATPase from kidney medulla. A suspension of membrane fragments in diluted buffer was frozen rapidly without cryoprotection in Freon-22 cooled with liquid nitrogen. The specimen was fractured at −100°C, and the fracture face was etched for 3 min before platinum/carbon evaporation at an angle of 45°. Fig. 17.8D: × 200,000; Fig. 17.8E: × 250,000.

The membrane fragment in Fig. 17.8D shows a circular area limited by a shallow step from a rather smooth outer surface. Within the circular area there are a few intramembrane particles. The direction of shadowing is from below. The lower part of Fig. 17.8E shows a fracture face with numerous intramembrane particles. In places chains of intramembrane particles seem to continue as elevated ridges on the surface in the upper part of the micrograph.

COMMENTS

Ice crystal formation is extensive in specimens manually plunge-frozen without proper cryoprotection. Ice crystals outside the cells are separated by elevated eutectic ridges. When the glycerol concentration is 15% crystal formation inside the cell is largely avoided. Etching of a fracture surface is not possible if the specimen is cryoprotected with 30% glycerol. However, if the glycerol percentage is low or if glycerol is absent, as in Figs. 17.8D and 17.8E, the ice will rapidly sublime if the vacuum is high and there is a cooler object, e.g., the knife used for fracture, placed above the specimen. Etching results in exposure of membrane surfaces previously covered by ice. Thus in Fig. 17.8D the circular area with few particles is interpreted as an E face whereas the surrounding area is interpreted as the exposed inner surface of the Na,K-ATPase membrane. Figure 17.8E shows the P face with abundant intramembrane particles consisting of Na,K-ATPase protein units. The upper part of the micrograph, however, is interpreted as the outer surface of the membrane where the outer portions of the membrane proteins are exposed and cause the irregular elevations.

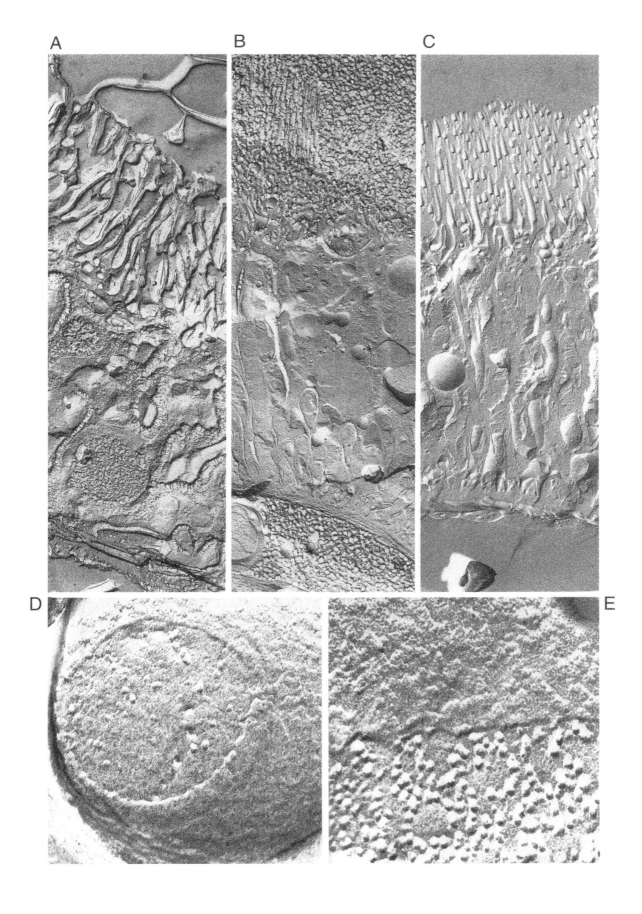

9. Quick-Freeze Deep Etching

FIGURE 17.9A Part of a demembranated sperm tail from a honeybee rapidly frozen by impact onto a liquid helium-cooled copper block in a laboratory-manufactured "quick-freeze" device built by P. Lupetti. Quick-frozen samples were then freeze-fractured in a Balzer freeze-etch machine and deep etched for 4 min at −100°C, thereby exposing the structural components of the sperm tail. These were then rotary replicated with a 2-nm-thick layer of platinum evaporated from an angle of 24° above the horizontal and finally supported with a 10-nm film of pure carbon. The replicas were cleaned by floating overnight in concentrated chromic acid, washed in water, picked up on 75 mesh Formvar-coated copper grids, and examined in an electron microscope. This micrograph was then taken from a portion of the replica judged to be sufficiently superficial not to be deformed by ice crystal formation. × 120,000.

There are no obvious specimen preparation artifacts. The patterns seen in the replica are hence the same as those seen by other preparation methods. From left to right: four accessory microtubules of the axoneme, the limiting membrane of the mitochondrial derivative, and the crystal of the mitochondrial derivative.

FIGURE 17.9B Same preparation as in Fig. 17.9A, although this micrograph was taken from a part of the replica that derives from a deeper zone in the frozen tissue block. × 120,000.

Numerous ice crystals obscure most of the biological structures. The sperm tail is largely hidden.

COMMENTS

Impact freezing on a pure copper mirror ensures a very rapid freezing rate. Upon this quick freezing, the water in the biological specimen will become amorphous ice and will not deform the biological structures. Upon sublimation, "etching," the specimen will be uncovered, and the various ultrastructural details seem to remain essentially as they were in the native state. The various components of the sperm tail seen in Fig. 17.9A are all assumed to derive from the biological specimen. The replica method shows three-dimensional configurations better than a thin section. The longitudinal 4-nm periodicity of the tubulin monomers can thus be seen in the microtubules, as can the regular and tight packing of intramembrane particles in the separating membrane, or the dumbbell-shaped subunits in the crystal of the sperm mitochondrion to the right. However, only a narrow superficial zone will be optimally frozen. At a few microns below the surface, freezing will be slower and ice crystals will form. Such ice crystals that remain after etching will be replicated and are seen in Fig. 17.9B.

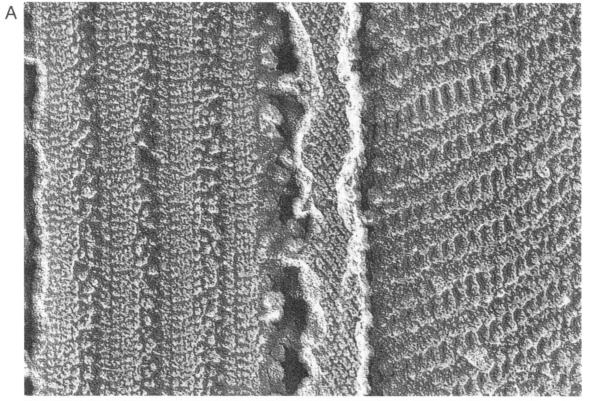

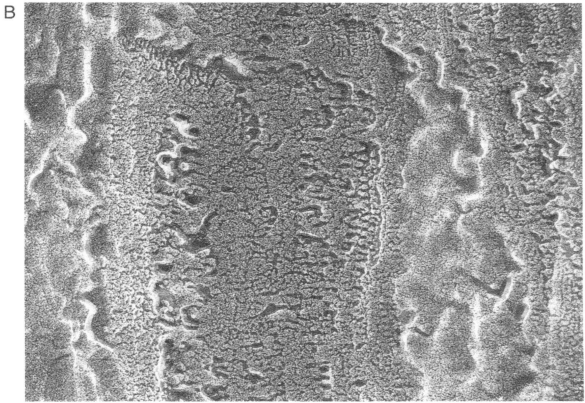

10. Identification of Transport Molecules

FIGURE 17.10A A freeze-fracture replica of liposomes that were formed during dialysis of a phosphatidylcholine solution dissolved in buffered sodium cholate (Skriver *et al.*, 1980). The liposomes were concentrated by centrifugation, and the pellet was resuspended and equilibrated with 20% glycerol. Vesicle aliquots were frozen in Freon-22 and fractured at −100°C. × 150,000.

The freeze-fractured liposomes show either convex or concave fracture faces. None of the fracture faces exhibit intramembrane particles.

FIGURE 17.10B Liposomes prepared as in Fig. 17.10A except that during dialysis the lipid mixture also contained purified Na,K-ATPase (669 μg of protein/10 mg of lipid). × 150,000.

Both convex and concave fracture faces carry some intramembrane particles.

FIGURE 17.10C A freeze-fracture replica of liposomes formed from *Escherichia coli* phospholipids dissolved in octylglucopyranoside and prepared according to Zeidel *et al.* (1994). × 90,000.

The fracture faces of the liposomes are either convex or concave. There are no intramembrane particles.

FIGURES 17.10D–17.10F Liposomes prepared as in Fig 17.10C and reconstituted with purified aquaporin-1. The lipid-to-protein ratio was 125 : 1 in Fig. 17.10D. In Figs. 17.10E and 17.10F the protein concentration was 2.5 and 5 times greater respectively. Liposomes were concentrated by centrifugation, cryoprotected in glycerol and after freezing in Freon-22 fractured and shadowed at an angle of 45°. × 90,000.

The liposomes in Figs. 17.10D–17.10F show an increasing density of intramembrane particles on the fracture faces.

COMMENTS

Liposomes reconstituted with purified Na,K-ATPase show the ability to transport sodium and potassium. After addition of ATP to the incubation medium, sodium ions from the medium accumulate inside the vesicles and potassium ions are extruded from the vesicle interior to the medium. The capacity for active cation transport is proportional to the frequency of intramembrane particles over a range of 0.2–16 particles per vesicle (Skriver *et al.*, 1980). Liposomes without ATPase do not transport ions. Correlation of ion transport data and densities therefore demonstrate that the particles in the vesicle membrane represent Na,K-ATPase molecules.

Liposomes reconstituted with increasing amounts of aquaporin-1 (Figs. 17.10D–17.10F) show increasing osmotic water permeabilities whereas liposomes without aquaporin-1 are impermeable (Zeidel *et al.*, 1994). Quantitative estimates of particle densities in freeze-fractured liposomes show that the intramembrane particles correspond to aquaporin-1 transport protein, primarily arranged as tetramers. Thus aquaporin-1 liposomes, as well as Na,K-ATPase liposomes, provide well-defined systems for biophysical analysis of transmembrane water or ion movements.

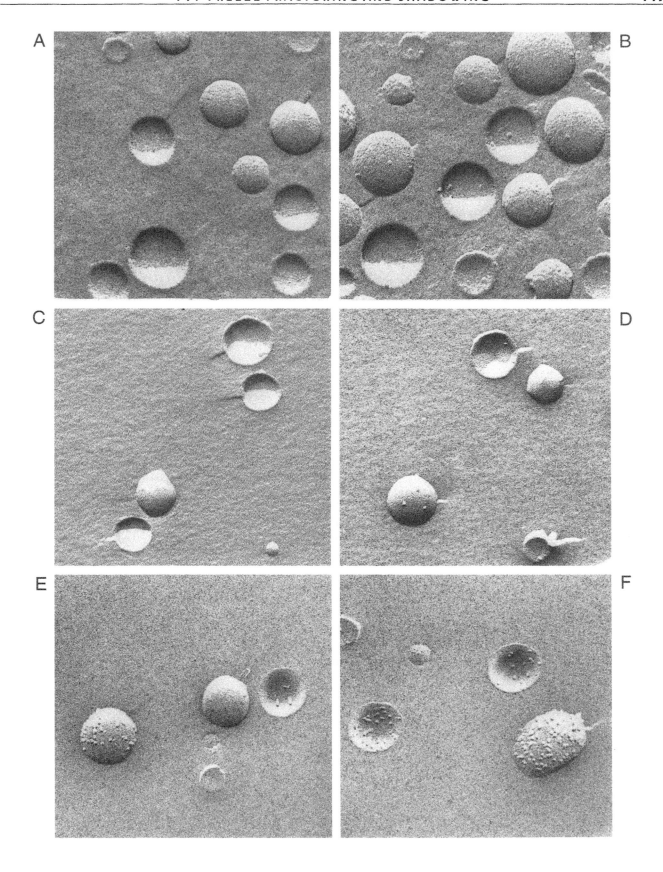

11. Contamination

FIGURE 17.11A A freeze-fracture replica of a plunge-frozen small droplet of pure water. Evaporation of platinum/carbon at an angle of 45° was initiated a few seconds after the last cut with the kinife. × 24,000.

The fracture surface shows elongated crystals oriented in all directions. There are small confluent bumps between, and in part on, the crystals.

FIGURE 17.11B Microvilli from a kidney cell fixed in glutaraldehyde, cryoprotected in glycerol, freeze-fractured at a temperature of −100°C, and evaporated with platinum/carbon at an angle of 45° about 1 min after the last cut of the knife. The direction of shadowing is from below. × 60,000.

In the tubule lumen, in the lower part of the micrograph, there are numerous small elevations with diameters of 20–60 nm. Similar elevations are also located adjacent to and on the microvilli in the upper part of the micrograph.

FIGURE 17.11C A higher magnification of a specimen similar to that in Fig. 17.11B. The direction of shadowing is from below. × 120,000.

The P faces of microvilli show numerous small particles as well as large particulate elevations. In several places there is a gradual transition from small to large particles. The fracture surface between microvilli is smooth.

COMMENTS

Water contamination of the fracture surface occurs if the vacuum is poor and the water pressure close to the specimen exceeds the tendency for water molecules to sublime from the fracture surface. It also occurs if the specimen temperature is too low.

Figures 17.11A–17.11C illustrate various aspects of contamination. In Fig. 17.11A, large crystals as well as irregular areas of granular deposits form on the fractured surface. In Fig. 17.11B, numerous large particles form on part of the specimen where there are no biological structures and also on the fractured microvilli. In Fig. 17.11C, similar particles are present on fractured P faces, which then show a range of particles ranging from intramembrane particles of normal appearance to large artificial particles. Evaluations of particles in such replicas are therefore hazardous. To avoid contamination the vacuum in the chamber should be high and the fractured surface protected by, e.g., the cooled knife. A simple means of reducing contamination is to start the platinum/carbon evaporation immediately before the last cut is made with the knife.

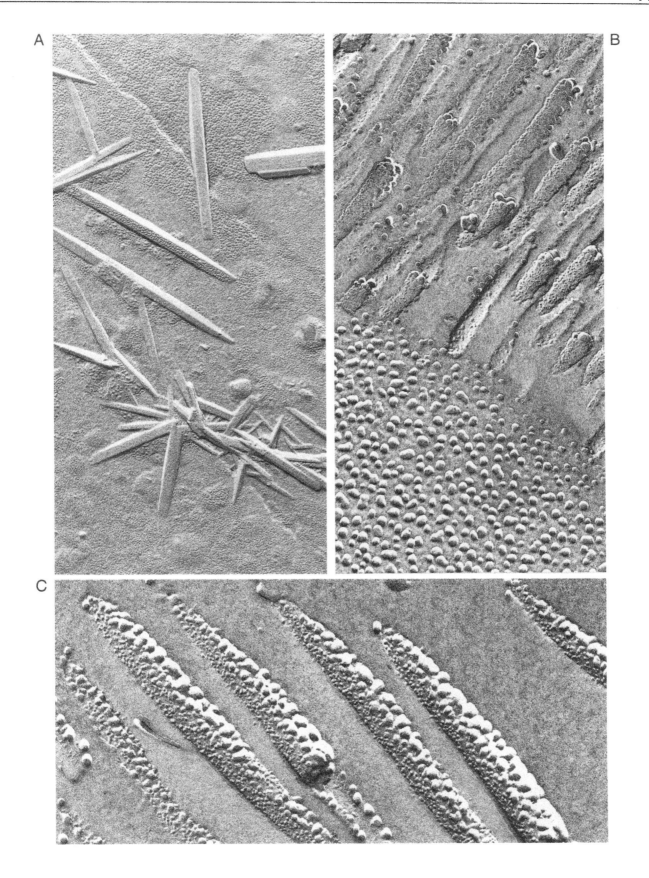

12. Plastic Distortion

FIGURE 17.12A A freeze-fracture replica of glycerol solution containing polystyrene particles with a diameter of 0.88 μm. The suspension of polystyrene particles was frozen in Freon-22 cooled with liquid nitrogen and conventionally fractured and shadowed at $-110°C$. The replica was cleaned in hypochlorite. $\times$ 30,000.

This micrograph shows an intact polystyrene particle to the right and a concave depression in the replica to the left. In the middle there is an irregular structure projecting out of the replica. The shadowing direction is from below.

FIGURES 17.12B–17.12D Higher magnifications of similar objects as in Fig. 17.12A and originating from an electron micrograph in a similar replica. The direction of shadowing is from below. $\times$ 70,000.

Figure 17.12B shows the intact polystyrene particle. Figure 17.12C shows a deformed polystyrene particle that is drawn out and gives a shadow. Figure 17.12D shows a concave surface approaching the diameter of a polystyrene particle. Note small particles inside the concave fracture face.

FIGURES 17.12E AND 17.12F Freeze-fracture replicas of microvilli of a kidney proximal tubule. The tissue was conventionally cryoprotected with 30% glycerol and frozen in Freon-22 cooled with liquid nitrogen. Platinum/carbon was evaporated 2 min after fracture. The direction of shadowing is from below. Fig. 17.12E: $\times$ 60,000; Fig. 17.12F: $\times$ 150,000.

Microvilli show normal fracture patterns. There are numerous holes in the replica between microvilli, each of which in the upper edge has a dense attachment appearing like a small piece of the replica. In Fig. 17.12F the dense attachments to the upper rims of the holes in the replica are in places displaced over the fractured membrane faces (arrows), giving the impression that they have "fallen backwards" over the replica.

COMMENTS

During fracture, objects may become deformed greatly, as illustrated in Figs. 17.12A and 17.12C. Some polystyrene particles are stretched into irregular shapes as in Fig. 17.12C. Other particles remain intact but leave a depression, which often shows a surface with small particles, almost resembling those in fractured biological membranes. The intact particles become attached to the underside of the replica during the cleaning procedure. The freeze-fracture replica illustrated in Figs. 17.12E and 17.12F shows defects of unclear origin. It is possible that holes in the replica and the superposition of small objects onto membrane fracture faces develop during thawing and subsequent cleaning of the replica. Whatever the origin, it illustrates complex patterns that are sometimes observed in freeze-fracture replicas.

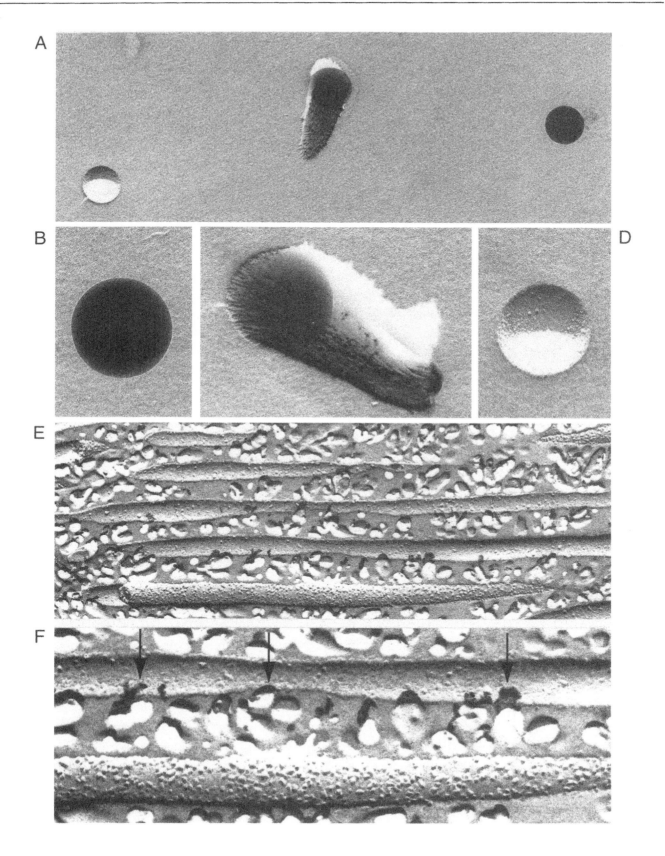

13. Replica Defects

FIGURE 17.13A A freeze-fracture replica of a liver cell. The tissue was aldehyde fixed and conventionally prepared for freeze fracture. The replica was shadowed with platinum/carbon using rotary shadowing at 45°. × 8000.

The replica has a normal appearance except that it has a cleft in the middle. The two edges of the cleft follow approximately parallel courses.

FIGURE 17.13B Same preparation as in Fig. 17.13A. × 8000.

In the lower part of the micrograph there is a large electron-dense area with sharp edges.

FIGURE 17.13C A freeze-fracture replica of a kidney tubule epithelium shadowed at an angle of 45°. The shadow direction is from below. × 5000.

Most of the replica shows normal conditions but in some places there are irregular, dark regions. In addition, small angular objects are evident in places, such as those seen in the upper left corner of the micrograph.

COMMENTS

Freeze-fracture replicas may acquire several defects during the various preparatory steps after the platinum/carbon shadowing until they are safely deposited on the grid. If the carbon layer applied after metal shadowing is too thin, the replica will be brittle and fall into pieces; the same thing will happen if the backing carbon is evaporated at too low a temperature, i.e., below about −80°C. If the replica breaks into pieces, it may in places fold upon itself as in Fig. 17.13B. Another common defect is incomplete removal of the biological specimen from the replica, which results in an irregular electron-dense deposit as in Fig. 17.13C.

A

B

C

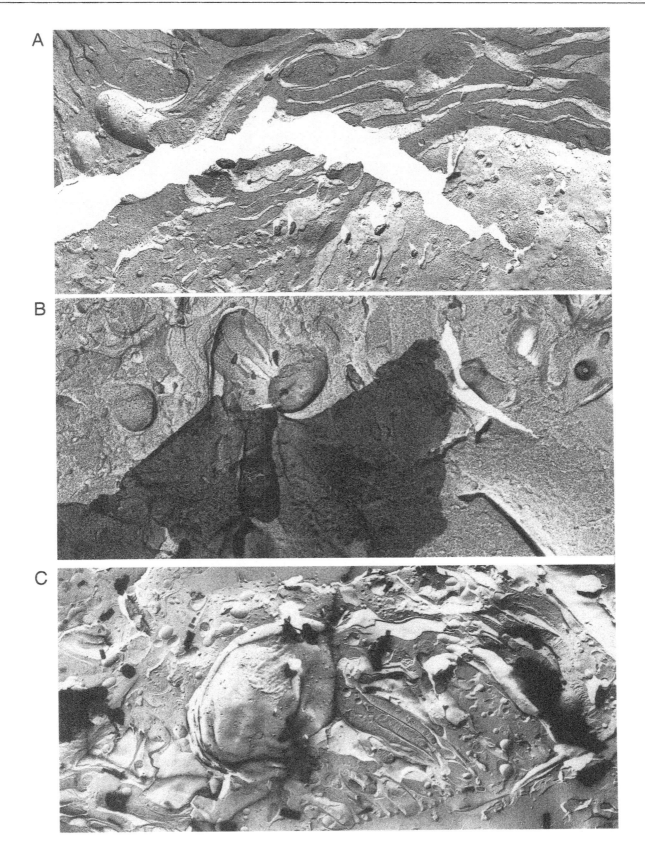

References

Afzelius, B. A., Dallai, R., and Callaini, G. (1989). Spermiogenesis and spermatozoa in *Melanopsis* (Mesogastropoda, Mollusca). *J. Submicrosc. Cytol. Pathol.* **21**, 187–200.

Böhler, S. (1975). "Artefacts and Specimen Preparation Faults in Freeze Etch Technology." Balzers Ag, Liechtenstein.

Branton, D., Bullivant, S., Gilula, N. B., Karnovsky, M. J., Moor, H., Mühlethaler, K., Northcote, D. H., Packer, L., Satir, B., Satir, P., Speth, V., Staehlin, L. A., Steere, R. L., and Weinstein, R. S. (1975). Freeze-etching nomenclature. *Science* **190**, 54–56.

Danielsen, C. C. (1987). Thermal stability of human-fibroblast-collagenase-cleavage products of type-I and type-III collagens. *Biochem. J.* **247**, 725–729.

Deguchi, N., Jørgensen, P. L., and Maunsbach, A. B. (1977). Ultrastructure of the sodium pump: Comparison of thin sectioning, negative staining, and freeze-fracture of purified, membrane-bound (Na⁺,K⁺)-ATPase. *J. Cell Biol.* **75**, 619–634.

Fisher, K., and Branton, D. (1974). Application of the freeze-fracture technique to natural membranes. *In* "Methods in Enzymology" (S. Fleischer and L. Packer, eds.), Vol. 32, pp. 35–44. Academic Press, New York.

Fujimoto, K. (1997). SDS-digested freeze-fracture replica labeling electron microscopy to study the two-dimensional distribution of integral membrane proteins and phospholipids in biomembranes: Practical procedure, interpretation and application. *Histochem. Cell Biol.* **107**, 87–96.

Fujimoto, K., Møller, J. V., and Maunsbach, A. B. (1996). Epitope topology of Na,K-ATPase α subunit analyzed in basolateral cell membranes of rat kidney tubules. *FEBS Lett.* **395**, 29–32.

Fujimoto, K., Noda, T., and Fujimoto, T. (1997). A simple and reliable quick-freezing/freeze-fracturing procedure. *Histochem. Cell Biol.* **107**, 81–84.

Gross, H. (1987). High resolution metal replication of freeze-dried specimens. *In* "Cryotechniques in Biological Electron Microscopy" (R. A. Steinbrecht and K. Zierold, eds.), pp. 205–215. Springer-Verlag, Berlin.

Hawes, C., and Martin, B. (1995). Freeze-fracture deep-etch methods. *In* "Methods in Plant Cell Biology" (D. W. Galbraith, H. J. Bohnert, and D. P. Bourque, eds.), Vol. 49, pp. 33–43. Academic Press, San Diego.

Heuser, J. E. (1989). Protocol for 3-D visualization of molecules on mica via the quick-freeze, deep-etch technique. *J. Electr. Microsc. Techn.* **13**, 244–263.

Heuser, J. E., Reese, T. S., Dennis, M. J., Jan, Y., Jan, L., and Evans, L. (1979). Synaptic vesicle exocytosis captured by quick freezing and correlated with quantal transmitter release. *J. Cell Biol.* **81**, 275–300.

Hui, S. W. (ed.) (1989). "Freeze-Fracture Studies of Membranes." CRC Press, Boca Raton, FL.

Jørgensen, P. L. (1974). Purification and characterization of (Na⁺-K⁺)-ATPase. III. Purification from the outer medulla of mammalian kidney after selective removal of membrane components by sodium dodecyl sulphate. *Biochim. Biophys. Acta* **356**, 36–52.

Kleinschmidt, A. K. (1968). "Methods in Enzymology" (L. Grossman and K. Moldave, eds.), Vol. 12B, p. 361. Academic Press, New York.

Kleinschmidt, A. K., and Zahn, R. K. (1959). Über Desoxyribonuclein-säure-Molekeln in Protein-Mischfilmen. *Naturforsch.* **14b**, 770–779.

Knoll, G. (1995). Time-resolved analysis of rapid events *In* "Rapid Freezing, Freeze Fracture and Deep Etching" (N. J. Severs and D. M. Shotton, eds.), pp. 105–126. Wiley-Liss, New York.

Margaritis, L. H., Elgsaeter, A., and Branton, D. (1977). Rotary replication for freeze-etching. *J. Cell Biol.* **72**, 47–56.

Maunsbach, A. B., Skriver, E., and Jørgensen, P. L. (1979). Ultrastructure of purified Na,K-ATPase membranes. *In* "Na,K-ATPase: Structure and Kinetics" (J. C. Skou and J. G. Nørby, eds.), pp. 3–13, Academic Press, London.

Moor, H., and Mühlethaler, K. (1963). Fine structure in frozen-etched yeast cells. *J. Cell Biol.* **17**, 609–628.

Müller, M., Meister, N., and Moor, H. (1980). Freezing in a propane jet and its application in freeze-fracturing. *Mikroskopie (Wien)* **36**, 129–140.

Niedermeyer, W., and Wilke, H. (1982). Quantitative analysis of intramembrane particle (IMP) distribution of biomembranes after freeze-fracture preparation by a computer-based technique. *J. Microsc.* **126**, 259–273.

Pinto da Silva, P., Parkison, C., and Dwyer, N. (1981). Freeze-fracture cytochemistry: Thin sections of cells and tissues after labeling of fracture faces. *J. Histochem. Cytochem.* **29**, 917–928.

Pinto da Silva, P., Torrisi, M. R., and Kachar, B. (1981). Freeze-fracture cytochemistry: Localization of wheat-germ agglutinin and concanavalin A binding sites on freeze-fractured pancreatic cells. *J. Cell Biol.* **91**, 361–372.

Rasch, J. E., and Hudson, C. S. (eds.) (1979). "Freeze Fracture: Methods, Artifacts and Interpretations." Raven Press, New York.

Robards, A. W., and Sleytr, U. B. (1985). Low temperature methods in biological electron microscopy. *In* "Practical Methods in electron Microscopy" (A. M. Glauert, ed.), Vol. 10. Elsevier, Amsterdam.

Severs, N. J., and Shotton, D. M. (eds.) (1995). "Rapid Freezing, Freeze Fracture and Deep Etching." Wiley-Liss, New York.

Shotton, D. M. (1998). Freeze fracture and freeze etching. *In* "Cell Biology: A Laboratory Handbook" (J. E. Celis, ed.), 2nd Ed., vol. 3, pp. 310–322. Academic Press, San Diego.

Shotton, D. M., and Severs, N. J. (1995). An introduction to freeze fracture and deep etching. *In* "Rapid Freezing, Freeze Fracture and Deep Etching" (N. J. Severs and D. M. Shotton, eds.), pp. 1–30. Wiley-Liss, New York.

Sjöstrand, F. S. (1979). The interpretation of pictures of freeze-fractured biological material. *J. Ultrastruct. Res.* **69**, 378–420.

Sjöstrand, F. S., and Kreman, M. (1979). Freeze-fracture analysis of structure of plasma membrane of photoreceptor cell outer segments. *J. Ultrastruct. Res.* **66**, 254–275.

Skriver, E., Maunsbach, A. B., and Jørgensen, P. L. (1980). Ultrastructure of Na,K-transport vesicles reconstituted with purified renal Na,K-ATPase. *J. Cell Biol.* **86**, 746–754.

Sleytr, U. B., and Robards, A. W. (1977). Freeze fracturing: A review of methods and results. *J. Microsc. (Oxford)* **111**, 77–100.

Sommerville, J., and Scheer, U. (eds.) (1987). "Electron Microscopy in Molecular Biology: A Practical Approach." IRL Press Limited, Oxford.

Steinbrecht, R. A., and Zierold, K. (eds.) (1987). "Cryotechniques in Biological Electron Microscopy." Springer-Verlag, Berlin.

Ting-Beall, H. P., Burgess, F. M., and Robertson, J. D. (1986). Particles and pits matched in native membranes. *J. Microsc.* **142**, 311–316.

Verkleij, A. J., and Leunissen, J. L. M. (eds.) (1989). "Immuno-Gold Labeling in Cell Biology." CRC Press, Boca Raton, FL.

Verkleij, A. J., and Ververgaert, P. H. J. T. (1978). Freeze-fracture morphology of biological membranes. *Biochim. Biophys. Acta* **515**, 303–327.

Vollenweider, H. J., Sogo, J. M., and Koller, Th. (1975). A routine method for protein-free spreading of double- and single-stranded nucleic acid molecules. *Proc. Natl. Acad. Sci. USA* **72**, 83–87.

Zeidel, M. L., Nielsen, S., Smith, B. L., Ambudkar, S. V., Maunsbach, A. B., and Agre, P. (1994). Ultrastructure, pharmacologic inhibition and transport selectivity of aquaporin channel-forming integral protein in proteoliposomes. *Biochemistry* **33**, 1606–1651.

SAMPLING AND QUANTITATION

An ultrathin section through a biological tissue reveals an enormous amount of structural information when imaged in the transmission electron microscope. The way of sampling the area to be analyzed in detail is obviously of importance for the final outcome of the study, and hence for the validity of the conclusions. In fact, the sampling procedure is crucial in any electron microscope study.

The range of sampling methods varies from simple strategies such as the search for a certain biological object (e.g., a virus particle, a centriole, or a protein crystal) to the systematic sampling of a tissue for a representative evaluation of its condition (e.g., the prevalence of apoptosis, the occurrence of basement membrane changes, lysosomal abnormalities). Other factors to be considered in the sampling procedure include decisions regarding the number of experimental animals, sections to be prepared, and electron micrographs to be recorded. Special strategies will have to be evoked when, for example, analyzing subcellular fractions isolated by centrifugation.

In many electron microscope studies the observations are qualitative and involve a characterization of the structural features of molecules, organelles, cells, and tissues. In other investigations the dimensions of normal or pathological objects are of interest. Some of these studies therefore require a precise knowledge of the magnification of the micrographs, as well as knowledge of artifactual distortions of the specimen. Measurements of distances, thicknesses, or areas may be performed either on printed electron micrographs or on digitally recorded images in the electron microscope in conjunction with computer image processing systems. Morphometry aims at a quantification of structures in a two-dimensional plane, e.g., the plane of the section, whereas stereology aims at three-dimensional quantitative estimates of structures on the basis of two-dimensional observations.

For many years, biological electron microscopy was based on qualitative evaluations of specimens. A few investigators, however, notably in the laboratories of Francis O. Schmitt and Fritiof Sjöstrand, already at the end of the 1940s and early 1950s performed careful measurements of, for example, periodicities of myelin and collagen and thicknesses of cellular membranes. A more general attention to problems of sampling and quantitation came only later. A turning point was the symposium "Quantitative Electron Microscopy," which was arranged in 1964 by Gunther Bahr and Elmar Zeitler. At that occasion electron microscopy experts from different fields of physics and life sciences met and presented their respective approaches to problems of sampling and quantitation. From thereon the attention to quantitative aspects in electron microscopy increased steadily and resulted in several major contributions in stereology, such as those by Ewald Weibel (1980) and Hans Jørgen Gundersen *et al.* (1988).

Bahr, G. F., and Zeitler, E. H. (eds.) (1965). *Lab Invest.* **14,** 729–1340.
Gundersen, H. J., *et al.* (1988). *APMIS* **96,** 379–394; 857–881.
Weibel, E. R. (1980). "Stereological Methods," Vols. I and II. Academic Press, New York.

1. Calibration of Magnification

FIGURE 18.1A Electron micrograph of a 1000 mesh specimen grid (= 1000 openings per inch or 25.4 mm). The magnification as given by the electron microscope was 200 times and the negative was enlarged 3 times, i.e., nominally to × 600.

The magnification as calculated from the average of the grid periodicity is 583 times.

FIGURE 18.1B Grating replica (Agar Scientific, Stansted, England) with a stated periodicity of 51,000 lines per inch (= period 0.498 μm) recorded in the microscope at the nominal magnification 12,500 times and photographically enlarged 3 times, i.e., totally 37,500 times.

Based on the measured average spacing the magnification of the micrograph would be 34,100 times.

FIGURES 18.1C and 18.1D Electron micrograph of latex particles (Ernst Fulham Inc., Schenectady, New York) with a stated diameter of 0.5 μm (Fig. 18.1C) and 0.312 μm (Fig. 18.1D), both recorded at a nominal value of 8000 times and enlarged 3 times, i.e., 24,000.

The magnification as calculated on the basis of the average size of the latex particles in the figures (12 and 7 mm, respectively) and their stated diameter is 24,000 (Fig. 18.1C) and 22,400 times (Fig. 18.1D).

FIGURE 18.1E Catalase crystal negatively stained with phosphotungstic acid. The photographic negative was recorded at a magnification of 26,000 times and enlarged photographically 5.7 times, thus totally 148,000 times.

Assuming that the periodicity of the catalase crystal is 8.6 nm (Wrigley, 1968) the magnification of this micrograph based on measurements of the average spacing is calculated to be 143,000 times.

COMMENTS

Measurements of ultrastructural objects require knowledge of the magnification in the microscope and the photographic enlargement. The nominal magnification of the microscope is displayed and is usually also printed on the plate. For many purposes the displayed magnification is sufficiently accurate. However, in many microscopes the displayed magnification may have an error of 5–10% or even more. More accurate measurements require a check of the actual magnification, such as those illustrated here. At low and medium magnifications, fine-mesh grids of grating replicas are the most suitable standards; at high magnifications the spacings of collagen fibers (periodicity 66 nm), tobacco mosaic virus, or catalase crystals are useful. The magnification of a micrograph is also modified if the height of the specimen differs from that in its normal position, e.g., if the specimen grid is slightly concave or convex or the specimen holder deviates from its proper position. The periodicities of the standards may also be modified, e.g., by shrinkage within the electron beam or by plastic deformation of the replicas.

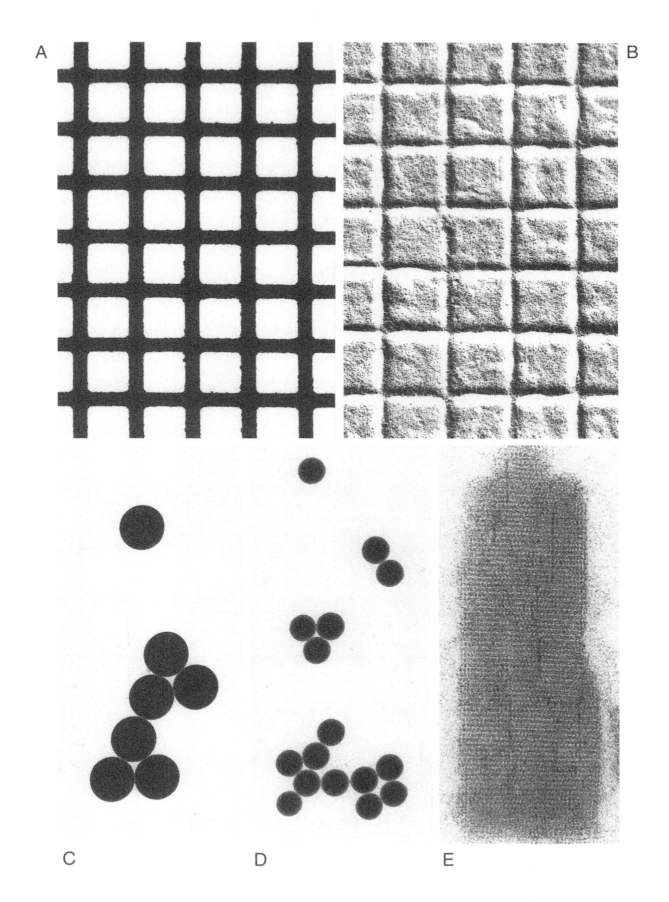

A

B

C D E

2. Sampling and Object Variability

FIGURE 18.2A Part of two skeletal muscle cells from a rat leg. The tissue was fixed in glutaraldehyde, postfixed in osmium tetroxide, and embedded in Epon. Section staining with uranyl acetate and lead citrate. × 10,000.

Mitochondria are few in the left cell but are frequent and large in the right cell. In the left cell the Z band in the middle of the light I band is thin and stained weakly whereas the Z band in the right cell is more prominent.

FIGURE 18.2B Cross-sectioned renal tubules from a perfusion-fixed rat kidney. × 3000.

A comparison between the proximal tubules labeled S1 and S2 shows a generally similar structure of the epithelium, although microvilli of the brush border are longer in S1 than in S2 tubules. Furthermore, the electron-dense lysosomes in the S2 tubules are more prominent than those in S1 epithelium. The distal nephron segment (D) is completely different from both S1 and S2 tubules.

COMMENTS

These two examples illustrate the necessity to consider the normal structural variability within a tissue of seemingly identical cells. Thus, skeletal muscle contains at least two morphologically distinct cell types: type II fibers (left cell in Fig. 18.2A) and type I fibers (right cell in Fig. 18.2A). In the kidney tubule the fine structure of the epithelium of the proximal tubule changes gradually from the glomerulus to the end of the proximal tubule, but due to the winding course of the tubule, different segments may be located immediately adjacent to each other in a thin section as illustrated here.

In quantitative studies in particular it matters a great deal if the measurements are performed on type I or type II muscle fibers or on proximal tubule segments S1 or S2. Also, in qualitative studies it is important to recognize the existence of different cell types in a tissue. An implicit conclusion is that survey electron micrographs should be recorded in studies of such complex tissues. With a limited field of view and a high magnification there is even a risk to confuse the cytoplasmic components, e.g., mitochondria, from a cell in the distal nephron (D in Fig. 18.2B) with those in proximal tubules (S1 or S2).

A

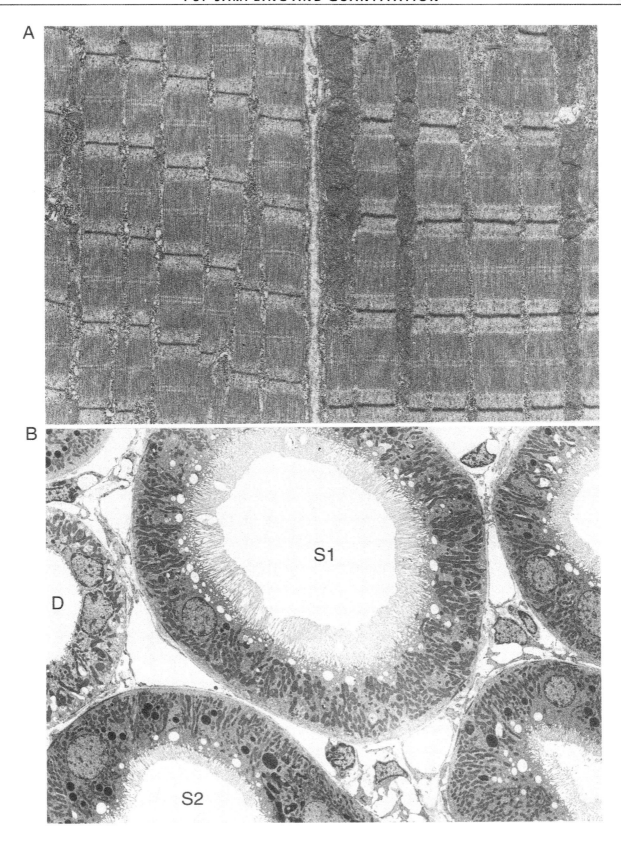

B

3. Sampling of Pellets: Differential Centrifugation

FIGURES 18.3A–18.3C Three electron micrographs taken at different levels in a pellet isolated by differential centrifugation from rat kidney cortex according to Maunsbach (1966). Cortical tissue was homogenized in 0.3 M sucrose and the nuclear fraction was removed by low-speed centrifugation. The supernatant was sedimented by centrifugation at 10,000 $\times$ g in an angle rotor. The resulting pellet had a brown bottom layer, a thick yellow middle layer, and a light yellow/white upper layer. It was fixed with glutaraldehyde and osmium tetroxide, embedded in Epon, and sectioned at each of the three levels.

FIGURE 18.3A Upper layer of pellet. $\times$ 18,000.

This layer contains round or elongated membrane profiles, sometimes originating from a central structure (*).

FIGURE 18.3B Middle layer of pellet. $\times$ 10,000.

This part of the pellet contains almost exclusively mitochondria of different shapes and diameters.

FIGURE 18.3C Bottom layer of pellet. $\times$ 10,000.

Here there are predominantly lysosomes but also some mitochondria and small vesicular structures. The boxed area contains only lysosomes.

COMMENTS

Subcellular fractions obtained by differential centrifugations are invariably layered. The bottom layer may be distinctly different from the upper or middle layers. In order to analyze the contents of such a pellet by electron microscopy it is necessary to cut sections at right angle to the surface of the pellet and to sample the pellet at different levels. A random section of the pellet may be very misleading with respect to the contents of the pellet. To obtain a representative cross section of a pellet it is advantageous that the pellet is thin, i.e., it should not be centrifuged from a solution with a very high protein content. Pellets suitable for electron microscopy can be obtained from suspensions of subcellular particles containing as little as 25–100 μg protein, provided that centrifugation is performed using a small centrifuge tube. In order to evaluate the representativeness of the content of a particular layer in a pellet it is necessary to record the section at sufficiently low magnification. If only a few cell organelles are recorded, the representativeness may be quite misleading. For example, in Fig. 18.3C it is possible to pick out a biased, small area (boxed) showing lysosomes exclusively, despite the fact that many mitochondria are also present in this layer of the pellet.

A

B

C

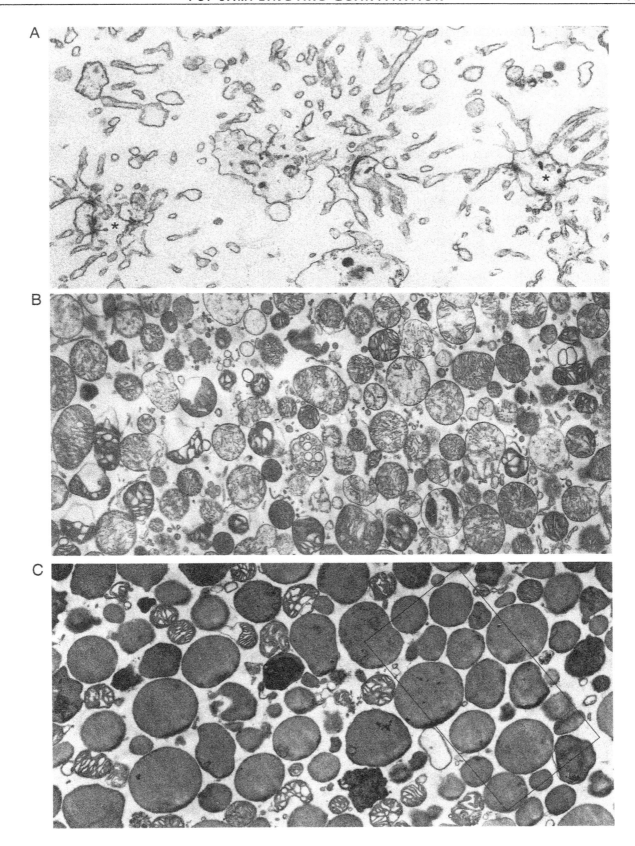

4. Sampling of Pellets: Gradient Centrifugation

FIGURES 18.4A–18.4C Micrographs of fractions obtained from a sucrose gradient used to purify a lysosomal pellet similar to that shown in Fig. 18.3C. A semipurified lysosomal fraction as in Fig. 18.3C was resuspended and purified on a continuous sucrose gradient (Maunsbach, 1966). Two bands formed in the sucrose gradient, an upper band and a lower band. The bands were diluted with sucrose, sedimented by centrifugation, and fixed and prepared for electron microscopy.

FIGURE 18.4A Content of the upper band of sucrose gradient. × 24,000.

This layer in the gradient contains mitochondria exclusively.

FIGURE 18.4B Lower band in sucrose gradient. × 10,000.

This fraction shows lysosomes exclusively.

FIGURE 18.4C Complete cross section of lysosomal fraction purified as in Fig. 18.4B. In this case the suspended lysosomes were sedimented onto the surface of a flat disc that was mounted in a swinging bucket rotor. The section was oriented at a right angle to the flat pellet and extends from its bottom to its upper surface. × 7000.

The pellet contains lysosomes almost exclusively.

COMMENTS

Pellets obtained from bands in density gradients usually show very uniform contents as illustrated in Figs. 18.4A and 18.4B. As is the case after differential centrifugations, such pellets may be thick and the contents difficult to evaluate quantitatively by electron microscopy. The organelles should then be sedimented onto a flat surface and the resulting pellet sectioned in such a way that both the bottom and the top of the pellet are included in the section, as in Fig. 18.4C. Such a pellet permits a morphometric analysis of the contents of the fraction.

A

B

C

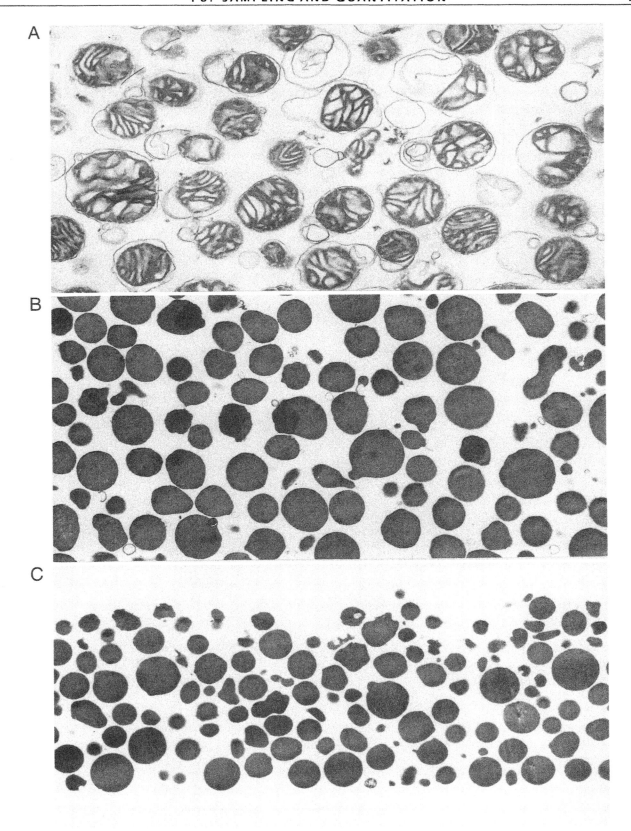

5. Micrograph Montages

FIGURE 18.5A Photograph of a montage of 18 electron micrographs recorded from one large section from a rat kidney cortex, which was perfusion fixed with 1% glutaraldehyde and postfixed with osmium tetroxide, embedded, and the thin section finally recorded in the electron microscope at an original magnification of 1500 times. The three times enlarged micrographs were cut carefully to avoid overlapping and then mounted two and two on nine pieces of cardboards, which were finally mounted together to form an area of 150 × 100 cm. This corresponds to 500 × 325 μm on the thin section, which is reproduced here at a magnification of about 300 times. This photograph was taken in February 1998 but the montage was originally made in 1964.

The photograph shows Karen Thomsen (left) and Else-Merete Løcke (right) in front of the montage, which is hanging in the laboratory. The separate micrographs form a continuous image, but the borders between separate micrographs can be discerned faintly in many places.

FIGURE 18.5B A montage of micrographs from a cross sectioned kidney tubule that was microperfused *in vitro* and then prepared for electron microscopy. Each micrograph is recorded at a magnification of 2600, enlarged three times, and the enlargements then mounted together with transparent tape and finally attached to a board. The vertical diameter of the tubule in the montage is 90 cm.

The photograph shows a typical montage used for overall evaluation or various stereological measurements. In such a routine preparation differences between micrographs with respect to contrast and density or even sectioning scratches are acceptable and do not appreciably disturb the analysis.

FIGURE 18.5C A montage of micrographs from cells in a cross-sectioned tissue culture that was fixed and processed for electron microscopy. Enlargement of micrographs and photographic reproductions are the same as in Fig. 18.5B.

By combining the micrographs the distribution of the pronounced structural differences between adjacent cells in the culture become very apparent.

COMMENTS

When analyzing complex tissues it is usually helpful to study low magnification micrograph or montages of micrographs covering large areas. In Fig. 18.5A, none of the individual micrographs is unusual in itself, but when all 18 micrographs are mounted together they give more complete information about the general architecture of the tissue. In fact, this particular montage on the wall in the laboratory has been used repeatedly as an inspiration for thoughts and speculations about tissue components and their fine structure, interrelationships, and functional significance.

The magnification in Figs. 18.5B and 18.5C (original 2600 times, final 7800 times) is sufficiently high to reveal considerable cellular information. Montages such as these are helpful by providing an overview of the tissue as a basis for more firm conclusions. Other types of sampling, such as recording every third field in the *x* and *y* direction, may be necessary in stereological studies, but may not provide the same general qualitative understanding of tissue fine structure as continuous montages.

A

B

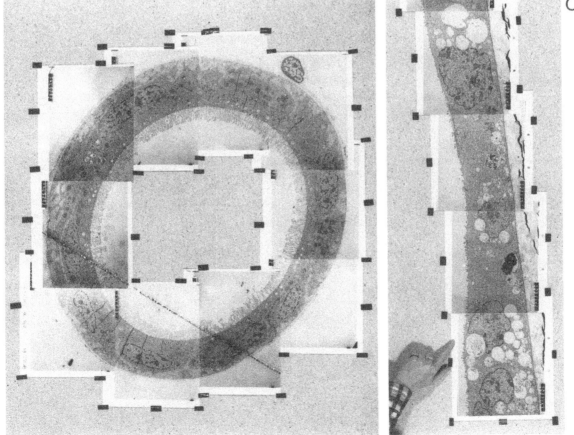

C

6. Automated Digital Montages

FIGURE 18.6A Gallery of six individual but overlapping electron micrographs from liver tissue that was fixed and prepared for electron microscopy. The micrographs were recorded with a Proscan CCD camera in a Zeiss 912 electron microscope using an automatic digital montage function of the AnalySIS program (Soft Imaging System, Münster, Germany). After deciding the parameters of the recordings (numbers of micrographs, magnification, amount of overlap, and exposure conditions), all six recordings were made automatically without manual interference. The recordings were printed on a Kodak 8650 PS printer. × 5000.

Adjacent micrographs show overlaps corresponding to approximately one-fifth of the micrograph width.

FIGURE 18.6B Electron micrograph composed of the six micrographs in Fig. 18.6A. The composition was made automatically with the aid of the AnalySIS program and the micrograph was printed with the same printer as in Fig. 18.6A. × 6000.

The micrograph shows liver cell cytoplasm without evidence of borders between the six individual micrographs. Cellular components recorded on two or more micrographs are completely continuous.

COMMENTS

Each micrograph in Fig. 18.6A was recorded with 1024 × 1024 pixels. After reduction for the overlap, the micrograph as illustrated contains 1848 × 2672 pixels. The definition and resolution in digital micrographs are limited by the number of pixels. Thus a 1024 × 1024 micrograph will invariably show less resolution than a 1848 × 2672 micrograph. The several electron micrographs shown in Fig. 18.5A were recorded individually on plates from a section and joined to form a photographic montage covering a large area of the object. The same result can be obtained by recording a series of fields digitally and then fusing these images digitally into a single micrograph. This method offers many possibilities with respect to the number of areas and magnifications and constancy of the exposure and overlap. It also provides a composite micrograph with a larger pixel content than the original individual recordings.

A

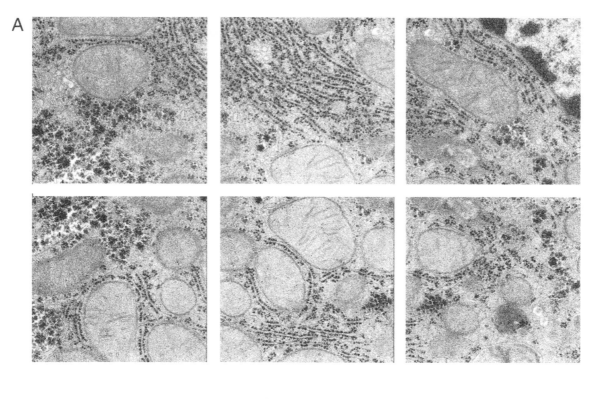

B

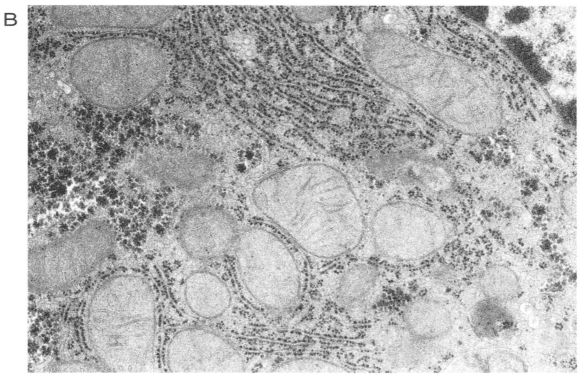

7. Resolution of Digital Montages

FIGURE 18.7A A single digital recording of the same area that was covered by the montage on the previous plate (Fig. 18.6B). The width of the micrograph corresponds to 1024 pixels. × 6000.

At this magnification the micrograph appears quite similar to that in Fig. 18.6B.

FIGURE 18.7B A higher magnification of part of the micrograph in Fig. 18.7A. The inset shows part of the right margin of the micrograph enlarged photographically to an even higher magnification. × 12,000 and 36,000, respectively.

The micrograph is composed of quadratic pixels that are particularly evident in the inset.

FIGURE 18.7C Corresponding area to that in Fig. 18.7B but enlarged from the composite in the previous plate (Fig. 18.6B). The inset shows the right edge of the micrograph enlarged photographically. × 12,000 and 36,000, respectively.

In these micrographs, pixels are also observable, although they are smaller and less evident than in Fig. 18.7B.

COMMENTS

Micrographs that are composed of several digitally recorded images contain more pixels than digitally recorded single micrographs of the same area. The difference is not obvious when the micrographs are inspected visually. However, it becomes evident if the images are enlarged electronically (compare Figs. 18.7B and 18.7C). The difference becomes even more evident if corresponding areas are enlarged photographically. A practical consequence of using digital montages is that large areas of an object can be recorded at higher magnifications (in the individual micrographs) than in a single recording of the same area. Therefore the resolution in digital montages is much higher than in single digital recordings of the same area.

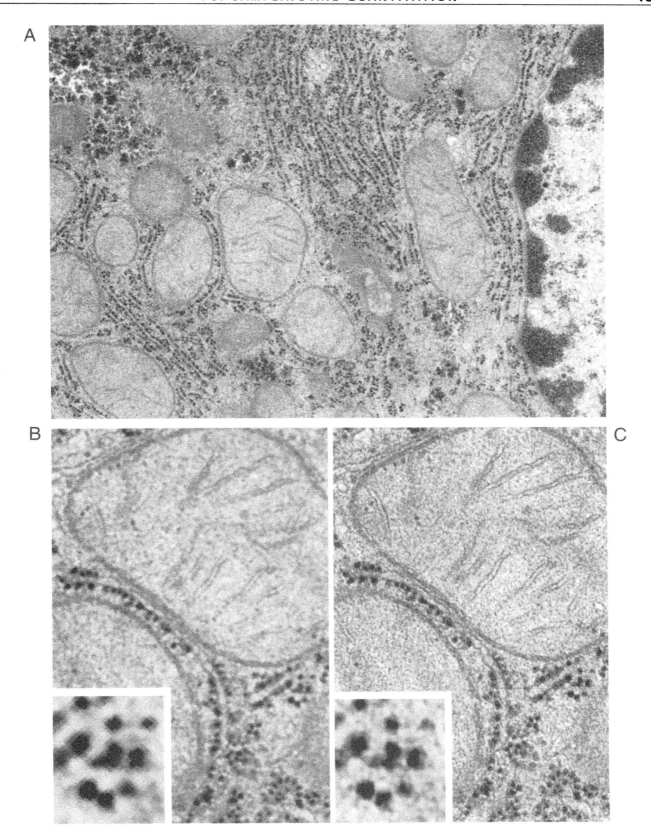

8. Measurements on Digital Images

FIGURE 18.8A Capillary wall of a rat kidney perfusion fixed with glutaraldehyde and tannic acid in phosphate buffer. Following Epon embedding the section was double stained with uranyl acetate and lead citrate. The micrograph was recorded with a Proscan CCD camera in a Zeiss 912 electron microscope and printed with a Kodak 8650 PS printer. × 90,000.

Normal glomerular capillary wall showing the basement membrane between epithelial foot processes (top) and endothelial cells (below). Arrows point at the slit membranes between the epithelial foot processes.

FIGURE 18.8B Distances between adjacent foot processes were measured by marking the endpoints for the measurements on the computer screen using the mouse of the AnalySIS system. The distances between the points were then automatically recorded.

The micrograph shows the marked points and interconnecting lines that are measured and printed automatically.

FIGURE 18.8C Each profile (numbered 1–6) of the epithelial foot processes was identified using a threshold and binarized, while the basement membrane and endothelial profiles were eliminated.

The software program automatically determined various parameters of the profiles, their areas, heights, circumferences, and shape factors and printed these parameters with the chosen statistics, including means and standard deviations.

COMMENTS

Digital images can be analyzed with respect to structural parameters using any of several software programs. The main problem is to define the lines and the areas to be measured while the actual calculations are instantaneous. However, areas may be difficult to define if they differ only slightly in contrast from the surroundings and the periphery difficult to define unequivocally. It is important to calibrate the system with respect to microscope, CCD camera, and measuring system. Points to be measured may also be difficult to locate precisely (Fig. 18.8B). Thus a certain amount of subjective judgement is usually unavoidable in these types of measurements and some measurements may therefore be time-consuming. For this reason the material to be analyzed may in some cases be smaller in number than when the same parameters are determined by stereological methods, which in well-designed projects can be based on large statistical samples. The direct measurements of parameters, illustrated here, may therefore be both more time-consuming and statistically inferior to simple stereological point or intersection counting.

A

B

C

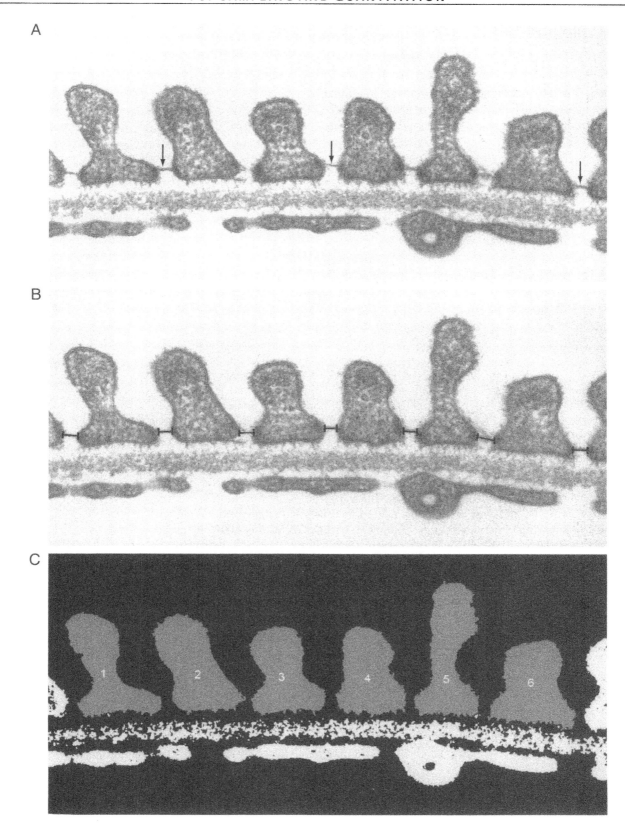

9. Stereological Grids

FIGURE 18.9A A test grid for counting of points and intersections overlying a section of liver cell cytoplasm. The maximum number of points is 54, with subsets containing 12 and 6 points. Lines are suitable for counting intersections of frequent membranes provided the section has a random orientation.

Frequent points are suitable for counting rare objects such as peroxisomes (lower right) or Golgi regions whereas the least frequent subset is suitable for nuclei. The subset of 12 points (ends of the lines) can be used for mitochondria.

FIGURE 18.9B A counting frame for unbiased counting (Gundersen, 1977) overlying a section of a liver cell. The counting frame consists of a solid line extending above and below to infinity and a dashed line. $\times$ 25,000.

Two-dimensional objects are counted if they fall fully inside the frame or cut the dashed "acceptance" line, provided they do not at the same time cut the solid "forbidden" line. Objects cut by the forbidden line are not counted.

COMMENTS

Several new stereological methods have been developed for the analysis of cells and tissues in microscopic sections (Gundersen *et al.*, 1988a,b; Howard and Reed, 1998). These methods allow unbiased stereological determinations of volumes, areas, and numbers of structures in cells and tissues. It is a basic requirement in these methods that the sampling is uniform and random in order to exclude systematic bias.

Figure 18.9B illustrates unbiased counting with a sampling frame. A crucial aspect in stereological studies is to design a sampling scheme that is not only unbiased but also optimizes sampling efficiency. This is illustrated in Fig. 18.9A where the volume of rare objects can be estimated with the most frequent points, whereas large structures are estimated with the least frequent points. The same principle can be used for estimating surface areas using systematically distributed short lines as in Fig. 20.9B instead of a dense network of lines. When initiating a stereological study, it is time well spent to perform a pilot study and to evaluate the method carefully with respect to possible systematic bias and efficiency.

A

B

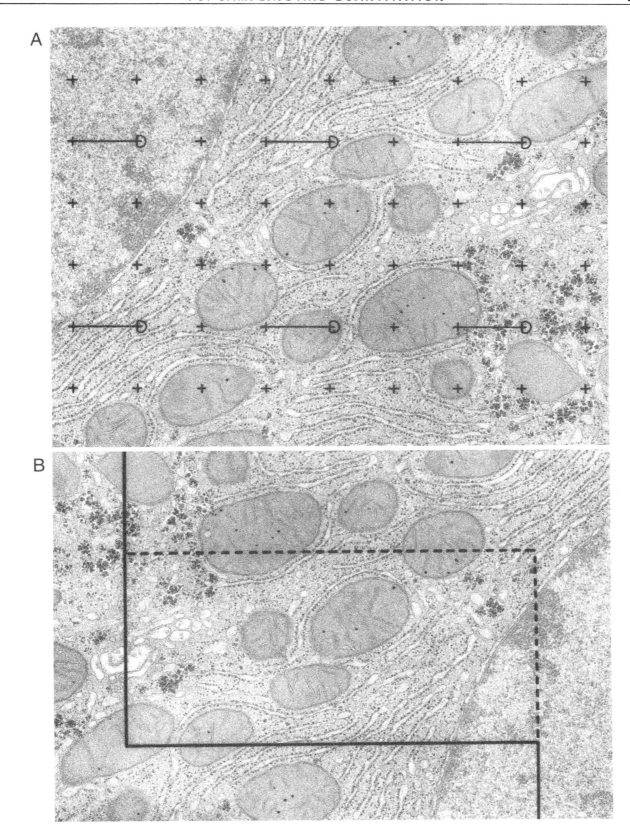

10. Cycloid Test System

FIGURE 18.10A A survey electron micrograph illustrating the sampling from a so-called vertical section (Baddeley *et al.*, 1986) through the renal medulla of a rabbit kidney (Mathiasen *et al.*, 1991). The section is oriented parallel to the vertical axis, which in turn is parallel to the axis of the tubular structures. The rotation of the section is random around the vertical axis. Using a rotary stage in the microscope the section was then aligned with the vertical direction parallel to the long side of the photographic frame. Micrographs were sampled for every fourth field on the fluorescence screen and in every second row. The long edge of the frame corresponds to 10 μm.

Although 15 areas are sampled in the section, only 6 contain parts of thick ascending limb epithelium (3 in the second row, 2 in the third row, and 1 in the fifth row). Five frames contain no cellular structures at all.

FIGURE 18.10B Sampled field in a vertical section similar to that in Fig. 18.10A with a superimposed cycloid test system. The long side of the micrograph is parallel with the vertical axis of the section. Oblique lines are not used for counting but facilitate the counting of intersections and points. × 16,000.

Intersections between basolateral membranes and cycloid arcs are counted with every second row of cycloid arcs, whereas the basal lamina is counted with all arcs. The volume of the reference space is obtained by counting the encircled test points, which hit the tubule cytoplasm.

COMMENTS

The cycloid test system makes it possible to obtain unbiased estimates of anisotropic membrane areas. If sections are cut perpendicular to the axis of a distal tubule, estimates of the membrane areas using other test systems will be underestimated and biased, as the basolateral membranes are arranged in sheaths that radiate toward the basal lamina and are oriented perpendicular to the tubular axis. Only if the cell surface is isotropic, as in some proximal tubules, is it possible to obtain a realistic value of the surface areas in sections oriented at a right angle to the tubule axis.

A

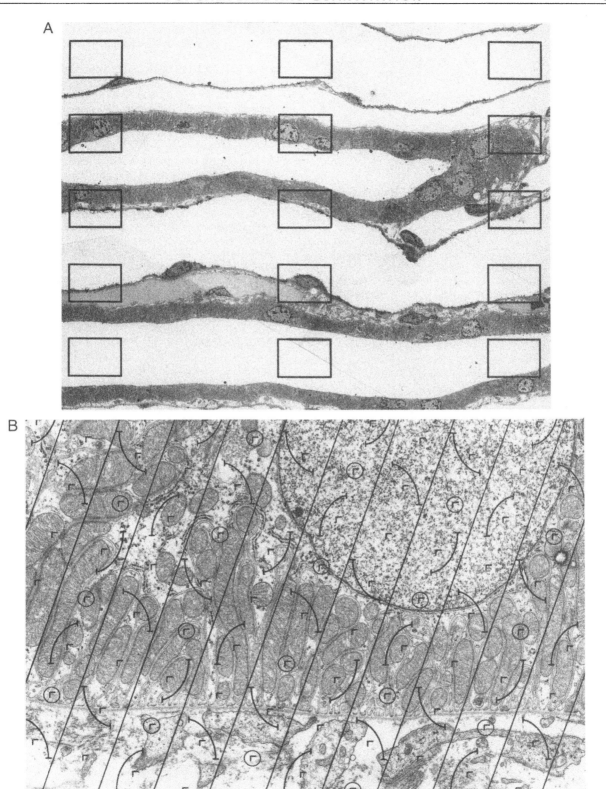

B

References

Agar, A. W., and Keown, S. R. (1978). Measurements from micrographs by optical diffraction. *Proc. Roy. Microsc. Soc. (London)* **13**, 147–158.

Backus, R. C., and Williams, R. C. (1948). Some uses of uniform sized spherical particles. *J. Appl. Phys.* **19**, 1186–1187.

Baddeley, A. J., Gundersen, H. J. G., and Cruz-Orive, L. M. (1986). Estimation of surface area from vertical sections. *J. Microsc. (Oxford)* **142**, 259–276.

Bahr, G. F., and Zeitler, E. H. (eds.) (1965). Quantitative electron microscopy. *Lab. Invest.* **14**, 729–1340.

Bohman, S.-O., Deguchi, N., Gundersen, H. J. G., Hestbech, J., Maunsbach, A. B., and Olsen, S. (1979). Evaluation of a procedure for systematic semiquantitative analysis of glomerular ultrastructure in human renal biopsies. *Lab. Invest.* **40**, 433–444.

Cermola, M., and Schreil, W.-H. (1987). Size changes of polystyrene latex particles in the electron microscope under controlled physical conditions. *J. Electr. Microsc. Techn.* **5**, 171–179.

Gundersen, H. J. G. (1977). Notes on the estimation of the numerical density of arbitrary profiles: The edge effect. *J. Microsc. (Oxford)* **111**, 219–223.

Gundersen, H. J. G., Bagger, P., Bendtsen, T. F., Evans, S. M., Korbo, L., Marcussen, N., Møller, A., Nielsen, K., Nyengaard, J. R., Pakkenberg, B., Sørensen, F. B., Vesterby, A., and West, M. J. (1988a). The new stereological tools: Disector, fractionator, nucleator and point sampled intercepts and their use in pathological research and diagnosis. *APMIS* **96**, 857–881.

Gundersen, H. J. G., Bendtsen, T. F., Korbo, L., Marcussen, N. Møller, A., Nielsen, K., Nyengaard, J. R., Pakkenberg, B., Sørensen, F. B., Vesterby, A., and West, M. J. (1988b). Some new, simple and efficient stereological methods and their use in pathological research and diagnosis. *APMIS* **96**, 379–394.

Gundersen, H. J. G., and Jensen, E. B. (1987). The efficiency of systematic sampling in stereology and its prediction. *J. Microsc. (Oxford)* **147**, 229–263.

Gundersen, H. J. G., and Østerby, R. (1981). Optimizing sampling efficiency of stereological studies in biology or 'Do more less well!' *J. Microsc. (Oxford)* **121**, 65–73.

Hennig, A., and Elias, H. (1963). Theoretical and experimental investigations on sections of rotary ellipsoids. *Z. Wissensch. Mikr. Mikrosk. Techn.* **65**, 133–145.

Howard, C. V. (1990). Stereological techniques in biological electron microscopy. *In* "Biophysical Electron Microscopy: Basic Concepts and Modern Techniques" (P. W. Hawkes and U. Valdrè, eds.), pp. 479–508. Academic Press, London.

Howard, C. V., and Reed, M. G. (eds.) (1998). "Unbiased Stereology: Three-Dimensional Measurement in Microscopy." BIOS Scientific Publishers, Oxford.

Karamanta, D. (1971). Polystyrene spheres in electron microscopy. *J. Ultrastr. Res.* **35**, 201–209.

Lickfeld, K. G., Menge, B., Wunderli, H., van den Broek, J., and Kellenberger, E. (1977). The interpretation and quantification of sliced intracellular bacteriophages and phage-related particles. *J. Ultrastr. Res.* **60**, 148–168.

Madsen, K. M., and Tisher, C. C. (1986). Structural-functional relationships along the distal nephron. *Am. J. Physiol.* **250**, F1–F15.

Mathiasen, F. Ø., Gundersen, H. J. G., Maunsbach, A. B., and Skriver, E. (1991). Surface areas of basolateral membranes in renal distal tubules estimated by vertical sections. *J. Microsc.* **164**, 247–261.

Maunsbach, A. B., and Boulpaep, E. L. (1984). Quantitative ultrastructure and functional correlates in proximal tubule of *Ambystoma* and *Necturus. Am. J. Physiol.* **246**, F710–F724.

Maunsbach, A. B., and Christensen, E. I. (1992). Functional ultrastructure of the proximal tubule. *In* "Handbook of Physiology, Section 8, Renal Physiology," Vol. 1, (E. E. Windhager, ed.), pp. 41–107. Oxford University Press, Oxford.

Østerby, R., and Gundersen, H. J. G. (1980). Fast accumulation of basement membrane material and the rate of morphological changes in acute experimental diabetic glomerular hypertrophy. *Diabetologia* **18**, 493–501.

Pagtalunan, M. E., Rasch, R., Rennke, H. G., and Meyer, T. W. (1995). Morphometric analysis of effects of angiotensin II on glomerular structure in rats. *Am. J. Physiol.* **268**, F82–F88.

Rasch, R., Jensen, B. L., Nyengaard, J. R., and Skøtt, O. (1998). Quantitative changes in rat renin secretory granules after acute and chronic stimulation of the renin system. *Cell Tissue Res.* **292**, 563–571.

Shay, J. (1975). Economy of effort in electron microscope morphometry. *Am. J. Pathol.* **81**, 503–512.

Sterio, D. C. (1984). The unbiased estimation of number and sizes of arbitrary particles using the disector. *J. Microsc. (Oxford)* **134**, 127–136.

Weibel, E. R. (1963). Principles and methods for the morphometric study of the lung and other organs. *Lab. Invest.* **12**, 131–155.

Weibel, E. R. (1980). "Stereological Methods," Vol. 1. Academic Press, New York.

Williams, M. A. (1977). Quantitative Methods in Biology. *In* "Practical Methods in Electron Microscopy," Vol. 6, (A. M. Glauert, ed.). North-Holland Publishing Co., Amsterdam.

Wrigley, N. G. (1968). The lattice spacing of crystalline catalase as an internal standard of length in electron microscopy. *J. Ultrastruct. Res.* **24**, 454–464.

IMAGE PROCESSING

The explosive development of computer technology has dramatically extended the possibilities to process and analyze electron micrographs. Whether recorded photographically and thereafter digitized or recorded by digital cameras or imaging plates, the micrographs can be further processed or analyzed by computer to reveal otherwise hidden information. Several possibilities exist: multiple images from the same type of object, e.g., a molecule, can be collected and averaged to increase ultrastructural information; the structural parameters of repeating elements in a two-dimensional crystal can be determined; and multiple image of the same object recorded at different tilt angles can be used to calculate the three-dimensional structures. Additionally, the gray scale of the electron micrograph image can be modified with respect to contrast and brightness and finally printed in black and white or pseudocolor without the involvement of photographic techniques. It is, alas, also possible to manipulate the image, e.g., by removing components or modifying or adding structural features. Such manipulations may range from simple removal of a scratch in a micrograph all the way down to the level of scientific fraud. The application of computer-based image processing techniques requires a certain measure of familiarity with the technology, but user-friendly programs are common. Furthermore digitized images can be archived systematically for future analysis and, not the least, they can be transmitted electronically to laboratories or journals in other parts of the world. This chapter illustrates some basic procedures in image processing and how they can be used on different biological objects.

Electron micrographs often contain hidden information not readily seen by visual inspection. Various strategies have been invented to retrieve such information. Thus, Roy Markham, Simon Frey, and Graham J. Hills (1963) described two "methods for the enhancement of image detail and accentuation of structure in electron microscopy." Their micrographs displayed symmetries that were enhanced by a photographic superposition technique. They had been inspired by the paper by Francis Galton, who in 1878 published a "compositive portrait" of several criminals, a study performed in a search for the characteristic facial traits of a villain. To Galton's surprise this "average criminal" was better looking than any of its ingredient portraits. The paper by Markham and colleagues in its turn inspired Aaron Klug and J. E. Berger (1964) to use optical diffractometers for the analysis of periodicities or the computer equivalent to diffractometers. Using such studies, image resolution improved and macromolecular shapes become accessible to study. Aaron Klug received the 1982 Nobel prize in chemistry "for his development of crystallographic electron microscopy and for his structural elucidation of biologically important nucleic acid–protein complexes."

Galton, F. (1878). Nature 18, 97–100.
Klug, A., and Berger, J. E. (1964). J. Mol. Biol. 12, 565–569.
Markham, R., Frey, S., and Hills, G. J. (1963). Virology 20, 88–102.

1. Digital Contrast Changes

FIGURE 19.1A Part of a liver cell following standard preparation (glutaraldehyde and osmium tetroxide fixation, Epon embedding, and section staining with uranyl acetate and lead citrate) recorded on an imaging plate at an original magnification of × 24,000. The imaging plate was processed in a Fujifilm FDL 5000 IP recorder and electronically enlarged and printed at a final magnification of × 70,000.

The cellular components are shown with good contrast and have sharp outlines.

FIGURES 19.1B–19.1D These three images derive from the same imaging plate recording as that used for Fig. 19.1A and have been electronically modified in the software program of a computer connected to the imaging plate reader. × 70,000.

In Fig. 19.1B the black-and-white scale is reversed, in Fig. 19.1C the contrast is enhanced relative to that in Fig. 19.1A, and in Fig. 19.1D the contrast is decreased.

COMMENTS

These four images illustrate that changes in contrast and density can be imposed on digitized images. The contrast can be modified over a much wider range than when using photographic material and indeed the black-and-white scale can be reversed (Fig. 19.1B). Although the images are made up of pixels, the low electronic enlargements (about three times here) from the imaging plate do not lead to noticeable decreases in image sharpness. For most purposes the image is fully compatible with a corresponding image recorded on photographic film. Digitized images obtained from imaging plates, or from scanned micrographs recorded on film, can be stored electronically for later use.

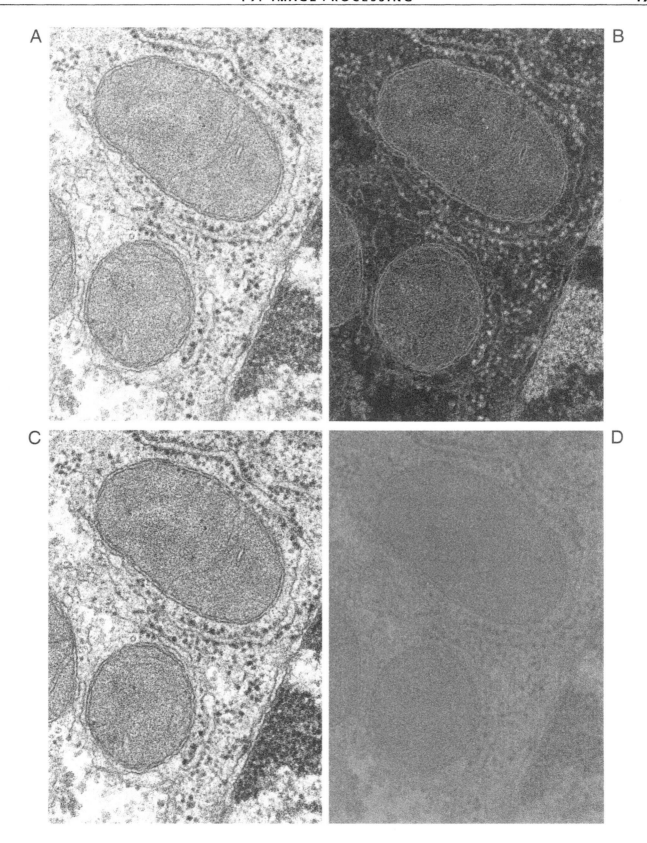

2. Processing of Scanned Image

FIGURE 19.2A An electron micrograph of liver cytoplasm after conventional fixation, embedding, and staining. The image was recorded on Agfa Scientia EM film. × 40,000.

The micrograph shows a sharp and contrasty image of cell structures.

FIGURE 19.2B The electron micrograph in Fig. 19.2A was scanned at 400 dpi with a UMAX VISTA-S8 scanner and then printed with the aid of a Kodak 8650PS color printer.

There is no distinct difference in crispness between Figs. 19.2A and 19.2B.

FIGURE 19.2C The black-and-white scale of the scanned image in Fig. 19.2B was digitally reversed and then printed as described previously.

This image shows reversed contrast of all cell components.

FIGURE 19.2D Same scanned image as in Fig. 19.2B and printed in the same way except that it was digitally reduced before printing. × 30,000.

Except for the magnification, this image is similar to that in Fig. 19.2B.

COMMENTS

Electron microscope images can be recorded in digital form either directly in the electron microscope by means of CCD cameras (see Chapter 11) or scanned from photographic negatives or enlargements. In both cases the image can be processed in a number of ways, e.g., printed with altered contrast or at a different magnification, as illustrated in Figs. 19.2C and 19.2D. Digitally scanned and printed images resemble closely the original photographic image as illustrated by Figs. 19.2A and 19.2B. Images scanned from electron micrograph negatives or enlargements can therefore be used for image processing in the same way as images digitally recorded already in the electron microscope. For many purposes, scanning and digital printing is an efficient way to modify or improve micrographs for publication.

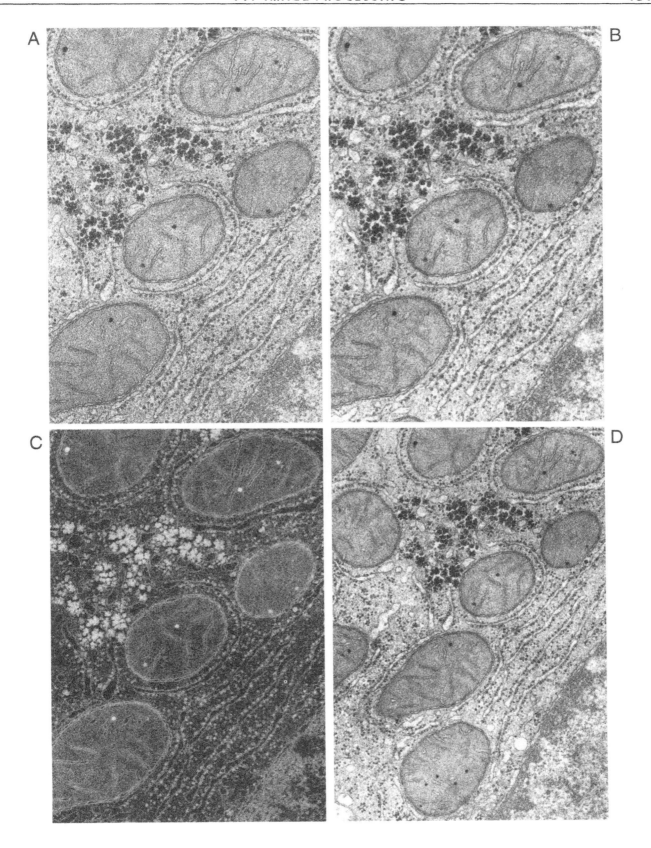

3. Translational Image Enforcement

FIGURE 19.3A An electron micrograph of closely packed latex particles with a diameter of 0.26 μm on a Formvar film. $\times$ 46,000.

The particles form a monolayer and are arranged in closely packed hexagonal arrays.

FIGURES 19.3B–19.3D Photographic images obtained from the same negative as in Fig. 19.3A through six consecutive exposures of the negative. In Fig. 19.3B the photographic paper was shifted a distance corresponding to one particle diameter between each exposure and in a direction parallel to one of the three directions of particle arrays. In Fig. 19.3C the shift corresponded to half the diameter of the particle and in Fig. 19.3D to one-third of the particle diameter. $\times$ 46,000.

The particles in Fig. 19.3B appear as particles in Fig. 19.3A, although slightly smaller. Each particle image represents the average of six latex particles. In Figs. 19.3C and 19.3D each particle image corresponds in shape to one-half and one-third of an original particle, respectively.

COMMENTS

The translational technique by Markham and colleagues (1963) was introduced to increase the signal-to-noise ratio in micrographs of periodic objects, such as collagen fibers, bacterial walls, or rod-shaped virus particles. It corresponds to the technique of optical diffraction, although it makes use of images in real space rather than in inverse space. Minute details, which are barely or not at all discerned in the original micrograph, will appear more distinct if the correct periodic shift distance is used in the darkroom. There is a risk, however, that false details will appear and these may be due to either a wrong shift distance (such as the fraction of the true translational period; compare those used for Figs. 19.3C and 19.3D) or an accidental irregularity, which influences the pattern in multiple units, which would be six in Fig. 19.3B. It is hence recommended that the longest possible periodic shifts that give the expected pattern should be used and that several negatives be analyzed to check for the reproducibility of the results.

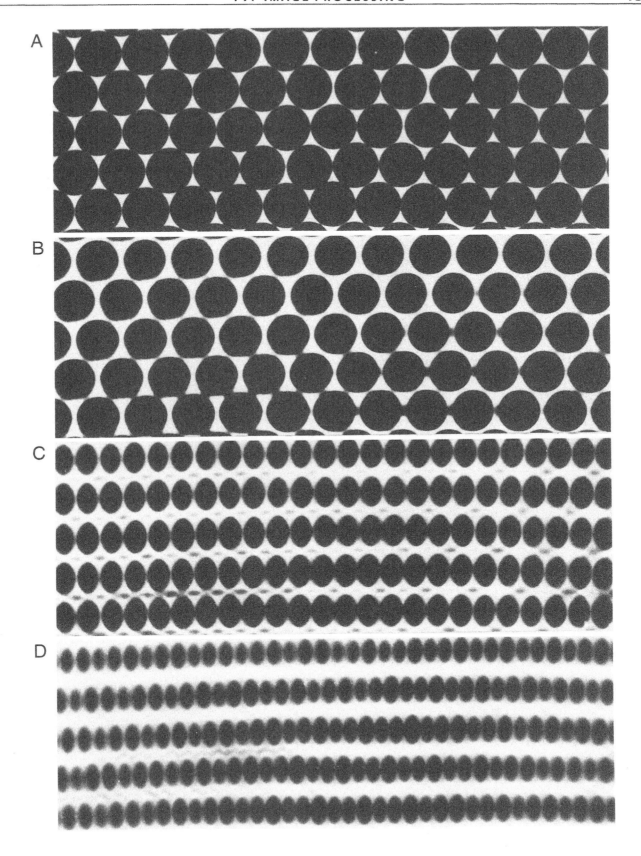

4. Averaging of Macromolecular Assemblies

FIGURE 19.4A An electron micrograph of purified Na,K-ATPase in a membrane fragment where the protein has aggregated to form linear arrays. The enzyme was incubated in 1 mM NH$_4$VO$_3$ and 3 mM MgCl$_2$ in imidazole–HCl buffer (Skriver *et al.*, 1981; Söderholm *et al.*, 1988). × 430,000.

The protein units are arranged in linear arrays of paired particles. The ribbons vary in lengths and are separated by stain-filled grooves. In places the ribbons run parallel, although they are not straight.

FIGURES 19.4B AND 19.4C Figure 19.4B shows Na,K-ATPase membrane similar to that in Fig. 19.4A with one ribbon outlined. In Fig. 19.4C the boxed ribbon is reconstructed by correlation averaging methods (Saxton and Baumeister, 1982; Hegerl and Altbauer, 1982). Figure 19.4B × 430,000, Fig. 19.4C × 3,000,000.

The plot in Fig. 19.4C illustrates an average calculated from 10 significant unit cells in the ribbon in Fig. 19.4B.

FIGURES 19.4D–19.4F Two-dimensional crystals induced as in Fig. 19.4A with sodium monovanadate in purified membranes of Na,K-ATPase and negatively stained with uranyl acetate. Well-ordered crystalline arrays were densitometered at 20-μm intervals and projection maps were calculated using the Fourier transform amplitudes and phases (Hebert *et al.*, 1982). Magnification of electron micrographs (a) × 520,000. In the diffraction patterns (b), 1 mm corresponds to 1.8 × 10^{-3} Å^{-1}. In the reconstructed images (c), 1 mm corresponds to 5.4 Å; a and b are the axes in the unit cell, γ the angle between the axes, and d_A and d_C are the diagonals in the unit cells

Three different crystal forms are illustrated. Each crystal form (D, E, F) is illustrated with an electron micrograph (a), the corresponding diffraction pattern (b), and the computer-reconstructed image from the same crystal (c). In the reconstructed images (c), the protein-rich regions (positive regions) are drawn with unbroken contour lines whereas negative stain regions have dashed lines.

COMMENTS

Two-dimensional crystals of Na,K-ATPase vary with respect to the type of crystallization and unit cell dimensions. The earliest stage in the formation of the crystal consists of the formation of linear arrays of protein units as illustrated in Fig. 19.4A. Figures 19.4B and 19.4C demonstrate that even short arrays can be used to reconstruct the projected shapes of the units and their interrelationships. When the crystals are fully developed, the protein units form confluent areas (Figs. 19.4Da–19.4Fa) that show unit cells of different dimensions and conformations, which depend on several factors, including the composition of the medium used for crystallization and duration of storage after induction of crystallization (Maunsbach *et al.*, 1991). The resolution in computer-averaged images of two-dimensional crystals depends primarily on the degree of order in the crystals, but is also limited by the negative stain itself. To improve resolution large and well-ordered crystals may be embedded in glucose (Unwin and Henderson, 1975) or ideally analyzed frozen hydrated in the cryoelectron microscope when resolutions below 3 Å are obtainable (Henderson, 1995).

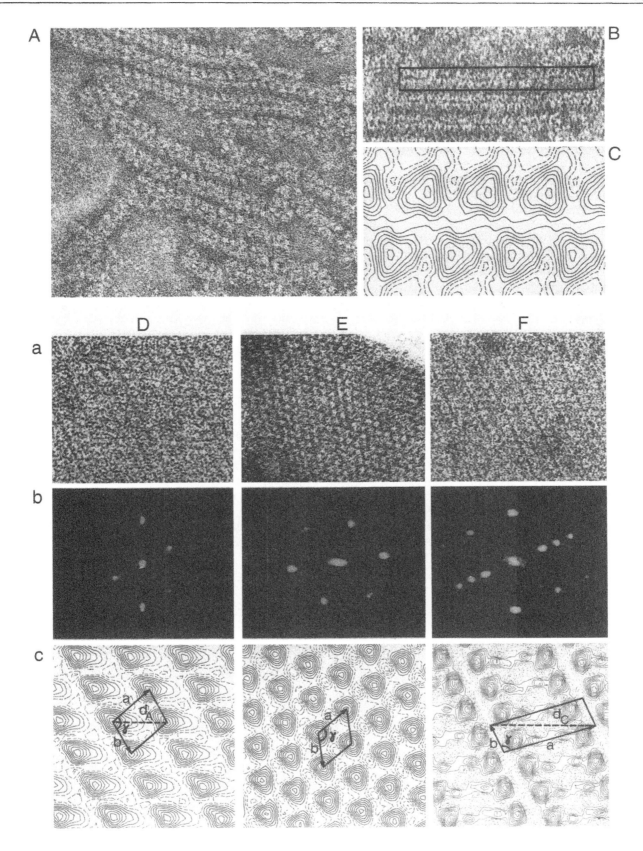

5. Rotational Image Enforcement

FIGURE 19.5A An electron micrograph of seven closely packed latex particles with a diameter of 0.26 μm and located on a supporting Formvar film. $\times$ 92,000.

The latex particles appear completely opaque due to their relatively large size.

FIGURES 19.5B–19.5F Patterns made from Fig. 19.5A with the rotational enforcement technique according to Markham *et al.* (1963). In each of Figs. 19.5B–19.5F the photographic paper was rotated around the center of the projected picture of Fig. 19.5A. In Fig. 19.5B the angle was 72° (360°/5) and five consecutive multiple exposures were made on the same photographic paper. In Fig. 19.5C the angle between each exposure was 1/6 of 360°, in Fig. 19.5D 1/7, in Fig. 19.5E 1/12, and in Fig. 19.5F 1/18 of 360°. $\times$ 92,000.

Figure 19.5C closely resembles the nonrotated original micrograph in Fig. 19.5A, whereas Figs. 19.5B and 19.5D do not show any evidence of the seven closely packed latex particles. Figures 19.5E and 19.5F show a repeat pattern somewhat similar to that of Fig. 19.5A, although the number of repeat units, in the circle is two or three times that of the original. Each unit has a noncircular shape. The white spot in the center of Figs. 19.5B–19.5F is the hole of the needle that impaled the photographic papers and around which the photographic papers were rotated.

COMMENTS

The Markham rotational technique is often used to enhance the contrast of significant details in structures with rotational symmetry, e.g., cilia, virus particles, or cross-cut microtubules. In the artificial example illustrated here, there is no uncertainty about the sixfold symmetry seen in Fig. 19.5A, and a sixfold multiple exposure as expected gives a pattern identical to the original one. If the pattern in Fig. 19.5A is rotated $n - 1$ or $n + 1$, the original symmetry is lost.

When the rotation is an even multipel of n, there is again a symmetrical pattern, but the individual "subunits" appear in increased numbers and have changed geometries.

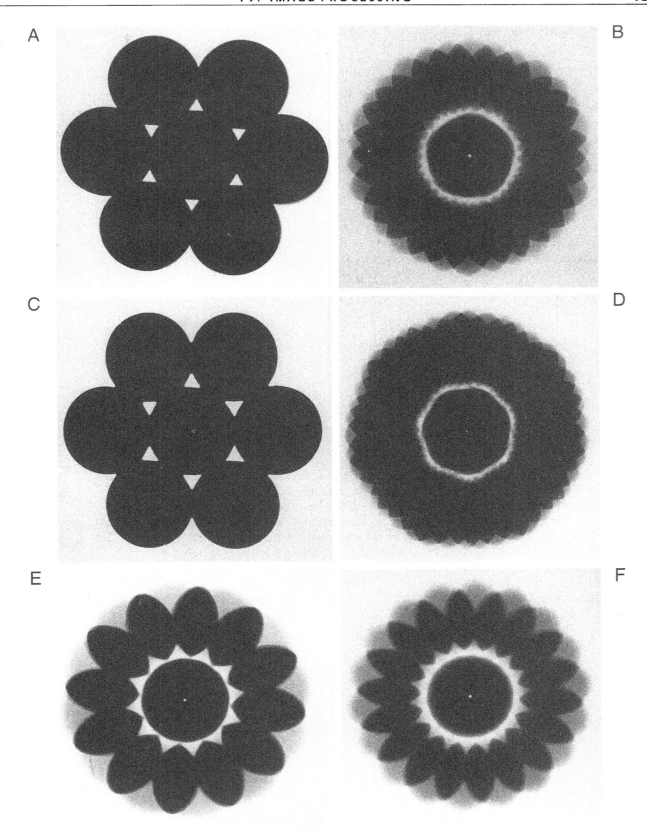

6. Photographic versus Computer Averaging

FIGURE 19.6A Cross section of a sperm axoneme from the stick insect *Baculum* sp. fixed with a mixture of glutaraldehyde and tannic acid, block stained with uranyl acetate, and embedded in Epon before sectioning, section staining, and recording. Electron optical magnification: $\times$ 50,000. Final magnification: $\times$ 600,000.

About half the axoneme is visible with five of its nine accessory microtubules and four of its nine microtubular doublets.

FIGURE 19.6B The sperm axoneme seen after application of Markham's photographic rotational technique (compare with Fig. 19.5). Nine exposures were thus made with a 40° rotation around the center between each exposure.

The microtubular doublets and the accessory tubules had such a fixed location relative to the center that equivalent units fall on each other on rotation.

FIGURE 19.6C A similar axoneme where all structures with a ninefold rotation symmetry have been superimposed by computer methods (see Afzelius *et al.*, 1990; Lanzavechia *et al.*, 1991).

Image details are well defined, including the individual protofilaments of the microtubular doublets, and so are the club-like radiating spokes and the material between the accessory tubules.

COMMENTS

A prerequisite for the photographic Markham rotational technique is that the individual elements have a strictly uniform radial arrangement around the center. In this example the rotational technique provides enhanced definition, particularly of the structures between the microtubules. In addition, the computer-based analysis makes it possible to reveal the detailed arrangements of the protofilaments within the microtubules. This is achieved by applying different types of computer correction programs for minor deviations from the ninefold symmetry of the axoneme. A further advantage of the computer-aided technique is that the images from a number of axonemes can be superimposed to provide an even more detailed image. Such structural elements that have no fixed positions relative to the central axis will not become visible, e.g., the eight or seven globules inside the four accessory tubules.

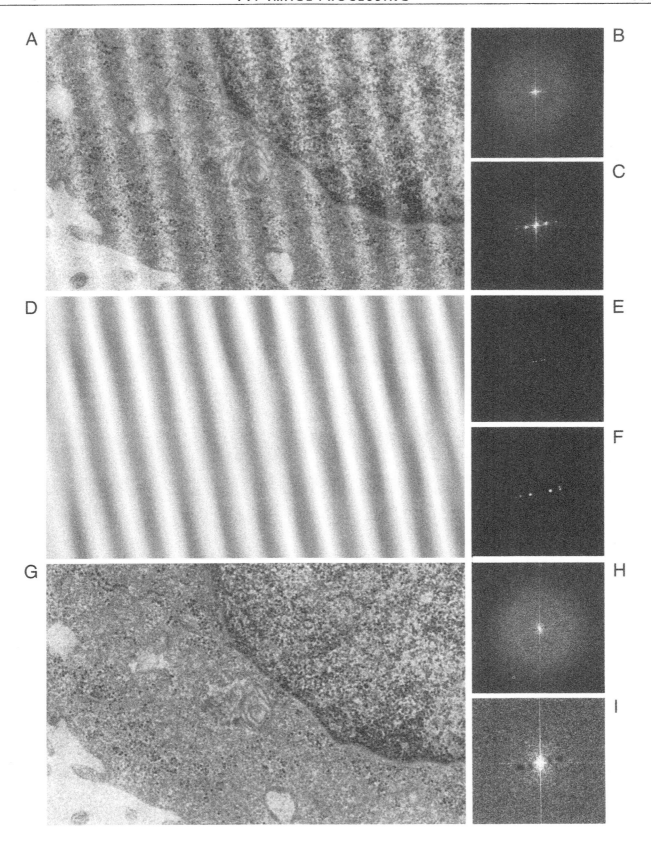

8. Removal of Image Defects

FIGURES 19.8A AND 19.8B A section of cells in a freeze-substituted tissue slice. The micrograph was recorded digitally with a CCD camera in an electron microscope. Figure 19.8B is the same micrograph as in Fig. 19.8A after digital removal of irregular densities using the AnalySIS program. Each linear scratch density was eliminated digitally as were the scattered dense particles. Small areas of adjacent cytoplasm were then copied into the empty holes. × 6000.

The original section shows vertical dark lines from knife scratches as well as contamination in the form of dense particles of different dimensions. The section in Fig. 19.8B is almost free from section scratches and contaminating irregular particles.

FIGURE 19.8C Digital image of cell cytoplasm containing lysosomes. × 15,000.

Arrows point at three knife scratches, each with a dark and a light component.

FIGURE 19.8D Figure 19.8C after image processing. The diffractogram of Fig. 19.8C contained elongated spots at a right angle to the scratch lines. Figure 19.8D is obtained after removal of these spots. × 6000.

The cell cytoplasm now is essentially free of sectioning defects.

FIGURES 19.8E AND 19.8F A proximal tubule cell after freeze-substitution and immunolabeling of actin. Using the AnalySIS program the density of the dark part of the image in Fig. 19.8E is reduced. × 15,000.

The image in Fig. 19.8E varies in density; the image is lighter above the dashed line than below. After processsing the micrograph (Fig. 19.8F) has a uniform density.

COMMENTS

Digitally recorded electron micrographs can be modified in a number of ways with several existing image processing software packages. The modified images in Figs. 19.8B, 19.8D, and 19.8F are examples of how nonbiological contamination, knife scratches, and variations in image density can be selectively removed or modified. The illustrated technical faults may not distort the biological information in the micrograph but are unaesthetic. As long as the modifications do not modify the biological contents of the images and only concern the purely technical aspects of the image they can be regarded as acceptable. However, there is obviously also a gray zone where the investigator must be very careful not to interfere with the biological contents of the images.

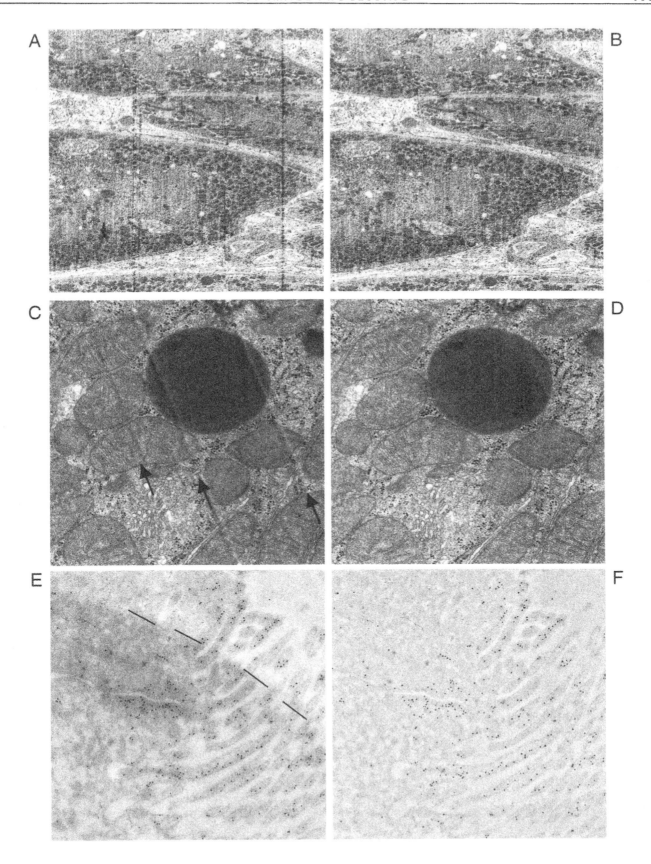

9. Scientific Fraud: Removal of Objects

FIGURE 19.9A Part of liver cell cytoplasm following conventional glutaraldehyde/osmium tetroxide double fixation and staining. The micrograph was recorded in digitized form in a Zeiss 912 Omega electron microscope with a CCD camera and printed with a Kodak 8650PS printer. × 30,000.

The micrograph shows normal liver cytoplasm with mitochondria and endoplasmic reticulum.

FIGURE 19.9B The same recording after image processing using the SIS image program. One mitochondrion has been removed and a correspondingly large cytoplasmic area copied into the empty space. × 30,000.

The replacement of the lower mitochondrion by cytoplasm containing rough surfaced endoplasmic reticulum in 19.9B is difficult to detect unless the original micrograph (Fig. 19.9A) is also shown.

FIGURE 19.9C Part of a lysosome-enriched pellet isolated from a rat kidney cortex, fixed, stained, and recorded in an electron microscope. The enlargement of the negative was scanned with an UMAX VISTA-S8 scanner and printed with a Kodak 8650PS printer. × 15,000.

The micrograph shows lysosomes but also many small cell fragments.

FIGURE 19.9D AND 19.9E The digital image in Fig. 19.9C was processed using the Adobe PhotoShop image processing program. All structural components except typical lysosomes were removed to give Fig. 19.9D. This new image was further processed in such a way that the lysosomes were moved in different directions in order to obtain a more even distribution of the lysosomes. × 15,000.

Figure 19.9D shows exclusively lysosomes against an empty background. The additionally manipulated image in Fig. 19.9E could be taken for "a highly purified fraction of lysosomes with a uniform distribution in the pellet."

COMMENTS

The image processing performed in Figs. 19.9B, 19.9E, and 19.9F represents striking examples of what must be called scientific fraud. Application of present-day image processing programs on digitized electron micrographs, whether they are derived from digital cameras or from scanned conventional prints make possible a number of modifications of the original image. In addition to modifications of micrograph contrast, density, brightness, irregularities, and application of different colors, these programs have made it possible to change the scientific content of micrographs. The manipulations shown here are almost impossible to detect. In fact, Fig. 19.9E may falsely be claimed to represent an unusually highly purified lysosomal fraction.

These new possibilities of manipulating microscope images raise serious problems, particularly for reviewers and editors of journals and books, should someone attempt to publish a falsified image and draw scientific conclusions from it. This serious question should be considered in the scientific community, including committees for scientific misconduct.

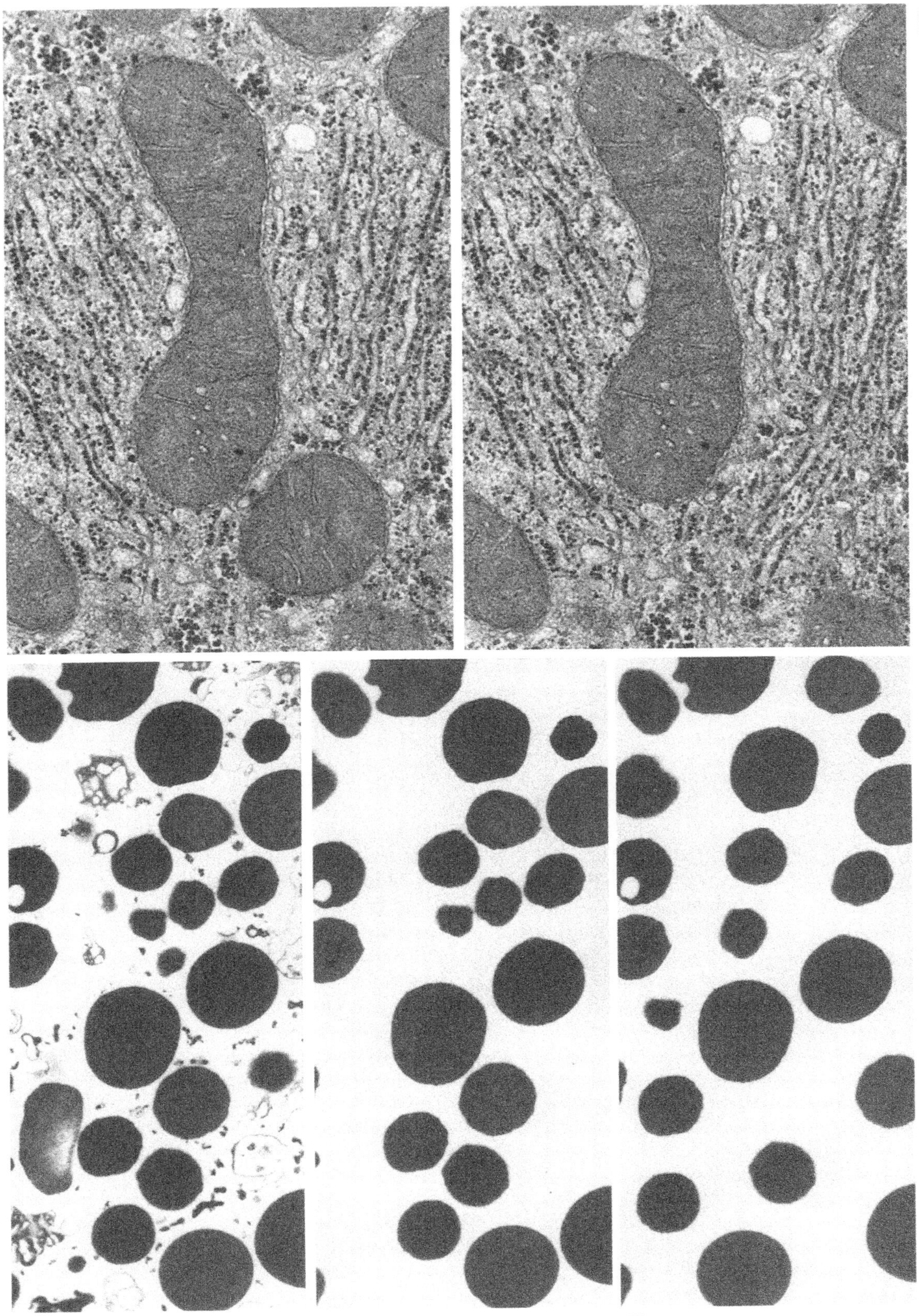
A
B
C
D
E

10. Scientific Fraud: Manipulation of Labeling

FIGURE 19.10A An electron microscope image of epithelial cell from the descending thin limb of the loop of Henle in a human renal biopsy. The tissue was fixed with 4% formaldehyde and freeze-substituted into Lowicryl HM20, and the ultrathin section was then labeled with rabbit anti-aquaporin-1 antibodies, which were detected with anti-rabbit IgG conjugated to 10-nm colloidal gold. The print of the negative was scanned with an UMAX VISTA-S8 scanner and the image printed with a Kodak 8650PS printer. × 60,000.

Both the luminal (upper) and the basal (lower) cell membranes of this cell are labeled heavily for aquaporin-1. There is only a negligible background labeling in the cytoplasm or outside the cell. The small dense particles in the mitochondria to the left represent an unspecific stain precipitate rather than colloidal gold particles.

FIGURE 19.10B The same digital image as in Fig. 19.10A modified with the Adobe PhotoShop image processing program. First all colloidal gold particles associated with the apical cell membrane were removed. A few background colloidal gold particles were also erased. In a second step, dense dots were placed over the basal cell membrane corresponding to the right half of the micrograph. The image was then printed with a Kodak 8650PS printer. × 6000.

After image processing, all apical cell membrane labeling and background labeling are absent. Instead the basal membrane appears to be more strongly labeled than in Fig. 19.10A.

COMMENTS

Image processing as applied in Fig. 19.10B is absolutely unacceptable. Attempt to publish such a modified micrograph would represent a serious example of scientific fraud as it completely distorts the scientific content of the original image in Fig. 19.10A. Readers of scientific literature must be aware that manipulations of this kind are technically easy to perform with equipment that is commonly available. Forged computer processed pictures of subcellular structures may appear—and have indeed already been published.

A

B

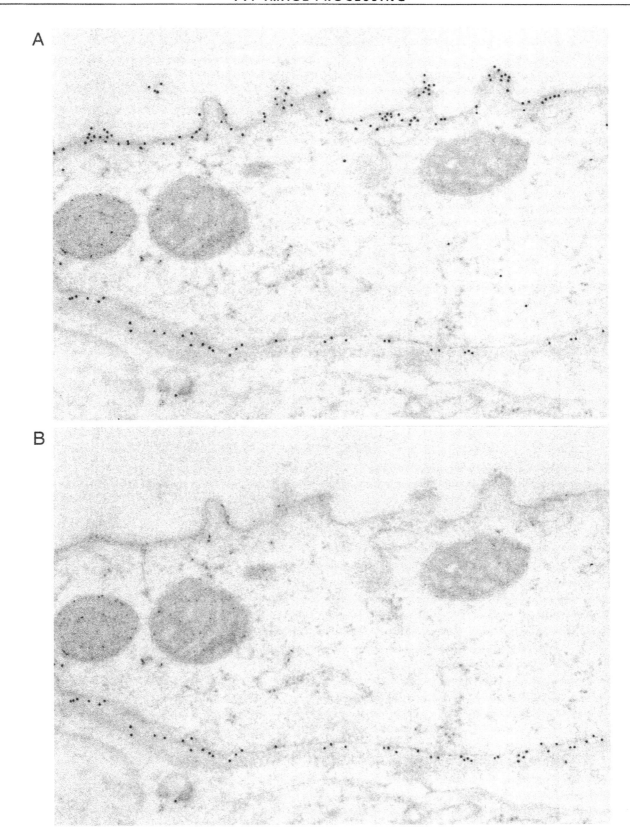

References

Aebi, U., Fowler, W. E., Buhle, E. L., and Smith, P. R. (1984). Electron microscopy and image processing applied to the study of protein structure and protein-protein interactions. *J. Ultrastruct. Res.* **88,** 143–176.

Afzelius, B. A., Bellon, P. L., and Lanzavecchia, S. (1990). Microtubules and their protofilaments in the flagellum of an insect spermatozoon. *J. Cell Sci.* **95,** 207–217, 1990.

Afzelius, B. A., Dallai, R., Lanzavecchia, S., and Bellon, P. L. (1995). Flagellar structure in normal human spermatozoa and in spermatozoa that lack dynein arms. *Tissue Cell* **27,** 241–247.

Bellon, P. L., and Lanzavecchia, S. (1992). Pattern reconstruction in ultrastructural morphology. *J. Microsc. (Oxford)* **168,** 33–45.

Carragher, B., and Smith, P. R. (1996). Advances in computational image processing for microscopy. *J. Struct. Biol.* **116,** 2–8.

Crowther, R. A., and Amos, L. A. (1971). Harmonic analysis of electron microscope images with rotational symmetry. *J. Mol. Biol.* **60,** 123–130.

Crowther, R. A., and Klug, A. (1975). Structural analysis of macromolecular assemblies by image reconstruction from electron micrographs. *Annu. Rev. Biochem.* **44,** 161–182.

Friedman, M. H. (1970). A reevaluation of the Markham rotation technique using model systems. *J. Ultrastruct. Res.* **32,** 226–236.

Gilev, V. P. (1979). A simple method of optical filtration. *Ultramicroscopy* **4,** 323–335.

Glaeser, R. M. and Downing, K. H. (1992). Assessment of resolution in biological electron microscopy. *Ultramicroscopy* **47,** 256–265.

Hebert, H., Jørgensen, P. L., Skriver, E., and Maunsbach, A. B. (1982). Crystallization patterns of membrane-bound $(Na^+ + K^+)$-ATPase. *Biochim. Biophys. Acta* **689,** 571–574.

Hebert, H., Skriver, E., and Maunsbach, A. B. (1985). Three-dimensional structure of renal Na,K-ATPase determined by electron microscopy of membrane crystals. *FEBS Lett.* **187,** 182–186.

Hegerl, R., and Altbauer, A. (1982). The "EM" program system. *Ultramicroscopy* **9,** 109–116.

Hegerl, R. (1992). A brief survey of software packages for image processing in biological electron microscopy. *Ultramicroscopy* **47,** 417–423.

Henderson, R. (1995). The potential and limitation of neutrons, electrons, and X-rays for atomic resolution microscopy of unstained biological molecules. *Rev. Biophys.* **28,** 171–193.

Klug, A. (1983). From macromolecules to biological assemblies: Nobel Lecture, 8 December 1982. *Biosci. Rep.* **3,** 395–430.

Klug, A., and DeRosier, D. J. (1966). Optical filtering of electron micrographs: Reconstruction of one-sided images. *Nature* **212,** 29–32.

Lanzavecchia, S., Bellon, P. L., and Afzelius, B. A. (1991). A strategy for the reconstruction of structures possessing axial symmetry: Sectioned axonemes in sperm flagella. *J. Microsc. (Oxford)* **164,** 1–11.

Markham, R., Frey, S., and Hills, G. J. (1963). Methods for the enhancement of image detail and accentuation of structure in electron microscopy. *Virology* **20,** 88–102.

Maunsbach, A. B., Skriver, E., and Hebert, H. (1991). Two-dimensional crystals and three-dimensional structure of Na,K-ATPase analyzed by electron microscopy. *In* "The Sodium Pump: Structure, Mechanism, and Regulation" (J. H. Kaplan and P. De Weer, eds.), pp. 159–172. Rockefeller Univ. Press, New York.

Ottensmeyer, F. P., Andrew, J. W., Bazett-Jones, D. P., Chan, A. S. K., and Hewitt, J. (1977). Signal to noise enhancement in dark field electron micrographs of vasopressin: Filtering of arrays of images in reciprocal space. *J. Microsc. (Oxford)* **109,** 259–268.

Russ, J. C. (1995). "The Image Processing Handbook." 2nd Ed. CRC Press, Boca Raton, FL.

Saxton, W. O., and Baumeister, W. (1982). The correlation averaging of a regularly arranged bacterial cell envelope protein. *J. Microsc.* **127,** 127–138.

Skriver, E., Maunsbach, A. B., and Jørgensen, P. L. (1981). Formation of two-dimensional crystals in pure membrane-bound Na^+,K^+-ATPase. *FEBS Lett.* **131,** 219–222.

Stewart, M. (1988). Introduction to the computer image processing of electron micrographs of two-dimensionally ordered biological structures. *J. Electr. Microsc. Techn.* **9,** 301–324.

Söderholm, M., Hebert, H., Skriver, E., and Maunsbach, A. B. (1988). Assembly of two-dimensional membrane crystals of Na,K-ATPase. *J. Ultrastruct. Mol. Struct. Res.* **99,** 234–243.

Unwin, P. N. T. (1974). Electron microscopy of the stacked disk form of tobacco mosaic virus. II. The influence of electron irradiation on stain distribution. *J. Mol. Biol.* **87,** 657–670.

Unwin, P. N. T., and Henderson, R. (1975). Molecular structure determination by electron microscopy of unstained crystalline specimens. *J. Mol. Biol.* **94,** 425–440.

THREE-DIMENSIONAL RECONSTRUCTIONS

Whereas a scanning electron micrograph gives immediate information on the topography of the investigated specimens, a transmission electron micrograph only provides a two-dimensional (*x* and *y* axes) image of the object. The image does not immediately contain information on the *z* axis. Because the third dimension is lacking, a three-dimensional (3D) object will appear as a two-dimensional area, a two-dimensional area projected parallel to the beam will appear as a one-dimensional line, and a one-dimensional line projected parallel to the beam will appear as a zero-dimensional dot.

The trained investigator will automatically reconstruct the lost dimension to micrographs of sectioned material, as the same type of structure, e.g., an organelle, may appear in numerous copies, although seen in different projections. Such intuitive interpretation is far from trivial, however, and often leads to erroneous conclusions. Information on the true third dimension can, in many instances, be obtained by recording the specimen at different tilt angles, by analysis of reconstructions based on serial sections, or by metal shadowing as in the freeze-fracturing technique (Chapter 17).

Recording of macromolecular objects at different angles provides the basis for computer-based reconstructions of their three-dimensional structure. The fact that such reconstructions have reached or approached the atomic level in a few ideal objects represents a major breakthrough in biological electron microscopy.

This chapter contains examples of different methods for obtaining 3D information about biological objects, ranging from cells to molecules.

The search for the third dimension in electron micrographs has been present almost from the very beginning of electron microscopy and the problem has been attacked in different ways: stereo recordings, serial sectioning, shadowing, high-voltage microscopy, and, last but not least, scanning electron microscopy. Thus already in 1942 Manne Siegbahn designed a tilting device that allowed stereo recordings with the electron microscope he had designed. Fritiof Sjöstrand (1958) was the first to demonstrate the usefulness of reconstructions based on ultrathin serial sections in his analyses of the synaptology of the retinal receptors. At the molecular level crystallographic methods proved their fundamental importance in studies on the 3D structure of the T4 phage tail by David DeRosier and Aaron Klug (1968) and of the bacterial purple membrane by Richard Henderson and Nigel Unwin (1975).

DeRosier, D. J., and Klug, A. (1968). *Nature* **217**, 130–134.
Henderson, R., and Unwin, P. N. T. (1975). *Nature* **257**, 28–32.
Siegbahn, M. (1942). *Nordisk Tidskrift Fotografi* **26**, 205–206.
Sjöstrand, F. S. (1958). *J. Ultrastruct. Res.* **2**, 122–170.

1. Comparison between Transmission and Scanning Electron Microscopy

FIGURE 20.1A A low magnification electron micrograph of cross-sectioned mucosa of the cecum from a germ-free rat. The tissue was perfusion fixed through the heart with 3% glutaraldehyde subsequent to a short rinse with Tyrode solution. Postfixation was in osmium tetroxide and embedding in Epon. Sections were cut at a right angle to the mucosa. × 600.

The epithelial cell lining of the mucosa shows an irregular course in this section. Two crypt-like openings are seen, but the three-dimensional appearance of the mucosal surface is not evident from this section.

FIGURE 20.1B A scanning electron micrograph of the cecal mucosa from a germ-free rat. The fixation procedure was the same as for Fig. 20.1A; the fixed tissue was then freeze-dried and coated in vacuum with gold and observed in a JEM scanning electron microscope at 10 kV. × 975.

This "aerial" view of the intestinal mucosa shows that the surface has round or elongated "hills" separated by "valleys" of variable depths and widths. At the center of each hill there is a depression, representing the opening of a crypt of Lieberkühn.

COMMENTS

These two micrographs illustrate different aspects of the same tissue as studied by transmission (TEM) and scanning electron microscopy (SEM). In this case the TEM micrograph shows the interior of the mucosa, including its cytoplasmic contents, and the general distribution of epithelial cells on the surface, but it does not give a clear impression of the topological arrangements of the crypts. The SEM picture, however, shows the three-dimensional topography similar to an "aerial" view. It shows the openings of Lieberkühn's crypts on the "hills" and the surrounding "valleys." It should be noted that the cecal lining in a germ-free rat has a much more complex topography than that of normal rats; this difference is clearly illustrated by scanning electron microscopy (Gustafsson and Maunsbach, 1971). However, the scanning micrograph does not illustrate the structural organization or components of the mucosa below the surface. A combination of transmission and scanning electron microscopy is therefore in some situations a great advantage for the understanding the three-dimensional organization of the tissue.

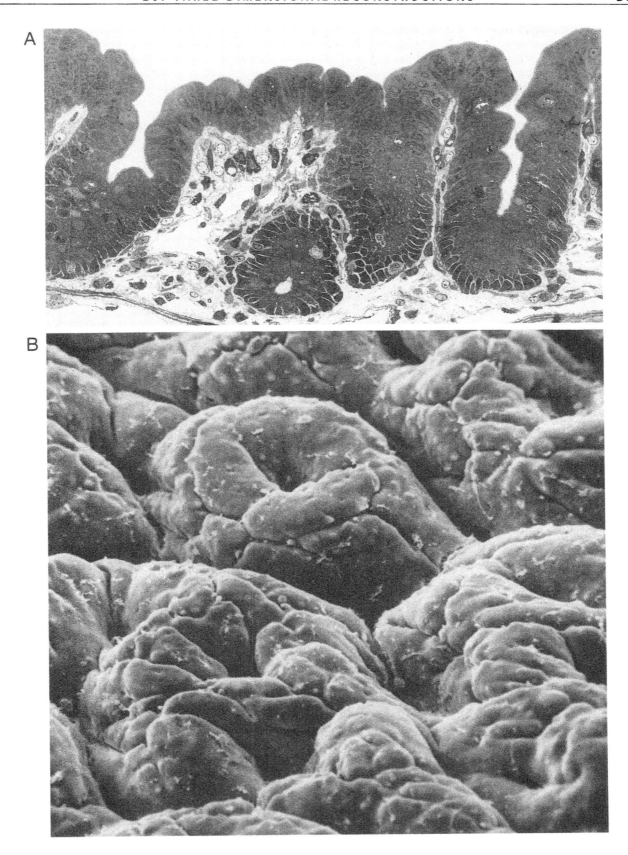

2. Serial Sectioning

FIGURES 20.2A–20.2D Parts of proximal tubule cells from the kidney of *Necturus maculosus* perfusion fixed with osmium tetroxide, stained *en bloc* with uranyl acetate, and embedded in Vestopal. The tissue was sectioned serially with a section thickness of 350 Å as estimated from spherical components. Every second or third section of this series of ribbon of sections is illustrated here as Figs. 20.2A–20.2D. × 33,000.

In the center of each of these micrographs the cell membranes are seen to interdigitate. A comparison between the micrographs shows that the interdigitating processes are not finger-like but correspond to folds, or membrane plica, oriented approximately perpendicular to the plane of the section.

FIGURES 20.2E–20.2H These four micrographs represent every second or third section in a similar series of serial sections. This part of the cytoplasm contains several mitochondrial profiles. × 30,000.

Figure 20.2E shows several mitochondrial profiles of which two (asterisks) can be also traced in the following sections. However, in Fig. 20.2F the two profiles seem to connect, as is better seen in Fig. 20.2G, whereas in Fig. 20.2H only a small part of the mitochondrion remains.

COMMENTS

The shape and extension of cytological structures can, to some extent, be estimated by an experienced observer from the shape and frequency of profiles in single sections. However, the true shape of a given structure can only be determined if it has been sectioned serially. For example, many types of cells seem to contain in single sections numerous mitochondria, but serial sections reveal that many profiles in fact belong to a smaller number of organelles (Bergeron et al., 1980) or even one single, large mitochondrion (Hoffman and Avers, 1973). The use of serial sectioning for ultrastructural analysis becomes increasingly demanding with increasing numbers of consecutive, ultrathin sections. However, for a morphological analysis of the nervous system, such investigations are of great importance and represent the only way to clarify the microcircuitry patterns of this tissue (Sjöstrand, 1958, 1974, 1998).

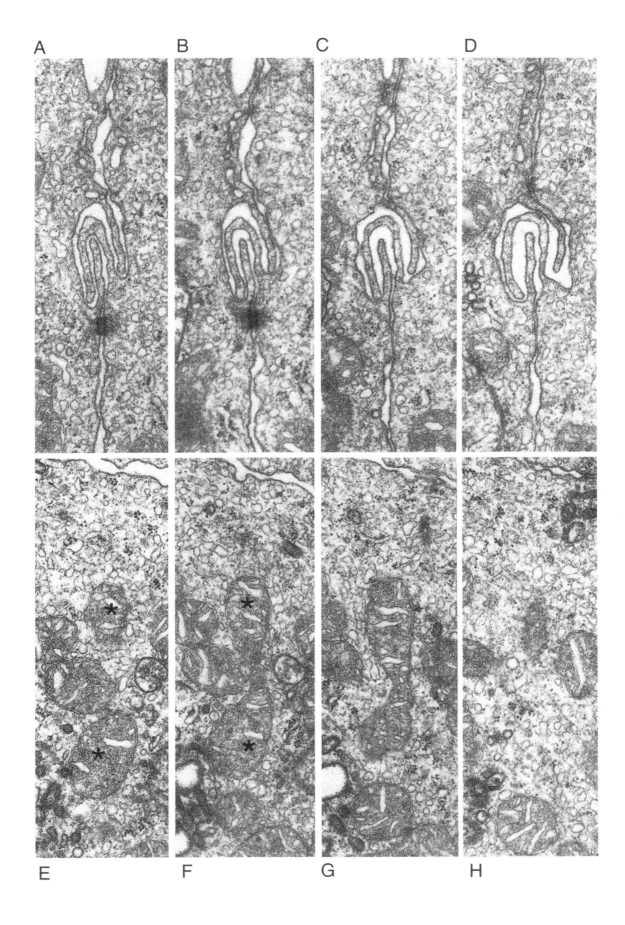

3. Large Three-Dimensional Objects

FIGURES 20.3A AND 20.3B A rat renal cortex perfusion fixed with 1% glutaraldehyde, postfixed in 1% osmium tetroxide, and embedded in Epon. Starting at the renal capsule, 1000 semithin (4 μm) sections were cut through the renal cortex toward the medulla. Every section was photographed in a light microscope (Fig. 20.3A) and used for computer-assisted 3D reconstructions of selected nephrons as described previously (Dørup *et al.*, 1994; Dørup and Maunsbach, 1997). Selected semithin sections were then reembedded (Maunsbach, 1978; compare Figs. 8.1) and sectioned for electron microscopy (Fig. 20.3B). Through this procedure, predetermined segments of the proximal tubule can be analyzed by electron microscopy. Fig. 20.3A: × 140; 20.3B: × 1500.

In the light micrograph (Fig. 20.3A) the asterisk indicates a tubule cross section, which is also shown by electron microscopy in Fig. 20.3B following reembedding and ultramicrotomy of the 4 μm section. The 3D reconstruction (Fig. 20.3C) demonstrated that the cross section is located 6350 μm from the glomerulus.

FIGURE 20.3C The convoluted part of the proximal tubule reconstructed on the basis of semithin (4 μm) Epon sections. The two lines represent a stereo plot of the axis of the tubule from the glomerulus (open rings) to the end of the convoluted part of the proximal tubule. For stereo viewing the tubule was rotated 4° around the *y* axis. × 124.

The stereo plot illustrates the complex course of a subcapsular proximal tubule. The convoluted tubule makes five contacts with the renal surface at the top of the reconstruction. Any predetermined segment of this reconstruction can be used for electron microscope analysis through reembedding of the relevant semithin sections.

FIGURES 20.3D–20.3G Brush border regions from different segments of the proximal tubule illustrated in Fig. 20.3C. The micrographs originate from cross sections located at 835, 4419, 6481, and 11,005 μm from the glomerulus as calculated along the tubule axis. × 25,000.

There are appreciable differences in the length and total surface area of the luminal brush border between the S1 segment (Fig. 20.3D), the S2 segment (Figs. 20.3E and 20.3F), and the S3 segment (Fig. 20.3C).

COMMENTS

Computer-assisted 3D reconstruction of large structures, on the basis of semithin sections in combination with electron microscopy, makes it possible to analyze preselected regions or segments of complex histological structures, such as kidney tubules, or seminiferous tubules, or very extended structures such as nerve processes, and to analyze their regional differences in fine structure. It is thereby possible to obtain information that is otherwise unobtainable. A prerequisite for the method is the ability to cut very long, uninterrupted series of semithin sections. By using 4-μm semithin sections as the basis for the reconstruction it is possible to cover much larger tissue dimensions than by using ultrathin serial sections, which normally would be in the range of 0.05–0.1 μm, e.g., some 40 times thinner.

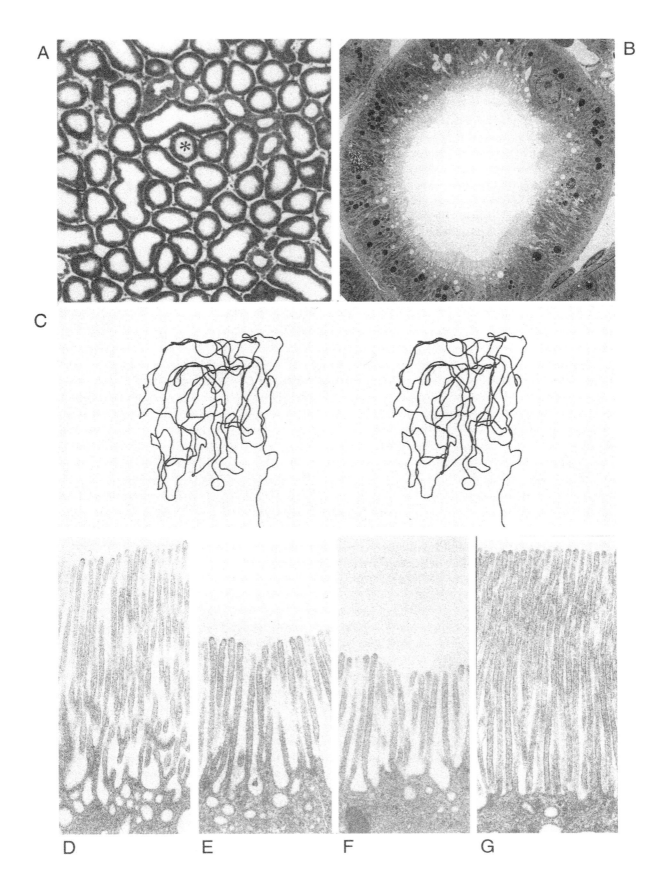

4. Tilting of Section: Cell Nucleus

FIGURES 20.4A–20.4F A nucleus in a rat liver cell fixed in glutaraldehyde, postfixed in osmium tetroxide, and embedded in Epon. The sections were stained with uranyl acetate and lead citrate. Micrographs were taken using a tilting stage. Figure 20.4A shows the untilted section before it is tilted 60° in the one (Fig. 20.4B) or the opposite direction (Fig. 20.4C). The grid was then rotated 90°, a new micrograph was taken (Fig. 20.4D), and was again tilted +60° (Fig. 20.4E) and −60° (Fig. 20.4F). Magnification × 18,000 in Figs. 20.4A and 20.4D.

The nucleus is cut at some distance away from its center rather than equatorially. Therefore the nuclear envelope is cut obliquely all around its circumference and has a fuzzy appearance in Figs. 20.4A and 20.4D. This is seen at higher magnifications in Figs. 20.5A–20.5F. In the tilted positions (Figs. 20.4B, 20.4C, 20.4E, and 20.4F), all cellular structures have been shortened by 50% in the tilting direction. Figures 20.4B and 20.4C appear quite similar superficially but distinct differences exist. In Fig. 20.4B the nuclear envelope is cut at a right angle on the left side of the nucleus but is still fuzzy in the right side, whereas in Fig. 20.4C the nuclear envelope is cut at a right angle on the right side. These areas are shown at higher magnifications in Figs. 20.5C and 20.5D. Figures 20.4E and 20.4F show the same features of the rotated and similarly tilted section.

COMMENTS

Tilting of the section can be used to analyze the orientation of membranes in the section. In the present example the interpretation would normally not present any problems, but in other cases an inspection of the structure from several different angles may be useful.

The entire micrographs of the tilted section are sharp at this magnification even though there is a height difference in the microscope between the edges amounting to about 8 μm. This fact illustrates the considerable depth of focus in the electron microscope.

A

D

B

E

C

F

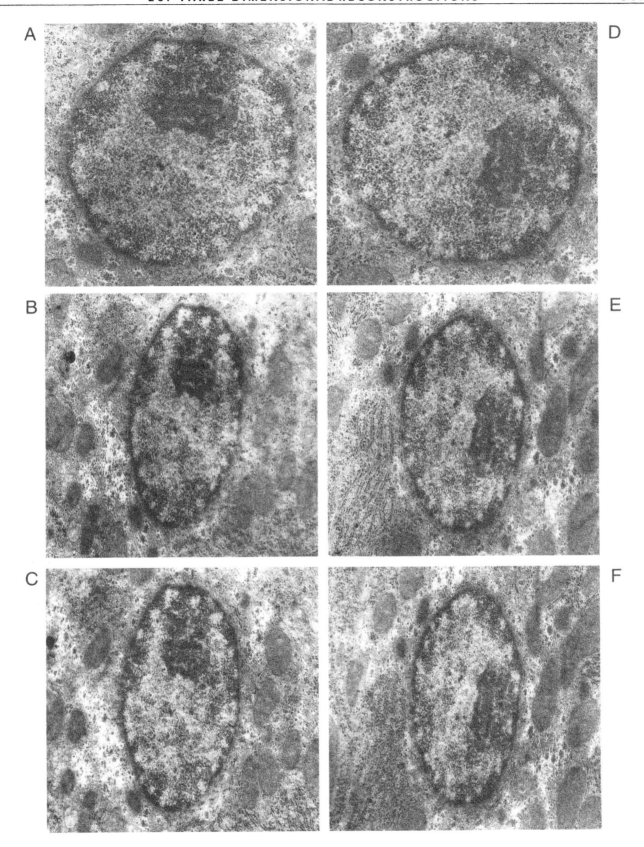

5. Tilting of Section; Nuclear Envelope

FIGURES 20.5A–20.5H Higher magnifications of the nuclear envelope from Figs. 20.4A–20.4F. The nuclear envelope in Figs. 20.5A, 20.5B, 20.5E, and 20.5F, where the section is in the horizontal plane, is shown in corresponding micrographs after 60° tilting in Figs. 20.5C, 20.5D, 20.5G, and 20.5H, respectively. Magnification of the nontilted section: × 65,000.

In the horizontal (e.g., nontilted) sections the nuclear envelope is oriented obliquely and does not show a sharp definition of its membranes (Figs. 20.5A, 20.5B, 20.5E, and 20.5F). Instead, some nuclear pore complexes are seen (arrowheads). In the tilted sections the membranes of the nuclear envelope are cut at right angle at several places, whereas the nuclear pore complexes are difficult to recognize. The membranes of the endoplasmic reticulum, which are distinct in Fig. 20.5G, cannot be seen before tilting (Fig. 20.5E). Note the different projections of the polyribosomes in the cytoplasm in these four figure pairs.

COMMENTS

This set of micrographs amplifies the comments made for Figs. 20.4A–20.4F and also illustrates that images of small components change appreciably upon tilting. This can be noted in, for example, the nuclear pore complexes and polyribosomes.

Tilting of the section is particularly useful in order to decide whether two adjacent structures are in true or only apparent continuity and is a powerful means of analyzing superposition effects in electron microscopy.

Pairs of micrographs taken at different tilt angles can be observed in a stereo viewer. This gives a three-dimensional image of the structures, but the tilt angle should then preferably be only 10–20°.

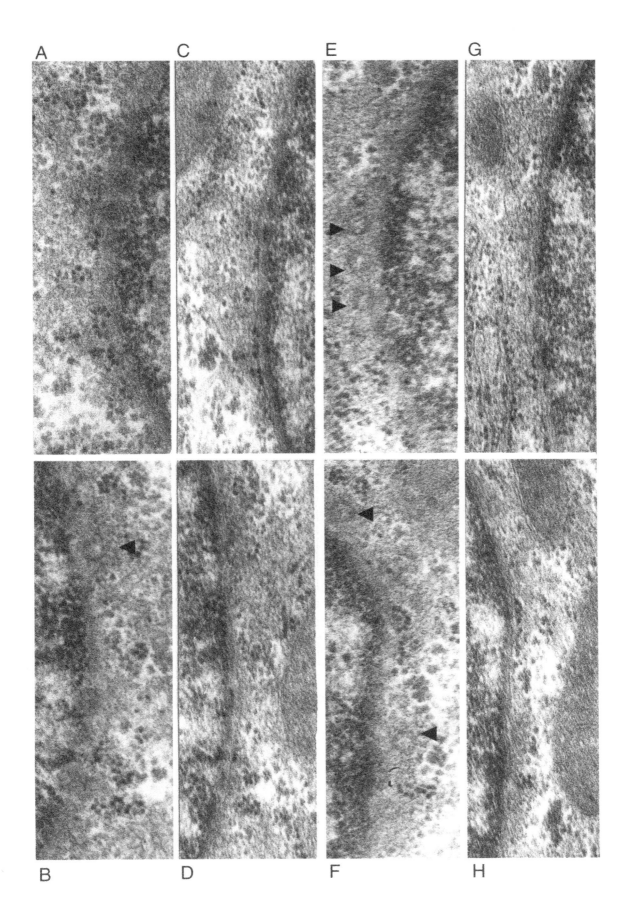

6. Helical Structures

FIGURES 20.6A–20.6D Shadows on photographic paper of helical structures placed on top of a glass plate (over the paper) and consisting of (from left to right) (1) a sinusoidal curve drawn with ink on the glass plate; (2) a left-handed metal helix constructed from a metal wire; (3) a metal wire, half of which is left-handed and the other half right-handed; and (4) a right-handed metal helix.

Figure 20.6A shows the shadow of this model in a untilted position. Figure 20.6B is the shadow of the model after tilting through an angle of 10° with the top of the field tilted toward the reader. In Fig. 20.6D the bottom of the field has been tilted toward the reader. Figure 20.6C is a photograph taken of the system. × 0.5.

It can be seen that the shadows in Fig. 20.6A from a left-handed helix and that of a right-handed one are indistinguishable and do not differ from a plane figure. The tilted helices will appear as a series of arcades or a series of catenaries depending on the sense of the helix. By contrast the shape of the flat sinusoidal curve remains unchanged.

FIGURE 20.6E Electron micrograph of a section through the testis of the leach *Piscicola geometra*. × 60,000.

The homogeneously dark structures in Fig. 20.6E are sections through the sperm nucleus.

COMMENTS

In a single, untilted shadowgram it is impossible to determine the absolute sense of a helix, i.e., its enantiomorphic form, that is the direction of its "handedness", or even if the object is helical at all. A single electron micrograph of helical structures in a negative-staining preparation or in a section does not provide information on the sense of the helix. Such information can be obtained by tilting the preparation and analyzing the changes in the images of the helices. When tilting the helices sidewise the sinusoidal pattern remains unchanged, except that positions of the flexure points will be shifted differently depending on the sense of the helix (Chretien *et al.*, 1996).

The nucleus in Fig. 20.6E has a helical shape but it is not directly clear whether this shape is due to one helical ridge (a single helix) or to two or more such ridges (double helix, triple helix, etc.). A section that passes obliquely through the nucleus shows that the ridges tilt in one direction in the anterior part of the nucleus and in the opposite direction in the posterior part. The number of ridges is seen to be different on opposite sides. The difference in number is equivalent to the number of ridges, thus three (compare the asterisks). In other words the nucleus appears as a triple helix. This is confirmed by surrounding cross-sectioned nuclei.

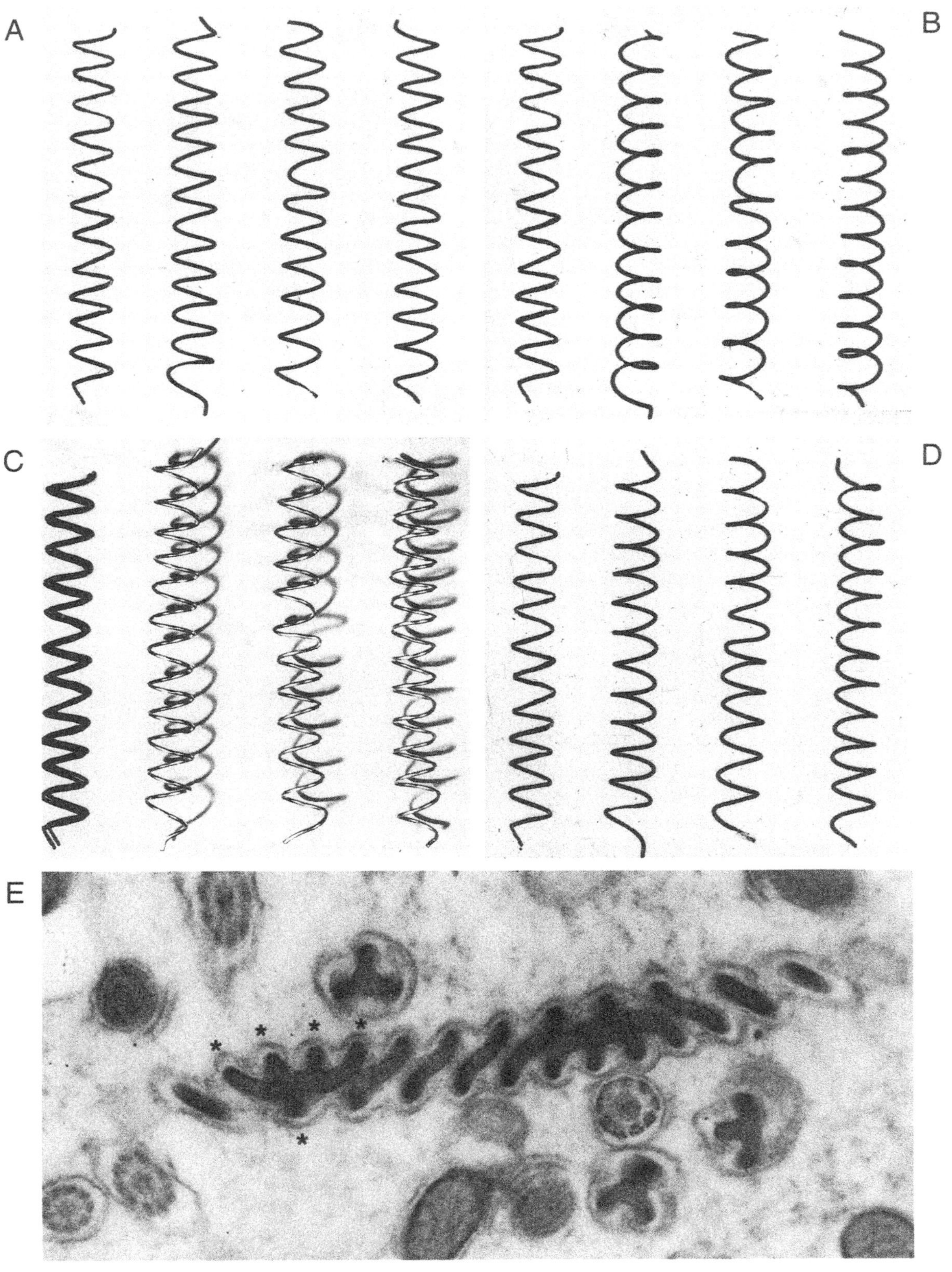
A
B
C
D
E
*
*
*
*
*
*

7. Computer-Analyzed Helices

FIGURES 20.7A–20.7C Sperm microtubule from the insect fungus gnat negatively stained with phosphotungstic acid. In a first step of computer analysis, this slightly curved microtubule segment (Fig. 20.7A) (and several others) has been straightened by an appropriate algorithm, which renders the image shown in Fig. 20.7B. Spectral analyses of its periodic densities are compatible with the assumption that the microtubule consists of subunits in a helical array. This fact can be used further in the preparation of filtered images of the front and back sides of the helix; one such side is shown in Fig. 20.7C. The image has been obtained by a two-dimensional filtering (for technical details, see Lanzavecchia *et al.*, 1994). × 440,000.

The periodicity of the microtubule is barely seen in the original micrograph (Fig. 20.7A) or in its scanned and computer-straightened equivalent (Fig. 20.7B), although some approximately longitudinal lines can be discerned by inspecting the figure at an oblique angle. In selecting a front (or back) side of the helix, a more distinct pattern can be discerned (Fig. 20.7C).

FIGURE 20.7D A computer average of several cross-sectioned sperm tails of the same insect as in Figs. 20.7A–20.7C, obtained as described in Dallai *et al.* (1994). × 260,000.

The periphery of the axoneme contains seven microtubules that have walls containing 16 protofilaments.

FIGURE 20.7E A three-dimensional view of the microtubule. The three-dimensional reconstruction has been obtained with customary methods of helical reconstruction from single projections (DeRosier and Moore, 1970). The helix has been arbitrarily fixed to be left-handed.

Eight of the 16 protofilaments of the microtubule are seen at the side facing the viewer. The protofilaments have a somewhat skew orientation along the microtubule. The subunits making up the protofilaments form pairs, which correspond to tubulin heterodimers.

COMMENTS

A micrograph of a negative-staining preparation may contain much information that is not seen readily with the naked eye, but which can be extracted either by the use of an optical diffractometer or, better still, by computer analysis of the scanned micrograph. Information on regularities of various kinds will appear as spots in an amplitude spectrum (not shown here), which can be regarded as the raw material of analysis. Although much information can be seen in the computer-reconstructed view of the microtubule, the handedness of the helix cannot be determined from a single projection but can be obtained by analyzing the same microtubule at different tilt angles. The true view of the microtubule of the fungus gnat sperm in fact may appear as the mirror image of Figure 20.7E.

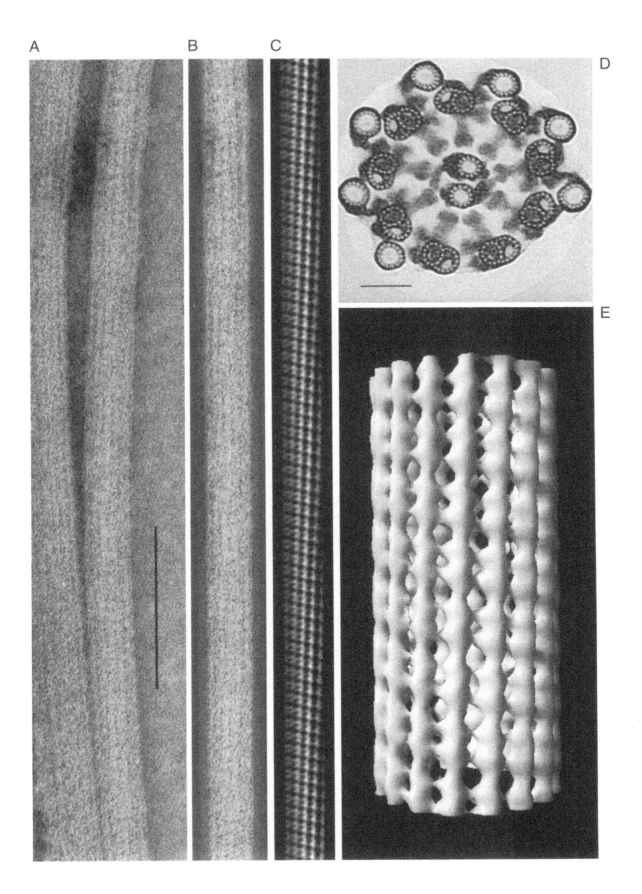

8. A Three-Dimensional Model of Na,K-ATPase

FIGURE 20.8A Electron micrographs of a membrane crystal of Na,K-ATPase. The enzyme was isolated from the outer medulla of a rabbit kidney in a membrane-bound form. The purified enzyme was crystallized with 3 mM ammonium vanadate and 3 mM magnesium chloride in 10 mM imidazole buffer at pH 7.0. Samples were negatively stained with 1% uranyl acetate, and the electron micrographs were recorded at a magnification of 50,000 using tilt angles from $-60°$ to $+60°$ with 6° intervals. The left micrograph shown here was not tilted and the right one was tilted 36°. $\times$ 200,000.

The two electron micrographs show the same Na,K-ATPase membrane crystal as observed at different angles.

FIGURE 20.8B As described previously (Hebert *et al.*, 1985), the electron micrographs were digitized and 3D reconstructed essentially according to Henderson and Unwin (1975) and Amos *et al.* (1982). The three illustrated density maps are 5-Å-thick sections oriented parallel to the membrane and located in the unit cell at 25, 0, and -15 Å relative to its center.

The density maps show that the unit cell contains two symmetrically related rod-like regions that run perpendicular to the membrane. A bridge-like connection is formed between adjacent rods in the lower drawing. The solid lines correspond to protein regions whereas negatively stained areas have dashed contours.

FIGURES 20.8C AND 20.8D A 3D model of the Na,K-ATPase dimer built from the density maps in balsa wood. $\times$ about 5,400,000.

The Na,K-ATPase dimer is seen in the direction of the c axis (Fig. 20.8C) and from the side (Fig. 20.8D). The vertical line in Fig. 20.8D indicates the level where the contrast variation has a minimum.

FIGURE 20.8E A computerized version of the Na,K-ATPase model shown in Fig. 20.8D rendered on the basis of the same density maps at 5-Å intervals. [*See also color insert.*]

This version of the dimer has a smoother surface but its shape and dimensions are the same. The cytoplasmic side is down.

COMMENTS

The contrast minimum observed at the level of the vertical bar in Fig. 20.8D is interpreted as the level of lipid bilayer. The protein protrudes approximately 40 Å on one side of the bilayer, probably the cytoplasmic side, and about 20 Å on the extracellular side. Na,K-ATPase can form different types of membrane crystals, which vary with respect to subunit interactions and unit cell dimensions (Maunsbach *et al.*, 1991). Thus 3D constructions have so far been performed on single membrane fragments. A few other membrane proteins, however, can be crystallized in extremely reproducible forms or are already in nature highly organized, as is the purple membrane (Henderson and Unwin, 1975). Electron microscope crystallography of such protein assemblies may reveal structural information down to, or even below, the 3-Å level. It is also possible to obtain 3D reconstructions of single macromolecular complexes such as ribosomes or large enzyme complexes (Frank, 1996; Frank and Radermacher, 1992), a line of research which has also provided much new information.

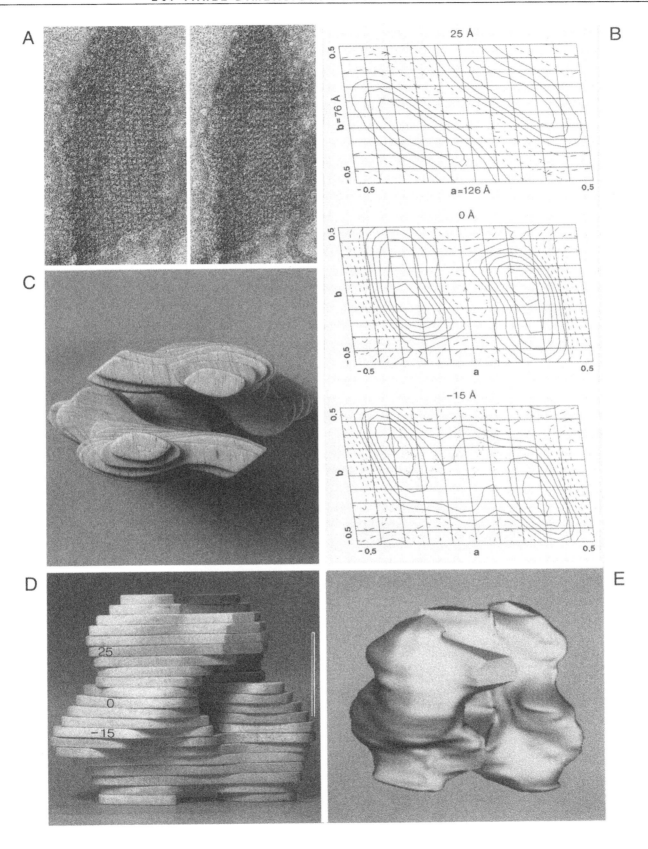

References

Afzelius, B. A., Dallai, R., and Callaini, G. (1989). Spermiogenesis and spermatozoa in *Melanopsis* (Mesogastropoda, Mollusca). *J. Submicrosc. Cytol. Pathol.* **21,** 187–200.

Amos, L. A., Henderson, R., and Unwin, P. N. T. (1982). Three-dimensional structure determination by electron microscopy of 2-dimensional crystals. *Progr. Biophys. Mol. Biol.* **39,** 183–231.

Barajas, L. (1970). The ultrastructure of the juxtaglomerular apparatus as disclosed by three-dimensional reconstructions from serial sections: The anatomical relationship between tubular and vascular components. *J. Ultrastruct. Res.* **33,** 116–147.

Bergeron, M., and Thiéry, G. (1980). Three-dimensional characteristics of the mitochondria of the rat nephron. *Kidney* **17,** 175–180.

Berthold, C.-H., and Rydmark, M. (1983). Electron microscopic serial section analysis of nodes of Ranvier in lumbosacral spinal roots of the cat: Ultrastructural organization of nodal compartments in fibres of different sizes. *J. Neurocytol.* **12,** 475–505.

Chrétien, D., Kenney, J. M., Fuller, S. D., and Wade, R. H. (1996). Determination of microtubule polarity by cryo-electron microscopy. *Curr. Biol.* **4,** 1031–1040.

Compans, R. W., Mountcastle, W. E., and Choppin, P. W. (1972). The sense of the helix of paramyxovirus nucleocapsids. *J. Mol. Biol.* **65,** 167–169.

Cui, S., and Christensen, E. I. (1993). Three-dimensional organization of the vacuolar apparatus involved in endocytosis and membrane recycling of rat kidney proximal tubule cells: An electron microscopic study of serial sections. *Exp. Nephrol.* **1,** 175–184.

Dallai, R., and Afzelius, B. A. (1994). Three-dimensional reconstructions of accessory tubules observed in the sperm axonemes of two insect species. *J. Struct. Biol.* **113,** 225–237.

DeRosier, D. J., and Klug, A. (1968). Reconstruction of three dimensional structures from electron micrographs. *Nature* **217,** 130–134.

DeRosier, D. J., and Moore, P. B. (1970). Reconstruction of three-dimensional images from electron micrographs of structures with helical symmetry. *J. Mol. Biol.* **52,** 355–369.

Dørup, J., Andersen, G. K., and Maunsbach, A. B. (1983). Electron microscope analysis of tissue components identified and located by computer-assisted 3-D reconstructions: Ultrastructural segmentation of the developing human proximal tubule. *J. Ultrastruct. Res.* **85,** 82–94.

Dørup, J., and Maunsbach, A. B. (1997). Three-dimensional organization and segmental ultrastructure of rat proximal tubules. *Exp. Nephrol.* **5,** 305–517.

Elias, H. (1971). Three-dimensional structure identified from single sections: Misinterpretation of flat images can lead to perpetuated errors. *Science* **174,** 993–1000.

Felluga, B. (1976). Precise guiding pattern for three-dimensional analysis of single cells by transmission electron microscopy. *J. Submicrosc. Cytol.* **8,** 193–204.

Frank, J. (1996). "Three-Dimensional Electron Microscopy of Macromolecular Assemblies." Academic Press, San Diego.

Frank, J., and Radermacher, M. (1992). Three-dimensional reconstruction of single particles negatively stained or in vitreous ice. *Ultramicroscopy* **46,** 241–262.

Gustafsson, B. E., and Maunsbach, A. B. (1971). Ultrastructure of the enlarged cecum in germfree rats. *Z. Zellforsch.* **120,** 555–578.

Hama, K., and Nagata, F. (1970). A stereoscopie observation of tracheal epithelium of mouse by means of the high voltage electron microscope. *J. Cell Biol.* **45,** 654–659.

Hawes, C. (1991). Stereo-electron microscopy. *In* "Electron Microscopy of Plant Cells" (J. L. Hall and C. Hawes, eds.), pp. 67–84. Academic Press, London.

Hebert, H., Skriver, E., Kavéus, U., and Maunsbach, A. B. (1990). Coexistence of different forms of Na,K-ATPase in two-dimensional membrane crystals. *FEBS Lett.* **268,** 83–87.

Hebert, H., Skriver, E., and Maunsbach, A. B. (1985). Three-dimensional structure of renal Na,K-ATPase determined by electron microscopy of membrane crystals. *FEBS Lett.* **187,** 182–186.

Henderson, R., and Unwin, P. N. T. (1975). Three-dimensional model of purple membrane obtained by electron microscopy. *Nature* **257,** 28–31.

Hoenger, A., and Aebi, U. (1996). 3-D reconstructions from ice-embedded and negatively stained biomacromolecular assemblies: A critical comparison. *J. Struct. Biol.* **117,** 99–116.

Hoffman, H.-P., and Avers, C. J. (1973). Mitochondrion of yeast: Ultrastructural evidence for one giant branched organelle per cell. *Science* **181,** 749–751.

Kanaya, K. (1988). Digital image processing of crystalline specimens examined by electron microscopy. *J. Electr. Microsc. Techn.* **10,** 319–367.

Klug, A., and DeRosier, D. J. (1966). Optical filtering of electron micrographs: Reconstruction of one-sided images. *Nature* **212,** 29–32.

Lanzavecchia, S., Bellon, P. L., Dallai, R., and Afzelius, B. A. (1994). Three-dimensional reconstructions of accessory tubules observed in the sperm axonemes of two insect species. *J. Struct. Biol.* **113,** 225–237.

Maunsbach, A. B., Skriver, E., and Hebert, H. (1991). Two-dimensional crystals and three-dimensional structure of Na,K-ATPase analyzed by electron microscopy. *In* "The Sodium Pump: Structure, Mechanism, and Regulation" (J. H. Kaplan and P. De Weer, eds.), pp. 159–172. Rockefeller Univ. Press, New York.

Mohraz, M., Simpson, M. V., and Smith, R. P. (1987). The three-dimensional structure of the Na,K-ATPase from electron microscopy. *J. Cell Biol.* **105,** 1–8.

Nogales, E., Wolf, S. G., and Downing, K. H. (1998). Structure of the $\alpha\beta$ tubulin dimer by electron crystallography. *Nature* **391,** 199–201.

Sjöstrand, F. S. (1958). Ultrastructure of retinal rod synapses of the guinea pig eye as revealed by three-dimensional reconstructions from serial sections. *J. Ultrastruct. Res.* **2,** 122–170.

Sjöstrand, F. S. (1974). A search for the circuitry of directional selectivity and neural adaptation through three-dimensional analysis of the outer plexiform layer of the rabbit retina. *J. Ultrastruct. Res.* **49,** 60–156.

Sjöstrand, F. S. (1998). Structure determines function of the retina, a neural center. 1. The synaptic ribbon complex. *J. Submicrosc. Cytol. Pathol.* **30,** 1–29.

Skriver, E., Maunsbach, A. B., and Jørgensen, P. L. (1981). Formation of two-dimensional crystals in pure membrane-bound Na$^+$,K$^+$-ATPase. *FEBS Lett.* **131,** 219–222.

Thalen, J., Spoelstra, J., van Breemen, J. F. L., Mellema, J. E., and van Bruggen, E. F. J. (1970). A high resolution tilting stage constructed for three-dimensional reconstruction of biological objects. *In* "Proceedings of the Seventh International Congress on Electron Microscopy, Grenoble," Vol. I, 439–440.

Turner, J. N., and Valdrè, U. (1992). Tilting stages for biological applications. *In* "Electron Tomography" (J. Frank, ed.), pp. 167–196. Plenum Press, New York.

Unwin, P. N. T., and Henderson, R. (1975). Molecular structure determination by electron microscopy of unstained crystalline specimens. *J. Mol. Biol.* **94,** 425–440.

APPENDIX:
PRACTICAL METHODS

Preparatory methods in biological transmission electron microscopy are innumerable and in steady increase. For each separate step in the sequence of preparation there are many alternative recipes. However, comparisons between related technical modifications are usually lacking. Often the importance of a certain modification of a protocol is greatly exaggerated.

In our laboratories we have over the years applied a large number of existing methods for a variety of tissues and have developed some methods ourselves, sometimes in collaboration with students, technical staff, or guest scientists. Some procedures have proven less useful, but what remains is a set of "standard" procedures that work well in our hands and that we therefore want to describe in some detail here.

1. Fixation

Adequate fixation is often obtained by simple immersion of small tissue pieces into the fixative solution. This is the only mode of application possible for small organisms, for isolated cells, for many types of animal or plant tissues, and for biopsies. However, a more rapid and uniform fixation is usually obtained in animals if the fixative solution is perfused via the vascular system through the heart, through the abdominal aorta, or in some cases through the venous system followed by immersion fixation.

A. Immersion Fixation

Immersion fixation of tissues for general ultrastructural studies is carried out with 1% buffered glutaraldehyde for small tissue blocks. If the size of the tissue in one dimension is about 0.5 mm or more, the concentration should be increased to 3%, which is recommended for renal biopsies. For large specimens, an alternative solution is 2% formaldehyde plus 2.5% glutaraldehyde in a 0.1 M cacodylate buffer. For immunocytochemical studies, 4% formaldehyde plus 0.1% glutaraldehyde fixative should be used. If the antigen is very sensitive to fixation, glutaraldehyde should be excluded.

Procedure

1. Cut out a piece of tissue from the organ under study and place it in a precooled, empty petri dish.
2. Hold the tissue gently with a forceps and cut thin slices with a thin razor blade using sawing movements. The slices should not exceed 0.5 mm in thickness. Great care should be taken not to strain the tissue mechanically. Areas where the forceps have touched the tissue should be discarded.
3. Trim the slices to less than 0.5 × 5 × 5 mm and immerse them into the fixative solution.
4. Swirl the vial occasionally during fixation to secure uniform penetration of the fixative from all sides into the tissue. Fix for at least 2 hr. In most cases the temperature of the fixative is not important. Initial

fixation can be carried out at room temperature and followed by fixation in the cold.

5. Trim down the dimensions of the tissue slices (e.g., to $0.5 \times 0.5 \times 1.0$ mm) while in the fixative solution in order to get small blocks suitable for embedding.

6. Rinse the tissue 2×30 min in 0.1 M sodium cacodylate buffer.

7. Postfix the tissue in 1% osmium tetroxide in 0.1 M cacodylate buffer for 1 hr in the cold. Swirl the vial occasionally to secure uniform penetration of the fixative. (Lipid-rich tissues may require 2% osmium tetroxide.)

8. Rinse 2×30 min in 0.1 M sodium cacodylate buffer. The tissue is now ready for dehydration and embedding.

Following immersion fixation it is important that the tissue analyzed in the electron microscope originates from the surface layers of the tissue block, as there is a gradient in the quality of fixation from the surface to the center of the block. In the center of tissue slices, where the fixative has arrived with some delay, cytoplasm and organelles will appear swollen. Thus, it is practical first to examine semithin sections (1 μm) of the tissue block by light microscopy in order to select the optimal location of the tissue for electron microscopy analysis.

B. Fixation of Tissue Cultures

Procedure

1. Gently decant the tissue culture medium.
2. Immediately add 2% glutaraldehyde in 0.1 M cacodylate buffer. Very gently swirl the fixative in the culture dish.
3. Fix for 2 hr.
4. Proceed as described earlier, steps 6–8.

C. Fixation of Cell Suspensions

Procedure

1. Mix one volume cell suspension rapidly with an equal volume of 2% glutaraldehyde in 0.1 M cacodylate buffer.
2. Fix for 2 hr.
3. Sediment the cells in a centrifuge tube by centrifugation at approximately $1000\,g$ for 5 min. Decant the supernatant.
4. Resuspend the cells in an excess of 0.1 M cacodylate buffer. Repeat step 3 after 15 min.
5. Add 1% osmium tetroxide in 0.1 M cacodylate buffer and resuspend the cells. Fix for 30 min. Repeat step 3.

6. Add 0.1 M sodium cacodylate buffer and resuspend the cells. Repeat step 3.

7. Resuspend the cells in a very small volume of buffer. Inject the suspension with a fine pipette into a small drop of 4% melted agar kept at 40°C. Allow the drop to cool, whereupon it will turn into a gel and can be treated as a tissue block during dehydration and further processing. If the cell suspension should be frozen and freeze-substituted the agar should be exchanged for 10% gelatine.

D. Perfusion Fixation through Abdominal Aorta

The following procedure (Maunsbach, 1966a) results in efficient fixation of kidney, liver, pancreas, and small intestines. In the kidney, which is very sensitive to variations in the mode of applying the fixative, it preserves open tubules and normal relationships between tubule cells. The procedure is described for rats but can also be adapted for other animals. For most tissues, 1% glutaraldehyde is sufficient for general ultrastructural studies. This concentration does not require a preceding rinse of the vascular system with a salt solution.

Procedure

1. Place the closed flask containing the fixative upside down and about 150 cm above the aortic level of the animal. Wear gloves.
2. Connect the flask to the perfusion needle via the administration set for intravenous solutions and ventilate the flask.
3. The size of the needle should be 1.3–1.5 mm in outer diameter for a 300-g rat and proportionally smaller for lighter animals. Bend the needle at a 45° angle with the beveled side out (i.e., down during perfusion).
4. Connect the perfusion needle via the infusion set to the flask containing the fixative. Check that there are no air bubbles in the tubing of the infusion set.
5. Fix the anesthetized animal onto the operating table with its back down. Artificial respiration is unnecessary.
6. Open the abdominal cavity by a long midline incision with lateral extension and move the intestines to the left side of the animal.
7. Carefully expose the aorta below the origin of the renal arteries and gently free the aorta from overlaying adipose and connective tissues.
8. Hold the wall of the aorta firmly with a forceps with fine claws about 0.5–1.0 cm from the distal bifurcation. Insert the bent needle close to the forceps toward the heart into the lumen of the aorta (with the beveled side of the tip down).

9. In very rapid succession (a) cut a hole in the inferior caval vein with fine scissors, (b) start the perfusion, and (c) clamp the aorta below the diaphragm but above the origin of the renal arteries. When performing these manipulations, accuracy and speed are essential and the fixation procedure should be preferably carried out by two persons. It is particularly important to clamp the aorta rapidly after the perfusion has been started. This is best done by compressing the aorta toward the posterior wall of the peritoneal cavity with a finger and then by a clamp. Finally, cut the aorta above the compression.

10. The tissue surface must blanch *immediately* (within less than a second), show a uniformly pale color, and harden quickly. The flow rate should be at least 60–100 ml/min for an adult rat. Perfuse for 3 min. If the fixative flow is compromised during perfusion fixation, a sufficient concentration of fixative is not obtained throughout the tissue and cells may undergo various abnormal alterations before they are fixed. Usually there is a good correlation among the speed of tissue blanching, fixative flow as observed in the drip chamber, absence of blood in dissected tissues, and the final quality of tissue preservation as observed in the electron microscope. Stop the perfusion and excise and trim the tissues in the fixative with a razor blade. Store the tissue in vials and immersion-fix in the same fixative as used for perfusion for 2 hr or more.

11. Proceed as described in Section A, steps 6–8.

E. Perfusion Fixation through the Heart

The following procedure provides fixation of most rat organs with 1% glutaraldehyde. For some organs the glutaraldehyde concentration should be increased (e.g., to 5% for the brain) and the fixative preceded by a brief rinse with a balanced salt solution such as Tyrode.

Procedure

1. Place the closed flask containing the fixative upside down about 150 cm above the animal.
2. Connect the perfusion needle via the infusion set to the flask containing the fixative and ventilate the flask. Check that there are no air bubbles in the tubing of the infusion set.
3. Fix the anesthetized animal onto the operating table with its back down.
4. Open the thoracic cavity of the animal without giving artificial respiration.
5. Grasp the heart close to its apex with a forceps. Cut a small hole in the wall of the left ventricle close to the apex with fine scissors. Rapidly insert a blunt syringe

needle (2.0–2.4 mm outer diameter for a 300-g rat and proportionally smaller for lighter animals) and move it into the ascending aorta. Place a clamp on the aorta to hold the needle.

6. Cut a hole in the right atrium of the heart and start the perfusion immediately.
7. Check the flow rate in the drip chamber and flask. The flow rate should be at least 150 ml/min for an adult rat. Perfuse for 3 min. Stop the perfusion and remove pieces of tissue. Subdivide the tissue and fix it additionally for 2 hr in the same fixative.
8. Proceed as described in Section A, steps 6–8.

F. Fixative Containing Tannic Acid

This fixation procedure is particularly suitable for studies of microtubules, microfilaments, plasma membranes, and the glycocalyx (see Chapter 2). The primary fixative contains 2% glutaraldehyde and 1% tannic acid, and the secondary fixative contains 1% uranyl acetate in distilled water (Afzelius, 1988). No osmium fixation is used. The primary fixative can be used for either perfusion or immersion fixation.

Procedure

1. Dissect the specimen into small fragments and leave them in the glutaraldehyde/tannic acid fixative for 2 or more days.
2. Rinse the specimens in several changes of 0.1 *M* phosphate buffer, then in distilled water.
3. Postfix the specimens in the 1% uranyl acetate solution for 1 hr.
4. Rinse again in distilled water.
5. Dehydrate and embed.

The final outcome of the fixation will depend on the dimensions and type of the specimen. Tannic acid is a group of high molecular compounds that only slowly penetrate into the tissue. Not all cells will be impregnated by tannic acid.

G. Comments

For immunocytochemistry the sensitivity of the antigen to aldehydes determines the composition of the fixative. As a rule of thumb, insensitive antigens can be fixed with 1% glutaraldehyde, sensitive antigens with 4% formaldehyde plus 0.1% glutaraldehyde, and very sensitive antigens with 4% or 8% formaldehyde only (see Chapter 16).

Cells tend to swell if the osmolality of the fixative vehicle (buffer) is low, whereas they shrink if the fixative solution has a high solute concentration (Maunsbach, 1966b). For this reason the osmotic composition of the fixative vehicle has to be adapted to the type of tissue. In the outer renal medulla, where the extracellular osmolality is high, the normal perfusion fixative (1% glutaraldehyde in modified Tyrode solution or in 0.1 M cacodylate buffer) should be supplemented with 0.2 M sucrose. Freshwater organisms tend to have a low osmolality, whereas marine ones (except teleosts) have a high one. For amphibian tissues the vehicle osmolality should be lower than for mammalian ones. (See Chapter 3).

The pH of aldehyde fixatives is normally 7.0–7.5, and fine adjustments of pH are not crucial in most ultrastructural studies. The choice of buffer in the fixative may influence the appearance of the tissue but in most tissues only to a moderate degree. A modified Tyrode solution can be used instead of cacodylate (less toxic and less expensive). Phosphate buffers alone are also useful but give a fine precipitate at concentrations around 0.1 M or more in some tissues (see Chapter 3).

In highly vascularized organs, such as the pancreas, perfusion fixation leads to a distension of the extravascular space. Such swelling can be prevented by the addition of 2% dextran (molecular weight around 40,000) to the perfusion solution (see Chapter 3).

Storage of cells and tissues in glutaraldehyde fixatives has very little influence on the final appearance of the tissue in the transmission microscope. Except for immunocytochemical studies, specimens can be kept for days, or even months, in the aldehyde fixative. Pieces of tissue can therefore be transported in glutaraldehyde between laboratories (see Chapter 2). Storage in osmium tetroxide, however, is deleterious to the tissue if prolonged for several hours.

H. Checklist of Chemicals and Equipment

Chemicals

Agar
Calcium chloride
Disodium hydrogen phosphate dihydrate ($Na_2HPO_4 \cdot 2H_2O$)
Glucose
Glutaraldehyde, 25% aqueous stock solution
Hydrochloric acid
Magnesium chloride
Osmium tetroxide
Paraformaldehyde powder
Potassium chloride
Sodium bicarbonate
Sodium chloride
Sodium cacodylate ($C_2H_6AsNaO_2 \cdot 3H_2O$)
Sodium dihydrogen phosphate monohydrate ($NaH_2PO_4 \cdot H_2O$)

Sodium hydroxide
Tannic acid (Mallinckrodt, St. Louis, or Carlo Erba, Milano)
Uranyl acetate

Equipment

Operating table
Anesthetic
Scissors, forceps, scalpels, and clamps
Small forceps with fine claws
Gauze swabs
Plastic petri dishes
Thin razor blades
5- to 10-ml vials with lids for specimens
Gloves
Short-beveled syringe needle for perfusion of aorta, length about 50 mm, outer diameter 1.3–1.5 mm
Blunt syringe needle for heart perfusion, length about 100 mm, outer diameter 2.0–2.4 mm
Perfusion set with drip chamber as used for intravenous blood infusions
Flask to which the perfusion set fits
10- to 15-cm-long syringe needle to ventilate the flask
Stand to hold the fixative flask upside down about 150 cm above the operating table

I. Solutions

0.2 M sodium cacodylate buffer, pH 7.2. Dissolve 21.4 g sodium cacodylate in 480 ml distilled water. Adjust pH to 7.2 with 1 N HCl. Complete with distilled water to 500 ml.

0.2 M phosphate buffer, pH 7.2. Dissolve 1.93 g $NaH_2PO_4 \cdot H_2O$ and 6.41 g $Na_2HPO_4 \cdot 2H_2O$ in 240 ml distilled water. Adjust pH to 7.2 and fill up to 250 ml.

1% glutaraldehyde in modified Tyrode solution (Maunsbach, 1966a). The fixation consists of 1% (about 0.1 M) glutaraldehyde made up in a Tyrode solution that contains only 75% of the regular amount of sodium chloride (6.0 g NaCl, 0.20 g KCl, 0.20 g $CaCl_2$, 0.05 g $NaH_2PO_4 \cdot H_2O$, 0.10 g $MgCl_2 \cdot 6H_2O$, 1.0 g $NaHCO_3$, 1.0 g glucose, and 40 ml 25% glutaraldehyde and distilled H_2O to 1000 ml; dissolve the $CaCl_2$ separately). The fixative has a total osmolality of about 354 milliosmols and a pH of 7.3–7.4.

1% glutaraldehyde in 0.1 M sodium cacodylate buffer. Mix 250 ml of 0.2 M sodium cacodylate buffer and 20 ml of 25% aqueous glutaraldehyde and add distilled water to about 480 ml. Adjust pH to 7.2 with 0.1 N NaOH or 0.1 N HCl. Complete with distilled water to 500 ml.

3% glutaraldehyde in 0.1 M sodium cacodylate buffer. Mix 50 ml of 0.2 M sodium cacodylate buffer and 12 ml of 25% aqueous glutaraldehyde and add distilled water to about 90 ml. Adjust pH to 7.2 with 0.1 N NaOH or 0.1 N HCl. Complete with distilled water to 100 ml.

2% glutaraldehyde–1% tannic acid in 0.1 M phosphate buffer. Make a 0.2 M phosphate buffer (pH 7.2) and add 3.6% sucrose and 2% tannic acid. When the tannic acid dissolves it should give a clear yellow solution. Mix one part of this solution with one part of 4% glutaraldehyde.

1% uranyl acetate in distilled water. Do not adjust pH.

20% stock solution of formaldehyde. Mix 20 g of paraformaldehyde powder with 80 ml of distilled water in a glass flask. Heat to 60°C while gently agitating the milky solution. Add 1 N sodium hydroxide dropwise until the solution clears up. Complete to 100 ml with distilled water. This procedure should be carried out in a well-ventilated hood and with protection for the face. Always wear gloves. The solution can be stored for a few months in the refrigerator.

4% formaldehyde plus 0.1% glutaraldehyde in 0.1 M sodium cacodylate buffer for immunocytochemistry. Mix 100 ml of a 20% formaldehyde stock solution, 2 ml of 25% glutaraldehyde, and 250 ml of 0.2 M sodium cacodylate buffer. Add distilled water to about 480 ml. Adjust pH to 7.2 with 0.1 N NaOH or 0.1 N HCl. Complete with distilled water to 500 ml. For sensitive antigens, omit glutaraldehyde.

2% formaldehyde plus 2.5% glutaraldehyde in 0.1 M sodium cacodylate buffer, often referred to as "half-strength Karnovsky's fixative" (Karnovsky, 1965). Mix 50 ml of a 25% aqueous glutaraldehyde stock solution, 50 ml of a 20% formaldehyde stock solution, 250 ml of 0.2 M sodium cacodylate buffer; add distilled water to about 480 ml and adjust pH to 7.2 with 0.1 N NaOH or 0.1 N HCl. Complete with distilled water to 500 ml.

2% osmium tetroxide stock solution. Crystalline osmium tetroxide is usually obtained preweighed (1 g) in closed glass ampoules. Wash the ampoule carefully, remove the label, and score it around its perimeter with a diamond or a fine file. Break the ampoule cautiously and empty the crystals and the ampoule (some crystals attach to the glass) into 50 ml of distilled water in a glass vial with a tight lid. Because the crystals dissolve very slowly, the solution should be prepared at least the day before use. Shaking and/or ultrasonic treatment speeds up the process. All steps should be carried out in a well-ventilated hood. Always wear gloves when handling osmium tetroxide solutions. The solution is stable if kept in the cold and protected from strong light. Note that rubber stoppers should not be used.

1% buffered osmium tetroxide. To make 10 ml of fixative solution, mix 5 ml of 0.2 M cacodylate buffer and 5 ml of 2% osmium tetroxide stock solution.

4% agar in 0.1 M sodium cacodylate buffer. Dissolve during stirring 4 g of agar in 100 ml of 0.1 M sodium cacodylate buffer. Heat to close to 100°C until dissolved. Cool down to about 40°C before use.

I. Warning

Chemicals used in fixation for electron microscopy, notably aldehydes, osmium tetroxide, uranyl acetate, and cacodylate, are toxic and should be handled with adequate safety precautions, including gloves and good ventilation in a fume hood. Exposure to formaldehyde may lead to allergic reactions.

References

Afzelius, B. A. (1988). Microtubules in the spermatids of stick insects. *J. Ultrastruct. Mol. Struct. Res.* **98,** 91–102.

Bohman, S.-O., and Maunsbach, A. B. (1970). Effects on tissue fine structure of variations in colloid osmotic pressure of glutaraldehyde fixatives. *J. Ultrastruct. Res.* **30,** 195–208.

Glauert, A. M. (1975). Fixation, dehydration and embedding of biological specimens. *In* "Practical Methods in Electron Microscopy" (A. M. Glauert, ed.), Vol. 3, Part 1. Elsevier, Amsterdam.

Griffiths, G. (1993). "Fine Structure Immunocytochemistry." Springer-Verlag, Berlin.

Karnovsky, M. J. (1965). A formaldehyde-glutaraldehyde fixative of high osmolality for use in electron microscopy. *J. Cell Biol.* **27,** 137A.

Maunsbach, A. B. (1966a). The influence of different fixatives and fixation methods on the ultrastructure of rat kidney proximal tubule cells. I. Comparison of different perfusion fixation methods and of glutaraldehyde, formaldehyde and osmium tetroxide fixatives. *J. Ultrastruct. Res.* **15,** 242–282.

Maunsbach, A. B. (1966b). The influence of different fixatives and fixation methods on the ultrastructure of rat kidney proximal tubule cells. II. Effects of varying osmolality, ionic strength, buffer systems and fixative concentration of glutaraldehyde solutions. *J. Ultrastruct. Res.* **15,** 283–309.

Mizuhira, V., and Futaesaku, Y. (1972). New fixation for biological membranes using tannic acid. *Acta Histochem. Cytochem.* **5,** 233–235.

Pihakaski-Maunsbach, K., Soitamo, A., and Suoranta, U.-M. (1990) Easy handling of cell suspensions for electron microscopy. *J. Electr. Microsc. Techn.* **15,** 414–415.

Sabatini, D. D., Bensch, K., and Barrnett, R. J. (1963). Cytochemistry and electron microscopy: The preservation of cellular ultrastructure and enzymatic activity by aldehyde fixation. *J. Cell. Biol.* **17,** 19–58.

2. Dehydration and Embedding

A. Ethanol Dehydration and Epon Embedding

The protocol for ethanol dehydration in connection with epoxy resin embedding can be varied considerably without great effects on the final result. In most cases, short dehydration times seem preferable, but storage of specimens overnight at 4°C in 95 or 100% ethanol is usually without problems (see Chapter 5). Acetone is often used as an alternative to ethanol for dehydration with similar results. Both ethanol and acetone dehydration lead to extraction of tissue lipids and shrinkage of tissue dimensions. These effects may be slightly different in different protocols but cannot be eliminated. Most commercial epoxy resin kits give embeddings of similar hardness and cuttability.

Procedure

1. Rinse tissue fixed in aldehydes and/or osmium tetroxide for 2 × 30 min in the same buffer as used for fixation. Keep the tissue at 0–4°C.
2. Dehydrate for 2 × 15 min in 70% ethanol at 0–4°C. As in all following steps up to step 10, remove fluid with a plastic Pasteur pipette (broken off fragments from glass pipettes may damage expensive diamond knives) before adding new fluid. The tissue must never be allowed to dry.
3. Dehydrate for 2 × 15 min in 90% ethanol. This and the following steps of dehydration and infiltration are carried out at room temperature.
4. Dehydrate for 2 × 15 min in 95% ethanol.
5. Dehydrate for 2 × 15 min in absolute ethanol.
6. Place the tissue for 2 × 15 mm in propylene oxide (alternatively acetone or limonene). Take particular care that the tissue does not dry, as propylene oxide evaporates very rapidly. Because propylene oxide is toxic and very volatile, this and all subsequent steps should be carried out in a well-ventilated hood. Always use gloves.
7. Infiltrate the tissue for at least 60 min in a 1:1 mixture of propylene oxide and a complete mixture of epoxy resin (thus also containing the accelerator).
8. Transfer the specimens to the surface of the epoxy resin in a clean vial containing 100% resin. Use a fine forceps or a wooden stick to handle the specimen. Propylene oxide diffuses out of the tissue blocks when they sink through the resin. Leave the specimens in the epoxy resin overnight at room temperature.
9. Fill a flat embedding mold with epoxy resin. The molds should preferably contain a thin bottom layer of already polymerized resin; in this way the specimen will become totally surrounded by resin, rather than bordering to the mold surface, which may lead to a soft block. Place a small piece of paper with the identification number of the specimen in the resin. If necessary, adjust the location of the tissue block and/or the paper with a wooden stick.
10. Polymerize the specimens at 60°C for 2 days.

Difficulties in sectioning tissue embedded in epoxy resin are often due to soft resin blocks, which may originate from one or more deviations from the dehydration/infiltration protocol: (a) wrong epoxy resin composition, (b) insufficient stirring of the components of the resin, (c) too short infiltration time in pure resin, (d) too large specimen (all dimensions exceeding 1 mm), (e) too short polymerization time, (f) incomplete dehydration, (g) incomplete removal of ethanol, and (h) too low polymerization temperature.

B. En Bloc Staining/Fixation with Uranyl Acetate

After step 1 in the previous procedure, rinse the specimen for 2 × 15 min with sodium maleate buffer, pH 5.2. Stain for 60 min in 0.5% uranyl acetate in maleate buffer, pH 5.2. Rinse for 2 × 15 min in sodium maleate buffer. Continue to step 2 in Procedure 2A.

C. Checklist of Chemicals and Equipment

Chemicals

Epoxy resin kit containing epoxy resin, dodecenyl succinic anhydride (DDSA), methyl nadic anhydride (MNA). The catalyst BDMA (N-benzyldimethylamine) is less viscous than DMP 30 and can be substituted for it and used at a concentration of 3% (see Glauert, 1991).
Ethanol
Maleic acid
Uranyl acetate dihydrate
Propylene oxide

Equipment

Glass vials (5–10 ml) with lids
Latex gloves
Disposable beaker for mixing resin
Disposable plastic Pasteur pipettes
Flat embedding molds; alternatively, gelatin capsules
Fine forceps
Wooden sticks

D. Solutions

0.05 M maleate buffer. Dissolve 0.58 g maleic acid in about 80 ml water and adjust pH to 5.2 with 1 N NaOH. Fill up to 100 ml with water.

0.5% uranyl acetate in 0.05 M sodium maleate buffer. Dissolve 0.58 g maleic acid in 80 ml water and adjust pH to 6.0 with 1 N NaOH. Dissolve 0.5 g uranyl acetate dihydrate in this solution and adjust (if necessary) pH to 5.2 with NaOH. Fill up to 100 ml with water.

Epoxy resin. To make 100 g resin, mix 48 g Epon 812, 19 g DDSA, and 33 g MNA. Stir continuously for 5 min. Add 2 g DMP 30 or 3 g BDMA and stir continuously for another 5 min. The complete epoxy mixture should be used for initial infiltration within the next few hours as it will slowly start to polymerize at room temperature. The freshly mixed complete resin can be stored in the freezer (e.g., $-20°C$) for months in closed vials. The vials must attain room temperature before being opened and used for embedding. The hardness of the polymerized epoxy blocks can be modified by changing the ratio of DDSA/MNA. Thus, an increase of DDSA gives softer blocks, whereas an increase in MNA gives harder blocks (Luft, 1961; Glauert, 1975). Adjustment of the anhydride : epoxide ratio is unnecessary, although previously considered important for controlling the properties of the resin. Whereas polymerization is normally carried out at 60°C, a higher curing temperature (80–100°C) can be used for very rapid polymerization, although this may lead to structural derangement of membranes.

E. Warning

Chemicals used during dehydration and embedding in epoxy or acrylic resins are toxic (mutagenic, allergenic, and, in some cases, perhaps carcinogenic) and should be handled with adequate safety precautions (Ringo et al., 1982). Work in a well-ventilated hood and use gloves. Note that propylene oxide and resins can penetrate most types of gloves within a short time.

References

Glauert, A. M. (1975). Fixation, dehydration and embedding of biological specimens. In "Practical Methods in Electron Microscopy" (A. M. Glauert, ed.), Vol. 3, Part 1. North-Holland, Amsterdam.

Glauert, A. M. (1991). Epoxy resins: An update on their selection and use. Eur. Microsc. Anal. September 15–20.

Luft, J. H. (1961). Improvements in epoxy resin embedding methods. J. Biophys. Biochem. Cytol. 9, 409–414.

Maunsbach, A. B. (1998). Embedding of cells and tissues for ultrastructural immunocytochemical analysis. In "Cell Biology: A Laboratory Handbook" (J. E. Celis, ed.), 2nd Ed. Vol. 3, pp. 260–267. Academic Press, San Diego.

Ringo, D. L., Brennan, E. F., and Cota-Robles, E. H. (1982). Epoxy resins are mutagenic: Implications for electron microscopists. J. Ultrastruct. Res. 80, 280–287.

3. Low Temperature Embedding

A. Cryofixation and Lowicryl Embedding

Cryosectioning and freeze substitution followed by Lowicryl embedding are alternative procedures in immunoelectron microscopy. Cryosectioning and immunolabeling provide a higher sensitivity for most antigens. However, the equipment for freeze-substitution is less expensive and easier to handle than equipment for cryoultramicrotomy. Furthermore, Lowicryl blocks are stable and can be stored at 4° or −20°C and repeatedly resectioned and immunolabeled.

For immunoelectron microscopy the tissue is usually fixed for a short time (e.g., 30 min–2 hr) with 2, 4, or 8% paraformaldehyde or 4% paraformaldehyde plus 0.1% glutaraldehyde, but should never be postfixed in osmium tetroxide. Tissue blocks should be small and completely infiltrated with 2.3 M sucrose before freezing. Substitution in 0.5% uranyl acetate in methanol (Schwarz and Humbel, 1989) results in good ultrastructural preservation combined with retained tissue antigenicity. The size of the specimen for freeze-substitution must be small to allow complete removal of water and of methanol. Remaining water or methanol will result in white blocks and holes in the tissues. Substitution times may require adjustment according to the tissue studied.

Procedure

1. Transfer the tissue directly from the aldehyde fixative to the buffered sucrose–paraformaldehyde solution. The tissue blocks should not exceed 0.5 mm in any direction. Place the specimens in the upper layer of the sucrose–paraformaldehyde and stir the solution. Infiltrate for 1 hr at room temperature.
2. Cryofix the tissue in liquid nitrogen. Hold the specimen gently with fine forceps provided with cold-insulated handle. Dip the specimen quickly into the liquid nitrogen, where it may be left for weeks or months or processed immediately. Carefully follow the safety regulations when handling liquid nitrogen. It is necessary that the specimen is frozen ("cryofixed") without ice crystal formation, which will leave holes in the tissue.
3. Transfer the frozen tissue very rapidly with a precooled forceps from the liquid nitrogen to the methanol/0.5% uranyl acetate solution, which is kept at −85−−90°C in the capsules in the freeze-substitution unit. During this transfer great care must be taken not to warm the specimen. For this purpose the vessel with the liquid nitrogen must be placed adjacent to the substitution unit. In order to avoid recrystallization of tissue water during freeze-substi-

tution, it is important that the temperature of the specimen does not increase in connection with fluid changes. Fluids must be properly temperature equilibrated in the freeze-substitution unit before being added to the samples.
4. After 4–8 hr, withdraw most of the substitution fluid from the capsules with a polyethylene Pasteur pipette. The diameter of the tip of the Pasteur pipette should be smaller than the size of the specimens as the specimens are difficult to observe at this step and may otherwise be removed. Fill the capsules with temperature-equilibrated methanol/0.5% uranyl acetate solution and raise the temperature to −80°C for 24 hr. Rinsing and infiltration periods are adjusted in the following to suit regular working hours.
5. Rinse the specimens three times with pure methanol at −70°C over a period of 8 hr.
6. Rinse the specimens three times with pure methanol at −45°C over a period of 20 hr.
7. Infiltrate the specimens with a 2 : 1 mixture of methanol and Lowicryl HM20 for 6 hr at −45°C.
8. Infiltrate the specimens with a 1 : 1 mixture of methanol and Lowicryl HM20 for about 14 hr at −45°C.
9. Infiltrate the specimens with pure Lowicryl HM20 at −45°C for 8 hr with three changes.
10. Infiltrate the specimens with pure Lowicryl HM20 at −45°C for 24 hr.
11. Fill up the capsules with fresh Lowicryl HM20 and close the lids. Polymerize with indirect UV light at −45°C for 48 hr. Capsules should be completely filled with Lowicryl and the lids completely closed to exclude oxygen during polymerization. Oxygen in the resin interferes with polymerization and results in soft blocks.
12. Increase the temperature to 0°C and continue UV polymerization for 24 hr. The specimens are now ready for conventional ultramicrotomy at room temperature.

Embedding in Lowicryl K4M or LR White can be performed with a similar fixation procedure and provides blocks and sections that are also suitable for immunoelectron microscopy, but the embedding in the resin should be performed at −30°C.

B. Checklist of Chemicals and Equipment

Chemicals

Disodium hydrogen phosphate dihydrate ($Na_2HPO_4 \cdot 2H_2O$)

Liquid nitrogen

Lowicryl HM20 kit containing HM20 resin monomer E, HM20 cross-linker D, and HM20 initiator C

Methanol
Paraformaldehyde (dry powder)
Sodium dihydrogen phosphate monohydrate (NaH$_2$PO$_4$ · H$_2$O)
Sodium chloride
Sodium hydroxide
Sucrose
Uranyl acetate dihydrate

Equipment

Disposable beaker for mixing Lowicryls
Fine forceps with cold-insulated shaft
Freeze-substitution apparatus with regulation between −85° and 0°C: Balzers FSU 010 freeze-substitution unit (Bal-Tec), Leica EM AFS automatic freeze-substitution system, Reichert freeze-substitution unit (Leica AG), or an equivalent (e.g., customer-built) freeze-substitution apparatus
Polyethylene capsules with pyramid shape and hinged lids for low-temperature embedding
Polyethylene Pasteur pipettes with extended fine tips
UV lamp for polymerization with 350-nm UV light (unless built into the freeze-substitution apparatus)

C. Solutions

Lowicryl HM20. To make about 20 g, gently mix 3.0 g cross-linker D and 17.0 g monomer E. Bubble dry nitrogen gas into the mixture for 5 min to exclude O$_2$ which inhibits Lowicryl polymerization. Add 0.1 g initiator C and mix gently until it is dissolved. Avoid making air bubbles.

Methanol : Lowicryl HM20 mixtures in proportions 2 : 1 and 1 : 1.

2.3 M sucrose in PBS-buffered 2% paraformaldehyde. To make a 100-ml solution, weigh out 0.038 g sodium dihydrogen phosphate monohydrate, 0.128 g disodium hydrogen phosphate dihydrate, 0.877 g sodium chloride, and 78.7 g sucrose. Add water and 10 ml of 20% paraformaldehyde to about 95 ml. Use a magnetic stirrer until the sucrose is dissolved, which usually requires several hours. Adjust pH to pH 7.2 and fill up with water to 100 ml.

20% stock solution of paraformaldehyde. To make a 100-ml solution, mix 20 g of paraformaldehyde powder with 80 ml of distilled water in a glass flask. Heat to 60°C while gently agitating the milky solution. Add 1 N sodium hydroxide dropwise until the solution clears up. Complete to 100 ml with distilled water. This procedure should be carried out in a well-ventilated hood and with protection of the face. Always wear gloves. The solution can be stored for a few weeks in the refrigerator.

0.5% uranyl acetate in methanol. Dissolve 0.5 g uranyl acetate dihydrate in 100 ml methanol.

References

Carlemalm, E., Garavito, R. M., and Villiger, W. (1982). Resin development for electron microscopy and an analysis of embedding at low temperature. *J. Microsc.* **126,** 123–143.

Carlemalm, E., Villiger, W., Hobot, J. A., Acetarin, J.-D., and Kellenberger, E. (1985). Low temperature embedding with Lowicryl resins: Two new formulations and some applications. *J. Microsc.* **140,** 55–63.

Maunsbach, A. B. (1998). Embedding of cells and tissues for ultrastructural and immunocytochemical analysis. *In* "Cell Biology: A Laboratory Handbook" (J. E. Celis, ed.), 2nd Ed., Vol. 3, pp. 260–267. Academic Press, San Diego.

Newman, G. R., and Hobot, J. A. (1993). "Resin Microscopy and On-Section Immunocytochemistry." Springer-Verlag, Berlin.

Schwarz, H., and Humbel, B. M. (1989). Influence of fixatives and embedding media on immunolabelling of freeze-substituted cells. *Scan. Microsc.* Suppl. 3, 57–64.

4. Support Films

The three most commonly used support films are made of Formvar, carbon, and collodion. Formvar films are strong and elastic; carbon films are thin but strong and stable; and collodion films are easy to make but have to be strengthened by an evaporated carbon layer as they are fragile and tend to drift in the electron beam; they lose more than half of their mass by electron bombardment.

A. Formvar Films

A Formvar film (polyvinyl formal) is a replica of a glass surface, e.g., a 3″ × 1″ microscope slide.

Procedure

1. Dissolve 0.35% Formvar in pro analysi chloroform (or redistilled ethylene dichloride). This is done the day before making the support film, as Formvar dissolves slowly.
2. Pour the Formvar solution into a narrow glass jar with a tight-fitting lid to a height of about 4 cm.
3. Clean a light microscopical glass slide carefully, using a liquid laboratory detergent or acetone; rinse in deionized water and dry with optical lens tissue. If polished too energetically, however, it will become electrostatically charged and attract dust particles.
4. Immerse the glass slide (or a glass strip) in the Formvar solution for a few seconds; lift the slide out of the fluid but keep it inside the glass jar for half a minute. Then lift the slide and touch the bottom edge lightly with a filter paper. Excess solution will evaporate rapidly. The thickness of the Formvar film depends on the concentration of the Formvar solution and how much is drained off the slide. Steps 1–3 should be performed in a ventilated hood, the remaining ones on a laboratory bench.
5. Inspect the Formvar-coated slide for irregularities. If acceptable, continue, otherwise try another glass. Then score the Formvar film along the edges of the glass slide, using a sharp metal point such as a forceps prong or a razor blade.
6. Float the Formvar film off from the glass slide and onto a clean water surface. This is done by lowering the slide into the water at an angle of about 30°, which will result in one film, or at a right angle to the surface, which (usually) results in two film. It is particularly important that the water be kept clean; fingers must never come into the water.
7. When the film has come off and is floating on the water surface it should be inspected. A thin film should be almost invisible in reflected light. Put grids onto the good parts of the film and avoid areas where there are wrinkles and irregularities.
8. Pick up the film with its grids using Parafilm or a nonsoaking paper.
9. Before proceeding to make more Formvar films, inspect a few of the first ones in the electron microscope. If they are acceptable, continue making more films, if not (contamination, dust, wrinkles) try another slide.
10. Evaporate a thin layer of carbon on the Formvar to strengthen the film. This will also greatly diminish the tendency of Formvar film to drift in the electron microscope.

B. Carbon Films

Mica is used as the substrate onto which carbon is deposited. Freshly cleaved mica has the advantages of being evenly planar down to atomic dimensions and by having a hydrophilic surface, which simplifies stripping of the film.

Procedure

1. Cut a piece of mica with a scalpel into the shape of rectangles with a side length of a few centimeters, then cleave, and, with the fresh surface up, put into the carbon sputter apparatus using carbon filaments.
2. Evaporate about 1–2 nm of carbon onto the mica. The adequate amount can be monitored by the use of a white porcelain chip with a drop of glycerol located beside or at a distance somewhat closer to the carbon source than the mica. With some experience the amount of carbon can be estimated easily from the color difference between the porcelain and the glycerol, which stays white.
3. Place a number of clean grids on a filter paper of similar (or smaller) dimensions as the piece of mica and submerge the filter paper in the water used to float off the carbon film. Then float the evaporated carbon film off (with the same method as described for Formvar film manufacture). Defects will be detected in reflected light on the water surface.
4. Lower the water surface and maneuver at the same time the carbon film over the grids. This step is easy if the water is kept in a funnel with regulated outlet. Let the grids dry.

C. Collodion Films

Procedure

1. Place a dish with a diameter of 20 cm or more on a tray with a black background. (A black plastic sheet between the dish and the tray can be used and will enable the operator to see any dust on the water surface or support film.) Fill the tray with filtered or deionized water until the water surface is convex. Clean the water surface by sweeping a clean glass rod over the surface.
2. Place a drop of a 2% solution of collodium (e.g., Parlodion or some other nitrocellulose) in amylacetate onto the water surface. Amylacetate evaporates rapidly and leaves a round film of collodion floating on the surface.
3. Check the collodion film for wrinkles, dust, and irregularities using light reflections of a strong lamp.
4. Place a number of support grids on the floating collodion film.
5. Place a piece of Parafilm on top of the film, and pick up and put the grids with their adhering collodion film into a large petri dish or in another dust-free space and allow to dry. Store in petri dish.
6. Strengthen the collodion-coated grid by sputtering a thin layer of carbon onto it.

D. Checklist for Chemicals and Equipment

Chemicals

Chloroform, pro analysi
Collodion (nitrocellulose, a trade name is Parlodion)
Deionized or Milli-Q-filtered water
Ethylacetate
Ethylene dichloride, double distilled
Formvar (polyvinyl formal)

Equipment

Carbon sputter
Dish with a diameter of 20 cm or more placed on a tray; a piece of black plastic is placed on the tray under the dish
Filter paper
Glass jar with a tight-fitting lid
Glass rod
Glass slides
Grids
Mica
Optical lens tissue
Parafilm
Petri dishes
Watchmakers' forceps (antimagnetic)

5. Ultramicrotomy

Ultramicrotomy of resin-embedded biological objects is a basic method in most laboratories for biomedical electron microscopy. Ultramicrotomes have been perfected gradually over the years and it is now possible to cut routinely sections that are both thin and large. Different aspects of ultramicrotomy have been thoroughly discussed by Reid and Beesley (1991), and the characteristics of the ultramicrotomes and their handling are described in detail in the instrument manuals and will not be discussed here. The following only comments on a few important steps in the preparation of thin resin sections.

Ultrathin sections can also be cut from tissue, which has been aldehydefixed, cryoprotected, and frozen. However, cryoultramicrotomy is more demanding than resin ultramicrotomy and it is advised to learn it in a laboratory where it is practiced or at technical courses; some of which are given by the manufacturers.

A. Cutting Semithin Sections

Remove the specimen from the embedding mold or BEEM capsule. If the specimen is embedded in a gelatine capsule, the capsule should be peeled off. Trim off excess resin above and on the sides of the embedded tissue. Slice off enough on the top to expose the tissue specimen.

Semithin sections (1 μm) provide good overview of the specimen in the light microscope. A glass knife is used to shave off the sectioning surface using the "bad" side of the knife, then to cut the semithin sections on the "good" part of the knife. The semithin sections can be cut dry and lifted with a watchmaker's forceps and placed on water drops on a microscope slide. Alternatively, they can be cut with water in the trough and picked up with a wire loop. After drying, staining with toluidine blue, and mounting, the sections are used to identify the interesting part of the specimen.

B. Trimming for Ultrathin Sectioning

The block face is trimmed down to the shape of a tetraeder ("mesa"); two sides of the block face should be parallel to each other and parallel to the knife edge during sectioning, and the other two sides may be perpendicular to the knife edge during sectioning, or one side slightly slanted to give the sections a recognizable shape; this will help the investigator to orient the sections the same way in the electron microscope. Use a very sharp and clean razor blade to cut the final sides of the block face straight and smooth. If the ultrathin sections are collected on grids without a supporting film, a margin around the area of interest has to be left on all four sides. The mirror-smooth surface obtained by the cutting of semithin sections is an essential aid in the specimen-to-knife alignment in three dimensions before the start of sectioning (the block face can thus be seen from its front, from above, and from the side).

C. Glass and Diamond Knives

For the production of glass knives, the instructions of the manual accompanying the knife maker should be used. The glass knife should be inspected with a binocular to decide what part of the knife is free of irregularities.

For most specimens, diamond knives are preferable to glass knives and, in the long run, more economical. When purchasing a diamond knife, the edge length should be more than twice the length of the standard section width to be used: lengths of 2.0–2.5 mm are commonly used. This gives several separate cutting areas. Diamond knives are delivered with a recommended cutting speed and clearance angle. The resin hardness should be adjusted to the diamond knife, whereas the opposite is true for glass knives. A "soft" or badly polymerized resin will leave a thin deposit on the knife edge. When the deposit slowly polymerizes, the knife will appear dull. After the sections have been collected, the knife and trough should be rinsed with double-distilled water from a plastic wash bottle and the knife cleaned along the edge with a styrofoam or elderpit stick or a boiled wooden toothpick sharpened to a wedge with a razor blade and wetted with double-distilled water.

Diamond knife edges have a tendency to be hydrophobic, preventing total wetting of the knife; this can be remedied by leaving it for 20 min with a very high meniscus with double-distilled water.

D. Cutting Ultrathin Sections

It is of utmost importance that the water trough of the knife, the grids, the forceps, and other implements used for sectioning and section collection are very clean. The actual sectioning procedure is described in the microtome manual. Particular care should be paid to the thinness of the first section. If a specimen block tends to attract water from the trough while passing the knife, the water level in the trough should be lowered. Sometimes this is only possible with a retained wetted edge if *one* drop of ethanol is added to the water. (With stronger ethanol solutions, components of the glue from the trough of the diamond knife might dissolve and cause contaminations.)

E. Section Collection

There are different techniques for collecting sections from the water surface. One way is simply to lower a net grid with or without support film from above onto the sections on the water surface. However, this method usually results in sections with wrinkles. A better way is to lower the grid three-quarters into the water. With the aid of an eyelash hair attached to a toothpick, a row of sections is drawn to the grid and secured against it. The grid is then raised slightly and, in the case of small sections, one or two more rows of sections can be attached parallel to the first row. A third, and recommended, way is to pick up large sections using a one-hole grid (Galey and Nilsson, 1966). Lower the empty, cleaned one-hole grid onto the water surface so that the sections will appear in the grid opening. (Note that the sections should first be moved far away from the knife edge in order to avoid damaging the edge.) Lift the one-hole grid with its water droplet and sections and place it on a grid with support film. Very carefully remove some of the water from the sandwiched grids by touching the grid edge perpendicularly with a small piece of filter paper, let the grids dry, and carefully separate them. Sections picked up in this manner tend to have less wrinkles than those picked up by other methods.

References

Reid, N., and Beesley, J. E. (1991). Sectioning and cryosectioning for electronmicroscopy. *In* "Practical Methods in Electron Microscopy" (A. M. Glauert, ed.), Vol. 13. Elsevier, Amsterdam.

Galey, F. R., and Nilsson, S. E. G. (1966). A new method for transferring sections from the liquid surface of the trough through staining solutions to the supporting film of a grid. *J. Ultrastruct. Res.* **14,** 405–410.

6. Section Staining

Unstained sections of biological tissues, in general, show low contrast when examined in a transmission electron microscope. Almost invariably sections are therefore contrasted, or "stained," by solutions containing salts of heavy metals, such as uranyl acetate (Watson, 1958) and/or lead citrate (Reynolds, 1963). The staining serves to enhance the contrast of cellular components in the microscope, but from a chemical point of view it is rather unspecific.

In most staining procedures the grids are floated on a droplet of the staining solution, then rinsed in water and dried. This in principle is very simple, but there are, in addition to the previously mentioned stains, a number of other recipes for staining solutions and variations of the actual staining procedure (Afzelius, 1992; Lewis and Knight, 1992). This diversification is due to the fact that many factors influence the results. Even minor modifications may give unwanted results, such as precipitation or contamination on the specimen.

A. Section Staining with Uranyl Acetate

Uranyl acetate staining results in a uniform, but fairly weak, general staining of cellular components. There is a slight preference for an increased contrast of DNA and RNA.

Procedure

1. Filter a suitable volume of stain solution through a 0.2-μm filter immediately before use.
2. Place a series of drops on Parafilm or disposable petri dishes using a clean plastic Pasteur pipette or a syringe with a Millipore filter. Note that new glass Pasteur pipettes are usually dirty and have to be cleaned before use. The number of drops should equal the number of grids to be stained.
3. Place the grids section side down on the drops with an interval of 1 min between each grid.
4. Leave the grids on the stain drops for 10 min.
5. Remove and rinse the grids in the same order that they were placed on the staining solution. Lift grid with clean forceps and hold vertically. Rinse with a stream of 15–20 drops of redistilled or Millipore-filtered water from a Pasteur pipette onto the grid.
6. Remove the last drop of water by touching the edge of the grid with a filter paper. Place a small piece of filter paper between the prongs of the forceps to remove additional excess water.
7. While still clamped in the forceps, let the grid dry for 5 min.

B. Section Staining with Lead Citrate

Lead staining is used when a general survey of tissue ultrastructure is the goal. The following procedure results in a general increase in contrast of tissue components; membranes will stand out distinctly. Lead citrate staining is also used when the tissue has been stained previously *en bloc* in uranyl acetate according to Karnovsky (1967) before embedding in resin. Such sections usually provide optimum definition of fine structural details.

Procedure

1. Filter 10 ml of stain solution through filter paper into a 10-ml cylinder. Cover the cylinder with a piece of Parafilm. Handle the cylinder carefully without shaking it.
2. Withdraw some stain solution with a clean Pasteur pipette from the interior of the stain solution without blowing air bubbles into the stain.
3. To stain four grids, place 4 drops of stain on Parafilm. Do not stain more than four grids at a time as precipitate forms easily at the surface of the drops. (Surface precipitate can be minimized by placing NaOH tablets around the grids, supposedly to bind exhaled carbon dioxide, but in practice we do not find this necessary.)
4. Immediately place the grids on the drops with an interval of half a minute.
5. Stain the sections for 2 min (range: 15 sec to 20 min, depending on desired contrast).
6. Remove the grids with intervals of half a minute with fine forceps. Hold the grid vertically and squeeze gently 10–15 drops of redistilled or Millipore-filtered water from a Pasteur pipette onto the grid.
7. Touch the grid edge perpendicularly with a filter paper and leave the grid to dry in the forceps with a small piece of filter paper between the prongs of the forceps.

C. Double Staining with Uranyl Acetate and Lead Citrate

Double staining with uranyl acetate followed by lead citrate is applied when a general high tissue contrast is desired.

Procedure

1. Stain with uranyl acetate as described earlier.
2. Let the grid dry.
3. Stain with lead citrate as described earlier.
4. Let the grid dry.

D. Comments

The chemicals used for section staining are toxic and should be handled with adequate safety precaution. Staining should be performed in a well-ventilated hood using gloves. The results of section staining depend in part on the preceding fixation of the tissue and in part on the characteristics of the embedding medium and on the protocol of the staining process. Thus if uranyl acetate staining is carried out at 60°C instead of room temperature, contrast is enhanced greatly, which may reveal new tissue components but may overstain others. Various instruments and procedures have been worked out for the simultaneous staining of many grids, e.g., a full grid box. We have never adopted these procedures. It is better to lose only a few grids rather than a full grid box if the staining fails.

1. Dirt may contaminate the sections unless all glassware, Parafilm, and pipettes are *meticulously* clean and all solutions filtered. Contamination can also derive from the support film or from the water trough during sectioning.
2. Excess rinsing of lead stain may gradually remove stain and may make the staining more susceptible to beam damage in the electron microscope.
3. The surface of lead citrate drops rapidly acquires contamination due to interaction with carbon dioxide in the air. This will appear as contamination on the section. Therefore, drops must be placed on the Parafilm immediately before the grids are applied to the surface of the drops.
4. Some types of stain contamination may be prevented or removed by treating the sections with acids (see Kuo, 1980; Mollenhauer, 1987).

E. Checklist for Chemicals and Equipment

Chemicals

Lead citrate
Sodium hydroxide
Trisodium citrate dihydrate
Uranyl acetate dihydrate
Redistilled water

Equipment

Magnetic stirrer
Parafilm (or dental wax)
Disposable Pasteur pipettes

F. Solutions

Saturated solution of uranyl acetate. To make 100 ml, dissolve 7.69 g of uranyl acetate in 100 ml redistilled or Millipore-filtered water and stir for some hours. Store the solution at room temperature in the dark in a closed glass vial covered by aluminium foil.

Lead citrate solution. Boil about 50 ml redistilled or Millipore-filtered water for 5 min in a beaker. Dissolve 1.33 g lead nitrate and 1.76 g sodium citrate in 40 ml of the boiled water. Stir the solution rapidly for 30 min with a magnetic stirrer. Add 8.0 ml of 1 N sodium hydroxide freshly prepared with redistilled Millipore-filtered water. The solution clears after the addition of NaOH. Fill up to 50 ml with water. The solution should stand for 2–3 days before use. When stored at 4°C, the solution can be used for 6–12 months.

References

Afzelius, B. A. (1992). Section staining for electron microscopy using tannic acid as a mordant: A simple method for visualization of glycogen and collagen. *Microsc. Res. Techn.* **21**, 65–72.

Karnovsky, M. J. (1967). The ultrastructural basis of capillary permeability studied with peroxidase as a tracer. *J. Cell Biol.* **35**, 123–236.

Kuo, J. (1980). A simple method for removing stain precipitates from biological sections for transmission electron microscopy. *J. Microsc. (Oxford)* **120**, 221–224.

Lewis, P. R., and Knight, D. P. (1992). Cytochemical staining methods for electron microscopy. *In* "Practical Methods in Electron Microscopy" (A. M. Glauert, ed.), Vol. 14, pp. 1–321, Elsevier, Amsterdam.

Mollenhauer, H. H. (1981). Contamination of thin sections: Some observations on the cause and elimination of "embedding pepper." *J. Electr. Microsc. Techn.* **5**, 59–63.

Reynolds, E. S. (1963). The use of lead citrate at high pH as an electron-opaque stain in electron microscopy. *J. Cell Biol.* **17**, 208–212.

Watson, M. L. (1958). Staining of tissue sections for electron microscopy with heavy metals. *J. Biophys. Biochem. Cytol.* **4**, 475–478.

7. Microscopy and Image Recording

The procedure for the electron microscopic analysis of a grid depends of course on the brand of microscope used and will not be described in detail here. However, from our experience of using microscopes from several different manufacturers we may give a few general recommendations.

First of all check the alignment of the microscope and, in particular, make sure that the astigmatism is corrected. Make sure that there are no mechanical or electrical instabilities by using a test specimen or the actual specimen to be examined. A slight misalignment of the electron beam usually has only little influence on the image quality, whereas electrical instabilities due to, for example, contamination within the microscope column or mechanical instabilities causing stage movement, make it impossible to obtain useful micrographs.

Start microscopy of the specimen using low magnification (less than 5000 times) and try to find areas without obvious specimen defects, such as contaminations, section wrinkles, or holes in the supporting film. When a suitable area is available, define the aim of the investigation. If mainly survey micrographs are required, record these before inspecting details of the area at higher magnification, as there is a risk of uneven contamination or beam damage. However, if high magnification micrographs are required, these should be recorded first, before contamination obscures details. For stereological studies the investigator must design a strict sampling scheme, e.g., recording the upper left portion of each grid opening or some other predetermined area, rather than recording what appears interesting. Some microscopes are equipped with a system of making multiple recordings at predetermined distances.

Immediately before recording an area of the specimen the microscopist has to check that there is no specimen drift. This is done most effectively by comparing the positions of a point on the specimen screen and a neighboring object in the section, e.g., ribosome. Observation of these points for some seconds will show directly whether there is specimen drift, perhaps caused by movement of the supporting film. The remedy for this problem is to reduce beam intensity or to use more stable specimens, most easily by evaporating a thin carbon layer over the specimen.

The exposure dose should be adjusted in such a way that the resulting exposed and developed negatives have a sufficient density. However, the exposure time should not be too long; in general we never use exposure times over 1 sec, as some specimen drift may otherwise be disturbing. Different approaches can be made to achieve proper focus. If the instrument has a wobbler that is adjusted correctly, this is a helpful tool, particularly at magnifications less than 20,000 times. If the specimen contains contrasty edges, focusing is preferably done with the aid of these edges. In order to obtain a properly focused micrograph it may be useful to make a through focus series. With some practice, most microscopists learn to find a suitable focus level by observing the image pattern directly, for instance, the changes in granularity. the microscope must not be touched during the actual image acquisition.

Development of electron microscopical films should follow the instructions given by the manufacturer. A general rule is that the electron micrographs should not be underdeveloped, as they then lose in contrast. It is advisable to agitate the developer occasionally during the developing procedure.

8. Photographic work

A. Processing of Electron Microscope Negatives

Digital recording of images in the electron microscope is being used increasingly, but the classical recording on photographic material remains the standard method in most laboratories. Darkroom work with electron microscope negatives is similar to work with any other black-and-white negatives, as is the enlargement of negatives.

The developer should be freshly made (or given some refreshing medium). It should be protected from being oxygenized by the air, most simply done by using a floating lid on the tank with the developing solution. An aged or overused developer will give low-contrast negatives. When aged the developing solution will turn dark and turbid. It is a good idea to keep a log in the darkroom and note the date, when the developer is changed, and register how many negatives have been developed. Even if the developer is changed routinely, e.g., every Monday morning, an energetic investigator may unexpectedly develop several hundred negatives within 1 day, which will require a change of the developer. The temperature of the developer is important and should be controlled. Even a decrease of a few degrees below the recommended temperature will give underdeveloped negatives and will require an increased developing time.

The quality of the fixative solution should be checked by adding a drop of 10% sodium iodide to 1 ml of the solution; if the solution turns opaque it should be discarded; it would be a sign of the fixative being useless. Old fixative must not be poured down the drain as this harms the sewage treatment processes. After fixation the negative material should be rinsed for a minimum of 30 min in running tap water and then briefly dipped in tap water containing a small amount of detergent to prevent drying spots or stripes on the negatives.

In principle it is possible to use almost any negative material in the electron microscope, but photographic sheet films or photographic glass plates that are manufactured specifically for electron microscopy are preferable. Their emulsions are usually thinner than those of other films and are also relatively insensitive to visible light.

Instructions for the handling of film material are given on data sheets provided by the manufacturer.

B. Exposure of Photographic Paper

When focusing the electron microscope negative in the enlarger, a focusing aid in the form of a small focusing telescope is very helpful. In this step the diaphragm of the enlarger should be opened fully, then closed to f8 or, even better, f11 for making the prints. With a relatively long exposure (e.g., 15 sec or more) the operator is given time to selectively reduce the exposure of some parts of the photographic paper. If an in-built scale of the photographic enlargement is required, a transparent plastic ruler or some other scale may be laid on top of the negative.

C. Processing of Photographic Paper

The photographic paper is either processed in the classical way through developer, rinse, fixative, rinse, and then dried or processed in a printing machine. For printing routine electron micrographs, a printing machine is a good, time-saving device, but some investigators may want to print their best micrographs using open tanks with developer and fixative. One reason for this is that the paper used in some automatic printing machines appears to have a more narrow contrast scale than ordinary photographic paper and may, in addition, slowly acquire a yellow or brownish tone on storage or exposure to light. However, ordinary printing paper also seems to be produced with a more limited contrast scale, apparently due to a lowered silver content in the emulsion. Regardless of which brand of photographic paper is used, it is usually necessary to have access to soft, medium, hard, and extra hard paper. Usually the photographic paper for making prints of electron micrographs should be glossy. As a rather bright red light can be used in the darkroom, the photographer who develops in open tanks may be able to see the gradual darkening of the various motives and to stop or prolong the printing as seem suitable.

9. Negative Staining

Negative staining is a simple and rapid preparation procedure for isolated molecules or cell components. Different technical modifications exist (Bremer *et al.*, 1998; Harris, 1997) but they are in principle closely related. Several different stains can be used and 1% uranyl acetate and 2% sodium phosphotungstate are among the most commonly used. The properties of the support film are important; carbon films or Formvar films with a thin evaporated carbon layer are preferred. The surface of the films should preferably be made hydrophilic in a glow-discharge unit. The following two simple procedures are given for negative staining.

A. "Stain-on-Grid" Procedure

1. Hold the grid with a pair of clean forceps.
2. Place a 5-μl drop of sample solution on the grid.
3. After 60 sec blot off the sample by touching the edge of the grid to filter paper. The grid must not dry.
4. Immediately add about a 15-μl droplet of stain solution to the grid. After a few sec blot off stain and add a new stain drop.
5. After 60 sec blot off excess stain thoroughly.
6. Let the grid dry. It is now ready for the microscope.

B. "Grid-on-Stain" Procedure

1. Hold the grid with a pair of clean forceps.
2. Place a 5-μl drop of sample solution on the grid.
3. After 60 sec blot off the sample by touching the edge of the grid to filter paper. The grid must not dry.
4. Place the grid briefly on a drop of distilled water that has been placed on a sheet of Parafilm.
5. Blot off excess water. In some experiments step 4 can be repeated.
6. Place the grid on a drop of stain solution that has been placed on the Parafilm. After a few moments move the grid to a fresh drop of stain.
7. After 1 min blot off excess stain and let the grid dry.

For many samples a protein concentration of about 0.1 mg/ml is suitable. If a glow-discharge unit is not available the spreading of the negative staining can be improved by adding octylglucoside to the sample or stain solution at a concentration of about 2 mM or by adding 25 μg/ml bacitracin.

D. Checklist for Chemicals and Equipment

Chemicals

Bacitracin
n-Octyl-α-D-glycopyranoside (octylglucoside)
Phosphotungstic acid
Sodium hydroxide
Uranyl acetate dihydrate

Equipment

Double-distilled or deionized water
Filter paper
Forceps, antimagnetic
Micropipettes, 5 μl
Parafilm
Pasteur pipettes
Grids; 200 mesh with carbon or Formvar/carbon films

C. Solutions

2% sodium phosphotungstic acid. Dissolve 0.2 g phosphotungstic acid in 9 ml double-distilled water. Adjust the solution with 0.1 N sodium hydroxide to neutrality. Fill up with water to 10 ml.

1% uranyl acetate. Dissolve 0.1 g uranyl acetate dihydrate in 10 ml water. The pH of this solution will be ca. 4.5 and should not be adjusted. This solution is unstable above pH 5.0.

References

Aebi, U., and Pollard, T. D. (1987). A glow discharge unit to render electron microscope grids and other surfaces hydrophilic. *J. Electr. Microsc. Techn.* **7,** 29–33.

Bremer, A., Häner, M., and Aebi, U. (1998). Negative staining. *In* "Cell Biology: A Laboratory Handbook" (J. E. Celis, ed.), Vol. 3, pp. 277–284. Academic Press, San Diego.

Harris, J. R. (1997). "Negative Staining and Cryoelectron Microscopy: The Thin Film Techniques." Bios Scientific, Oxford.

Namork, E., and Johansen, B. V. (1982). Surface activation of carbon film supports for biological electron microscopy. *Ultramicroscopy* **7,** 321–330.

10. Autoradiography

In principle, electron microscope autoradiography (EMAR) is a simple method but requires attention to a number of practical steps. The following procedure has been used in our laboratories essentially unchanged for many years.

A. The Wire Loop Method

Ultrathin sections containing the radioactive isotope to be located are picked up on 200 mesh grids covered with Formvar supports films. Before applying the autoradiographic emulsion, the sections should be stained with lead citrate or double stained with uranyl acetate and lead citrate.

Procedure

1. Evaporate a thick carbon layer on the sections to prevent the developing and fixing solutions from dissolving the stain. The carbon layer should be thicker than that usually applied on grids to strengthen the support film.
2. All the following steps should be carried out in a darkroom. Make absolutely sure that the emulsion is not sensitive to the darkroom light.
3. Weigh out 4 g of Ilford L4 Nuclear Research Emulsion in a glass beaker; add 6 ml of distilled water.
4. Place the beaker in a water bath at 40°C. Stir occasionally until all emulsion is melted and the solution is homogeneous. This usually takes at least 30 min. Leave the melted emulsion in the water bath until all grids have been coated.
5. Dip the metal loop in the melted emulsion. When the loop is withdrawn a film of emulsion in sol form will appear in the loop.
6. Hold the grid in such an angle that the red light from the darkroom lamp is reflected from the film. Wait for 30–60 sec. The film in the loop will start to gel, which can be observed as an increase in the intensity of the reflected light.
7. After 30–120 sec, when the central part of the film has gelled (the periphery may still be in sol form) the film is ready for application on the grids with the sections, which have been coated with carbon in advance.
8. Place the grid with the sections facing upward on top of a rod (metal or wood) with a diameter slightly less than the grid.
9. Move the loop with the gelled emulsion onto and past the grid. In that way the emulsion film becomes attached to the grid. The outer non-gelled part of the emulsion film in the loop must not touch the grid.
10. Lift the grid from the rod with the forceps. Blow very gently on the emulsion side of the grid to make sure that the emulsion is in direct contact with the sections all over.
11. Keep the grids in a light-safe box at 4°C. After suitable exposure time (1 day–3 months, depending on the concentration and type of isotopes) develop the grids in D19 (90 sec), rinse briefly in distilled water (30 sec), fix in 25% sodium thiosulphate pentahydrate (120 sec), and carefully rinse twice in distilled water (60 sec). Developer and fixative should be filtered before use.
12. Dry the grid, which is now ready for the microscope.

Before applying emulsion to a series of grids the quality of the emulsion should be checked. For this purpose coat a few grids and observe them undeveloped in the electron microscope. The emulsion should be uniform without empty areas or regions with clumped silver grains (compare Figs. 14.1). In addition, one grid coated in the dark and not exposed to light or radioactivity should be developed to check the background of the emulsion (compare Fig. 14.2A). Very few developed grains should be detectable; preferably only a few per open square on the 200 mesh grid.

B. Checklist for Chemicals and Equipment

Chemicals

Distilled water
D19 developer
Ilford L4 Nuclear Research Emulsion
Sodium thiosulfate pentahydrate

Equipment

Balance (readable in the dark)
Darkroom with emulsion-safe light
Forceps
Lens paper
Metal (wood) rod, diameter about 2 mm; mounted vertically
Platinum wire loop, inner diameter 5 mm
Thermometer
Water bath

Reference

Maunsbach, A. B. (1966). Absorption of I[125]-labeled homologous albumin by rat kidney proximal tubule cells: A study of microperfused single proximal tubules by electron microscopic autoradiography and histochemistry. *J. Ultrastruct. Res.* **15**, 197–241.

11. Immunolabeling

Lowicryl resins (Carlemalm *et al.*, 1982) are well suited for electron microscope immunocytochemistry of formaldehyde or formaldehyde/glutaraldehyde-fixed tissues. Thin Lowicryl sections can be labeled with the appropriate primary antibody and the primary antibody then detected with a secondary antibody (or protein A) coupled to colloidal gold particles. The following protocol represents a standard procedure for the localization of several different antigens at the ultrastructural level.

A. Immunolabeling of Lowicryl Sections

Ultrathin sections of aldehyde-fixed and Lowicryl HM20 or K4M-embedded tissue are suitable for the immunocytochemical detection of many antigens. Depending on the sensitivity of the antigen, the tissue is fixed with either formaldehyde or mixtures of formaldehyde and glutaraldehyde (see Section 1). The tissue can be embedded in Lowicryl using the method of "progressively lowering temperature" (PLT) by Carlemalm *et al.* (1982) or by freezing and freeze-substitution into Lowicryl (Section 2). Tissue blocks can be conveniently stored, preferably in the cold room, and repeatedly resectioned. If there are holes in the sections when the blocks are resectioned, the first 30–50 ultrathin sections should be discarded before collection.

Procedure

1. Collect ultrathin sections of Lowicryl-embedded tissue on 200 mesh nickel grids with carbon-coated Formvar support film.
2. Place the grids section-side down on small drops (each 10–25 μl) of the appropriate solutions. Place the drops on a sheet of Parafilm (or dental vax) and transfer the grids successively from drop to drop. At each transfer remove most of the previous solution by touching the grid to a filter paper, but without allowing the grid to dry. Transfer the grid between the following drops (3–7):
3. Preincubation solution for 5–15 min at room temperature.
4. Solution of primary antibody diluted 1:50–1:5000 depending on the characteristics of antigen and antibody. Incubate for 1 hr at room temperature or at 4°C overnight in a moist chamber.
5. Rinsing solution three times, 5 min on each drop.
6. Secondary antibody (or protein A) conjugated to colloidal gold. Dilution is 1:50–1:200. Incubate for 1 hr at room temperature.
7. Rinsing solution three times, 5 min on each drop.
8. Rinse with water (redistilled and filtered) 5 min on each of three drops.
9. Remove excess water with filter paper and air-dry the grid.
10. Stain for 10 min on saturated and filtered uranyl acetate. Additional contrast can be obtained with lead citrate staining.
11. Rinse with 15–20 drops of redistilled water.
12. Air-dry the grid, which is now ready for examination in the electron microscope.

B. Immunolabeling of Cryosections

The procedure for labeling of ultrathin cryosections is the same as for ultrathin Lowicryl sections except that the procedure starts with three rinses on drops of rinsing solution without bovine serum albumin (BSA) or skimmed milk powder. After incubation with secondary antibody and rinses on buffer and on distilled water the grid is placed on a solution consisting of 9 parts of 2% methylcellulose and 1 part of 3% uranyl acetate. After 10 min the grid is lifted with a thin wire loop and the methylcellulose/uranyl acetate droplet is slowly drawn away with clean filter paper and the grid air-dried (Tokuyasu, 1986).

C. Comments

Electron microscope immunocytochemistry requires careful control of the experimental conditions. Immunoblotting will often—but not always—reveal if the antigen is present in the tissue and if the antibody specifically recognizes the antigen. For initial overview it is also recommended to apply immunofluorescence and/or immunoperoxidase labeling of 1-μm-thick cryosections for localization of the antigen at the light microscope level. Initial controls at the electron microscope level should include replacement of the primary antibody with buffer solution, with nonimmune serum or IgG from the same species, and, if at all available, preimmune serum. Another control is to preabsorb the antibody with the purified antigen (usually preabsorbed overnight in the cold room). Some of the more common problems in electron microscope immunocytochemistry include:

1. *Absence of labeling.* The antigen may be absent in the tissue or was destroyed during the tissue preparation. Absence of labeling may also be due to very low or no affinity of the antibody to the antigen. The antibody may be damaged, (aged, stored improperly, or repeatedly frozen and thawed) or too diluted. Check that the right secondary antibody is used.

2. *Weak labeling.* Damage of the antigen during fixation and/or embedding may lead to weak labeling. Suggested remedies include (a) decreasing the concentration of fixative; (b) using formaldehyde alone instead of glutaraldehyde or glutaraldehyde/formaldehyde; (c) shortening the fixation time; (d) using another type of Lowicryl, (e) reducing the salt concentration in the solutions used for the dilution of antibodies, (f) using smaller gold particles (e.g., 5 nm), (g) trying an etching or antigen retrieval procedure, or (h) turning to the alternative method of cryoultramicrotomy and labeling of cryosections.

3. *High background labeling.* Several factors may contribute to high background labeling: Insufficient rinsing of the grid before or after antibody labeling, insufficient blocking, or too high a titer of the primary antibody or the secondary antibody (or protein A). High background labeling may, in some cases, be prevented by increasing the sodium chloride concentration or the pH in the incubation solution for the primary antibody and/or the secondary gold conjugate. If the tissue contains high concentrations of free aldehyde groups following glutaraldehyde fixation, the sections should be preincubated with 0.5 mM ammonium chloride.

4. *Clustered colloidal gold particles.* The primary antibody or the colloidal gold conjugate has aggregated and should be renewed.

D. Checklist of Chemicals and Equipment

Chemicals

Bovine serum albumin (BSA)
Disodium hydrogen phosphate dihydrate
Gelatin (from cold water fish skin)
Glycine
Goat antirabbit IgG (or other appropriate antibody) conjugated to 10- or 5-nm colloidal gold particles
Methylcellulose
Polyethyleneglycol (PEG), molecular weight 20,000
Protein A conjugated to colloidal gold particles
Skimmed milk powder
Sodium azide (NaN$_3$)
Sodium chloride
Sodium dihydrogen phosphate monohydrate

Equipment

Disposable Pasteur pipettes
Filter paper, 0.2 μm
Fine forceps, antimagnetic
Magnetic stirrer
Nickel grids (for immunolabeling)
Parafilm (or dental vax)

E. Solutions

a. Rinsing solution. Consists of phosphate-buffered saline (PBS): 0.01 M sodium phosphate buffer containing 0.15 M sodium chloride with 0.1% skimmed milk powder (or 1% BSA). Mix 0.7 ml of 0.2 M NaH$_2$PO$_4$·H$_2$O, 1.8 ml of 0.2 M Na$_2$HPO$_4$·2H$_2$O, and 0.438 g NaCl. Adjust pH to 7.4 and fill up with redistilled H$_2$O to make 50 ml. Add 0.05 g skimmed milk powder (or 0.5 g BSA).

b. Preincubation solution. Rinsing solution with 0.05 M glycine. Add 0.375 g glycine per 100 ml solution.

c. Solution for dilution of primary antibody. Same as rinsing solution if skimmed milk powder is used. If BSA is used the concentration should be 0.1%.

d. Solution for dilution of gold-conjugated antibodies (or protein A). Same as solution for dilution of primary antibody but with the addition of 1.5 ml of 1% polyethyleneglycol and 0.555 g fish gelatine per 25 ml solution (to reduce aggregation of gold particles).

The stability of solutions for preincubation, rinsing, and dilution of antibodies can be improved by adding 0.02 M of sodium azide (NaN$_3$).

References

Carlemalm, F., Villiger, W., Hobot, J. A., Acetarin, J.-D., and Kellenberger, E. (1982). Resin development for electron microscopy and an analysis of embedding at low temperature. *J. Microsc. (Oxford)* **126,** 123–143.

Griffiths, G. (1993). "Fine Structure Immunocytochemistry." Springer-Verlag, Berlin.

Maunsbach, A. B. (1998). Immunolabeling and staining of ultrathin sections in biological electron microscopy. *In* "Cell Biology: A Laboratory Handbook" (J. Celis, ed.), 2nd Ed., Vol. 3, pp. 268–276. Academic Press, San Diego.

Newman, G. R., and Hobot, J. A. (1993). "Resin Microscopy and On-Section Immunocytochemistry." Springer-Verlag, Berlin.

Roth, J. (1986). Post-embedding cytochemistry with gold labelled reagents: A review. *J. Microsc. (Oxford)* **143,** 125–137.

Schwarz, H., and Humbel, B. M. (1989). Influence of fixatives and embedding media on immunolabelling of freeze-substituted cells. *Scann. Microsc. Suppl. 3,* **57,** 64.

Tokuyasu, K. T. (1986). Application of cryoultramicrotomy to immunocytochemistry. *J. Microsc. (Oxford)* **143,** 139–149.

12. Freeze Fracture

Freeze fracturing provides information about cells and tissues that is different than that obtained by thin-sectioning approaches. The characteristics of observations are closely related to the freeze-fracture technique, particularly to the characteristics of the instruments used. For this reason the following procedure focuses only on the principle steps in the procedure, whereas the details are dependent on the particular instrument type.

Procedure

1. Fix the tissue in 1–3% glutaraldehyde or formaldehyde/glutaraldehyde mixtures.
2. Rinse the tissue in buffer and trim it into 1- to 2-mm^3 blocks.
3. Immerse the blocks in 30% glycerol for 1 hr.
4. Place the blocks individually on the specimen holders fitting the freeze-fracture unit. Freeze the specimen by rapidly plunging the holder with the specimen into liquid Freon 22 that has been cooled with liquid nitrogen.
5. The frozen specimen can be stored in liquid nitrogen for future use or mounted directly in the freeze-fracture unit.
6. Mount the main gun of the instrument with platinum/carbon and the other gun with carbon according to the manual of the instrument.
7. Pump the instrument to a vacuum of 10^{-6}–10^{-7} hPA and cool to −150°C.
8. Insert the specimen.
9. When the vacuum is again stabilized, increase the temperature to −110°C.
10. Fracture the specimen by cutting sections with the built-in knife.
11. Immediately before the last knife fracture, activate the platinum/carbon gun and evaporate platinum onto the specimen at a 45° angle. The thickness of the platinum/carbon layer is monitored in most instruments with a quartz crystal monitor to a thickness of 1.5–2.5 nm.
12. To stabilize the replica, evaporate carbon for 1–2 sec with the second gun, which is oriented at an angle of 85–90°.
13. Increase the temperature of the specimen to −40°C and evaporate carbon again for 3 sec.
14. The specimen is removed from the freeze-fracture unit and cleaned by floating the tissue block, replica side up, on 40% chromic acid for 24 hr. If the specimen consists of suspensions of cells or particles it may be cleaned on sodium hypochloride (NaOCl) for 2 hr.
15. Rinse the replica 3 times on distilled water.
16. Collect the replica on 300-mesh copper grids.

Reference

Shotton, D. M. (1998). Freeze fracture and freeze etching. *In* "Cell Biology: A Laboratory Handbook" (J. E. Celis, ed.), 2nd Ed., Vol. 3, pp. 310–322. Academic Press, San Diego.

AUTHOR BIOGRAPHIES

A native of Sweden, Arvid B. Maunsbach received his M.D. and Ph.D. from the Karolinska Institute in Stockholm. As a second-year medical student he became interested in electron microscopy and joined the Anatomy Department at the Institute as a junior instructor in parallel with his studies. He then worked for almost 4 years in the United States, in both the Pathology and Zoology Departments at the University of California, Los Angeles, and in the Biophysical Laboratory at Harvard Medical School. Returning to Sweden he became a docent at the Karolinska Institute. In 1970 he was appointed professor and chairman of the Department of Cell Biology at the University of Aarhus, Denmark, where he initiated a laboratory for biological ultrastructure research and built a research group focusing on renal ultrastructure and function. He has been secretary and then president of the Scandinavian Society for Electron Microscopy and also chairman of the Nordic Society for Cell Biology. Later he became general secretary of the International Federation of Societies for Electron Microscopy (IFSEM) and in 1994 president of the Federa-

tion. He has also served as chairman of the Danish Medical Research Council and as dean of the Faculty of Health Sciences in Aarhus.

Arvid Maunsbach has authored 160 scientific papers and coedited four books. He clarified the physiological pathway for endocytosis in renal proximal tubule cells, where he also identified the apical membrane recycling process and investigated functions of lysosomes. He published the first electron microscope analyses of purified Na,K-ATPase, including its two-dimensional crystalization. Other research lines focus on the ultrastructure and function of epithelial transport routes for salt and water, and the segmentation of the kidney tubule. In connection with some of these studies he has developed new preparatory methods for fixation, autoradiography, and cell fractionation. Arvid Maunsbach has been invited speaker at numerous scientific meetings, and in addition to training graduate students in his own laboratory, he has given postgraduate courses in several other countries, notably within Scandinavia, in Portugal, and in Brazil.

Björn A. Afzelius was born and grew up in Stockholm, Sweden. He was introduced to his future field, electron microscopy of cells and tissues, at the Anatomy Department of the Karolinska Institute, where he was a guest for 5 years. He got a Ph.D. degree at Stockholm University before moving to the Johns Hopkins University, Baltimore, to work as a research fellow with faculty status. After his American year he returned to Sweden to become docent, then professor, at Stockholm University from 1959 until the present, all the time engaged in various ultrastructural studies. He received an M.D.h.c. from the Karolinska Institute and a Ph.D.h.c. from the University of Siena, Italy. He has acted as chairman of the Scandinavian Society for Electron Microscopy and as chairman of the Nordic Society for Cell Biology. He has

enjoyed being a teacher in various courses in electron microscopy in several countries. As author, editor, or coeditor he has published four books on spermatology and cell biology and has authored or coauthored 250 scientific papers. Some of his papers deal with the structure of mitochondria in Luft syndrome, i.e., a rare mitochondrial disease with defective respiratory control, or the demonstration of a general occurrence of mitochondrial DNA. Other papers deal with spermatozoa or cilia, the acrosome reaction of sea urchin sperm, the proposal that the sperm flagellum or the cilium moves through a sliding of the microtubules driven by dynein arms, and the organization of protofilaments in the ciliary axoneme. He also recognized a human disease where dynein arms are lacking, which was named immotile-cilia syndrome.

ACKNOWLEDGMENTS FOR REPRODUCTION OF FIGURES

Special thanks go to José David-Ferreira for letting us publish the following, previously unpublished, micrographs taken by him: FIGURES 2.7C, 2.11A, 4.7A–4.7D, 4.8A–4.8C, 5.2A–5.2E, 5.3A–5.3C, and 7.3A–7.3C; to Bjørn V. Johansen for recording the insets to FIGURES 10.16; to Lars Kihlborg for helping us with FIGURES 10.1B–10.1D; and Gunna Christiansen for the preparations for FIGURES 17.1A–17.1E.

Several micrographs in this volume originate from collaborations with our collegues or students. These micrographs and others published by us as follows:

Chapter 1

FIGURES 1.4A and 1.4B. From Maunsbach, A. B. (1966). *J. Ultrastr. Res.* **14,** 167–189, with permission from Academic Press, San Diego.

FIGURE 1.4E. From Nass, M. M. K., Nass, S., and Afzelius, B. A. (1965). *Expl. Cell Res.* **37,** 516–539, with permission from Academic Press, San Diego.

FIGURE 1.7A. From Maunsbach, A. B., Marples, D., Chin, E., Ning, G., Bondy, C., Agre, P., and Nielsen, S. (1997). *J. Am. Soc. Nephrol.* **8,** 1–14, with permission from Williams & Wilkins, Baltimore.

FIGURE 1.8A. From Afzelius, B. A. (1979). *Inter. Rev. Expl. Pathol.* **19,** 1–43, with permission from Academic Press, San Diego.

FIGURE 1.8B. From Afzelius, B. A., and Eliasson, R. (1979). *J. Ultrastr. Res.* **69,** 43–52, with permission from Academic Press, San Diego.

FIGURE 1.8H. From Malmqvist, E., Ivemark, B. I., Lindsten, J., Maunsbach, A. B., and Mårtensson, B. (1971). *Lab. Invest.* **25,** 1–14, with permission from Williams and Wilkins, Baltimore.

FIGURE 1.9A. From Maunsbach, A. B. (1966). *J. Ultrastr. Res.* **15,** 197–241, with permission from Academic Press, San Diego.

FIGURE 1.9B. From Maunsbach, A. B. (1976). *In* "International Review of Physiology: Kidney and Urinary Tract Physiology" (K. Thurau, ed.), Vol. II, pp. 145–167, with permission from University Park Press, Baltimore.

FIGURE 1.10B. From Miller, A. and Maunsbach, A. B. (1966). *Science* **151,** 1000–1001, with permission from American Association for the Advancement of Science, Washington, D.C.

FIGURES 1.11A and 1.11B. From Deguchi, N., Jørgensen, P. L., and Maunsbach, A. B. (1977). *J. Cell Biol.* **75,** 619–634, with permission from The Rockefeller University Press, New York.

FIGURES 1.11C and 1.11D. From Maunsbach, A. B., Skriver, E., Petersen, K. D., and Hebert, H. (1991). *In* "International Symposium on Electron Microscopy" (K. Kuo and J. Yao, eds.), pp. 136–150, with permission from World Scientific Publishing Co. Pte. Ltd., Singapore.

Chapter 2

FIGURES 2.2A, 2.2B, and 2.2C. From Maunsbach, A. B. (1966). *J. Ultrastr. Res.* **15,** 242–282 and 283–309, with permission from Academic Press, San Diego.

FIGURES 2.8A and 2.8B. From Christensen, E. I., and Maunsbach, A. B. (1979). *Kidney Int.* **16,** 301–311, with permission from Blackwell Science, Inc., Malden, MA.

FIGURES 2.8C and 2.8D. From Cotta-Perreira, G., Rodrigo, F. G., and David-Ferreira, J. F. (1976). *Stain Technol.* **51,** 7–11, with permission from Williams & Wilkins, Baltimore.

FIGURES 2.12A, 2.12C, 2.13A, 2.13B, and 2.13D. From Maunsbach, A. B. (1966). *J. Ultrastr. Res.* **15,** 242–282, with permission from Academic Press, San Diego.

Chapter 3

FIGURES 3.3A, 3.3B, 3.3C, 3.4A, 3.4B, 3.5A, 3.5B, and 3.5C. From Maunsbach, A. B. (1966). *J. Ultrastr. Res.* **15,** 283–309, with permission from Academic Press, San Diego.

FIGURES 3.7A, 3.7B, 3.7C, 3.8A, and 3.8B. From Bohman, S. O., and Maunsbach, A. B. (1970). *J. Ultrastr. Res.* **30,** 195–208, with permission from Academic Press, San Diego.

Chapter 4

FIGURE 4.1A. From Maunsbach, A. B. (1966). *J. Ultrastr. Res.* **15,** 242–282, with permission from Academic Press, San Diego.

FIGURES 4.2A and 4.2B. From Maunsbach, A. B., and Boulpaep, E. I. (1980). *Kidney Int.* **17,** 732–748, with permission from Blackwell Science, Inc., Malden, MA.

FIGURE 4.3A. From Maunsbach, A. B., Madden, S. C., and Latta, H. (1962). *J. Ultrastr. Res.* **6,** 511–530, with permission from Academic Press, San Diego.

FIGURES 4.6A–4.6D. From Maunsbach, A. B. (1994). *In* "Cell Biology: A Laboratory Handbook" (J. E. Celis, ed.), pp. 117–125, with permission from Academic Press, San Diego.

FIGURES 4.6A and 4.6B. From Maunsbach, A. B. (1966). *J. Ultrastr. Res.* **15,** 242–282, with permission from Academic Press, San Diego.

FIGURES 4.10A and 4.10B. From Maunsbach, A. B. (1979). *In* "Electron Microscopy in Human Medicine" (J. V. Johannesen, ed.), with permission from McGraw-Hill International Book Company, London.

Chapter 6

FIGURE 6.5. From Maunsbach, A. B. (1994). *In* "Cell Biology: A Laboratory Handbook" (J. F. Celis, ed.), Vol. 2, pp. 117–125, with permission from Academic Press, San Diego.

Chapter 8

FIGURES 8.1A–8.1D. From Maunsbach, A. B. (1978). *In* "Proceedings of the Ninth International Congress on Electron Microscopy, Toronto," Vol. II, pp. 80–81, with permission from Microscopical Society of Canada.

FIGURE 8.9B. From Maunsbach, A. B. (1966). *J. Ultrastr. Res.* **16,** 13–34, with permission from Academic Press, San Diego.

Chapter 10

FIGURES 10.26A, 10.26D, and 10.26E. From Overgaard, S., Lind, M., Josephsen, K., Maunsbach, A. B., Bünger, C., and Søballe, K. (1998). *J. Biomed. Mater. Res.* **39,** 141–152, with permission from John Wiley & Sons, Inc., Chichester.

FIGURES 10.28A and 10.28B. From Maunsbach, A. B., Hebert, H., and Kavéus, U. (1992). *Acta Histochem. Cytochem.* **25,** 279–285, with permission from the Japan Society of Histochemistry and Cytochemistry, Kyoto.

Chapter 14

FIGURES 14.1A, 14.1B, and 14.4C. From Maunsbach, A. B. (1966). *J. Ultrastr. Res.* **15,** 197–241, with permission from Academic Press, San Diego.

FIGURES 14.3A and 14.3C. From Maunsbach, A. B. (1966). *Nature* **212,** 546–547, with permission from McMillan Magazines, London.

FIGURE 14.4A. From Dohlman, G. F., Maunsbach, A. B., Hammarström, L., and Appelgren, L.-E. (1964). *J. Ultrastr. Res.* **10,** 293–303, with permission from Academic Press, San Diego.

Chapter 15

FIGURES 15.1B and 15.5C. From Maunsbach, A. B., Skriver, E., Söderholm, M., and Hebert, H. (1986). *In* "Proceedings of the Eleventh Congress on Electron Microscopy, Kyoto" pp. 1801–1806, with permission from The Japanese Society of Electron Microscopy, Tokyo.

FIGURES 15.3A–15.3F, 15.5A, 15.5B, 15.6C, and 15.7D. From Maunsbach, A. B. (1966). *J. Ultrastr. Res.* **16,** 197–238, with permission from Academic Press, San Diego.

FIGURE 15.4A. From Ottosen, P. D., and Maunsbach, A. B. (1973). *Kidney Int.* **3,** 315–326, with permission from Blackwell Science, Inc., Malden, MA.

Chapter 16

FIGURE 16.13A. From Pihakaski-Maunsbach, K., Griffith, M., Antikainen, M., and Maunsbach, A. B. (1996). *Protoplasma* **191,** 115–125, with permission from Springer-Verlag, Berlin.

FIGURES 16.17A, 16.17B, 16.17D, 16.17E, and 16.17F. From Ning, G., and Maunsbach, A. B. (1994). *Acta Histochem. Cytochem.* **27,** 347–356, with permission from the Japan Society for Histochemistry and Cytochemistry.

FIGURES 16.18A and 16.18B. From Fujimoto, K., Møller, J. V., and Maunsbach, A. B. (1996). *FEBS Lett.* **395,** 29–32, with permission from FEBS Letters, Elsevier Science BV, Amsterdam.

FIGURES 16.20A–16.20E. From Maunsbach, A. B., Marples, D., Chin, E., Ning, G., Bondy, C., Agre, P., and Nielsen, S. (1997). *J. Am. Soc. Nephrol.* **8,** 1–14, with permission from Williams & Wilkins, Baltimore.

Chapter 17

FIGURES 17.6A and 17.6B. From Maunsbach, A. B., Skriver, E., and Jørgensen, P. L. (1979). *In* "Na,K-ATPase. Structure and Kinetics" (J. C. Skou and J. G. Nørby, eds.), with permission from Academic Press, London.

FIGURES 17.7A and 17.7B. From Afzelius, B. A., Dallai, R., and Callaini, G., *J. Submicr. Cytol. Pathol.* **21,** 187–200, with permission from Editrici Compositori, Bologna.

FIGURES 17.8D and 17.8E. From Deguchi, N., Jørgensen, P. L., and Maunsbach, A. B. (1977). *J. Cell Biol.* **75,** 619–634, with permission from The Rockefeller University Press.

FIGURES 17.10A and 17.10B. From Skriver, E., Maunsbach, A. B., and Jørgensen, P. I. (1980). *J. Cell Biol.* **86,** 746–754, with permission from The Rockefeller University Press.

FIGURES 17.10C–17.10F. From Zeidel, M. L., Nielsen, S., Smith, B. L., Ambudkar, S. V., Maunsbach, A. B., and Agre, P. (1994). *Biochemistry* **33,** 1606–1615, with permission from the American Chemical Society, Washington, D.C.

Chapter 18

FIGURES 18.10A and 18.10B. From Mathiesen, F. Ø., Gundersen, H. J., Maunsbach, A. B., and Skriver, E. (1991). *J. Microscopy (Oxford)* **164,** 247–261, with permission from the Royal Microscopical Society, Oxford.

Chapter 19

FIGURES 19.4A–19.4C. From Söderholm, M., Hebert, H., Skriver, F., and Maunsbach, A. B. (1988). *J. Ultrastr. Mol. Struct. Res.* **99,** 234–243, with permission from Academic Press, San Diego.

FIGURES 19.4D–19.4F. From Hebert, H., Jørgensen, P. I., Skriver, E., and Maunsbach, A. B. (1982). *Biochim. Biophys. Acta* **689,** 571–574, with permission from Elsevier Science BV, Amsterdam.

Chapter 20

FIGURE 20.1A. From Gustafson, B. E., and Maunsbach, A. B. (1991). *Zeitschr. Zellforsch. Mikrosk. Anat.* **120,** 555–578, with permission from Springer-Verlag, Berlin.

FIGURE 20.3C–20.3G. From Dørup, J., and Maunsbach, A. B. (1997). *Expl. Nephrol.* **5,** 305–317, with permission from Karger AG, Basel.

FIGURE 20.7A. From Lanzavecchia, S., Bellon, P. L., Dallai, R., and Afzelius, B. A. (1994). *J. Struct. Biol.* **113,** 225–237, with permission from Academic Press, San Diego.

FIGURES 20.8A–20.8D. From Hebert, H., Skriver, E., and Maunsbach, A. B. (1985). *FEBS-Lett.* **187,** 182–186, with permission from Elsevier Science BV, Amsterdam.

INDEX

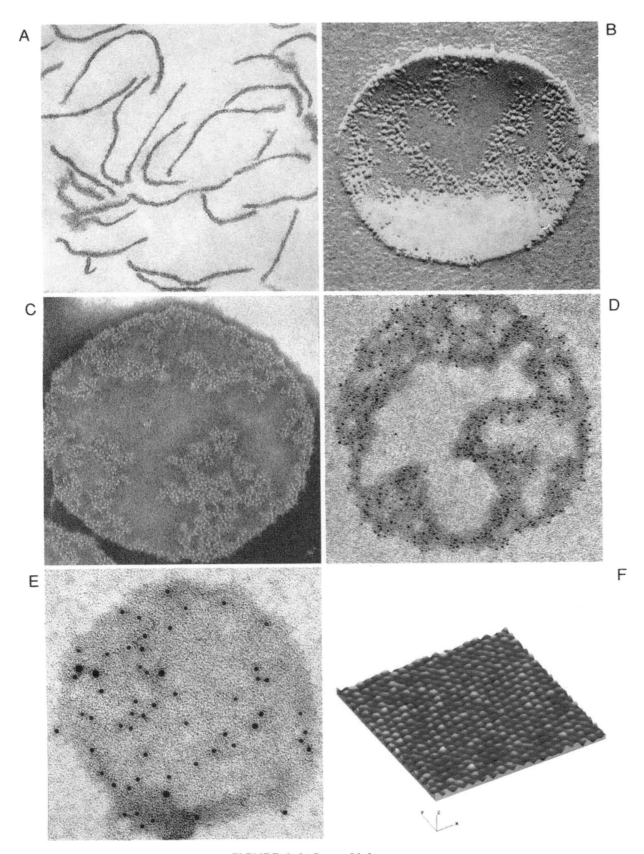

FIGURE 1.11 See p. 22 for text.

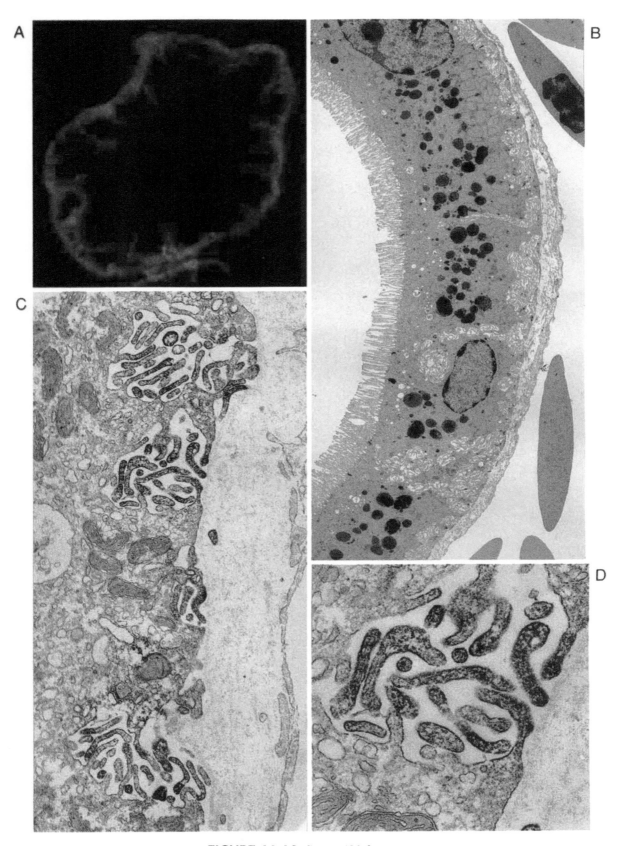

FIGURE 16.19 See p. 420 for text.

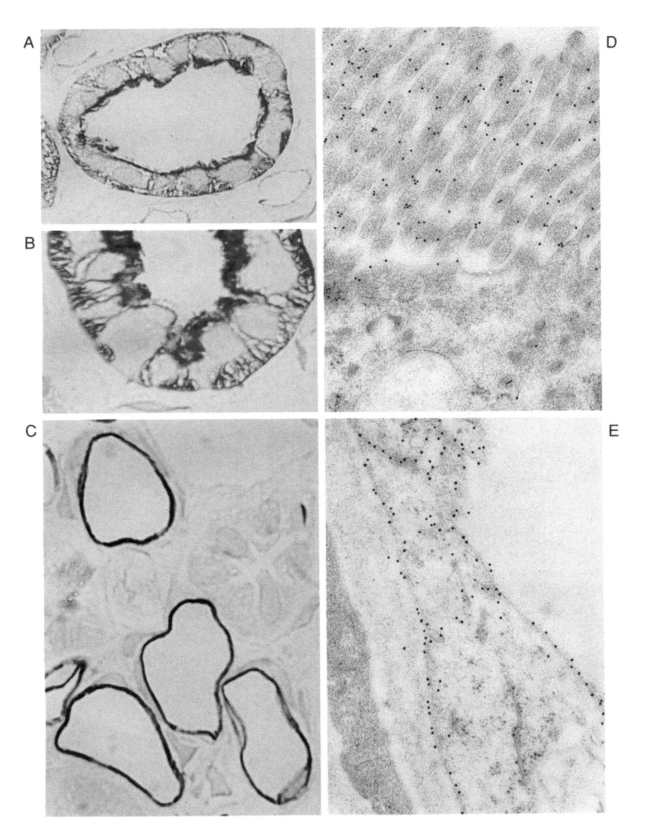

FIGURE 16.20 See p. 422 for text.

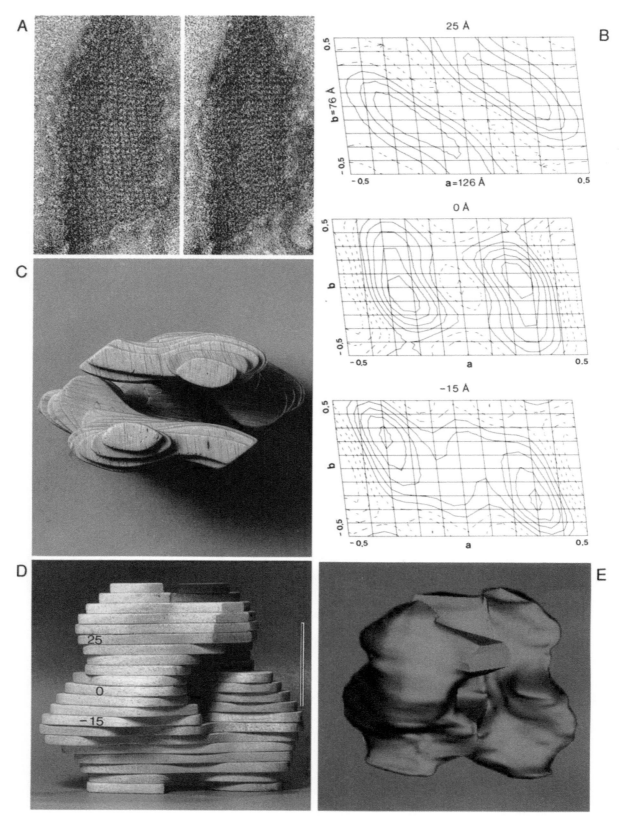

FIGURE 20.8 See p. 514 for text.

Printed and bound by CPI Group (UK) Ltd, Croydon, CR0 4YY

08/05/2025

01865029-0004